Peptides 1986

Peptides 1986

Proceedings of the
19th European Peptide Symposium
Porto Carras, Chalkidiki, Greece
August 31 – September 5, 1986

Editor
Dimitrios Theodoropoulos

W DE G Walter de Gruyter
Berlin · New York 1987

Editor
Dimitrios Theodoropoulos
Professor of Organic Chemistry
Laboratory of Organic Chemistry
University of Patras
Greece

Library of Congress Cataloging in Publication Data

European Peptide Symposium (19th : Chaklidikě, Greece)
Peptides, 1986.
Bibliography: p.
Includes indexes.
1. Peptides--Sythesis--Congresses. 2. Peptides--Congresses.
I. Theodoropoulos, Dimitrios, 1926- . II. Title.
QP552.P4E9 1986 660'.63 87-5413
ISBN 0-89925-310-5

CIP-Kurztitelaufnahme der Deutschen Bibliothek

Peptides ... : proceedings of the ... Europ. Peptide Symposium. -
Berlin ; New York : de Gruyter
Auf dem Umschlag: ... EPS
NE: EPS
19. 1986. Porto Carras, Chalkidiki, Greece, August 31 - September 5, 1986. - 1987.
ISBN 3-11-010687-6

 Printing: Gerike GmbH, Berlin. Binding: D. Mikolai, Berlin. - Printed in Germany.

Dedicated to the Memory
of Leonidas Zervas

Preface

Almost thirty years ago, in September 1958, a small group of chemists met in Prague, with Professor Joseph Rudinger as the spiritus agen, and decided to concentrate their scientific efforts on peptide synthesis and the rapidly emerging new disciplines based on peptides. The following meetings, annual at the beginning and at about two years intervals later on, developed quickly into highly esteemed specialized meetings which found response from North American and Far Eastern chemists as well. Nearly 23 years ago the European Peptide Symposium was organized in Athens by the late Professor Leonidas Zervas to whose memory this 19th EPS is dedicated.

Professor Zervas, after his pioneering work with Max Bergmann in Germany and later on at the Rockefeller Institute with M. Bergmann and J. Fruton, decided to return home. Starting from scratch, Professor Zervas built the foundations of chemical research in Greece and continued to contribute immensely to modern peptide chemistry.

The principal event of the 19th EPS, held at Porto Carras, Chalkidiki, Greece, from August 31st to September 5th, 1986, was the first Rudinger Memorial Lecture, endowed by Ferring Pharmaceuticals to commemorate Professor Joseph Rudinger's role in the foundation of the European Peptide Symposia and the diverse contributions he made to peptide chemistry. Professor Robert Schwyzer was appointed by the European Peptide Committee to be the first Joseph Rudinger Memorial Lecturer.

The Organizing Committee was particulary pleased to welcome their guests, 250 scientists and 100 accompanying persons coming from 26 countries.

The 19th EPS retained the main characteristics of previous Symposia: its aim was to highlight new, original research in the methodology of synthesis, including polymer supported peptide synthesis, coupling, enzymatic synthesis, semi-synthesis, recombinant synthesis, general methodology (including labelling, large scale synthesis, analysis and purification), physical methods, biologically active peptides and miscellaneous questions (including structural investigation). All contributions included in the volume are based on a five day scientific meeting of the Programme Committee held in Athens, March 1986. Over 150 oral and poster presentations provided a forum for discussing the latest developments. The current approaches to the design and synthesis of biologically active peptides were highlighted by R. Hirschmann and R. Geiger, who chaired a general discussion session.

I would personally like to thank my colleagues in the Organizing Committee for their cooperation during nearly two years of preparations, the Secretary Miss E. Logiadou for her exellent work beyond the call of duty, Mrs Aleka Miliotis for her assistence, and my daughter Melpo for designing the emblem and the poster of the Symposium.

This Symposium could not take place without economic support from many sources. I express my gratitude to all concerned.

At the European Peptide Committee Meeting at Porto Carras, a decision was taken to increase the number of participants while retaining the number of published contributions at about the same number. Participation of new scientists will also be encouraged.

The publisher of the Proceedings of this Symposium is Walter de Gruyter, Berlin. I am grateful to Mrs E. Glowka for her cooperation on the completion of this book.

Patras, December 1986 *Dimitrios Theodoropoulos*

Committees

EUROPEAN PEPTIDE COMMITTEE

Austria:	E. Haslinger
Belgium:	A. Loffet
Bulgaria:	B. Aleksiev
Czechoslovakia:	K. Bláha
Denmark:	K. Brunfeldt
France:	P. Fromageot
FRG:	G. Jung
GDR:	H. Niedrich
Greece:	D. Theodoropoulos
Hungary:	K. Medzihradszky
Israel:	A. Patchornik
Italy:	R. Rocchi
Netherlands:	G. I. Tesser
Norway:	J. Bøler
Poland:	G. Kupryszewski
Portugal:	M. J. S. A. Amaral Trigo
Spain:	E. Giralt
Sweden:	U. Ragnarsson
Switzerland:	C. H. Schneider
United Kingdom:	R. C. Sheppard
USSR:	Y. Ovchinnikov

ORGANIZING COMMITTEE

P. Cordopatis
A. Cosmatos
C. Sakarellos
G. Stelakatos
D. Theodoropoulos, Chairman

PROGRAMME COMMITTEE

K. Bláha
D. Brandenburg
K. Medzihradszky
U. Ragnarsson
C. Sakarellos
D. Theodoropoulos
G. T. Young

The 19th European Peptide Symposium was supported by

The Ministry of Culture
The Ministry of Northern Greece
The Greek National Tourist Organisation
The University of Athens
The University of Ioannina
The University of Patras

the following Companies:

Achaia Clauss
Anelor, S.A.
Bacacos, S.A.
Bachem AG
Bachem Inc.
Bayer AG
Beecham Pharmaceuticals
Biomega
Boxites S.A.
Bristol
Carlo Erba
C.E.A. Saclay
Chrotex, S.A.
Credit Bank
Degussa AG
Diamalt AG
Dow Chemicals (Greece)
Farmacia
Fluka
Glaxo
Hoechst
Hoffmann-La Roche Inc.
Hoffmann-La Roche & Co.
I C I
Institut Choay
Merck Sharp & Dohme
Merck
Novabiochem AG
Novo
Nyco
Olympic Airways
Opregen
Peninsula Laboratories
Peptides International, Inc.
Propeptides
Rhone Poulenc
Sandoz
Syntex
U C B
UpJohn

and the Leonidas Zervas Foundation

Contents

A. METHODOLOGY OF SYNTHESIS

POLYMER SUPPORTED PEPTIDE SYNTHESIS

ENZYMATIC SYNTHESIS. SEMISYNTHESIS. RECOMBINANT SYNTHESIS

GENERAL METHODOLOGY: INCLUDING LABELLING,
LARGE SCALE SYNTHESIS, ANALYSIS AND PURIFICATION

B. PHYSICAL STUDIES ON PEPTIDES

C. BIOLOGICALLY ACTIVE PEPTIDES

NEUROPEPTIDES

PANCREATIC PEPTIDES

IMMUNOPEPTIDES

D. MISCELLANEOUS: INCLUDING STRUCTURAL INVESTIGATION

Karel Jošt
/1932 - 1986/

RNDr Karel Jošt, DrSc, senior scientist of the Institute of Organic Chemistry and Biochemistry, Czechoslovak Academy of Sciences in Prague, died suddenly on June 25, 1986. He left prematurely at the age of 54 years, in the time the results of his long years' standing and intensive scientific work began to ripen. If we use to say that some people devoted their life to work, we certainly have to count the deceased Karel Jošt among them. He didn't have any hobbies, he didn't take any holidays, he concentrated solely on his activity in chemistry.

Born on April 20, 1932 in Nový Bydžov (North-East Bohemia) he also visited the high school in this town in 1951 - 1956. Already at that time he was deeply interested in natural sciences, along with it in chemistry. The studies at the Faculty of Mathematics and Physics, Charles University in Prague (1958 - 1961) focussed his interest on the interdisciplinary area between organic chemistry and biological sciences, known today as bio-organic chemistry. His Ph.D. studies under the leadership of Joseph Rudinger at the Institute of Organic Chemistry and Biochemistry, Czechoslovak Academy of Sciences, brought him to peptide chemistry. His Ph.D. thesis entitled "Oxytocin analogs, protected nona-, deca-, and dodecapeptides" stemmed from topics characteristics for the

Prague Laboratory - from the chemistry of peptide neurohypophyseal hormones oxytocin and vasopressin. Here, Karel Jošt met with scientific problems he has never left. In his work he was developing the idea of structural modification of natural hormones with the aim to achieve compounds with better properties for medical application whether human or veterinary. This was the ultimate intention of his study of relationships between chemical structure and biological activity of oxytocin and vasopressin which is reflected also by a short statistics concerning his scientific production: He published about 150 original scientific communications, delivered a number of lectures and prepared 35 patent applications, a number not common for a scientist in an academical institution. In an attempt to attain a realization on the industrial scale he personally devoted much of his time to the detailed work-up of "unfailing" production procedures. And he really attained his goal - some of the compounds he designed and prepared are successful commercially accessible pharmaceuticals.

Dr. Jošt was a distinctive representative of the classical conception in peptide chemistry - in the synthesis, analysis and characterization, in the investigation of properties and also in the final interpretation. The word "classical" does not mean negation of new more effective methods in this case. It rather demands their working-out on such a level of reliability which the existing methods permit. This is proved best by his interest in exploiting the enzymatic methods in peptide synthesis which motivated his recent work.

The central point of his attention remained the relation between structure and biological activity as regarded by an organic chemist. Here he attained the most expressive new findings appreciated by the doctor-of-science degree in 1977. In this field he became an internationally acknowledged specialist. The heighth of his literary work will be the Handbook of Neurohypophyseal Hormone analogs, published by CRC Press, U.S.A. As the senior editor, he made full use of his broad knowledge, with his coworkers he collected all the data and prepared a comprehensive manuscript. Unfortunately, he did not live long enough to see its printing. Just on the day when we learned about his death the first proofs came to our laboratory.

His scientific results have been awarded with many approvals, inter alia also the Czechoslovak State Prize in 1981. He was member of many scientific bodies connected with scientific education and planning of science. In all these instances his expert level has been esteemed. He trained a number of Ph.D. students who are now developing the peptide chemistry in scientific or industrial institutions. It would be very difficult to estimate the time he devoted to local and foreign students and to all who came to ask his advice.

His physical condition did not allow him lately a more frequent and long travelling. Therefore he relatively seldom attended our EPS, however, he was in live contacts with most of the significant laboratories in the whole world. In the family of peptide chemists his name will continue to live. In publicity the trace of his results will exist although he himself will stay nameless. The Czechoslovak chemistry is loosing in Karel Jošt one of the distinguished representatives and our laboratory a good coworker and friend.

/Karel Bláha/

THE JOSEF RUDINGER MEMORIAL LECTURE: AN INTRODUCTORY NOTE

It must be a matter of great satisfaction to all European peptide chemists that Josef Rudinger, the founder of the European Peptide Symposia, is now to be commemorated by a Memorial Lecture.

It was in the nature of Josef Rudinger that he wished to meet and discuss with colleagues in his chosen field; indeed, it might be said that in later years he inspired and to some extent co-ordinated research in centres that he visited. It was in 1949, shortly after the publication of my first brief note on the synthesis of γ-glutamyl peptides, that he wrote to me saying that he was working along similar lines and suggesting that we should correspond. This we did, exchanging details of procedures and the scripts of papers before publication. In the middle fifties the climate for East-West scientific meetings became more favourable, and Josef wrote to say that he believed there was a possibility of organising in Prague a small symposium of those actively engaged in peptide synthesis in Europe. He was disappointed at that time, and I offered to propose a Chemical Society meeting in England instead. However, in 1957 he told me that the Czechoslovak Academy of Sciences had agreed to organise a symposium in the following year. Some thirteen participants from outside Czechoslovakia met with the hosts in Prague in September 1958. The value of such direct and detailed discussions at a time when the subject was beginning to grow rapidly was immediately appreciated, and at the close of the meeting it was agreed to meet again regularly; so began the European Peptide Symposia, the first of their kind.

Some time ago, Dr. Frederik Paulsen, the founder of Ferring AB, Sweden, indicated to Dr. Ulf Ragnarsson that his company might be willing to establish a prize of some kind in peptide chemistry. The proposal was generally welcomed, and during the meeting of the Programme Committee in Uppsala in 1984 I suggested to Dr. Ragnarsson that it would be appropriate if such a prize were to be in the form of a Josef Rudinger Memorial Lecture, thus commemorating our founder. Josef Rudinger had in fact had some contact with Ferring in research projects, and this suggestion was accepted both by Ferring and by the European Peptide Committee at its meeting at Djurönäset in 1984. Regulations governing the Lecture were approved at that meeting, with the following preamble:

"The Josef Rudinger Memorial Lecture was established by the generosity of Ferring Pharmaceuticals in 1984, as a commemoration of Josef Rudinger's role in the foundation of the European Peptide Symposia and the diverse contributions he made to peptide chemistry, both by his own work and indirectly through the stimulation of others in many countries".

G.T. Young

PREDICTION OF POTENCY AND RECEPTOR SELECTIVITY OF REGULATORY PEPTIDES: THE MEMBRANE COMPARTMENT CONCEPT
FIRST JOSEF RUDINGER MEMORIAL LECTURE

Robert Schwyzer
Institut für Molekularbiologie und Biophysik, Eidgenössische Technische Hochschule, CH-8093 Zürich, Switzerland

Appointment by the European Peptide Committee to be the First Josef Rudinger Lecturer is a very great honour, indeed. In accepting the choice, I remain aware of the fact that many other scientists would have equally or better deserved such a distinction. It shall be not only my duty, but my pleasure to speak in memory of my dear friend with whom I was so closely connected for many years. Let me also remind you of his family: his wife, his daughter, and his mother, a strong personality with an impressive attitude towards life born of a deep sense of responsibility, honesty, and courage. Traits also typical of her son.

My friendship with Joe dates from the late 'fifties when he and I served together with Joseph Fruton and Geoffrey Young on the IUPAC-IUB Joint Commission on Biochemical Nomenclature and met for the first time in Zürich. Later, an official invitation to Prague gave me the opportunity of seeing him amidst his coworkers, of meeting his wife, Edita, and his superior, Fritz Šorm, President of the Czechoslovak Academy of Sciences. At a lunch for two, a sincere and friendly atmosphere was immediately established between Šorm and myself. This was to become important. During the next four years, Joe visited my newly founded Department at the ETH a couple of times, and a closer collaboration was envisaged. In June 1968, I was alarmed by signs of a beginning collaps of the political reforms in Czechoslovakia in which Joe, according to his personality, was enthusiastically involved. I wrote to Šorm, inviting Joe as a temporary visiting scientist to my laboratory. Šorm acted promptly and granted Joe a leave of absence which allowed him (and his family) to stay in Zürich for one year. However, as things developed, Joe decided to remain with me, and I succeeded in procuring for him an appointment as full professor at the ETH. He held it from 1970 to his untimely death in 1975. We two shared my personal space and facilities in the new building on the Hönggerberg without curtailing my junior colleagues, Dr. Zuber and Dr. Wüthrich, whom I had also engaged for the new enterprise of Molecular Biology and Biophysics at the ETH.

Joe and I shared the same enthusiasm for peptide chemistry and its use as a tool in molecular pharmacology, especially for the study of structure-activity relationships, the basis for the prediction of potency and receptor selection of regulatory peptides. His fascination is well-documented by scholarly articles such as 'The Design of Peptide Hormone Analogs' (1). He was an ideal partner, and our interests in research and teaching were complementary as we approached the matter from different angles, according to our scientific heritage.

Receptor requirements. My scientific 'grandfather' was Paul Ehrlich (1854-1915), founder of chemotherapy (2), with whom my scientific 'father' Paul Karrer (1889-1971) (3), had passed some of his happiest postdoctoral years. Ehrlich summarized his insights with the words: "Corpora non agunt nisi fixata", molecules do not act (upon living cells) unless they are bound. Binding of toxins and drugs proved to be specific and was thought to be caused by a 'lock-and-key' fit between cell surface receptors and the agonists. The receptor theory was independently proposed by John Newport Langley (1852-1925) (4) and appealed to me particularly because I had taught dyestuff, textile, and surface chemistry during my years with Karrer.

In early work with peptide antibiotics and hormones (1952-1963), I was intrigued by the question of how a flexible peptide, like a soft-boiled spaghetti, could serve as a 'key' to open a 'lock'. The amino acid sequence of a peptide agonist determines both its specific biologic activity and its conformational space, whereas its immediate environment determines its preferred conformation. I expected the receptor, as a specialized solute in the environment, to induce a preferred conformation on the peptide. With the proper amino acid sequence, the peptide could adapt snugly to a complementary surface of the receptor discriminator site and meet the manifold 'receptor requirements' for specific binding and triggering of the response. In this process, mutual forces, actio et reactio, were expected to impose conformational changes on both agonist and receptor, the latter perhaps serving as a 'stimulus' for the biologic effect, see Figure 1, adapted from (5).

While teaching peptide synthesis to Cho Hao Li's group in Berkeley (1958), I discovered that, in flexible peptides, distinct segments were responsible for distinct parts of the biologic effect (6). I considered 'message' segments to stimulate the receptors, 'potentiator' segments to enhance the effect, and 'address' segments to change the specific affinity of one and the same message among different receptors or receptor subtypes (7). This concept of 'sychnologic' organization was later refined and extended to other peptides (8,9).

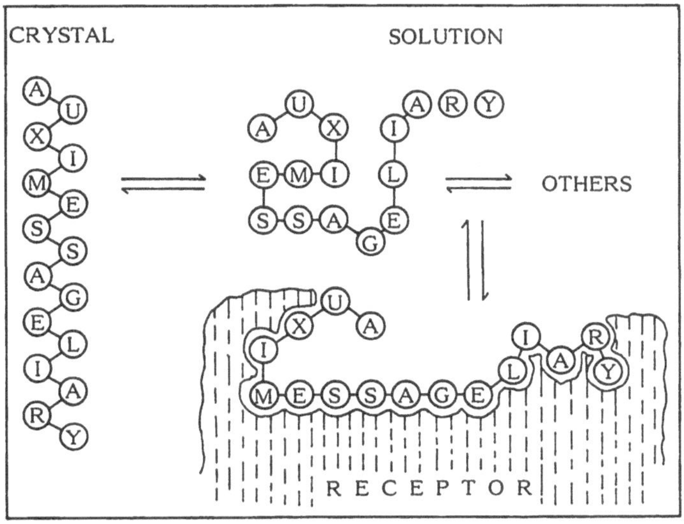

Fig. 1. Adaptation according to receptor requirements.

Today, the search for receptor requirements and their correlation with potency and receptor selectivity has become very intensive. The once fashionable idea (which I never cherished) that crystal structures or solution conformations of flexible peptides could provide valid clues is losing its appeal, but the mimicking of receptor-affine and receptor-triggering message conformations by restriction of conformational space through cyclization is proving its worth, for example with somatostatin (10), melanotropin (11), the enkephalins (12), and peptides related to Substance P (13). In these cases, it appears that certain receptors or their subtypes require folded conformations for optimal selection, interaction, and triggering. However, the rôle of address segments in chosing the proper receptor is still unclear: Do address segments interact classically with special surface elements of the receptors as proposed generally (5) and specifically for dynorphin (14)? Or do they in some unknown manner influence the conformation of the message segments (8,13)?

The proposed model of 1963 (5) also has its thermodynamic and kinetic difficulties. In the 'lock-and-key' concept, one out of many conformers present in solution must be selected for. The low probability of the peptide agonist being present as the proper conformer presents a high entropy barrier to the transition state of peptide--receptor interaction. 'Zipper' type and multistage reactions between peptide and receptor have been proposed to overcome this difficulty (15,16,17).

Finally, the question remains, why Nature has elaborated flexible peptides as hormones, neuromodulators, and neurotransmitters. A tentative answer applies genetic considerations (9). Sychnologic organization may aid in the expression of pleiotropic genes. We shall see that thermodynamic reasons must also be considered.

Quantitative Structure-Activity Relationships (QSAR). Josef Rudinger's scientific heritage was mainly British, as he had studied in Newcastle and fought in the RAF during World War II (18). His ideas about peptide biologic activity were strongly influenced by the work of Charles Ernest Overton (1865-1923) who, born in England, studied and worked in Zürich, and was Professor of Pharmacology in Lund from 1907--1923 (19). While still in Zürich, Overton proposed a theory of narcosis and analgesia in which he correlated potency of nonelectrolyte narcotics, such as chloroform and ether, with their partition coefficients between an aqueous and a hydrophobic phase (20). He postulated that the agonists dissolve in the membrane lipid phase and produce their narcotic effect by changing physico-chemical properties of cell membranes. The actions of basic organic electrolytes, such as alkaloids, did not fit into the scheme, and Overton postulated the formation of salt-like complexes between agonist and species- and effect-specific proteins, in other words: receptors. Overton's ideas appealed to many eminent chemists and pharmacologists, e.g. Pauling, Miller, Tanford, Hansch, and others. Considering salt-like and hydrophobic interactions between agonist and receptor led to quantitative estimates of structure-activity, i.e. structure-potency relationships (receptor selectivity excluded). In QSAR, eq. 1 (21), hydrophobic, steric, electronic, polarizability, and residual (unknown) molecular parameters are combined according to Hammett's linear free energy relationship.

$$\Delta G = \Delta G_{lip} + \Delta G_{elec} + \Delta G_{ster} + \Delta G_{pol} + \Delta G_{res} \qquad (1)$$

Despite its success with many drugs and its recent application to flexible peptides and their analogues with carboranyl-, adamantyl-, and (4-nitrophenyl)-alanine (22,23), QSAR is notoriously unsuccessful with peptides and, in its present form, cannot answer the question of receptor selection. We shall see that this is mainly due to the unequal distribution of hydrophobic and hydrophilic elements in many peptides which endows them with properties of surfactants.

Membrane-Catalyzed Receptor Selection. The discovery of specific, biologically significant interactions between regulatory peptides and artificial lipid membranes (24,25) offers a completely new approach to the problem of predicting potency and receptor selectivity of flexible peptides. Thus, the rôle of address segments becomes clear, thermodynamic and kinetic problems are alleviated, missing parameters in QSAR are provided, and certain advantages of flexible peptides over more rigidly folded proteins are suggested. Furthermore, new explanations for high capacity-low affinity and low capacity-high affinity binding sites are offered. The new approach is characterized by membrane catalysis of receptor interaction and selection (26,27,28,29). It has led to the elucidation of the molecular mechanism of opioid receptor selection

(30). This mechanism may be rather general, as suggested by investigations with adrenocorticotropin and melanotropin peptides, with substance P, eledoisin, and neurokinin B. Recent developments shall be described in the following chapters.

Membrane lipid phase as catalyst for peptide-receptor interactions (29). Dynorphin A-(1-13)-tridecapeptide ($dynorphin_{1-13}$) and adrenocorticotropin-(1-24)-tetracosapeptide ($ACTH_{1-24}$) react in aqueous solutions with neutral and anionic lipid membranes.

+YGGFLR(+)R(+)IR(+)PK(+)LK(+)- $Dynorphin_{1-13}$

+SYSME(-)HFR(+)WGK(+)PVGK(+)K(+)R(+)R(+)PVK(+)VYP- $ACTH_{1-24}$

Their accumulation on the membrane surface is driven by hydrophobic and electrostatic forces. On the membrane, they adopt preferred conformations and orientations: the N-terminal message segments, $dynorphin_{1-4}$ (14) and $ACTH_{1-10}$ (8), are inserted into the membrane hydrophobic phase as parts of quite perpendicularly oriented α-helical domains; the C-terminal address segments, $dynorphin_{5-13}$ (14) and $ACTH_{11-24}$ (8), remain in the aqueous phase as random coil domains (25,31,32,33,34,35), Figure 2.

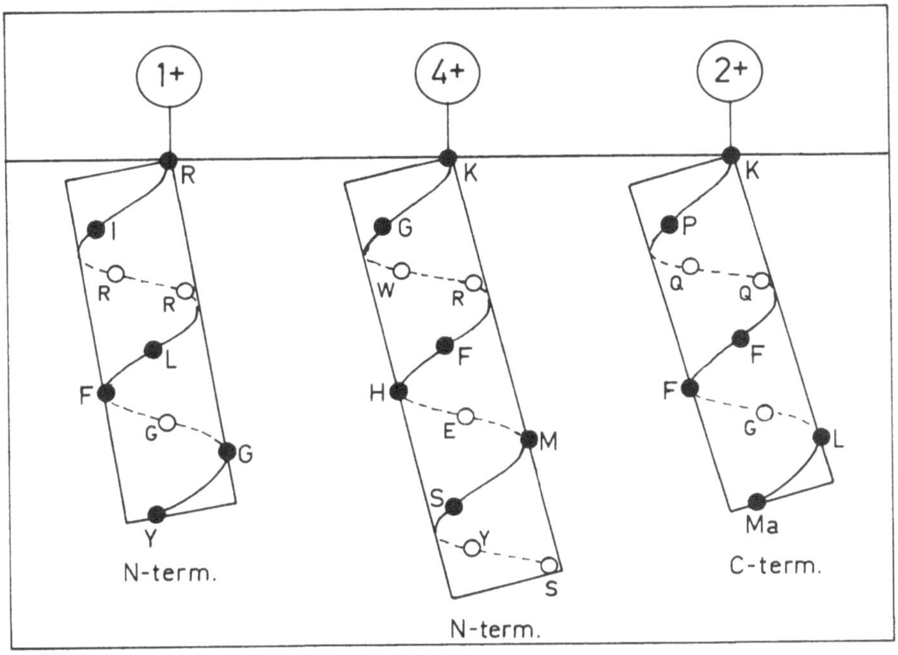

Fig. 2. Adaptation according to membrane requirements.
Left: Dynorphin-(1-13); centre: ACTH-(1-24); right: Substance P.

This situation suggests a catalytic action of the lipid phase of the target cell membrane on peptide-receptor interactions in much the same manner as detergent micelles catalyze bimolecular chemical reactions (26,27,28). "A catalyst decreases the free energy and activity coefficient of the transition state of the catalyzed reaction relative to the reactants in a given volume of solution" (36). In our cases, the binding energy between the peptide and the lipid membrane may be utilized to overcome the entropy requirements in bringing the reacting groups together. This is accomplished by accumulation and orientation of the peptides on the membrane surface (enhancement of the local concentration and reduction of the degrees of translational and rotational freedom). Furthermore, the free energy from the binding interaction between the membrane and part of the peptide may be utilized to force the reacting portion (message) into an environment in which it is destabilized and can reach the transition state more easily. Clearly, if a receptor requires interaction with a perpendicularly oriented helix and with charged or H bond forming groups, this condition is fulfilled. The consequences of this concept were examined theoretically together with my colleague, Dr. David F. Sargent (29).

In our model (29) a direct ligand-receptor interaction is replaced by multiple sequential steps including surface accumulation of charged ligands, ligand-membrane interactions and ultimately binding to the receptor itself. By dividing the total free energy of binding among several steps, the energy per step, including the intrinsic receptor interaction, is kept to moderate values. The model thereby yields simple explanations for the large apparent association constants, the high association and dissociation rates, and the heterogeneity of binding sites so frequently found with pharmacolcogical and biochemical ligand-receptor interactions. Furthermore, the measured apparent association constant is a function of the whole system rather than just the receptor. The same, fully functional receptor may show different binding characteristics in different surroundings, such as in different membrane compartments, another tissue or in a reconstituted system.

In summary, the expected catalysis of peptide-receptor interactions may be caused by (i) induced association of peptide with receptor, (ii) message destabilization, and (iii) effect of sequential steps. These mechanisms are expected to influence both the reaction rate and the equilibrium constant (36). The influence of two-dimensional diffusion of the agonist on the membrane surface (25) was not considered, as crucial parameters such as local viscosity are unknown.

Estimation of peptide membrane structures. I have suggested a simple method to estimate the preferred conformation, orientation, and accumulation of peptides on a membrane, using a pocket calculator. It is based on the premise that the most stable conformation of a relatively hydrophobic peptide segment in hydrophobic surroundings is the α-helix (37). It gives excellent values for the number of residues embedded in the hydrophobic phase, the hydrophobic association constant, and the orientation on the membrane surface of dynorphin$_{1-13}$ (38) and ACTH$_{1-24}$ (39), in agreement with the experimental observation. Values for electrostatic interactions may be calculated on the basis of assumed membrane surface potentials; their relative values are indicative of relative electrostatic accumulation on the membranes. Four parameters are important for estimating conformation, orientation, and accumulation of peptides on a membrane surface, Figure 2.

Hydrophobic association. The Gibbs free energy of hydrophobic association, $\Delta G°_{ass}(m)$, through m residues at the more hydrophobic end of a peptide chain is calculated from the free energy of transfer, $\Delta G°_{tr}(i)$, of the individual residues from their random-coil conformation in water to their helical conformation in a hydrophobic phase (40). The relations are:

$$\Delta G°_{tr}(m) = \Sigma_{i=1}^{i=m} \Delta G°_{tr}(i) + \Delta G°_{tr}(end) \tag{2}$$

$$\Delta G°_{ass}(m) = \Delta G°_{tr}(m) + \Delta G°_{t+r} \tag{3}$$

where $\Delta G°_{tr}(end)$ accounts for unsatisfied H bonds at the helix ends, and $\Delta G°_{t+r}$ is the free energy change caused by the loss of two degrees of rotational and one degree of translational freedom of the peptide bound to the membrane (41). The hydrophobic association constant and the length of the helix, m, are determined from the position of the energy minimum.

Amphiphilic Moment. Segregation of charged and uncharged amino acid residues into hydrophilic and hydrophobic domains endows peptides with an amphiphilic character. Such peptides will tend to accumulate on aqueous-hydrophobic interphase boundaries and orient themselves in the direction of minimum free energy. The segregation of hydrophobic and hydrophilic properties may be measured in analogy to the helical hydrophobic moment (42) by the molecular amphiphilic moment (38). In the helical hydrophobic moment the direction vectors originate on the helix axis and are perpendicular to it, whereas in the amphiphilic moment the direction vectors originate from the centre of the helix. The molecular amphiphilic moment is defined as

$$\vec{A} = \Sigma_{i=1}^{i=m} \Delta G°_{trh}(i)\vec{R}_i \tag{4}$$

where $\Delta G°_{trh}(i)$ is the signed numerical value of the Gibbs free energy change for the transfer of the ith residue in its helical conformation from water to a hydrophobic phase (values from (43)). $\vec{R}_i$ is the position vector from the helix centre to the α-carbon of the ith residue measured in units of helix radius, 0.188 nm. Random coil segments are assumed to exert their action at the helix end. The amphiphilic moment of a peptide located in a hydrophobic gradient produces a torque that tends to orient $\vec{A}$ perpendicular to the surfaces of equal hydrophobicity in the surrounding medium. The greater the scalar magnitude,

A, the less pronounced the thermal tumbling of the peptide molecules. Usually, a value of A of at least 150 arbitrary units is necessary to produce biologically relevant membrane associations (30).

Electric Dipole Moment. Alignment of peptide bond dipoles (44) and asymmetric arrangement of charged amino acid residues endow peptide helices with a molecular dipole moment. Such helices will tend to orient themselves in the direction of minimum free energy within the surface dipole layer of membranes. The molecular dipole moment was estimated in analogy to the molecular amphiphilic moment, eq. (4), by

$$\mu = \Sigma_{i=1}^{i=m} z_i \vec{R}_i \qquad (5)$$

where z_i is the charge number of the ionic side chain (assumed to be ±1 e) of the ith residue, or of the partial charges assigned to the helix ends (assumed to be ±0.63 e (44) located on the end residues, i = 1 and i = m). R_i is the position vector from the helix centre to the α-carbon of the ith residue, measured in units of 0.1 nm.

Net Charge. Charged peptides will be attracted or repulsed by the fixed charge layer of a membrane surface according to a Boltzmann distribution:

$$c_b/(c_{bmax} - c_b) = c_o \exp(-zFV_{gc}/RT) \qquad (6)$$

where c_b is the surface concentration of the peptide (number of adsorbed molecules per unit area), c_{bmax} the surface concentration at saturation, c_o the molar peptide concentration in the bulk phase, z the net charge, V_{gc} the Gouy-Chapman fixed charge potential, F the Faraday constant, R the universal gas constant, and T the absolute temperature in °K. Biologic membranes usually contain excess negatively charged lipid in such an amount that we may assume a characteristic V_{gc} of at least -40 mV for their lipid phase.

Positively charged peptides will accumulate in an aqueous compartment close to the fixed charge layer. Negatively charged peptides will accumulate in the bulk aqueous compartment, whereas neutral peptides will be indifferent. The two compartments caused by the fixed charge layer are characterized by the Debye--Hückel length:

$$l_D = (\varepsilon_0 \varepsilon RT/2c_e)^{1/2}/F \qquad (7)$$

which is the distance, x, from the fixed charge layer at which the electric potential, $\phi(x)$, has decreased from its maximum, ϕ_0, to ϕ_0/e. It depends on the dielectric constant, ε, of the phase (ε_0 is the electric field constant) and the electrolyte concentration, c_e. At physiologic conditions (simulated by c_e about 0.1M of a 1:1 electrolyte), l_D is about 1 nm or roughly 10 diameters of a water molecule, or the length of an α-helix with 6-7 residues.

<u>Molecular mechanism of opioid receptor selection</u> (30). Opioid peptides offer excellent examples of receptor subtype selectivity (45), an important phenomenon in nervous and endocrine regulation. Thus, [Leu]- and [Met]enkephalin prefer opioid δ-receptors, β-endorphin reacts about equally well with μ- and δ-sites, adrenorphin (metorphamide) is one of the most potent and selective, naturally occurring μ-agonists, while dynorphin A is a typical κ-agonist.

```
+YGGFL-                                        [Leu]enkephalin
+YGGFM-                                        [Met]enkephalin
        -+         +    +   ++ -
+YGGFMTSEKSQTPLVTLFKNAIIKNAYKKGE-              β-Endorphin
      ++
+YGGFMRRVamide                                 Adrenorphin
      ++ + + + -
+YGGFLRRIRPKLKWDNQ-                            Dynorphin A
```

Opioid peptides have a common N-terminal tetrapeptide sequence, Tyr-Gly-Gly-Phe-, which is the 'message' segment triggering the receptor responses, but they have different C-terminal 'address' segments which are responsible for their receptor subtype preferences.

I estimated the interactions of 26 synthetic and naturally occurring opioid peptides with anionic lipid membranes and compared the parameters with pharmacologic and binding data taken from the literature (Chavkin & Goldstein (14); Kosterlitz & Paterson (45)). The peptide amphiphilic moment (which measures the orientation of a peptide on an aqueous-hydrophobic interphase boundary) and the Boltzmann accumulation and repulsion on the anionic membrane surface turned out to be the principal determinants for opioid receptor selectivity and potency. Interaction with μ-receptors is governed by the net positive charge effective at the message domain of the agonist peptides, z(eff), as the Boltzmann term $e^{z(eff)}$ that determines the relative accumulation on anionic biologic membranes, eq. (6). Selection for δ-receptors decreases with increasing positive charge, and correlates with $e^{-z(eff)}$. Selection for κ-receptors is governed by the peptide amphiphilic moment, $\vec{A}$. A pronounced scalar magnitude, A, and almost perpendicular orientation of the N-terminal message domain as an α-helix are favourable for κ-selection. Potencies and binding affinities correlate with $A \cdot e^{z(eff)}$. This indicates that the κ-site recognizes those agonist molecules that are accumulated on the hydrophobic side of the negative surface layer of the membrane <u>and</u> have the proper orientation and conformation.

Classical site selectivity caused by <u>receptor requirements</u> for complementary and 'productive' (17) fit of the agonist to the discriminator sites of opioid receptors (e.g. requirements for Tyr-Gly-Gly-Phe- as N-terminal message for all opioid receptors, and folded conformations for μ-sites (12)) is thus crucially complemented by a set of membrane requirements. Membrane requirements are defined by a list of peptide structural features necessary for 'productive' membrane interaction. Productive membrane interaction means a membrane interaction specifically facilitating productive binding to a particular receptor. The ensuing selection mechanism is shown in Figure 3. In this model, the δ-site is exposed to the aqueous compartment surrounding the target cell at a distance comparable to or greater than the Debye-Hückel length, and is in a cationic vicinity. The μ-site is exposed to the anionic fixed charge compartment of the membrane in aqueous surroundings. The κ-site is buried in a more hydrophobic membrane compartment close to the anionic fixed charge layer. The relative accumulation of the opioid message domains in these compartments is determined by the address domains and constitutes a major part of the site selection mechanism.

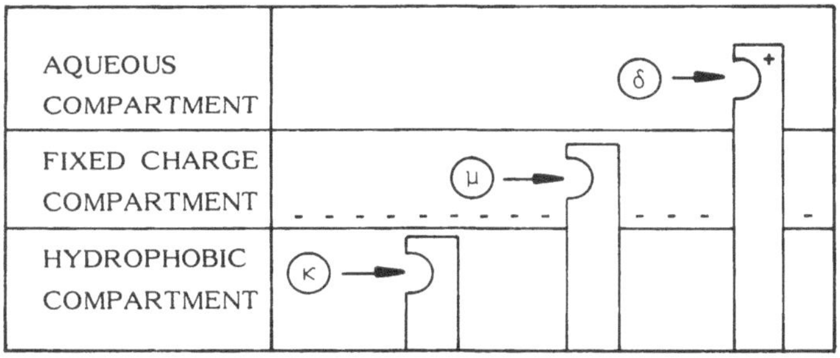

Fig. 3. Membrane Compartment Concept.
Molecular Mechanism of Opioid Receptor Selection.

This means that the δ-receptor may interact directly with peptides present in the aqueous phase, preferably with peptides carrying zero (or negative?) net charge. It will impose its conformational requirements on such peptides according to the classical concept. The μ-sites will encounter mainly those peptides that carry a positive net charge but do not insert their message domains into the hydrophobic compartment. Adsorption to the anionic surface may, without intervention of the receptor, impose folded structures on the peptides. It is known from cyclization experiments (46) and from IR investigations (47) that adsorption to surfaces restricts the conformational space of flexible molecules and favours β-type structures. The κ-sites will interact with those peptide message segments that intrude into a relatively hydrophobic compartment (which membrane layer this is, hydrocarbon or ester bond, is unknown at present). The fact that almost perpendicular orientation of a helical structure correlates with receptor binding and triggering may indicate that the κ-site requirement is for a perpendicularly inserted helix.

This new molecular mechanism for opioid receptor selection readily explains the facts that replacement of Gly^2 in dynorphins by D alanine destroys κ-site activity and enhances μ-interaction, and that the site specificity of enkephalins (zero net charge) is shifted from $\delta > \mu$ to $\delta < \mu$ in the enkephalin amides (net charge 1+). The proposed topological arrangement of the δ-site resembles that of the acetyl choline receptor (48) with its discriminator site about 4 nm distal from the lipid head groups. A somewhat shorter distance, about 3 nm or 3 times the Debye-Hückel length may easily be bridged by β-endorphin in its conformation postulated by Kaiser & Kézdi (49): the address positive charges may remain in contact with the fixed charge compartment while the message domain could reach the δ-site. An alternative δ-site is the surface of a separate 'δ-cell' with a positive Gouy-Chapman potential (30).

Preferred conformation, orientation, and accumulation of Substance P on lipid membranes. The estimated membrane structure of Substance P (A) has a quite perpendicularly oriented C-terminal α-helix, extending from Lys^3 to Met^{11} amide (Schwyzer, Erne, Rolka (50)). The molecular amphiphilic moment and dipole moment vectors are strong, 338 arbitrary units and 124 Debye, respectively. They point towards the N-terminus, 15° and 9° away from the helix axis. The estimated molar dissociation constants for the equilibrium between membrane-bound and free Substance P in aqueous solution are about 46mM for neutral membranes and 0.43mM to 36nM for anionic membranes with 10% to 100% molar ratios of anionic to neutral lipid, respectively. Thus, membrane interactions with negatively charged anionic target cell membranes are predicted to be biologically significant (29). These predictions are consistent with experimental data obtained with circular dichroism (51) and infrared (47) techniques. In particular, the formation of helical structures on anionic, but not on neutral liposomes is confirmed. The predictions for [Leu^9]substance P (stronger hydrophobic association) and for Substance P-(2-11)-decapeptide (weaker electrostatic interaction) are also consistent. Infrared band shape analysis with quantum-mechanical parameters shows that the helix length present in trifluoroethanol solution corresponds exactly to the value calculated for an interphase boundary. The Chassaing model with α-helical and folded structures (52) is predicted to be less stable in hydrophobic surroundings than our model by some 31 kJ/mol and has considerably weaker amphiphilic and dipole moments, excluding biologically significant membrane interactions. We propose that Substance P, adsorbed on anionic lipid membranes, interacts with the membrane in principally the same manner as $Dynorphin_{1-13}$ and $ACTH_{1-24}$; the only difference is that the message domain that is inserted into the hydrophobic phase is C-terminal in Substance P and N-terminal in the other peptides.

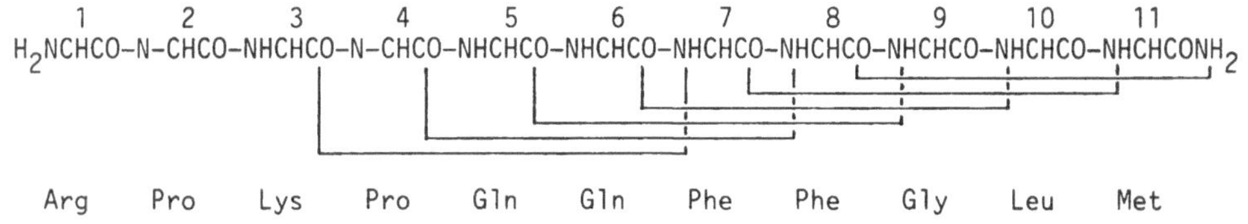

A: Hydrogen bonding system of Substance P on membranes (estimated) and in trifluoroethanol (IR band shape analysis).

Biologically inactive [D-Leu^9]substance P was shown neither to form an α-helix at the C-terminus, nor to interact with anionic membranes because of the lack of hydrophobic interaction without the helix (K. Rolka, to be published).

Predicted molecular mechanism of Substance P receptor selection. Two subtypes of Substance P receptor are found to exist in rat brain cortex cell membranes. They have been convincingly demonstrated and defined by their relative affinities for Substance P (SP), Eledoisin (EL), and Neurokinin B (NKB) (13). One subtype, which I shall call the SP-site, reacts with SP about 33 times more strongly than with EL. This subtype is present in rat parotid cells as a practically homogeneous population, and appears to be responsible for the sialogogic effect of SP. The other subtype, which I shall call the EL-site, reacts preferentially with EL, about 7 times more strongly than with SP. Neurokinin B, found in mammalian brain, reacts 60 times more strongly with EL-sites than with SP-sites. The cyclic peptide IA that mimicks a folded structure of the C-terminal hexapeptide message segments of the natural products shows a behaviour towards the two subsites that is very similar to that of EL and NKB in selectivity and potency, but not to that of SP.

+Arg(+)-Pro-Lys(+)-Pro-Gln-Gln-Phe-Phe-Gly-Leu-Met-NH_2	Substance P, SP (3+)
Pyr-Pro-Ser-Lys(+)-Asp(-)-Ala-Phe-Ile-Gly-Leu-Met-NH_2	Eledoisin, EL (0)
+Asp(-)-Met-His(+)-Asp(-)-Phe-Phe-Val-Gly-Leu-Met-NH_2	Neurokinin B, NKB (0)
Pyr-Phe-Phe-NHC*HCO-Leu-Met-NH_2	IA (0) (* = R)

The authors (13) exclude the exchange of Phe^8 in Substance P for Ile and Val in Eledoisin and Neurokinin B as the cause for receptor selection. They find, however, that the EL-site has a conformational requirement for a folded conformation similar to that in IA. This requirement is not readily met by Substance P. They conclude that: "...unknown determinants in the amino-terminal sequences of substance P must strongly contribute to the carboxyl-terminal peptide selectivity and conformation."

These determinants are revealed by estimation of the membrane structures of SP, EL, and NKB. We have seen that SP is able to interact with anionic membranes in a conformation- and orientation-specific manner. This is due to the net charge and the amphiphilic and dipole moments supplied by the N-terminal Arg-Pro-Lys-segment. The amphiphilic moment and the dipole moment prevent thermal tumbling and provide the proper orientation so that the C-terminal domain can assume an α-helical structure providing hydrophobic binding energy in addition to the electrostatic binding energy. This orientation and conformation allows the message domain to interact with the SP-site which, according to this model, must be exposed to a hydrophobic compartment, very much like the opioid κ-receptor. The unbound Substance P in the aqueous phase which is in equilibrium with the membrane-bound species may interact with the EL-sites and meet their requirement for a folded structure of the message. This interaction must, again according to this model, take place at an EL-site exposed to water. This explains semiquantitatively the 12-17% potency of SP at the EL-site compared to EL. EL and NKB lack not only the net charge, but also a sufficient hydrophobic interaction energy to become associated with the membrane in a manner similar to that of SP. Furthermore, their amphiphilic moments are smaller and less aligned to the helix axes than that of SP. EL and NKB must, therefore interact with an EL-site exposed to water. Whether this site is similar to the opioid δ-site or to the μ-site is hard to decide with the biologic data at hand.

Selection for steroidogenic, melanotropic, and neuromodulatory receptors (39). Peptides of the opiomelanocortin family act as hormones and neuropeptides. The main action of adrenocorticotropin is on steroidogenic receptors in the adrenal cortex. A well-known action of the melanotropins is on melanophore receptors. Both peptides (and the endorphins) also act in the central nervous system, probably as neuromodulators. As mentioned above, the predicted and experimentally observed membrane-structures of adrenocorticotropin-(1-24)-tetracosapeptide ($ACTH_{1-24}$) agree excellently. The peptide interacts strongly with neutral membranes ($K_d = 5x10^{-5}M$) and inserts its message, contained in residues 1-11, into the membrane hydrophobic phase as an almost perpendicularly oriented α-helix. Interaction with anionic membranes, $V_{gc} = -40mV$, is expected to be very strong, K_d about $4x10^{-9}M$. The 'free' message, ACTH-(1-10)-decapeptide, is estimated and experimentally shown not to interact with neutral and anionic membranes, mainly because of unfavorable amphiphilic and dipole moments, and lack of net charge. Alpha-melanotropin (α-MSH), Ac-Ser-Tyr-Ser-Met-Glu-His-Phe-Arg-Trp-Gly-Lys-Pro-Val-NH_2, shares its amino acid sequence with the N-terminus of $ACTH_{1-24}$, but is acetylated and amidated. Acetylation results in reduction of hydrophobic interaction ($K_d = 8x10^{-4}$) compared to $ACTH_{1-24}$. Its molar dissociation constant from anionic membranes, $V_{gc} = -40mV$, was estimated as $1x10^{-4}$. The calculated amphiphilic and dipole moments are not very favourably oriented and probably too weak to keep the molecules oriented on an interphase boundary. Dr. Shunsaku Kimura in my laboratory investigated α-MSH-membrane interactions experimentally and found that the conformations and orientations typical of $ACTH_{1-24}$ were not adopted on neutral and anionic membranes (report for Helv. Chim. Acta in preparation). It appears from an analysis of the available date on biologic action, that the topological arrangement of the steroidogenic, melanotropic, and neuromodulatory receptors resembles that of the opioid κ-, μ-, and δ-sites, respectively. For a detailed discussion, see (39).

The Membrane Compartment Concept (Conclusion). A regulatory peptide approaching its target cell from the intercellular fluid by diffusion may encounter receptors protruding their discriminator sites into the aqueous phase surrounding the cell at a distance equal to or greater than the Debye-Hückel length where the surface potential of the cell membrane is practically ineffective. These discriminators in the aqueous compartment will recognize and interact with those peptides that can meet their specific 'receptor requirements'. If these receptors carry a charge near the discriminator site, the charge on the peptide will be an important component of the 'receptor requirement'.

Peptides carrying positive charges will accumulate in the aqueous phase close to the negative fixed charge compartment of the membrane according to a Boltzmann distribution. The population of conformers of these pepides is probably almost identical with that present in the aqueous compartment, although restrictions may be imposed by effects of charge or reduction of water activity by the head groups. Receptors exposing their discriminator sites into this compartment will experience a peptide concentration higher than that in the aqueous compartment. Those accumulated peptides that can meet the requirements of one of these receptors will interact with it and trigger the specific response. However, a very long peptide with positive charges in the address domain that are remote from the message may have its address domain accumulating in the fixed charge compartment and its message domain interacting with a discriminator site in the aqueous compartment, particularly if the message and its connection to the address are sufficiently hydrophilic (see β-endorphin).

Peptides accumulating in the fixed charge compartment will interact with the hydrophobic compartment of the membrane if they meet certain 'membrane requirements'. The interaction is regioselective, exposing the peptide message into the hydrophobic compartment. It is conformation-selective, imposing a helical structure on the message-containing peptide domain with a maximum of internal H bonds and Coulomb interactions. It is also orientation-selective, orienting the helical message domain on the membrane surface according to the amphiphilic and dipole moments. Those receptors exposing their discriminator sites into the hydrophobic compartment will interact with those inserted peptides that meet their 'receptor requirements'. These may be closely related to the 'membrane requirements'. Where, exactly, the hydrophobic compartment is located, is not clear: it may be mainly the membrane layer containing the lipid ester bonds and the outer portions of the lipid hydrocarbon chains (report with Dr. Shunsaku Kimura in preparation).

Acknowledgement. I wish to thank my former colleagues, Prof. U.-P. Fringeli, Dr. H.-U. Gremlich, Dr. B. Gysin, Dr. K. Nakajima (Osaka), and Dr. P. Schoch, as well as my present colleagues, Drs. J. Bean, D. Erne, S. Kimura (Kyoto), K. Rolka (Gdansk), D. F. Sargent, S. D. Sharma (Lucknow), and A. Tun-Kyi, and my secretary, Mrs. M. Schütz, for their expert contributions and friendship.

References

1. Rudinger, J. 1971. In: Drug Design (E. J. Ariëns, ed.). Academic Press, Vol. 2, pp. 319-419.

2. Schwyzer, R. 1954. Angew. Chem. 66, 345-348.

3. Isler, O. 1978. Biographical Memoirs of Fellows of the Royal Society 24, 245-321.

4. Parascandola, J. 1981. In: Towards Understanding Receptors (J. W. Lamble, ed.). Elsevier North Holland Biomedical Press, p. 1-7.

5. Schwyzer, R. 1963. Pure Appl. Chem. 6, 265-295.

6. Schwyzer, R., C. H. Li. 1958. Nature (London) 182, 1669-1670.

7. Schwyzer, R. 1973. In: Peptides 1972 (H. Hanson and H. D. Jakubke, eds.), North Holland Publ. Co., pp. 424-436.

8. Schwyzer, R. 1977. Annals N. Y. Acad. Sci. 297, 3-26.

9. Schwyzer, R. 1980. Proc. R. Soc. Lond. B210, 5-20.

10. Veber, D. F., F. W. Holly, W. J. Palaveda, R. F. Nutt, S. J. Bergstrand, M. Torchiana, M. S. Glitzer, R. Saperstein, R. Hirschmann. 1978. Proc. Natl. Acad. Sci. USA 75, 2636-2640; Veber, D. F., R. M. Freidinger, D. S. Perlow, W. J. Palaveda, F. W. Holly, R. G. Strachan, R. F. Nutt, B. H. Arison, C. Homnick, W. C. Randall, M. S. Glitzer, R. Saperstein, R. Hirschmann. 1981. Nature (London) 292, 1-3.

11. Sawyer, T. K., W. L. Cody, J. J. Knittel, V. J. Hruby, M. E. Hadley, M. D. Hirsch, T. L. Donohue. 1983. In: Peptides, Structure and Function (V. J. Hruby and D. H. Rich, eds.), Pierce Chemical Co., Rockford, Ill., pp. 323-331.

12. Schiller, P. W. 1984. In: The Peptides (E. Gross and J. Meienhofer, eds.), Academic Press, N. Y., Vol. 6, 219-268; Schiller, P. W., T. M.-D. Nguyen, C. Lemieux, L. A. Maziak. 1985. J. Med. Chem. 28, 1766-1771; Schiller, P. W. 1986. In: Medicinal Chemistry and Molecular Pharmacology of Opiate Peptides and Opiates (S. R. Rapaka, R. L. Hawks, and G. Barnett, eds.), Government Printing Press, Rockville, MD, in press.

13. Cascieri, M. A., G. G. Chicchi, R. M. Freidinger, C. D. Colton, D. S. Perlow, B. Williams, N. R. Curtis, A. T. McKnight, J. J. Maguire, D. F. Veber, T. Liang. 1986. Molecular Pharmacol. 29, 34-38.

14. Chavkin, C., A. Goldstein. 1981. Proc. Natl. Acad. Sci. U.S.A. 78, 6543-6547.

15. Laiken, N., G. Némethy. 1971. Biochemistry 10, 2101-2106.

16. Burgen, A. S. V., G. C. K. Roberts, J. Feeney. 1975. Nature (London) 253, 753--755.

17. Franklin, T. J. 1980. Biochem. Pharmacol. 29, 853-856.

18. Schwyzer, R. 1975. Chimia (Aarau) 29, 376-377.

19. Lipnick, L. R. 1986. Trends Pharmacol. Sci. (TIPS) 7, 161-164.

20. Overton, E. 1901. Studien über die Narkose, zugleich ein Beitrag zur allgemeinen Pharmakologie. Gustav Fischer, Jena.

21. Hansch, C., A. Leo. 1979. Substituent Constants for Correlation Analysis in Chemistry and Biology. John Wiley& Sons.

22. Do, K. Q., J.-L. Fauchère, R. Schwyzer, P. W. Schiller, C. Lemieux. 1981. Hoppe-Seyler's Z. Physiol. Chem. 362, 601-610.

23. Fauchère, J. L. 1982. J. Med. Chem. 25, 1428-1431.

24. Schoch, P., D. F. Sargent, R. Schwyzer. 1979. Biochem. Soc. Trans. 7, 846-849; Schoch, P. 1980. Kapazitätsminimalisierung: eine neue Methode zur Messung von Oberflächenpotentialen und ihre Anwendung auf Lipid-Peptid Wechselwirkungen. Dissertation ETH Nr. 6699, Zürich.

25. Schwyzer, R., H.-U. Gremlich, B. Gysin, D. F. Sargent, H.-P. Fringeli. 1983. In: Peptides 1982 (K. Bláha and P. Maloň, eds.). Walter de Gruyter & Co., pp. 55-71.

26. Schwyzer, R. 1985. In: Peptides, Structure and Function (C. M. Deber, V. J. Hruby, and K. D. Kopple, eds.). Pierce Chemical Co., Rockford, Ill., pp. 3-12.

27. Schwyzer, R. 1985. In: Natural Products Chemistry 1984 (R. I. Zalewski and J. J. Skolik, eds.). Elsevier Science Publishers B.V., pp. 591-598.

28. Schwyzer, R. 1986. In: Natural Products and Biological Activities (H. Imura, T. Goto, T. Murachi, and T. Nakajima, eds.). University of Tokyo Press, pp. 197-207.

29. Sargent, D. F., R. Schwyzer. 1986. Proc. Natl. Acad. Sci. U.S.A. (in press).

30. Schwyzer, R. 1986. Biochemistry 25, in press.

31. Schwyzer, R., H.-U. Gremlich, B. Gysin, D. F. Sargent, U.-P. Fringeli. 1983. In: Peptides, Structure and Function (V. J. Hruby and D. H. Rich, eds.). Pierce Chemical Co., Rockford, Ill., pp. 657-664.

32. Gremlich, H.-U., U.-P. Fringeli, R. Schwyzer. 1983. Biochemistry 22, 4257-4264; 1984, ibid. 23, 1808-1810.

33. Gysin, B., R. Schwyzer. 1984. Biochemistry 23, 1811-1818.

34. Gysin, B., R. Schwyzer. 1983. Arch. Biochem. Biophys. 225, 467-474.

35. Erne, D., D. F. Sargent, R. Schwyzer. 1985. Biochemistry 24, 4261-4263.

36. Jencks, W. P. 1975. Adv. Enzymol. Relat. Areas Mol. Biol. 43, 219-410.

37. Henderson, R. 1979. Soc. Gen. Physiol. 33, 3-15.

38. Schwyzer, R. 1986. Biochemistry 25, 4281-4286.

39. Schwyzer, R. 1986. Helv. Chim. Acta 69, in press.

40. Von Heijne, G., C. Blomberg. 1979. Eur. J. Biochem. 97, 175-181.

41. Janin, J., C. Chothia. 1978. Biochemistry 17, 2943-2948.

42. Eisenberg, D., R. M. Weiss, T. C. Terwilliger. 1982. Nature (London) 299, 371--374; 1984. Proc. Natl. Acad. Sci. USA 81, 140-144.

43. Von Heijne, G. 1981. Eur. J. Biochem. 97, 175-181.

44. Wada, A. 1976. Adv. Biophys. 9, 1-63; Hol, W. G. J. 1985. Prog. Biophys. Mol. Biol. 45, 149-195.

45. Kosterlitz, H. W., S. J. Paterson. 1985. Phil. Trans. R. Soc. Lond. B 308, 291--297.

46. Schwyzer, R., J.-P. Carrión, B. Gorup, H. Nolting, A. Tun-Kyi. 1963. Helv. Chim. Acta 47, 441-464.

47. Erne, D., K. Rolka, R. Schwyzer. 1986. Helv. Chim. Acta 69, in press.

48. Conti-Tronconi, B. M., M. A. Raftery. 1982. Ann. Rev. Biochem. 51, 491-530.

49. Kaiser, E. T., F. J. Kézdi. 1983. Proc. Natl. Acad. Sci. USA 80, 1137-1143.

50. Schwyzer, R., D. Erne, K. Rolka. 1986. Helv. Chim. Acta 69, in press.

51. Rolka, K., D. Erne, R. Schwyzer. 1986. Helv. Chim. Acta 69, in press.

52. Chassaing, G., O. Convert, S. Lavielle. 1986. Eur. J. Biochem. 154, 77-85.

FROM PEPTONES TO PEPTIDES - A HISTORICAL RETROSPECT

Joseph S. Fruton
Yale University, New Haven, Connecticut 06520

I thank the Organizing Committee not only for the honor of participating in this Symposium, but also for the opportunity to acknowledge a debt incurred a half-century ago. In June 1934, I joined the laboratory of Max Bergmann at the Rockefeller Institute for Medical Research in New York. A few months later, Leonidas Zervas arrived there from Dresden, and was installed in the laboratory where I had begun to work. This association, which lasted two years, changed my perception of the art and rigor of chemical research. I am proud to have been one of Professor Zervas's first students, for I consider him to have been the true pioneer of modern peptide synthesis.

I have been asked to speak about the historical development of peptide chemistry. The period I shall try to cover spans nearly one hundred years, and may be divided roughly into three stages: the nineteenth-century work on the products of the enzymatic digestion of proteins; then, the contributions of Emil Fischer, Theodor Curtius and others during the first three decades of this century; and finally, the period 1930-1940, whose highlight was of course the Bergmann-Zervas carbobenzoxy method.

It is customary to begin an account of the history of peptide chemistry with the work of Emil Fischer [1852-1919], and this is justified to the extent that he was the first to explore the field systematically, and that he introduced the word peptide into the language of chemistry. The story properly begins, however, in the middle of the nineteenth century, with the study by physiologists of the digestion of dietary proteins in the gastrointestinal tract of man and higher animals. After the extensive studies by Theodor Schwann [1810-1882] on pepsin (1), discovered by Johann Nepomuk Eberle [1798-1834], during the 1840's Louis Mialhe [1807-1886] examined the action of gastric juice on a variety of protein materials. He was followed by Carl Gotthelf Lehmann [1812-1863], who introduced the term peptone to denote the products formed in this process (2). Throughout the nineteenth century, the

Peptides 1986

study of protein metabolism was dominated by the so-called Pepton-lehre, according to which blood proteins were considered to be derived directly from absorbed peptones formed in the digestive tract. Leading physiological chemists, notably Willy Kühne [1837-1900] and Franz Hofmeister [1850-1922], devoted much effort to the fractionation and characterization of peptones, and also of the so-called albumoses, thought to be intermediates in the formation of peptones from proteins (3).

The word peptide first appeared in print in 1902, at the end of a brief note by Emil Fischer (4). He proposed that "in analogy to the known designation of carbohydrates as disaccharides, trisaccharides, etc., the substances of the type glycyyl-glycine be named dipeptides and anhydride-like combinations of a greater number of amino acids be denoted tripeptides etc." In subsequent publications, Fischer also began to use the term polypeptide. Clearly, he derived the suffixes from the names he had introduced into the sugar field, but he also followed the ancient tradition that associated the digestion of proteins with pepsis.

The circumstances of the publication of Fischer's note are of some historical interest. During the week of September 21-27, 1902, the Society of German Scientists and Physicians held its 74th annual meeting in Karlsbad. The first of the principal scientific lectures was presented during the morning of September 22nd by Hofmeister, under the title Über den Bau des Eiweissmoleküls. On the afternoon of the same day, at the first session of the Section of Chemistry, Fischer gave a report Über die Hydrolyse der Proteinstoffe. In both talks, the theory was advanced that in proteins the constituent amino acids are joined to each other by amide bonds. In later historical accounts, these two lectures mark the appearance of the so-called Fischer-Hofmeister theory of protein structure.

In his lecture (5), Hofmeister attached special significance to the biuret reaction - the purple color given by proteins and peptones with alkaline copper sulfate (6). Among the synthetic materials, apart from biuret, that give this color test he cited the products obtained by Theodor Curtius [1857-1928] during the early 1880's either by the self-condensation of glycine ethyl ester or by the reaction of benzoyl chloride with silver glycinate (7). In the latter process, one of the isolated products was hippuryl-glycine, which in retrospect may be considered to represent the first synthetic peptide derivative. In favor of his theory, Hofmeister also offered evidence from physiological studies on the enzymatic cleavage of proteins.

In contrast to Hofmeister's lecture, the paper read by Fischer was not published. Instead, a brief Autoreferat prepared by Fischer appeared a few weeks later in the Chemiker-Zeitung (4). After describing his results on the isolation of amino acids and peptides from protein hydrolysates, he also wrote: "Finally the speaker discussed the coupling of the amino acids in protein molecules. The idea that amide groups play the principal role most readily comes to mind [liegt am nächsten], as Hofmeister also assumed in his general lecture this morning. The same conviction led him [Fischer] more than one and a half years ago to initiate experiments to effect the synthetic linkage of amino acids."

It should noted that the idea of amide linkages in proteins had appeared in the chemical literature before 1902. During the period 1860 to 1900, several chemists attempted to prepare peptone-like materials by subjecting amino acids to dehydration at high temperatures. One of these investigators was Édouard Grimaux [1835-1900]; in 1882, he published a paper (8) in which alanyl-alanine is depicted as an amide. In this paper he wrote: "If one succeeds in producing anhydrides containing the residues of aspartic acid, of leucine, of tyrosine, etc., one may thereby be able to transform them into amides that resemble more and more the nitrogenous colloids made by living organisms."

To return to Fischer's priority claim in his Autoreferat, this was based on his famous 1901 paper with Ernest Fourneau (9). They described the preparation of the first free peptide - glycyl-glycine - by partial hydrolysis of 2,5-diketopiperazine, made by Curtius during the 1880's. Also, they attempted to use the carboethoxy group to protect the alpha-amino group in condensation reactions by the acid-chloride method. Fischer hoped to remove the carboethoxy group selectively by treatment with ammonia, but this approach proved to be unsatisfactory. He chose instead to circumvent the need for an amino-protecting group by starting a peptide synthesis with an alpha-halogen acyl halide. By 1907, Fischer's associates had used this method to prepare over a hundred peptides, including the octadecapeptide H-Leu-$(Gly)_3$-Leu-$(Gly)_3$-Leu-$(Gly)_8$-Gly-OH (10). This formidable synthesis was completed by Fischer's assistant Walter Axhausen (b. 1882).

At each stage, Fischer called attention to the fact that, like the albumoses and peptones, the synthetic peptides gave the biuret reaction and that most of them were cleaved to amino acids by pancreatic enzymes. Indeed, he claimed that "...l-leucyl-triglycyl-l-tyrosine prepared artificially has all the properties of the albumoses. These observations are of importance in casting doubt on the view which formerly prevailed that, being intermediate products between proteins and peptones, the albumoses are substances of considerable molecular complexity" (11).

Also, Fischer repeatedly declined to accept the evidence showing that proteins have molecular weights above 15,000, and he set the maximum value at 4,000 to 5,000. For Fischer, the crystalline proteins isolated during the nineteenth century by Hofmeister and other physiological chemists were aggregates of such smaller molecules, and he hoped to synthesize a protein by mixing the appropriate peptides made in his laboratory. In a letter to his former teacher, Adolf von Baeyer [1835-1917], written in December 1905, Fischer stated: "My entire yearning is directed toward the first synthetic enzyme. If its preparation falls into my lap with the synthesis of a natural protein material, I will consider my mission fulfilled" (12). A month later, he delivered before the German chemical society a lecture (13) that evoked great attention in the popular press. Many people came to believe that the preparation of synthetic proteins was around the corner.

It is difficult to escape the impression that Fischer's insistence on the relatively low molecular weight of natural proteins came in large part from his unwillingness to admit publicly of the possibility that his method was inadequate for the synthesis of the large molecules advocated by the physiological chemists. In the excitement over Fischer's synthetic peptides, Hofmeister's name was forgotten, but it is worth noting that in 1908 he published a paper entitled Einiges über die Bedeutung und den Abbau der Eiweisskörper." The paper (14) began as follows: "That the end products of protein hydrolysis are largely linked to one another as amides was concluded by me on the basis of chemical and physiological evidence, and was securely established by E. Fischer through the preparation of such compounds - the peptides." Hofmeister then stated that the isolation of well-defined cleavage products was the most important requirement for the determination of protein structure. He summarized the efforts that had been made to isolate peptones in the form of acyl derivatives, and concluded that the precipitation methods used in this work could not yield chemically-homogeneous products. He did not mention, and probably did not know, that Mikhail Semenovich Tsvet [1872-1919] had begun to use adsorption chromatography to separate natural pigments; it was many years before anyone sought to apply this approach to the protein problem. Finally, Hofmeister noted that: "The rapid progress in the synthesis of polypeptides under the aegis of Fischer could lead to the view that one may expect the elucidation of protein structure to come entirely from this approach. This hope at once comes to naught if one considers the enormous number of synthetic possibilities." Indeed, Fischer's confidence in the power of the alpha-halogen acyl halide method proved to be short-lived. The strenuous efforts of his assistants and students produced much less than he had hoped for, and his disappointment may be inferred from the fact that after 1910 there were no further experimental publications on peptide synthesis from his laboratory.

Another name forgotten was that of Curtius. Stimulated by Fischer's entry into the protein field, Curtius resumed his work on peptides and in 1902 developed a coupling method involving the azides of acylamino acids. Some harsh words were exchanged, and each of these great chemists felt that the other had not cited his work properly (15). The public accolade went to Fischer, but the verdict of history has gone to Curtius. Today, peptide chemists have little occasion to use alpha-halogen acyl halides, but the azide coupling method is still valuable.

In the years between the end of the First World War and the invention of the carbobenzoxy method, there were only modest advances in the art of peptide synthesis. The Fischer method was used extensively only the laboratory of his disciple Emil Abderhalden [1877-1950], who also elevated to the status of a theory Fischer's suggestion in 1906 that diketopiperazines are structural elements of proteins (16). Apparent support for this view came from X-ray data for silk fibroin. The diffraction patterns indicated an ordered structure, and were consistent with either a long polypeptide chain or the non-covalent association of glycyl-alanyl diketopiperazine units. The latter alternative was favored, and it was until organic chemists began to accept the the idea advanced by Hermann Staudinger [1881-1965] that long-chain macromolecules could exist that they acknowledged the validity of the views advanced by Hofmeister and other biochemists that proteins are composed of long polypeptide chains.

During the 1920's, Max Bergmann [1886-1944] also proposed a theory that invoked the presence, in proteins, of unsaturated diketopiperazine units. In connection with this idea, he described (17) a method of peptide synthesis that involved the use of oxazolones of the kind studied earlier by Emil Erlenmeyer [1864-1921]. The reaction of the oxazolones with natural amino acids yielded acyldehydropeptides; hydrogenation gave a mixture of the two diastereoisomeric dipeptides, which had to be separated. After Bergmann came to New York in 1933, a large part of his personal research effort was devoted to the synthesis of dehydropeptides by the oxazolone method.

A small contribution of considerable historical interest was made during the 1920's by a young physician, Rudolf Schoenheimer [1898-1941], while working in the laboratory of Karl Thomas [1883-1969] at Leipzig. Schoenheimer took up the problem that Fischer had abandoned - the search for an amino-protecting group that could be removed without cleavage of the newly-formed peptide bond. He took advantage of the finding made in 1914 by Bergmann, shortly after becoming Fischer's assistant, that tosylamino acids can be cleaved by reduction with a mixture of hydriodic acid and phosphonium iodide. Schoenheimer used the azides of tosylamino acids to make a few simple peptides in this manner (18). As is well known, Schoenheimer later gained distinction for introducing the isotope technique into the study of intermediary metabolism.

Still another contribution of the 1920's was the work of Friedrich Wessely [1897-1967]. I mentioned before that in 1901 Fischer had attempted to use the carboethoxy group for amino-protection in peptide synthesis by the acid chloride method. A few years later, Hermann Leuchs [1879-1945] showed that the acid chlorides of carboethoxy-amino acids are readily converted into N-carboxy-anhydrides. Wessely (19) studied the polymerization of these so-called Leuchs anhydrides, and produced peptides that attracted some attention after the Second World War (20).

This brings us to the invention of the carbobenzoxy method, and its application during the decade before the war. The key step - the removal of a benzyl group by hydrogenolysis - had its origins during the early 1920's, when there appeared several papers on the hydrogenolysis of benzyl ethers and esters with palladium black as the catalyst, and in 1928 Karl Freudenberg [1886-1983] described the use of this method in the synthesis of sugars (21).

Three years later, there appeared a paper with Leonidas Zervas [1902-1980] as the sole author on the synthesis of 1-benzoylglucose by catalytic hydrogenolysis of its 4,6-benzylidene derivative (22). It is possible that the work on which this 1931 paper was based may have triggered the idea of using benzylchlorocarbonate in place of the ethylchlorocarbonate that Fischer and Fourneau had employed thirty years earlier in their attempt to find a suitable amino-protecting group for the synthesis of peptides.

In their famous 1932 paper (23), and in those published from the Dresden institute during the next two years, Bergmann and Zervas clearly established the carbobenzoxy method as the most promising approach yet devised to meet the challenge set by Fischer. They described the synthesis of many hitherto inaccessible peptides, including those containing amino acid residues with reactive side chains. The introduction of these residues into peptide chains involved the use of blocking groups that could be removed by catalytic hydrogenolysis after the chain had been assembled. Moreover, a very important advantage of using carbobenzoxy-amino acids in coupling reactions was found to be that, in contrast to acetyl- or benzoylamino acids, they are relatively resistant to racemization. For a summary of these papers, see (24).

In Dresden, the new peptides were used as substrates for intestinal peptidases, and when the Nazis obliged Bergmann to leave Germany, he brought this line of work to his New York laboratory. During my association with him until his untimely death in 1944, my principal and most rewarding effort was the search for synthetic peptide substrates for well-defined proteolytic enzymes, especially those that act on proteins. These were indeed provided by the carbobenzoxy method. The potential of the method also became evident during the 1930's from work in other laboratories. Charles Harington [1897-1972] used it to synthesize glutathione (with phosphonium iodide in acetic acid as the reducing agent), and an improved procedure was described soon afterward by Vincent duVigneaud [1901-1978], who applied the sodium-liquid ammonia method of reduction. Before 1940, the only other naturally-occurring peptide that had been synthesized was carnosine, also made by duVigneaud's group. All this was before the recognition that many hormones and antibiotics are peptides, and before the impetus these discoveries gave to the development of the art of peptide synthesis. After the war, the acid chloride coupling method described by Bergmann and Zervas was replaced by better procedures such as the use of mixed anhydrides, carbodiimides, and substituted phenyl esters, as well as the azide method introduced by Curtius. Although the carbobenzoxy (benzyloxycarbonyl) group still remains a valuable amino- and side-chain protecting group, and the benzyl group a useful carboxyl-protecting group, the more recent introduction of new ones (e.g., t-butyloxycarbonyl, trityl, fluorenyl-methoxycarbonyl) has greatly increased the scope and power of peptide synthesis.

Why did some of these new developments not come sooner, immediately after the publication of the Bergmann-Zervas 1932 paper? In my opinion, the principal reason was the continued uncertainty during the 1930's about the structure of proteins, and doubt about the validity of the Fischer-Hofmeister theory. The incentive that brought Fischer into the protein field with the intention of making long-chain polypeptides had apparently weakened, because serious attention was still given to the idea that native protein molecules contain covalent cyclic structures. Although the high-molecular weight of proteins was readily acknowledged, many people thought that only denatured proteins are long-chain polypeptides. As evidence for this interpretation, I cite the reaction to the finding of synthetic substrates for well-defined proteinases. For many years it was believed that pepsin attacks some linkage in proteins other than the peptide bond. With the aid of the carbobenzoxy method, I found the first synthetic peptide substrates not only for crystalline pepsin, but also for crystalline trypsin and chymotrypsin. In 1938, a very great physical biochemist, Kaj Linderstrøm-Lang [1896-1959], concluded from studies on protein denaturation that his data "provide sufficient basis for giving warning against the conclusion that genuine proteins contain peptide bonds because they are split by proteinases like trypsin" (25). Such doubts disappeared after Frederick Sanger determined the amino-acid sequence of insulin and used proteinases to produce cleavage products that were no longer called peptones but peptides.

In 1893, in an address delivered upon his admission to the Berlin Academy of the Sciences. Emil Fischer announced his intention to enter the field of protein chemistry, and added that the total analysis and synthesis of proteins might require the efforts of several generations of chemists. When he actually entered the field in 1899, he could have had no inkling that the main impetus to peptide chemistry would come, not from the synthesis of proteins, but from the discovery of peptide hormones and peptide antibiotics. This impetus was clearly evident at the meeting, in 1958, of the first European

Peptide Symposium in Prague, organized by our beloved friend Josef Rudinger [1924-1975], and the one ten years later of the first American Peptide Symposium at Yale. Since then, great advances have been made in the laboratory synthesis of proteins, but the undiminished vitality of these symposia indicates that much more remains to be done. I am certain that this continued effort will be inspired by the achievements of the great investigators who came before. Among these achievements, those of Leonidas Zervas are pre-eminent, because of the originality, scope, precision, and elegance of his work (26).

Acknowledgment

The preparation of this paper was aided by NEH grant RH-20739-86.

References

1. Hickel, E. 1975. Naturwiss. Rundsch. 28, 14.

2. Lehmann, C.G. 1855. Physiological Chemistry (translated from 2nd ed. by G.E. Day; edited by R.E. Rogers). Blanchard and Lea, Philadelphia, vol. 1, p. 450.

3. Fruton, J.S. 1972. Molecules and Life: Historical Essays on the Interplay of Chemistry and Biology. Wiley, New York, p. 431.

4. Fischer, E. 1902. Chem. Ztg. 26, 939.

5. Hofmeister, F. 1902. Naturwiss. Rundsch. 17, 529, 545.

6. Kurzer, F. 1956. Chem. Revs. 56, 95.

7. Curtius, T. 1881. J. prakt. Chem. [2] 26, 145.

8. Grimaux, E. 1882. Bull. Soc. Chim. [2] 38, 64.

9. Fischer, E., E. Fourneau. 1901. Ber. chem. Ges. 34, 2868.

10. Fischer, E. 1907. Ber. chem. Ges. 40, 1754.

11. Fischer, E. 1907. J. Chem. Soc. 91, 1749.

12. Letter from E. Fischer to A. von Baeyer, 5 December 1905. Fischer Collection, Bancroft Library, University of California, Berkeley.

13. Fischer, E. 1906. Ber. chem. Ges. 39, 530.

14. Hofmeister, F. 1908. Arch. exp. Path. Pharm. Suppl., 273.

15. Curtius, T. 1904. J. prakt. Chem. [2] 70, 57.

16. Fruton, J.S. 1979. Ann. N. Y. Acad. Sci. 325, 1.

17. Bergmann, M., F. Stern, C. Witte. 1926. Ann. Chem. 449, 277.

18. Schoenheimer, R. 1926. Z. physiol. Chem. 154, 203.

19. Wessely, F. 1925. Z. physiol. Chem. 146, 72.

20. Woodward, R.B., C.H. Schramm. 1947. J. Am. Chem. Soc. 69, 1551.

21. Fruton, J.S. 1982. TIBS 7, 37.

22. Zervas, L. 1931. Ber. chem. Ges. 64, 2289.

23. Bergmann, M., L. Zervas. 1932. Ber. chem. Ges. 65, 1192.

24. Fruton, J.S. 1949. Adv. Protein Chem. 5, 1.

25. Linderstrøm-Lang, K., R.D. Hotchkiss, G. Johansen. 1938. Nature 142, 996.

26. Katsoyannis, P.G., ed. 1973. The Chemistry of Polypeptides. Plenum Press, New York.

KNOWLEDGE-BASED DESIGN OF NOVEL PEPTIDES

T. L. Blundell

Laboratory of Molecular Biology, Department of Crystallography
Birkbeck College, London University, Malet Street, London WC1E 7HX

Introduction

In the coming years a knowledge of the chemistry and three-dimensional structures of protein receptors will transform the design of novel peptides. Recombinant DNA technology offers the possibility, not only of sequencing the proteins which are the receptors for drugs, herbicides and insecticides, but also of the production of large quantities for X-ray analysis and other biophysical studies. At the same time computer technology has developed at a rapid rate and offers data bases to store, retrieve and compare the known sequences and conformations, graphics to display and manipulate molecular models, receptors and new vector processors for simulation of their low energy conformers, normal modes and molecular dynamics. We can begin to define the molecular shape, charge, chemical functionality and flexibility necessary for binding of peptide ligands and to simulate complexes of potentially useful but novel molecules with their protein receptors. In this paper I briefly review these new technologies which provide a rational approach to peptide design.

Modelling protein receptors by homology

The first task is to define the three-dimensional structure of the receptor. If this has not been studied by X-ray analysis we may attempt to model it if the three-dimensional structures of homologous proteins have been defined. This is a consequence of the fact that tertiary structures tend to be strongly conserved in evolution.
Historically, modelling by homology has been achieved in several simple stages starting with alignment of the sequences, followed by creation of insertions, deletions and replacements in the known three-dimensional

Peptides 1986

structure of the homologous protein (see Blundell et al (1) for a review). Early work was achieved using physical models but subsequent attempts exploited computer graphics. The initial models have been refined using energy minimisation techniques to give a reasonable final structure. Thus kidney renin, which is an important target for drugs which lower blood pressure, was modelled in this way based on other aspartic proteinases (2,3). The bilobal enzyme with an extensive cleft with catalytic residues and specificity subsites is shown in Figure 1.

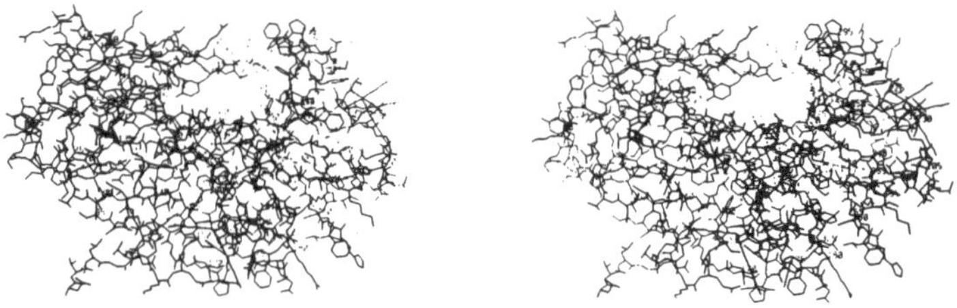

Figure 1 A stereoview of the model of human renin

However, with over 300 structures in the Brookhaven Data Bank, of which only ~100 are non-homologous, there may well be more than one structure available to serve as a basis for modelling.

A knowledge-based approach is to attempt to use simultaneously but selectively the information from all the known tertiary structures of the homologous family. For example, for modelling one domain of tissue plasminogen activator (Athwal, D., Harris, T. M. and Blundell, T. L., manuscript in preparation) there are five high resolution mammalian serine proteinases - trypsin, elastase, chymotrypsin, kallikrein and mast cell proteinase - of known tertiary structure. We first construct an imaginary "framework" which contains a virtual atom at points in three-dimensional space of the topologically equivalent residues common to the family. We then align the new sequence so that topologically equivalent residues of the "framework" are equivalent in the sequence alignment and select those fragments of each of the homologous proteins which are closest in sequence: for the modelling of tissue plasminogen activator serine proteinase domain we used seventeen fragments (Figure 2). New computer programs are being developed at Birkbeck to achieve this (Sutcliffe, M. and Blundell, T. L., unpublished results).

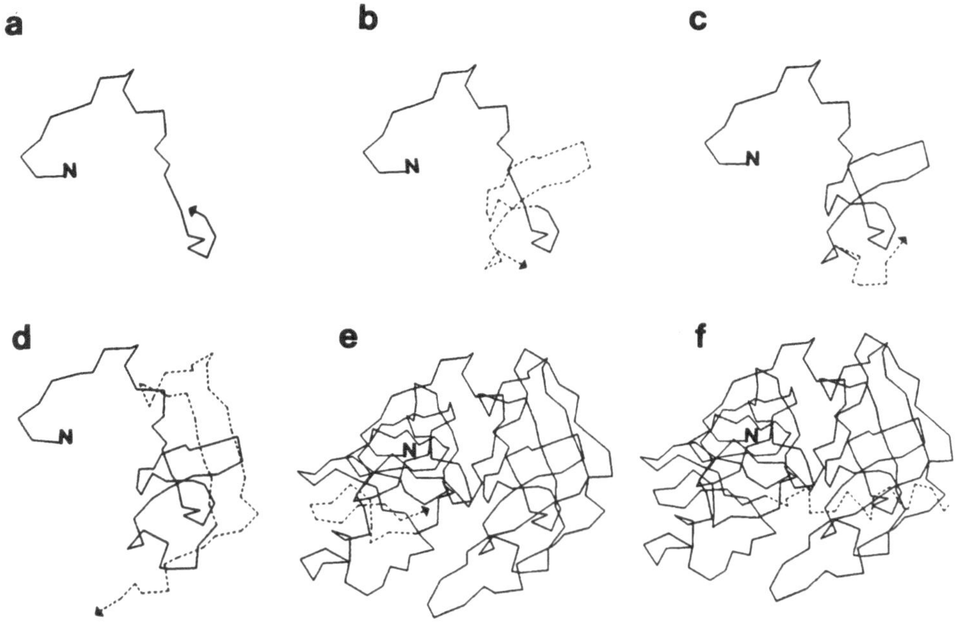

Figure 2
Modelling a protein from fragments of the three-dimensional structures of closely homologous proteins. The example is that of tissue plasminogen activator constructed on the basis of the structures of elastase, trypsin, chymotrypsin and kallikrein which have been defined by X-ray analysis. Views a to d show the addition of consecutive fragments. There were 17 in total. Views e and f show the penultimate and last fragments to give the full structure with the exception of four short insertions (D. Athwal, T. Harris and T. L. Blundell, unpublished results).

Since the majority of significant differences in protein structure occur in loop regions, these are the most difficult to construct. The choice of a conformation found in a loop of equivalent length in an homologous protein is often a good guide even if the sequence differs. However, the optimal approach is to explore conformations from other proteins in similar supersecondary structures. For example a detailed analysis of β-hairpin loops shows that in short loops certain structures with characteristic sequences recur many times in unrelated proteins (4). Although there is little sequence homology between loops certain glycines are often conserved and there is a preference for particular β-strand or α-helix forming residues at certain positions. These conformers can be thus used selectively in modelling.

Refinement using energy minimisation and molecular dynamics

Where the proteins have sequence homology of 50% or more, the models will probably be correct to better than 1Å although individual side-chains may be more in error (5). Errors can be decreased by minimisation of the potential energy, to give reasonable agreements (rms 0.5-1.0Å) with coordinates derived from X-ray analysis. However, X-ray analysis gives a time- and space-averaged structure in a crystalline lattice with perhaps 40% solvent, whereas energy minimisation gives the structure at a single minimum often in vacuo. To explore local changes requiring larger shifts torsion angles can be varied locally to locate conformations for subsequent energy minimisation. Alternatively, molecular dynamics may often be useful to explore a range of local conformations.

Extension of homology techniques to ab initio structure prediction

What can we do if no homology is found? The standard approach is to perform a conventional secondary structure prediction and calculate a hydropathy profile. However, the unknown structure is probably constructed from the structural motifs common to other globular proteins such as βαβ units or ββ hairpins. The problem is to identify these motifs from the sequence. Progress towards this goal will involve identification of the critical sequence patterns for each motif or "fingerprint". For example, the fingerprint of a residue β-hairpin (4) might be

β-β-β-β-Gly-Ser/Thr-β-β-β-β

This can be used as a template to search for such motifs. Larger supersecondary structures also have characteristic patterns. For example, a particular pattern of a glycine and a serine along with a few conservatively varied residues and hydrophobic regions has been used to identify a 40 amino acid Greek key structure in the surface protein S of the sporulating bacterium Myxococcus Xanthus even though there is no significant overall homology (6).

Data Bases

The information now available is so complex that its organisation in an easily accessible data base is becoming high priority. Birkbeck College and Leeds University, supported by ICI, Glaxo, CellTech, Sturge and the SERC, are collaborating in the development of a relational data base in which information concerning protein sequence, secondary structure, motifs, domains and general tertiary interactions are stored. Such a data base should make the design process accessible to a broader constituency of potential practitioners.

Peptide-protein receptor binding

Once the receptor model has been constructed the nature of the ligand interactions needs to be defined. Again, this is difficult to achieve using totally automated docking procedures or by interactive computer graphics modelling. The most useful approach appears to be to learn from other structural studies. Thus in defining renin-angiotensinogen interactions much can be gained from the study of other aspartic proteinase inhibitor complexes that have been studied by X-ray analysis. For example, most modified peptide inhibitors bind to aspartic proteinases, including renin, in a way that mimics the substrate-forming hydrogen bonds as shown in Figure 3 (7: Foundling, S. I., Watson, F. E., Cooper, J., unpublished results). Such X-ray analyses also define the specificity pockets. This knowledge provides a firm starting point for modelling angiotensinogen fragments into renin.

Peptide design

The detailed models obtained by such techniques then provide a rational basis for design. We begin by identifying the critical hydrogen bond interactions and ensuring that hydrogen bond donors such as peptide NH groups will have acceptors. We then generate the complementary surfaces of protein and ligand using computer graphics and investigate the possibility of closer and more extensive interactions perhaps by introducing synthetic amino acids. Thus peptide inhibitors of renin can certainly have larger side-chains at P_1 and P_1' than the leucine and valine found in angiotensinogen. Finally, we can introduce more rigid or cyclic groups which fix the conformer bound by the protein.

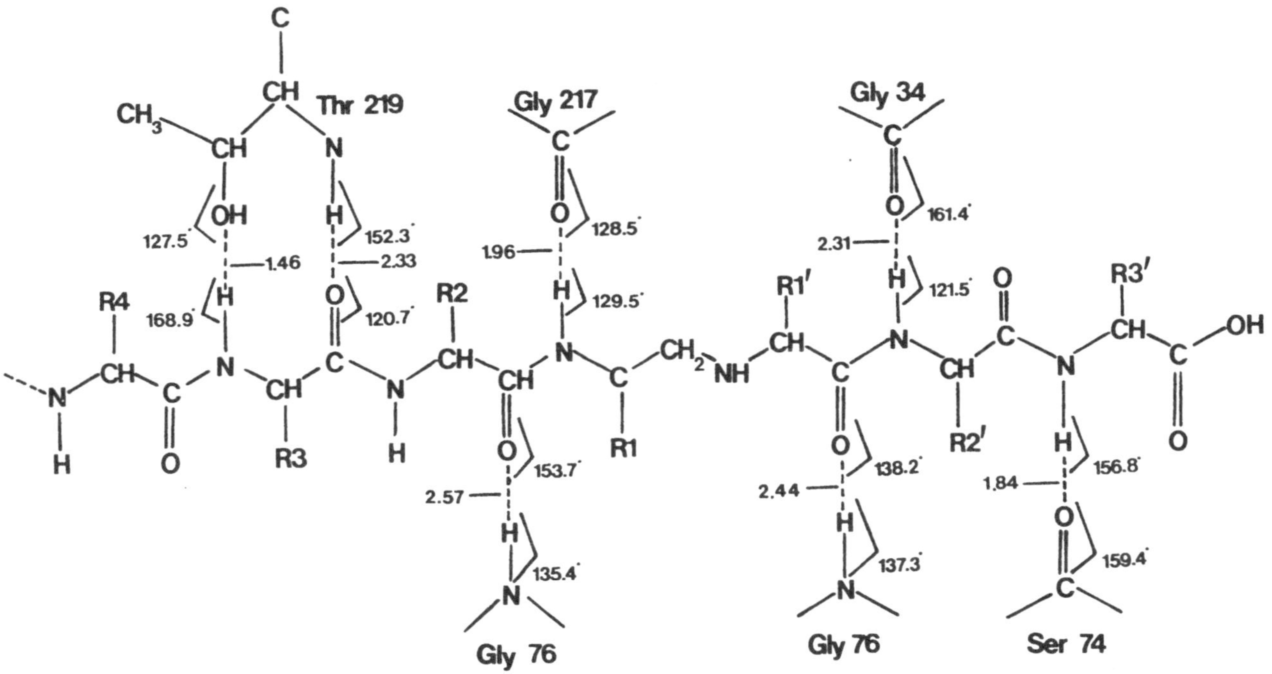

<u>Figure 3</u>
A schematic view of the hydrogen bonding of the renin inhibitor, H142, to the aspartic proteinase, endothiapepsin.
(S. I. Foundling, unpublished results)

We can assume that the sequences of almost all pharmaceutically important receptors will be defined by cloning and sequencing their complementary DNA in the next five years. Peptide chemists must exploit this information using such knowledge-based modelling procedures if they are to progress in the design of novel molecules.

Acknowledgements

I thank the following colleagues for stimulating discussion and active collaboration in this work :
Dee Athwal, Geoff Barton, David Barlow, Steve Bryant, Jonathan Cooper, Stephen Foundling, Mark Edwards, Fiona Hayes, Suhail Islam, Ilyas Haneef, Andrew Hemmings, Tim Hubbard, Lynn Sibanda, Mike Sternberg, Mike Sutcliffe, Janet Thornton, Ian Tickle, Willie Taylor, Annemarie Treharne, Frank Watson, Francis Whiteley, Marketa Zvelebil.

References

1. Blundell, T.L. et al. 1986. Phil. Trans. Roy. Soc. A317, 273.

2. Blundell, T.L., Sibanda, B.L. and Pearl, L.H. 1983. Nature 304, 273.

3. Sibanda, B.L., Blundell, T.L., Hobart, P., Chirgwin, J., Dorning, B. and Bindra, J. 1984. FEBS Letters 174, 103.

4. Sibanda, B.L. and Thornton, J.M. 1984. Nature 316, 170.

5. Chothia, C. and Lesk, A. 1986. EMBO J. 5, 823.

6. Wistow, G.J., Summers, L.J. and Blundell, T.L. 1985. Nature 316, 771.

7. Hallett, A. et al. 1985. In: Aspartic proteinases and their inhibitors (V. Kostka, ed.). Walter de Gruyter, Berlin, p. 467.

INSULIN-LIKE GROWTH FACTORS: STRUCTURE-ACTIVITY RELATIONSHIPS

P.G. Katsoyannis, G.P. Schwartz, G.T. Burke, S. Joshi
Department of Biochemistry, Mount Sinai School of Medicine of the City University of New York, N.Y. 10029

M.M. Rechler, L.Y-H. Tseng
Molecular, Cellular and Nutritional Endocrinology Branch, NIADDKD, National Institute of Health, Bethesda, MD 20205

It has been known for several years that a group of polypeptides circulate in the plasma which exhibit growth promoting effects, such as the ability to stimulate DNA synthesis, the incorporation of sulfate into cartilage and which also display insulin-like metabolic effects in extraskeletal tissues (1, 2, 3, 4). Serum concentration of these polypeptides are governed to a large extent by growth hormone secretion and it was proposed that the growth promoting effects of the growth hormone are indeed mediated through this group of polypeptides. The term somatomedin was hence widely used to describe them. In recent years it became clear that two major categories of these factors exist, and virtually all the published biological effects of the somatomedins can be ascribed to members of these groups of factors (2, 3). One group of factors, the plasma concentration of which is highly dependent on levels of growth hormone, appears to interact with a unique receptor, the Type I receptor. The second group, less dependent on growth hormone plasma levels, interacts with a distinctly different type of receptor, the Type II receptor (5, 6, 7). This rather confused picture regarding somatomedins was substantially clarified a few years ago by Rinderknecht and Humbel, who isolated from human plasma, purified to homogeneity and sequenced two polypeptides, one containing 70 amino acid residues and the other 67 amino acid residues (8, 9). These polypeptides exhibit striking homology to human proinsulin and they have been designated as Insulin-like Growth Factors I and II, abbreviated as IGF-I and IGF-II. It is now apparent that the somatomedins

Peptides 1986

described in the earlier literature are similar or identical to IGF-I and IGF-II (10, 11). Specifically the IGF-I which is a more growth hormone-dependent polypeptide interacts with high affinity with the Type I receptor which structurally is similar but not identical to the insulin receptor (6, 12); the IGF-II which is a less growth hormone-dependent polypeptide interacts preferentially with the Type II receptor, consisting of a single binding sub-unit which is distinctly different from the Type I and from the insulin receptor (13, 14).

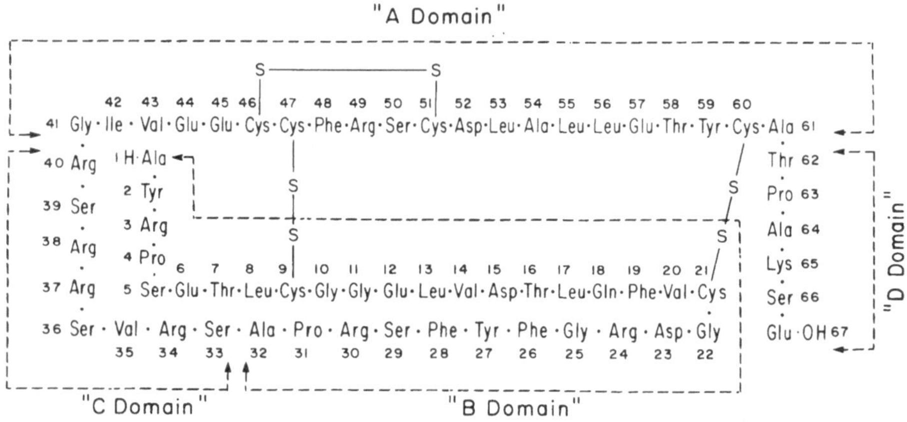

Fig. 1. Structure of human IGF-II.

Like proinsulin, IGF-II is a single-chain polypeptide containing 67 amino acid residues and cross-linked by three disulfide bridges (Fig.1)(9). The location of these bridges and the size of the two-ring systems formed is identical to those found in insulin. Like proinsulin, it contains an N-terminal region (sequence 1-32) homologous to the B chain of insulin, designated as the B domain, a shorter connecting peptide the C domain and a 21-amino acid sequence homologous to the A chain of insulin, designated the A domain. Unlike insulin or proinsulin IGFs contain at the C-terminus of the A domain an extension sequence, the D-domain which in IGF-II is a hexapeptide. A similar situation exists in IGF-I (Fig.2)(8). There is the B domain (sequence 1-29) and A domain (sequence 42-62) homologous to the B and A chains of insulin respectively, the C-domain (sequence 30-41) which is

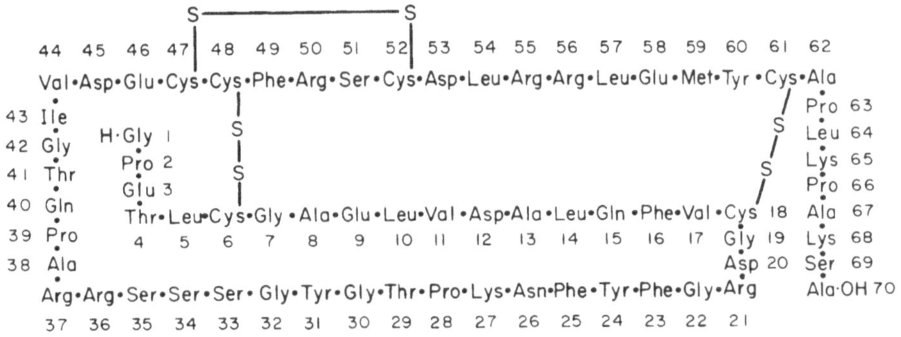

Fig. 2. Structure of human IGF-I.

dissimilar in size and sequence from the C-domain of IGF-II, and the D domain (sequence 63-70) which in IGF-I is an octapeptide. However, the D domain of IGFs contains a common tetrapeptide sequence (-Pro-Ala-Lys-Ser-). Over-all the A and B domains of IGF-I are ca 60% identical to the corresponding domains of IGF-II.

It is interesting to note that the two sets of basic amino acid residues at both ends of the C-peptide which serve as the cleaving sites in the conversion of proinsulin to insulin are absent in IGFs (-Thr-Gly-;sequences in IGF-I 41-42 and 29-30). This may explain the fact that IGF, in contrast to proinsulin, is not

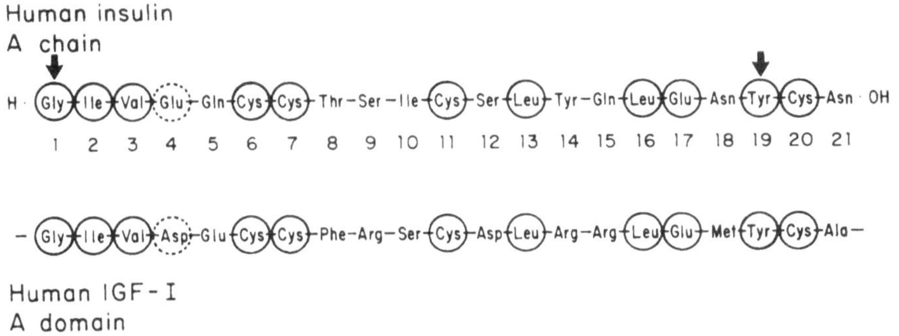

Fig. 3. Comparison of the sequences of human insulin A chain and human IGF-I A domain (IGF-I sequence 42-62). The numbering for insulin A chain is given between sequences. Identical residues are enclosed in solid circles while conservatively varied residues are enclosed in dotted circles.

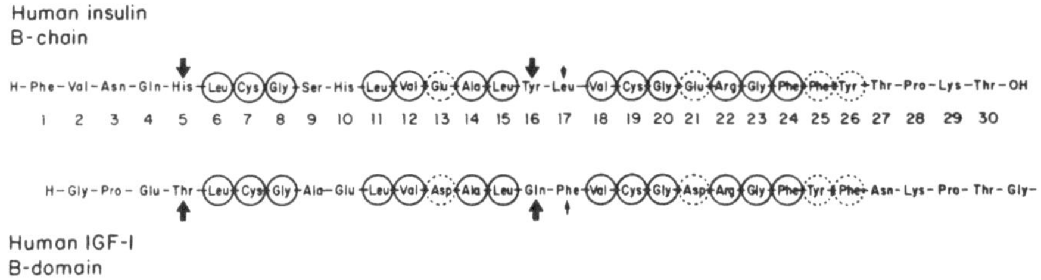

Fig. 4. Comparison of the sequences of human insulin B chain and human IGF-I B domain (IGF-I sequence 1-30). The numbering for insulin B chain is given between sequences. Identical residues are enclosed in solid circles while conservatively varied residues are enclosed in dotted circles.

converted, as far as we know, to a two-chain structure in vivo. The close homology of the IGF A and B domains to the A and B chains of insulin is shown in figures 3 and 4. As can be seen (Fig.3), of the twenty-one amino acid residues in the insulin A chain, twelve are identical or conservatively varied in the IGF A domain. Furthermore, of the three amino acid residues in the A chain of insulin, A^1 Gly, A^{19} Tyr and A^{21} Asn which are in the putative insulin receptor binding region or in the periphery of that region (15, 16), two, A^1 Gly and A^{19} Tyr are also present in the IGF A domain (Fig. 2; positions 42 and 60). In Fig. 4 the sequences of the human insulin B chain and IGF-I B domain are compared. Out of the thirty amino acid residues in the insulin B chain, seventeen are identical or conservatively varied in the IGF B domain. Furthermore, of the six amino acid residues of the insulin B chain thought to be in the putative insulin receptor binding region, or in the periphery of that region (15, 16), B^{12} Val, B^{13} Glu, B^{16} Tyr, B^{23} Gly, B^{24} Phe and B^{25} Phe, five are present or conservatively varied in the IGF B domain. A glaring difference is the replacement of B^{16} Tyr in insulin by Gln in IGF (Fig. 2, position 15). It was suggested by Blundell and Humbel (16) that the replacement of the B^{16} Tyr of insulin by Gln in IGF may be a contributing factor in the growth-promoting activity of IGFs. In this connection I may mention that we have reported recently the synthesis of an insulin analog in which the B^{16} Tyr is replaced by Gln (17). This analog, Glutamine B^{16} insulin, exhibits at least twice as much growth promoting potency as insulin-like potency.

On the basis of the X-ray structure of the 2 Zn crystalline insulin and from the known sequences of proinsulin, IGF-I and IGF-II Blundell and Humbel (16, 18), using computer graphics, suggested a strong correspondence of the conformation of insulin with that of proinsulin and IGFs. These computer generated models indicate that the IGFs can have an identical hydrophobic core and identical conformation of the A and B domains to that of insulin. On the basis of these models Blundell and Humbel explained some of the biological behavior of the IGFs (16). For example, the greatly reduced insulin-like activity of IGFs is attributed to the partial covering of the putative insulin-receptor binding region, which is substantially conserved in the IGFs, by the C and D regions of the IGFs.

The inability of IGFs to react with antibodies against insulin is also explained by these models. In general the antigenic determinants of insulin are clustered in two regions (19). One region involves the N- and C-terminal areas of the A chain and the C-terminus of the B chain. The other region involves the N-terminus of the B chain and the neighboring region of the A chain (A^8-A^{14}). The first region is partly covered in IGFs by the C and D domains and the second region in IGFs is sequentially quite different of that of insulin which explains the inability of IGFs to react with antiinsulin antibodies. However, our results which I will discuss momentarily indicate that some important differences between the conformation of insulin and IGF must exist, which are not apparent in these models, to account for the structure-activity relationships that we have demonstrated to prevail with the IGFs.

In view of their high degree of structural homology it is not surprising that insulin and IGFs have a comparable spectrum of biological activities. The relative potencies, however, vary greatly depending upon the tissue or cell type used for the assay, and upon the particular activity being measured. In general IGFs are more potent than insulin in promoting DNA synthesis in fibroblasts, whereas insulin is more potent than IGFs in stimulating short-term metabolic effects such as the stimulation of glucose transport in adipocytes (20, 21). This reflects cross-reactivity of insulin and IGFs with the same receptors but with different affinity. Insulin and IGFs in general exert their insulin-like effect by interacting with the insulin receptor and their mitogenic effects by interacting with the IGF receptors (22). IGF-I and IGF-II interact, although with quite different affinities, with both types of IGF receptors, namely, the Type I and Type II receptors, and insulin interacts with only the Type I receptor which is similar but not identical to the insulin receptor (6, 23, 24).

A unique property of the IGFs, not shared by insulin, is that they circulate in plasma bound to specific carrier proteins, which prolong the half-life of IGFs in circulation and which play a pivotal role in the delivery of the IGFs to their target tissues (4, 25). Insulin does not interact with carrier proteins.

For the past several years our laboratory has been involved in studies pertaining to the Insulin-like Growth Factors. One aspect of these studies concerns the synthesis of disulfide-linked two-chain insulin-like molecules in which the A and B chain moieties correspond either to the A and B domains of IGF or one chain corresponds to the A or B domain of IGF and the other chain is the B or A chain respectively of insulin (26, 27, 28, 29, 30, 31, 34). Biological evaluation of these compounds provided most interesting conclusions regarding structure-activity relationships in IGFs.

A schematic representation of the two-chain insulin-like molecules that we have synthesized is shown in the next two figures.

In Figure 5 the first two-chain compound is an insulin in which the A chain moiety is extended at the C-terminus with the D domain of IGF-II (Fig. 1, sequence 62-67). We refer to this compound as $A^{27}{}_{INS}$-"D"-IGF-II/$^{B}{}_{INS}$.

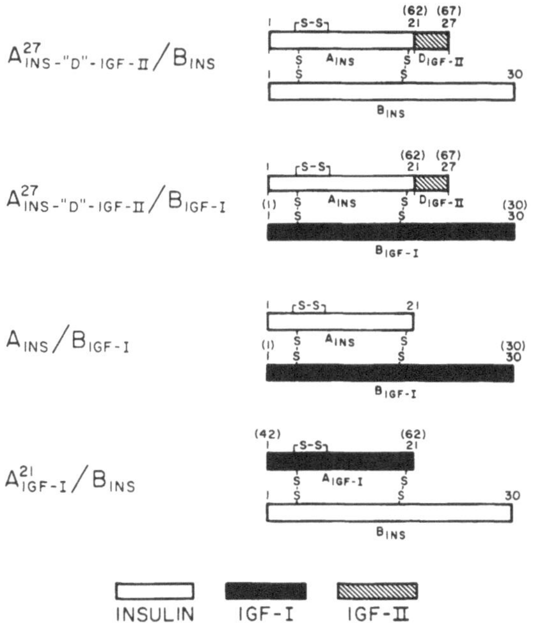

Fig. 5. Schematic representation of the two-chain insulin-like molecules. Sequences corresponding to insulin, IGF-I and IGF-II are indicated by open bars, solid bars and stippled bars respectively. The upper bar in each two-chain molecule represents the A chain (insulin or IGF) and the lower bar the B chain (insulin or IGF). Numbers in parenthesis indicate the actual sequences in the IGFs.

The second compound is an insulin-like molecule in which the A chain moiety is the A chain of insulin extended at the C-terminus with the D domain of IGF-II and the B chain moiety is the sequence corresponding to the B domain of IGF-I (Fig. 2, sequences 1-30). We refer to this compound as $A^{27}{}_{INS-"D"-IGF-II}/B_{IGF-I}$.

The third compound is an insulin-like molecule in which the A chain is that of insulin and the B chain moiety is the sequence corresponding to the B domain of IGF-I (Fig. 2, sequence 1-30). We refer to this compound as A_{INS}/B_{IGF-I}.

The fourth compound is an insulin-like molecule in which the A chain moiety is the sequence corresponding to the A domain of IGF-I (Fig. 2, sequence 42-62) and the B chain moiety is the B chain of insulin. We refer to this compound as $A^{21}{}_{IGF-I}/B_{INS}$.

In Figure 6 the A chain moiety of the first insulin-like compound is the sequence corresponding to the A domain of IGF-I extended at its C-terminus with the D domain of IGF-I (Fig. 2, IGF-I

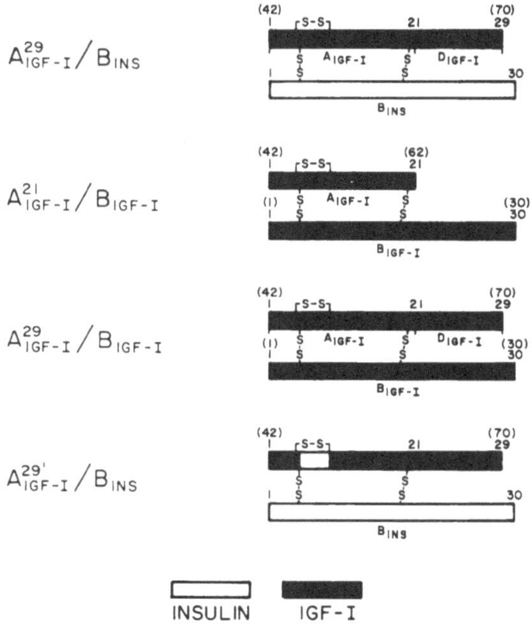

Fig. 6. Schematic representation of the two-chain insulin-like molecules. Sequences corresponding to insulin, IGF-I and IGF-II are indicated by open bars, solid bars and stippled bars respectively. The upper bar in each two-chain molecule represents the A chain (insulin or IGF). Numbers in parenthesis indicate the actual sequences in the IGFs.

sequence 42-70) and the B chain moiety is the B chain of insulin. We refer to this compound as $A^{29}{}_{IGF-I}/B_{INS}$.

The second compound is an insulin-like molecule in which the A chain moiety is the sequence corresponding to the A domain of IGF-I (Fig. 2, sequence 42-62) and the B chain moiety is the sequence corresponding to the B domain of IGF-I (Fig. 2, sequence 1-30). In other words this compound is related to IGF-I as insulin is related to proinsulin. We refer to this compound as $A^{21}{}_{IGF-I}/B_{IGF-I}$.

The third compound is an insulin-like molecule in which the A chain moiety is the sequence corresponding to the A domain of IGF-I extended at its C-terminus with the D domain (Fig. 2, sequence 42-70) and the B chain moiety is the sequence corresponding to the B domain of IGF-I (Fig. 2, sequence 1-30). We refer to this compound as $A^{29}{}_{IGF-I}/B_{IGF-I}$.

The last compound is an insulin-like molecule in which the B chain moiety is the B chain of insulin and the A chain moiety is the sequence corresponding to a modified version of the A domain of IGF-I extended at its C-terminus with the D domain (Fig. 2, sequence 42-70). The modification consists in the substitution of the intrachain sequence found in both IGFs, namely -Phe-Arg-Ser- (Fig. 1, sequence 48-50; Fig. 2, sequence 49-51), with the corresponding sequence found in the A chain of insulin, namely $-Ala^{8}-Gly^{9}-Val^{10}-$. We refer to this compound as $A^{29'}$IGF-I/^{B}INS.

The synthesis of all these two-chain insulin-like compounds was accomplished by the same procedures we have used for the synthesis of insulin and insulin analogs (32, 33). In other words we have synthesized the S-sulfonated derivatives of the polypeptide sequences corresponding to the A or the extended A domain and the B domain of IGF and have combined the appropriate S-sulfonated domains and S-sulfonated insulin chains to obtain the desired two-chain compounds. For the synthesis of the S-sulfonated derivatives the protected polypeptides (heptacosapeptide, heneicosapeptide, nonacosapeptide and triacontapeptide) embodying the amino acid sequences found in the respective IGF domains have been synthesized by solution methods, namely, stepwise elongation and fragment condensation. Removal of the protecting groups on exposure to liquid hydrogen fluoride or trifluoromethanesulfonic acid in the presence of appropriate scavengers and sulfitolysis of the resulting reduced chains led to the respective S-sulfonated derivatives (26, 27). Purification of the synthetic S-sulfonated chains was accomplished by ion-exchange chromatography. Specifically the $[A^{27}_{INS-\text{"D"}-IGF-II}]SSO_3^-$ and $[A^{21}_{IGF-I}]SSO_3^-$ were purified on a Cellex-E column with 0.1M Tris-HCl buffer (pH 7.0) and a linear NaCl gradient (26, 34); the $[A^{29}_{IGF-I}]SSO_3^-$ was purified on a Whatman DE-52 column with a 0.02M Tris-HCl buffer (pH 7.4) and a linear NaCl gradient (34); and the $[B_{IGF-I}]SSO_3^-$ was purified on a Cellex D column with a linear NH_4HCO_3 gradient (27).

The combination of the A and B chains to produce the two-chain molecules was accomplished by the same procedures we have used to combine the A and B chains of insulin or analogs in the synthesis of insulin or insulin analogs (32). These procedures consist of the interaction of the sulfhydryl form of the A chain with the S-sulfonated form of the B chain at pH 10.5 or, using the modification introduced by Chance and co-workers (35), of the interaction of the S-sulfonated A and B chains in the presence of dithiothreitol at pH 10.5. Isolation of the two-chain compounds from the combination mixture was accomplished in some cases by ion-exchange chromatography (26, 27) and in other cases by high-performance liquid chromatography (34).

The biological evaluation of these two-chain compounds is shown in the next two tables. In Table 1 two types of activities are recorded and the values are expressed as per cent of insulin. Insulin-like effects (first three columns) are measured by insulin-receptor binding in rat liver plasma membranes, lipogenesis in rat adipocytes and RIA with antibodies raised in guinea pigs

Table 1

BIOLOGICAL ASSAYS OF TWO-CHAIN COMPOUNDS
(ALL VALUES ARE EXPRESSED AS PERCENT OF INSULIN)

	INSULIN ASSAYS			GROWTH FACTOR ASSAYS	
COMPOUND	RECEPTOR BINDING	LIPOGENESIS	RIA	RECEPTOR BINDING	DNA SYNTHESIS
A^{27}INS-"D"-IGF-II/^{B}INS	35	40	18	36	48
A^{27}INS-"D"-IGF-II/^{B}IGF-I	< 0.3	0.3	0.05	9	13
^{A}INS/^{B}IGF-I	2	2	0.3	16	13
A^{21}IGF-I/^{B}INS	40	38	< 0.1	ca 200-300	ca 500-1,000
A^{29}IGF-I/^{B}INS	(11)*	21	0.2	ca 200-300	ca 500-1,000
A^{21}IGF-I/^{B}IGF-I	8	6	0.2	n.d.**	n.d.**
A^{29}IGF-I/^{B}IGF-I	0.7	0.3	n.d.	Preliminary assays indicate similar potencies, greater than insulin but less than IGF.	
A$^{29'}$IGF-I/^{B}INS	(15)*	12	n.d.		

*tentative values

**n.d.= not determined

against porcine insulin. Complete details of these assays have been described (36). Growth factor effects (last two columns) are measured by IGF-receptor binding in chick embryo and human fibroblasts and DNA synthesis are measured by [^{3}H] thymidine incorporation in chick embryo and human fibroblasts. These procedures are described in detail elsewhere (30).

In the first compound (A^{27}_{INS}-"D"-IGF-II/B_{INS}) (26, 30) it can be seen that extension of the A chain of insulin at its C-terminus with the D domain of IGF-II caused approximately a 60% loss of insulin-like activity which offers experimental verification of the suggestion by Blundell and Humbel (16, 18) that the D domain partially covers the putative insulin-receptor binding region. The growth factor assays show that this compound is less potent than insulin itself despite the presence of the D domain of IGF-II. This is the first indication that the D domain of IGF-II is not involved in growth-promoting activity.

The biological behavior of the second compound, (A^{27}_{INS}-"D"-IGF-II/B_{IGF-I}) (29), indicates that substitution of the B chain of insulin by the B domain of IGF-I with the simultaneous presence of the D domain of IGF-II not only results in a dramatic loss of insulin activity but fails to produce significant growth factor activity. The next compound, (A_{INS}/B_{IGF-I})(27), is related to the previous one by the absence of the D domain of IGF-II. As might be expected the absence of the D domain leads to an increase of insulin activity as compared to the previous compound but still does not produce significant growth factor activity. The biological behavior of these three compounds constitutes strong evidence that neither the B domain of IGF-I nor the D domain of IGF-II is importantly involved in the growth-promoting activities of IGFs.

In dramatic contrast to the results just presented the substitution of the insulin A chain with the A domain of IGF-I(A^{21}_{IGF-I}/B_{INS})(31, 34), produces a strong mitogen, ca 5 to 10 times more powerful than insulin in promoting DNA synthesis. It is of interest to note that the insulin-like properties of this compound are only modestly reduced relative to natural insulin. Apparently the high homology exhibited between the A chain of insulin and the A domain of IGF-I is sufficient to produce this insulin-like activity but the small divergence in homology is sufficient to endow the molecule with strong growth promoting properties. A similar situation exists in regard to the next compound ($A^{29}{}_{IGF-I}/B_{INS}$). The presence of the twenty-one amino acid residues corresponding to the A domain of IGF-I endows the molecule with mitogenic properties; the D domain is without obvious effects on these properties but as we have seen in the previous compounds is deleterious to insulin activity.

The growth-promoting properties of the final three compounds in the Table are still in the process of being determined. Preliminary results with the last two compounds indicate similar potencies, greater than insulin but less than IGF-I. All these data are consistent with the concept that the A domain of IGF is primarily responsible for conferring growth-promoting activity. It is of interest to note that the last compound, ($A^{29'}{}_{IGF-I}/B_{INS}$) (34), in which the intrachain sequence of the A domain of the IGFs (Phe-Arg-Ser) has been substituted by the corresponding sequence of insulin (Ala-Gly-Val) appears to be a mitogen and is similar in insulin activity to the compound containing the unsubstituted A domain of IGF-I i.e., $A^{29}{}_{IGF-I}/B_{INS}$. This might indicate that as in the case of insulin, this sequence is not important for insulin-like or growth-promoting activities.

We have already demonstrated that the B domain of IGF-I is in itself unlikely to be involved in the expression of growth-promoting activity. However another important concept emerges regarding the role of this domain when we compare several pairs of compounds which differ from each other in that one contains the B chain of insulin whereas the other contains the B domain of IGF-I. In every case the presence of the B domain of IGF-I is deleterious to insulin activity but indifferent towards growth-promoting activity, the latter being determined primarily by the A domain. This pattern persists whether the A moiety corresponds to the A chain of insulin (that is insulin versus A_{INS}/B_{IGF}), to the A^{21} domain of IGF-I (that is $A^{21}{}_{IGF-I}/B_{INS}$ versus $A^{21}{}_{IGF-I}/B_{IGF-I}$), or to the A^{29} domain of IGF-I (that is

$A^{29}{}_{IGF-I}/B_{INS}$ versus $A^{29}{}_{IGF-I}/B_{IGF-I}$). It thus appears likely that the B domain of IGF-I interferes with the interaction of IGF with the insulin receptor and that this interference is independent of the previously mentioned role of the D domain. These data might indicate that the B domain of IGF plays a role in determining receptor specificity.

With regard to recognition of these compounds by anti-insulin antibodies it is apparent from Table 1 that the insulin immunogenic determinants are largely suppressed when either the A or B chain of insulin is substituted by the corresponding domain of IGF-I. Further suppression of immunoreactivity accompanies the presence of the D domain of IGF-II.

Table 2 shows the properties of several compounds in interaction with IGF-carrier proteins and with antibodies raised against IGF (28, 31). Details of these assays are presented in ref. 28. The most important point is that insulin does not interact with either preparation and therefore all values are expressed as percent of IGF. It is apparent that all compounds in which the B moiety corresponds to the B domain of IGF-I regardless of the nature of the A moiety (i.e. insulin A chain of IGF A domain)

Table 2

BIOLOGICAL ASSAYS OF TWO-CHAIN COMPOUNDS
(ALL VALUES ARE EXPRESSED AS PERCENT OF IGF)

	GROWTH FACTOR ASSAYS	
COMPOUND	CARRIER PROTEINS BINDING	RIA
INSULIN	0	0
$A^{27}{}_{INS}$-"D"-$IGF-II/B_{INS}$	0	0
$A^{27}{}_{INS}$-"D"-$IGF-II/B_{IGF-I}$	ca 16**	n.d.*
A_{INS}/B_{IGF-I}	ca 16**	10
$A^{21}{}_{IGF-I}/B_{INS}$	0	0
$A^{29}{}_{IGF-I}/B_{INS}$	0	0

*n.d. = not determined ** carrier proteins from adult rat serum (28) interact with carrier proteins and anti-IGF antibodies. Compounds containing the B chain of insulin are without activity, regardless of the nature of the A moiety. Thus it is clear that carrier proteins and anti-IGF antibodies recognize determinants in the B domain of IGF.

To summarize, all our data lead us to the following tentative conclusions:

1. The A domain of IGF-I is primarily responsible for endowing the molecule with growth-promoting properties.

2. The B domain of IGF-I is not a major contributor to growth-promoting activity, but is crucial for the recognition of the molecule by IGF-carrier proteins and anti-IGF antibodies.

3. The D domain of IGF-II does not contribute to growth-promoting activity or to recognition by carrier proteins or antibodies, but participates in the suppression of insulin-like activity. Our data indicate a similar situation with the D domain of IGF-I.

4. The B domain of IGF-I, independent of the D domain, interferes with the interaction of the molecule with the insulin receptor, and may play a role in modulating specificity in receptor recognition.

From the data available thus far it is now possible to speculate regarding specific amino acid residues in IGFs and insulin which may underlie the observed differences in their biological behavior. The results with our two-chain molecules indicate that the A domain of IGF is primarily responsible for conferring growth-promoting properties. Thus when the insulin A chain is combined with the insulin B chain we obviously obtain insulin which is a weak growth factor. When instead the A domain of IGF is combined with the B chain of insulin the resulting molecule possesses modest insulin activity (ca 40% of insulin) but greatly enhanced growth-promoting activity (ca 5-10 fold that of insulin). The key to this divergence in activity must reside in the differences between the sequences of insulin A chain and the IGF-A domain. A comparison of the two sequences as shown in Fig. 7, reveals the following differences: A^4 Glu to Asp; A^5 Gln to Glu; A^8-A^{10} Thr-Ser-Ile to Phe-Arg-Ser; A^{12} Ser to Asp; A^{14}-A^{15} Tyr-Gln to Arg-Arg; A^{18} Asn to Met; and A^{21} Asn to Ala.

The conservative change A^4 Glu to Asp is of no consequence, since in IGF-II A^4 Glu is retained. The A^8-A^{10} Thr-Ser-Ile to Phe Arg-Ser change would appear to be of importance since IGF-II also has the IGF-I sequence. However, as was mentioned previously, the two-chain molecule $A^{29'}{}_{IGF-I}/B_{INS}$, in which the A^8 to A^{10} sequence has been substituted by the corresponding sequence in insulin appears to have comparable activity to the unsubstituted compound. This indicates that this sequence is not important in the expression of the biological activity of IGF.

It may be recalled that the A^8 to A^{10} sequence varies greatly in insulins from different species as well. The A^{14} - A^{15} change appears to be of little significance since IGF-I and IGF-II themselves differ considerably in this region (Arg-Arg to Ala-Leu).

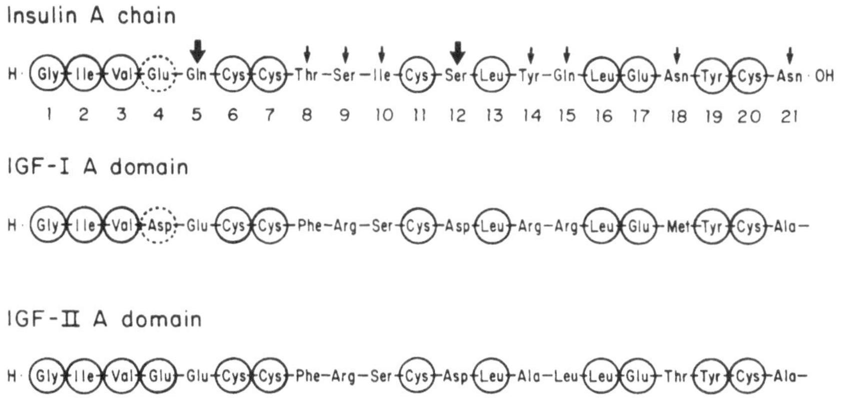

Fig. 7. Comparison of the sequences of human insulin A chain and human IGF-I and IGF-II A domains. The numbering for insulin A chain is given between sequences. Identical residues are enclosed in solid circles while conservatively varied residues are enclosed in dotted circles. The arrows refer to points of nohomology which are discussed in the text.

The same is true comparing the A^{18} change. The A^{21} change (Asn to Ala) also does not appear to be of any significance since several insulin analogs involving substitution of the A^{21} Asn have similar biological activity as insulin (34). The only two changes which appear to be of significance are the A^5 Gln to Glu and A^{12} Ser to Asp which are also noted in IGF-II. There are two, not mutually exclusive, speculations as to why the A^5 and A^{12} changes could be involved in the divergence of the biological behavior between insulin and the synthetic compounds. It could be assumed that the negative charges at the A^5 and A^{12} positions subtly alter the conformation of the molecules, from that of insulin, so as to greatly enhance their binding to the IGF receptor without seriously disturbing their putative insulin-receptor binding region. Alternatively the conformation of these two-chain molecules may be very similar to that of insulin, as the computer graphics models suggest, but the negative charges at A^5 and A^{12} positions are themselves responsible for enhancing the interaction of these molecules with the IGF receptor. We are currently investigating this problem by synthesizing appropriate insulin and two-chain IGF analogs. We are also about to initiate conformational studies by circular dichroism and time-resolved fluorescence spectroscopy and we will attempt to crystallize one of the two-chain IGF molecules for crystallographic studies.

References

1. Van Wyk, J.J. and Underwood, L.E. 1978. In: Biochemical Action of Hormones, Vol. V (G. Litwack, ed.). Academic Press, New York, pp. 101-148.

2. Humbel, R.E. 1984. In: Hormonal Proteins and Polypeptides, Vol. XII (C.H. Li, ed.). Academic Press, New York, pp. 57-79.

3. Van Wyk, J.J. 1984. In: Hormonal Proteins and Polypeptides, Vol. XII (C.H. Li, ed.) Academic Press, New York, pp. 81-125.

4. Nissley, S.P. and Rechler, M.M. 1984. In: Biochemical Action of Hormones, Vol. XII (C.H. Li, ed.). Academic Press, New York, pp. 127-203.

5. Kasuga, M.E., Van Obberghen, Nissley, S.P., Rechler, M.M. 1981. J. Biol. Chem. 256, 5305.

6. Massague, J. Czech, M.P. 1982. J. Biol. Chem. 257, 5038.

7. For a review see: Perdue, J.F. 1984. Can. J. Biochem. Cell Biol. 62, 1237.

8. Rinderknecht, E., Humbel, R.E. 1978. J. Biol. Chem. 253, 2769.

9. Rinderknecht, E., Humbel, R.E. 1978. FEBS Lett. 89, 283.

10. Klapper, D.G., Svoboda, M.E., Van Wyk, J.J. 1983. Endocrinology 112, 2215.

11. Marquardt, H. Todaro, G.J., Henderson, L.E., Oroszlan, S. 1981. J. Biol. Chem. 256, 6859.

12. Pilch, P.F., Czech, M.P. 1980. J. Biol. Chem. 255, 1722.

13. Oppenheimer, C.L., Czech, M.P. 1983 J. Biol. Chem. 258, 8539.

14. August, G.P., Nissley, S.P., Kasuga, M., Lee, L., Grunstein, L., Rechler, M.M. 1983. J. Biol. Chem. 258, 9033.

15. Pullen, R.A., Lindsay, D.G., Wood, S.P., Tickle, I.J., Blundell, T.L., Wollmer, A., Krail, G., Brandenberg, D., Zahn, H., Gliemann, J., Gammeltoft, S. 1976. Nature, 259, 369.

16. Blundell, T.L., Bedarkar, S., Rinderknecht, E. Humbel, R.E. 1978. Proc. Natl. Acad. Sci. USA 75, 180.

17. Schwartz, G.P., Wong, D., Burke, G.T., DeVroede, M.A., Rechler, M.M., Katsoyannis, P.G. 1985. J. Protein Chem. 4, 185.

18. Blundell, T.L., Bedarkar, S., Humbel, R.E. 1983. Fed. Proc. 42, 2592.

19. Arquilla, E.R., Bromer, W.W., Mercola, D.A. 1969. Diabetes 18, 193.

20. Zepf, J., Schoenle, E., Froesch, E.R. 1978. Eur. J. Biochem. 87, 285.

21. King, G.L., Kahn, C.R. 1981. Nature 292, 644.

22. For a review see: Czech, M.P., Oppenheimer, C.L., Massague, J. 1983. Fed. Proc. 42, 2598.

23. Rechler, M.M., Kasuga, M., Sasaki, N., DeVroede, M.A., Romanus, J.A., Nissley, S.P. 1983. In: Insulin-like Growth Factors/Somatomedins: Basic Chemistry, Biology and Clinical

Importance (E.M. Spencer, ed.). Walter deGruyter Co., New York. pp. 459-490.

24. Zapf, J., Froesch, E.R., Humbel, R.E. 1981. Curr. Top. Cel Regul. 19, 257.

25. Zapf, J., Mader, M., Waldvogel, M., Schalch, D.S., Froesch, E.R. 1975. Arch. Biochem. Biophys. 168, 630.

26. Ogawa, H., Burke, G.T., Katsoyannis, P.G. 1984. J. Protein Chem. 3, 327.

27. Joshi, S., Burke, G.T., Katsoyannis, P.G. 1985. Biochemistry 24, 4208.

28. DeVroede, M.A., Rechler, M.M., Nissley, S.P., Joshi, S. Burke, G.T., Katsoyannis, P.G. 1985. Proc. Natl. Acad. Sci. USA 82, 3010.

29. Joshi, S., Ogawa, H., Burke, G.T., Tseng, L. Y-H., Rechler, M.M., Katsoyannis, P.G. 1985. Biochem. Biophys. Res. Commun. 133, 423.

30. DeVroede, M.A., Rechler, M.M., Nissley, S.P., Ogawa, H., Joshi, S., Burke, G.T., Katsoyannis, P.G. 1986. Diabetes 35, 355.

31. Tseng, L. Y-H., Schwartz, G.P., Sheik, M., Chen, Z.Z., Joshi, S., Wang, J-F., Nissley, S.P., Burke, G.T., Katsoyannis, P.G., Rechler, M.M. 1986. Fed. Proc., 45, 1714.

32. Katsoyannis, P.G., Trakatellis, A.C., Zalut, C., Johnson, S., Tometsko, A., Schwartz., G., Ginos, J. 1967. Biochemistry 6, 2656.

33. For a review see: Katsoyannis, P.G. 1981. In: Structural Studies on Molecules of Biological Interest (G. Dodson, J.P. Glusker, D. Sayre eds.). Oxford University Press, London. pp. 454-486.

34. Unpublished data from this laboratory.

35. Chance, R.E., Hoffman, J.A., Kroeff, E.P., Johnson, M.G., Schrimer, E.W., Bromer, W.W. 1981. In: Peptides, Proceedings of the 7th American Peptide Symposium (D.H. Rich, E. Gross, eds.). Pierce Chemical Co., Rockford, Illinois. pp. 721-728.

36. Kitagawa, K., Ogawa, H., Burke, G.T., Chanley, J.D., Katsoyannis, P.G. 1984. Biochemistry 23, 1405.

Acknowledgments This work was supported in part by the United States Public Health Service, National Institutes of Health Research Grant No. AM 29988.

SELECTIVE REMOVAL OF THE 2-BROMOETHYL GROUP IN PEPTIDE SYNTHESIS

M. Joaquina S.A. AMARAL TRIGO and M. Isabel A. OLIVEIRA SANTOS
Centro de Investigação em Química, Faculdade de Ciências
Universidade do Porto, 4000 PORTO, PORTUGAL

The 2-bromoethyl esters of glycine[1] and L-phenylalanine were prepared as hydrochlorides by direct acid catalysed esterification of the aminoacids with 2-bromoethanol at 40°C. These esters were converted into six protected dipeptides by condensation with N-protected glycines. The selective removal of the 2-bromoethyl group from these dipeptides (I) was achieved by the action of sodium sulphide (scheme 1).

$$XNHCH_2COOH + Cl^-\{H_3\overset{+}{N}CHRCOOCH_2CH_2Br\}$$

$$\downarrow \quad DCCI/Et_3N$$

$$XHNCH_2CO\text{-}HNCHRCOOCH_2CH_2Br \quad (I)$$

$$\downarrow \quad 1)\ S^{2-} \quad 2)\ H_3O^+$$

$$XHNCH_2CO\text{-}NHCHRCOOH \quad (II)$$

X = Boc, Trt, Z

R = H, CH_2Ph

Scheme 1

Peptides 1986

Table 1. The Characteristics and Yields of the Synthesized Dipeptides

Fully Protected dipeptide (I)	N-Protected dipeptide (II)
Boc-Gly-Gly-OCH_2CH_2Br m.p. 60-62°C Yield 52%	Boc-Gly-Gly-OH[a] m.p. 135-136°C Yield 55%
Trt-Gly-Gly-OCH_2CH_2Br m.p. 139-140°C Yield 72%	Trt-Gly-Gly-OH[b] m.p. 180-181°C Yield 60%
Z-Gly-Gly-OCH_2CH_2Br m.p. 107-108°C Yield 65%	Z-Gly-Gly-OH[c] m.p. 178-179°C Yield 65%
Boc-Gly-Phe-OCH_2CH_2Br Oil Yield 76%	Boc-Gly-Phe-OH m.p. 140-142°C Yield 40%
Trt-Gly-Phe-OCH_2CH_2Br m.p. 73-74°C Yield 80%	Trt-Gly-Phe-OH[d] m.p. 211-215°C (dec.) Yield 59%
Z-Gly-Phe-OCH_2CH_2Br m.p. 88-91°C Yield 70%	Z-Gly-Phe-OH[e] m.p. 123-124°C Yield 68%

[a]Lit.[2] m.p. 135-136°C; [b]Lit.[3] m.p. 186°C; [c]Lit.[4] m.p. 178-179°C; [d]Lit.[5] m.p. 210-215°C (dec.); [e]Lit.[6] 125-126°C.

The experimental procedure used for the removal of the 2-bromoethyl group was similar to that described, previously, for the cleavage of the 2-chloroethyl groups[2]. However, the reactions with the 2-bromoethyl esters of the dipeptides were carried out, at room temperature, during about 1 $^{1}/_{2}$ hour for the dipeptides of glycine and 3 hours for the dipeptides of glycine and L-phenylalanine. The results obtained are summarised in Table 1.

Acknowledgement

This research was supported by the Instituto Nacional de Investigação Científica, Lisboa and, in part, by the C.I. nº 2/85 of the University of Porto.

References

1. Blinova, G.G., G.V. Erkina, E.G. Sochilin. 1975. Chem. Abs. 82, 43704t.

2. Amaral Trigo, M.J.S.A., M.I.M.R.E. Barbedo. 1983. In: Peptides 1982 (K. Bláha and P. Malon, eds.). Walter de Gruyter, Berlin, p. 133.

3. Hillmann-Elies, A., G. Hillmann, H. Jatzkewitz. 1953. Z. Naturforsch. 8b, 445.

4. MacLaren, J.A.. 1958. Austral. J. Chem. 11, 360.

5. Zervas, L., D.M. Theodoropoulos. 1956. J. Amer. Chem. Soc. 78, 1359.

6. Grassmann, W., E. Wünsch. 1958. Chem. Ber. 91, 449.

SELECTIVE PROTECTION OF CYSTEINE WITH FERROCENE DERIVATIVES

C.N.C. Drey, A.S.J. Stewart
School of Chemistry, Robert Gordon's Institute of Technology, St. Andrew Street, Aberdeen, AB1 1HG

Introduction

The present work arose from earlier studies with the protection of the N (1,4/5H) positions of the imidazole nucleus of histidine using alkylferrocene derivatives. The utility of these organo-metallic derivatives led us to consider their introduction into cysteine chemistry. We now report on the preparation and properties of S-ferrocenylmethyl(FCM)-L-cysteine (1) and its use in the synthesis of glutathione (GSH) (11).

Results

Protection of the thiol group of cysteine in peptide synthesis is normally essential. The benzyl group (1) was formerly the most frequently used, but severe conditions required for deprotection led to the development of alternative groups including, t-butyl (2), benzhydryl (3), trityl (4) and acetamidomethyl (5).

Application of ferrocenylalkyl derivatives in peptide chemistry has been exemplified by the 'four component condensation reaction' (6) and recently in the reductive N-ferrocenylalkylation of amino acids (7). We found that the trifluoracetic acid (TFA) catalysed reaction of ferrocenylmethanol with L-cysteine in aqueous acetone provided S-FCM-L-cysteine (1) in essentially quantitative yield. Cleavage reactions on (1) confirmed that cysteine was released by treatment with TFA or mercuric ions. Having established the compatability of (1) under typical peptide handling conditions we undertook an illustrative synthesis of GSH (11). The FCM derivative (1) could be readily converted into N-9-fluroenylmethyl(FMOC)-S-FCM and N-t-butyloxycarbonyl-(t-BOC)-S-FCM-L-cysteine (2) and (3) isolated as dicyclohexylammonium salts.

Peptides 1986

The scheme details a route in which stripping of protecting groups was achieved under acidolytic conditions. Thus coupling of (2) with glycine-t-butyl ester (4) by the dicyclohexylcarbodi-imide(DCI)/N-hydroxybenzotriazole(HBT) method provided a 97% yield of the protected dipeptide (6). Removal of the FMOC group proceeded in diethylamine at room temperature (8). The free base was extracted and coupled to the glutamate derivative (8) by the same method. Normal work up procedures afforded the fully protected GSH (9) which was isolated as an oil in 76% yield. Treatment of the tripeptide (9) for three hours with TFA containing 5% thiophenol removed all the protecting groups. The reaction was then quenched with ether and the precipitated GSH (11) retrieved by centrifugation.

Scheme

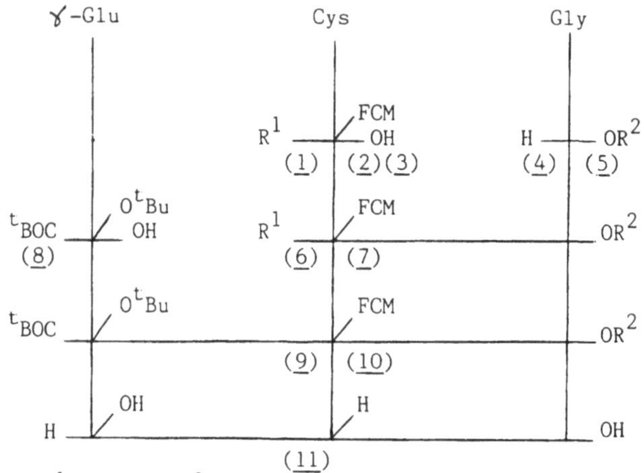

(1) = R^1 = H
(2) = R^1 = FMOC
(3) = R^1 = ^{t}BOC
(4) = R^2 = ^{t}Bu
(5) = R^2 = Et

(6) = R^1 = ^{t}BOC; R^2 = ^{t}Bu
(7) = R^1 = ^{t}BOC; R^2 = Et
(9) = R^2 = ^{t}Bu
(10) = R^2 = Et
FCM = Ferrocenylmethyl

Successive washings with ethanol and ether led to the isolation of analytically pure GSH (11) in 52% yield.

The scheme also illustrates an alternative route using the t-BOC cysteine derivative (3). Coupling of (3) to glycine ethyl ester (5) using the same procedures as previously, provided the fully protected dipeptide (7). Reaction with two equivalents of p-toluene sulphonic acid and (7) led to the selective cleavage of the N-t-BOC group. After isolating the free amine from (7) it was then coupled to the N-terminal glutamate (8) to give the fully protected GSH (10) recovered once more as an oil in 75% yield. Selective deprotection of (10) at the C-terminus using 1.1 equivalent of sodium hydroxide in methanol followed by treatment in TFA/5% thiophenol provided pure GSH (11) in 48%.

In all cases intermediates were fully characterised by elemental analysis and where appropriate by mass spectrometry.

Conclusion

The new S-FCM protecting group is readily accessible in high yield and may be removed by the action of TFA/thiophenol or mercuric ions. It is stable to conditions encountered in typical peptide reactions and work up procedures, including mild acids and bases. It is also lipophilic in character and is coloured providing a useful visual monitor. The utility of the S-FCM group has been demonstrated by the synthesis of GSH (11).

Acknowledgement

This work was supported by Robert Gordon's Institute of Technology (ASTS). We also wish to thank Dr. R. Wade and CIBA (Horsham) Ltd. for useful discussions and support.

References

1. Sifferd, R.M., Du Vigneaud: 1935. J.Biol.Chem. 108, 753.
2. Callahan, F.M., Anderson, G.W., Paul, R., Zimmerman, J.E.: 1963. J.Amer. Chem.Soc. 85, 201.

3. Zervas, L., Photaki, I.: 1962. J.Amer.Chem.Soc. 84, 3887.

4. Amiard, G., Heymes, R., Velluz, K.: 1956. Bull.Soc.Chem.Fr. 698; Zervas, L., Theodoropoulos, D.: 1956. J.Amer.Chem.Soc. 78, 1359.

5. Veber, D.F., Milkowski, J.D., Varga, S.L., Denkewalter, R.G., Hirschmann, R.: 1972. J.Amer.Chem.Soc. 94, 5456.

6. Marquarding, D., Hoffman, P. Heitzer, H. Ugi, I.: 1970. J.Amer.Chem.Soc. 92, 1969.

7. Eckert, H., Seidel, C.: 1986. Angew.Chem.Int.Edn.Engl. 25, 159.

8. Bodanszky, A., Bodanszky, M., Chandramouli, J., Kwei, J.Z., Martinez, J., Tolle, J.C.: 1980. J.Org.Chem. 45, 72.

ADVANTAGES AND DISADVANTAGES IN THE USE OF FORMIC ACID IN PEPTIDE SYNTHESIS

M. Galantino, E. Arlandini, R. Forino, G. Perseo, R. de Castiglione
Research & Development, Farmitalia Carlo Erba, Milan, Italy

Introduction

Formic acid is used in peptide chemistry for the acidolytic removal of the ter-butyloxycarbonyl group (1). Its main advantage, as compared to other deblocking agents, consists in the smooth and selective removal of the protecting group, particularly whenever acid sensitive amino acid, such as tryptophan and threonine, are present. Other important benefits are very good solvent properties, lower hygroscopicity of the resulting salts, commercial availability, moderate cost and easy industrial use.
Some side-reactions, however, can occur as a consequence of the use of this reagent. As a matter of fact, both O-formylation of Ser residues during prolonged treatment of proteins with anhydrous formic acid (2), and N-formylation during coupling with peptide formates by the DCC and activated ester methods (3,4) have been reported.

Results and Discussion

In the synthesis of dermorphin, eledoisin and tryptophyllins we have encountered both O-formylation of serine during formic acid removal of the Boc protecting group, and N-formylation by the formate ion during condensation by the azide and mixed anhydride procedures.
Analytical data of the identified by-products and their corresponding unformylated analogues are reported in Table 1.
The O-formyl derivatives Ia, IIa and IIIa were obtained in variable yields (20-50%) during treatment of the Boc-protected peptides with formic acid in the unusual conditions (1-3 hours at room temperature). Complete formylation was achieved in 12-15 hours in

Peptides 1986

the presence of a stoichiometric amount of HCl. The products could be isolated in a relatively pure form from ethyl acetate-diethyl ether.

Deformylation was observed to occur in the solvent system (n-butyl alcohol/acetic acid/water = 4/1/5) used for purification of the target compounds by counter-current distribution. In this system the deformylation of O-formyl peptides was $\geq$95% after 1-4 days at room temperature. To this end, the n-butyl alcohol can be replaced to advantage by methyl or ethyl alcohol.

Compounds IVa and Va are two examples of N-formylation by the formate ion during coupling by the mixed anhydride and the azide method, respectively. They formed in 50% yield, and were isolated by crystallization from diethylether after purification in flash chromatography.

O-formylation can be avoided by converting formate salts into tetrazolate salts, as recently suggested by Bodanszky (4).

In particular cases formic acid may prove very useful, as in the deblocking of Boc-Trp-Met-Asp-Phe-NH_2 (5), an intermediate in the synthesis of ceruletide, CCK and gastrin. We observed neither t-butylation of the Trp and Met residues (reported to possibly occur with other acidolytic agents), nor N-formylation in the subsequent coupling step. The deprotected tetrapeptide, in fact, was obtained from methyl alcohol as an internal salt because of the presence of the free β-carboxyl group of the aspartic acid, and consequently no formate ions were available for N-acylation.

Table 1. Analytical Data of O- and N-Formyl-Peptides (a) and their Corresponding Unformylated Analogues (b)

PEPTIDE	FD-MS* / FAB-MS**	^{1}H-NMR (diagnostic signals)	H.V.P.E.[1]	TLC[2] (Rf)	HPLC[3] ($R_T$min)
Ia	$[MH]^+$ 831*	8.20δ CH_2-O-CO-H	$E_{1.2}$ 0.51	Rf_A 0.45	10.50
Ib	$[MH]^+$ 803*	5.0 δ CH_2-OH	$E_{1.2}$ 0.51	Rf_A 0.40	9.57
IIa	$[MH]^+$ 571*	8.20δ CH_2-O-CO-H	$E_{1.2}$ 0.63	Rf_A 0.38	8.60
IIb	$[MH]^+$ 543*	5.0 δ CH_2-OH	$E_{1.2}$ 0.63	Rf_A 0.34	7.73
IIIa	$[MH]^+$ 1216**	8.17δ CH_2-O-CO-H	E_F 0.74	Rf_A 0.11	12.21
IIIb	$[MH]^+$ 1188**	5.1 δ CH_2-OH	E_F 0.74	Rf_A 0.11	11.47
IVa	$[M]^{+\cdot}$ 246*	{7.74δ NH-CO-H 4.34δ α-Trp	$E_{1.2}$ 0.11	Rf_B 0.62	12.38
IVb	$[M]^{+\cdot}$ 218*	3.98δ α-Trp	$E_{1.2}$ 1.09	Rf_B 0.08	8.33
Va	$[M]^{+\cdot}$ 377*	{7.91δ NH-CO-H 4.64δ α-Trp	$E_{1.2}$ 0	Rf_B 0.52	13.96
Vb	$[M]^{+\cdot}$ 349*	4.04δ α-Trp	$E_{1.2}$ 0.82	Rf_B 0.05	11.74

I) HCOOH · H-Tyr-D-Ala-Phe-Gly-Tyr-Pro-Ser(X)-NH_2;

II) HCOOH · H-Tyr-D-Ala-Phe-Gly-Ser(X)-NH_2;

III) HCOOH · H-<Glu-Pro-Ser(X)-Lys-Asp-Ala-Phe-Ile-Gly-Leu-Met-NH_2;

IV) X-Trp-OMe

V) X-Trp-Met-OMe (a, X = For; b, X = H)

1) $E_{1.2}$ or E_F, electrophoretic mobility at pH 1.2 or in HCOOH 80%, respectively, referred to glutamic acid

2) Thin layer chromatography on silica gel. Eluent system: A, n-BuOH/AcOH/H_2O 4/1/1; B, benzene/AcOEt/AcOH/H_2O = 10/10/2/1. Peptides Ia, Ib, IIa, IIb, IIIa, IIIb, IVb and Vb are ninhydrine-positive while IVa and Va are ninhydrine-negative.

3) Column RP-18, LiChrosorb (Merck), Mobile phase: A (0.02M KH_2PO_4, pH 3.5, buffer containing 10% v/v CH_3CN); B (0.02M KH_2PO_4, pH 3.5 buffer containing 70% v/v CH_3CN). Linear gradient from 10 to 90% of eluent B in 20 minutes. Detection: 210 nm.

References

1. Halpern, B., Nitecki, D.E. 1967. Tetrahedron Letters 31, 3031.

2. Narita, K. 1959. J. Am. Chem. Soc. 81, 1751.

3. Fletcher, G.A., Löw, M., Young, G.T. 1973. J . Chem. Soc. Perkin I, 1162.

4. Bodanszky, M., Bodanszky, A. 1984. Int. J. Peptide Protein Res. 24, 563.

5. Galantino, M., de Castiglione, R. 1979. U.K. Patent Application, GB 2016478A.

DIBENZYL DICARBONATE (Z_2O), A NEW CARBOBENZOXYLATING REAGENT

W. Graf [1], O. Keller [1], W. Keller [1], G. Wersin [1], E. Wünsch [2]

1) Research & Development Laboratories, FLUKA Chemical Co. Ltd., CH-9470 Buchs - Switzerland

2) Max-Planck-Institut für Biochemie, D-8033 Martinsried - FRG

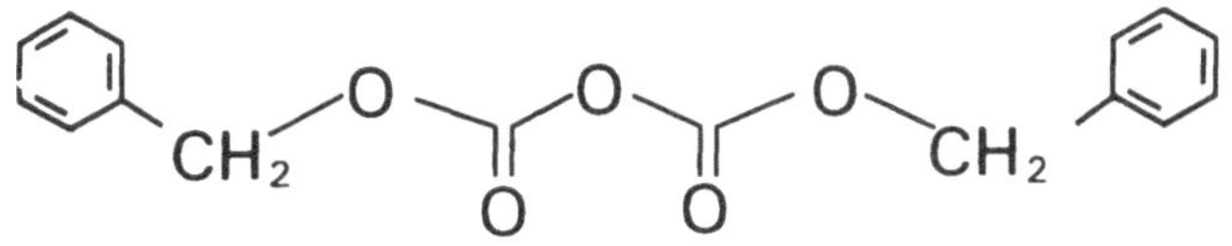

Introduction

The standard reagent used today to introduce the benzyloxycarbonyl group into organic molecules such as amino acids, peptides, alkaloids etc. is benzylchloroformate. This reagent by itself is thermally unstable even at room temperature and loses CO_2 yielding benzylchloride wich often gives rise to significant amounts of benzylated by-products. This fact led to the search for new, stable and highly reactive reagents for introducing the benzyloxycarbonyl group into organic molecules.

Results

It was found that dibenzyl dicarbonate can be prepared according to the procedure given in [1]. (Z_2O : white crystals, odourless, M.P. 25 - 30°, decompositions point ~ 50°, stable at 4°). First results obtained from a comparison of Z_2O and benzyl chloroformate as reagents in the synthesis of Z-Ser-OH, Z-Tyr-OH, Z-Thr-OH, Z-Trp-OH, Z-Phe-OH, Z-Pro-OH show that Z_2O reacts as fast as Z-Cl but less by-products are formed. E.g. Z-Trp-OH is obtained in 95 % yield with Z_2O, and in 70 % yield with Z-Cl [2].

[1] W. Kraatz in Houben-Weyl, Vol. E 4, p. 1000, Thieme-Verlag 1983

[2] W. Graf, O. Keller, W. Keller, G. Wersin, E. Wünsch, to be published 1986

APPLICATION OF FLUORENYLMETHYL ESTERS IN THE SYNTHESIS OF VIRAL PROTEIN FRAGMENTS

Alexander Kozhich, Andrey Ostrovsky

Shemyakin Institute of Bioorganic Chemistry, USSR Academy of Sciences, Moscow, USSR

Vladimir Pozdnev

Institute of Biological and Medical Chemistry, Academy of Medical Sciences of the USSR, Moscow, USSR

Preparation of the 9-fluorenylmethyl esters of 15 amino acids and their use in peptides synthesis is described. The 9-fluorenylmethyl esters originally introduced to peptide synthesis by Bodanszky (1, 2) and Kessler (3) were obtained by traditional methods in rather low yields. Considerably higher yields of these esters were obtained in this work with the aid of pyrocarbonate method (4). Mixture of Boc-amino acid, 0.2-1 equivalent of pyridine, 1.1-1.2 equivalent of 9-fluorenylmethanol and Boc_2O in the aprotic solvents (ethylacetate, tetrahydrofuran or dioxan) at concentration 0.5-1 M were stirred 6-12 hours at room temperature. Workup included extraction followed by open column chromatography on silica gel. After crystallization esters were obtained with 80-90% yields (Table 1). Most of them are crystalline solids, stable under prolonged storage. Tosylates of amino acid fluorenylmethyl esters obtained after Boc removal by trifluoroacetic acid or its mixture with acetic acid in the presence of toluene-4-sulfonic acid are also crystalline, non-gygroscopic solids.

Fluorenylmethyl esters of Ala, Glu(OBzl), Ser(Bzl) and Leu were used in the synthesis of a decapeptide from hepatitis A VP-1 coat protein and a dodecapeptide from pX protein of human T-leukemia virus (I). Peptides were synthesized by fragment condensation in solution. The N-terminal amino groups were temporary protected by Boc group. The C-terminal carboxyls were provided with benzyl or fluorenylmethyl moiety. The 9-fluorenylmethyl ester was removed by one hour treatment with 15% piperidine in DMFA (Schemes 1 and 2).

Peptides 1986

Table 1. Yields Properties and of Boc-AA-9-ffuorenylmethyl Esters and Their Tosylates

Amino acid (derivative)	Boc-AA-OFm			TsOH AA-OFm		
	Yield, %	m.p., ^{o}C	$\|\alpha\|_{D}^{23}$ c=1, MeOH	Yield, %	m.p., ^{o}C	$\|\alpha\|_{D}^{23}$ c=1, MeOH
Ala	96.5	76-78	-25.0	100	210-211	-17.7
Gly	93	72-75	-	100	161-162	-
Val	71	67-68	-33.1	100	179-181	-14.8
Leu	93	70-72	-34.9	97	211-213	-11.7
Ile	64	61-62	-24.9	95	177-179	-10.0
Pro	89	88.5-90	-44.0	100	oil	-28.7
Phe	89	127-128	-28.1	100	171-172	-43.5
Ser(Bzl)	86.5	oil	-10.0	85.8	167-168	-12.7
Glu(Bzl)	76	98-99	-25.6	100	183-185	-16.7
Trp	84	95-96	-5.2*	98	159-160	-17.2
Cys(Bu^t)	94	76.5-77.5	-20.7	100	188-190	-27.6
Tyr(Bzl)	81	110-111	-15.0	100	163-164	-34.4
Thr(Bzl)	88	oil	-10.5	92	145-147	-24.6
Lys(Z)	91	67-71	-24.1	96	140-142	-18.1
Phe(4-NO_2)	86	166-167	-7.0*	100	215-216	-41.6

* c=1, $CHCl_3$

The protected peptides were deblocked by liquid HF in the presence of p-cresole and were purified by preparative reverse phase HPLC (Fig, 3, 4). Peptides were obtained in 60-70% yield and characterized by amino acid analysis, NMR, and mass spectrometry.

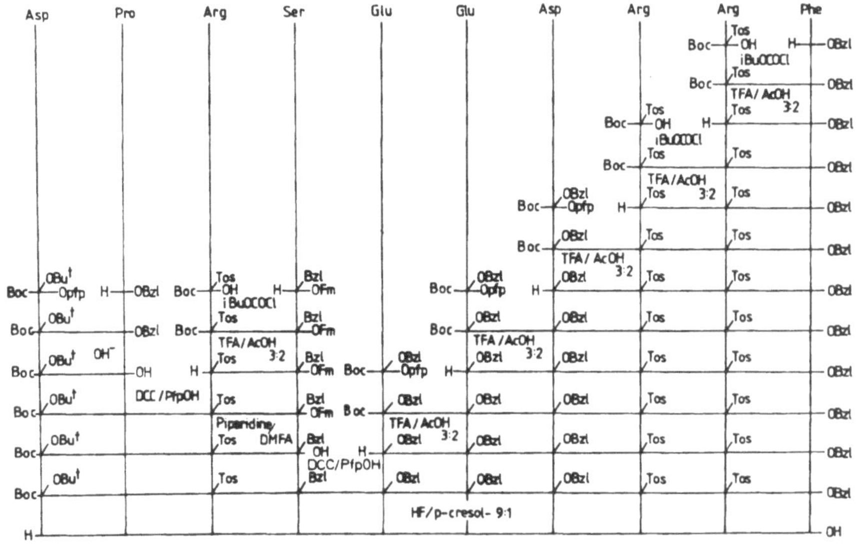

Fig. 1. Scheme for synthesis of decapeptide from hepatitis A VP-1 coat protein.

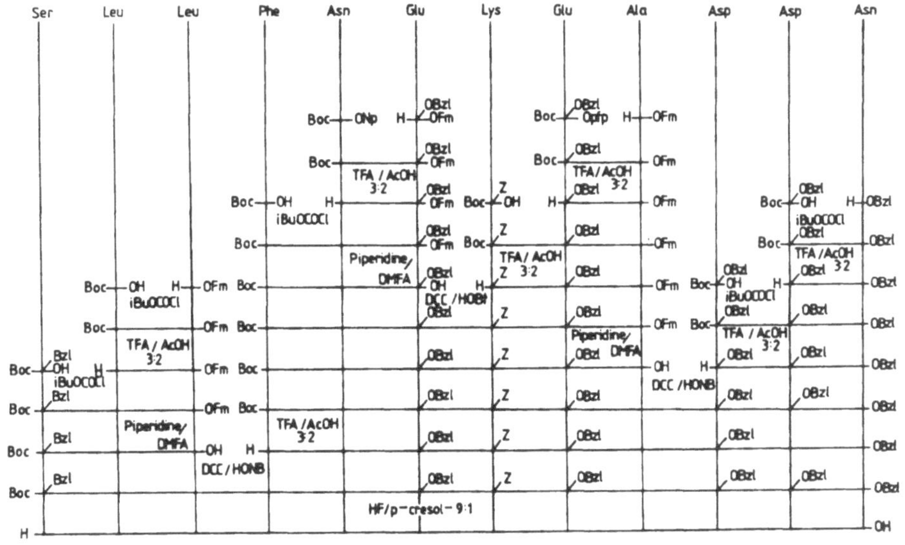

Fig. 2. Scheme for synthesis of dodecapeptide from pX protein of human T-leukemia virus (I).

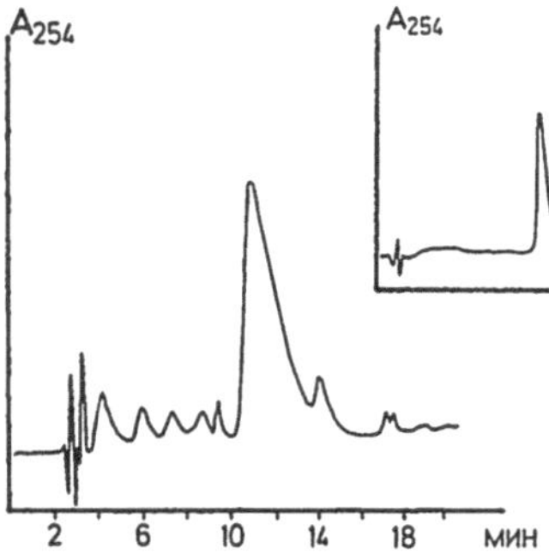

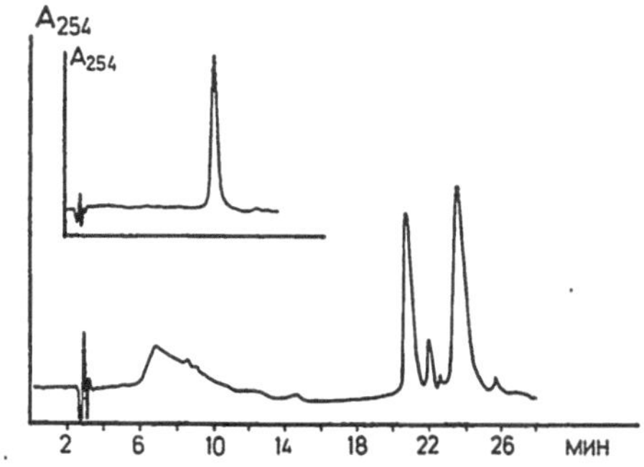

Fig. 3. Reverse phase HPLC of the deprotected 10-peptide from hepatitis A VP-1 protein (Zorbax ODS 9.4x250 mm, A-0.5% TFA/ H_2O, B-0.1% TFA/MeCN, gradient from 10% B to 35% B in 15 min). Insertion - analytical HPLC of purified peptide.

Fig. 4. Reserve phase HPLC of the deprotected 12-peptide from pX protein of human T-leukemia virus (I) (Zorbax ODS 9.4x250 mm, A-0.1% TFA, B-MeCN, gradient from 10% B to 30% B in 25 min). Insertion - analytical HPLC of purified peptide.

Conjugates of these peptides with BSA and KLH were prepared using glutaric aldehyde as crosslinking agent. Sera obtained after immunization of guinea pigs and mice had antipeptide titres about 10^3 but did not bind to respective inactivated viruses.

References

1. Bodanszky, M., M. Bednarek, A. Bodanszky, J.C. Tolle. 1981. In: Peptides, Proc. 16 Eur.Pept.Symp. (K. Brunfeld, ed.). Scriptor, Copenhagen, p. 93-98.

2. Bednarek, M.A., M. Bodanszky. 1983. Int.J.Pept.Prot.Res. 21, 196-201.

3. Kessler, H. 1983. Tetr.Lett. 24, 281-282.

4. Pozdnev, V.F. 1984. Bioorg.Khim. 10, 912-920.

ELECTROCHEMICAL CLEAVAGE OF PROTECTING GROUPS

H.L.S. Maia, M.J. Medeiros, M.I. Montenegro

Centro de Química Pura e Aplicada, Universidade do Minho, Largo do Paço, 4719 Braga Codex, Portugal

D. Pletcher

Department of Chemistry, The University, Southampton SO9 5NH, United Kingdom

Introduction

During the last two decades several authors have reported the use of electrochemical methods of cleavage of protecting groups (1). Some of these protecting groups are of common use in peptide synthesis but, although in many cases the yields reported were much better than 80%, these techniques have not received much attention from peptide chemists; for example, Young describes interesting and high yield reactions involving the picolyl and *N*-piperidino-oxycarbonyl groups but these groups are almost unknown to peptide chemists. The lack of application of electrochemical methods maybe due to a failure to rationalise experimental conditions as they varied widely from author to author, and also to absence of comments regarding either the reasons that led each author to choose a particular set of experimental conditions (e.g. solvents, supporting electrolytes, electrodes), or the need for such conditions in order to achieve satisfactory yields. Also in most cases model compounds were used instead of molecules of real interest to peptide chemists. This is most common in earlier work where even the potentials at which the cleavages were carried out were often not reported.

Wishing to familiarise organic chemists with electrochemical techniques, we have been carrying out kinetic and mechanistic studies on electrochemical cleavage of protecting groups commonly used in peptide synthesis. For this purpose we used the homogeneous redox catalysis technique (2) which enabled us to understand the kinetics and mechanisms and, thus, the effect of a given set of experimental conditions on the yields and also to set up the conditions required for the best results. Therefore, the cleavage of the tosyl group in monofunctional simple tosyl esters and primary and secondary amides was studied under several different experimental conditions (2,3); it was found that cleavage occurs at potentials close to −2.25 V in aliphatic esters and within the range −1.57 to −1.75 V in aromatic esters; amides cleaved within the range −2.3 to −2.4 V. These results were tested on *N*-tosyl-serine methyl ester, where the *N*-protecting group cleaved at −2.39 V, and on

Peptides 1986

N,O-ditosyl-serine methyl ester, where the *O*-protecting group cleaved at a less negative potential (-2.03 V) than that found for cleavage of the *N*-protecting group (-2.23 V). A proposal was made for the cleavage mechanism and it was concluded that the presence of a weak proton donor would be advisable for synthetic purposes. We also studied the splitting of the benzyloxycarbonyl group in primary and secondary simple aliphatic amides and in amino-acid and peptide esters of up to four amino-acid residues (4,5); in all compounds cleavage took place at a potential close to -2.8 V, although the compounds having proline at the *N*-terminus cleaved at potentials 0.05-0.13 V more negative than those required for cleavage of all other amino-acid derivatives. The splitting mechanism was elucidated and it was found that, the addition of acid catalyses this decarboxylation.

When scanning the literature concerned with removal of protecting groups of interest or of possible interest in peptide synthesis (1), one finds out that with a few exceptions, the groups that cleave under reductive conditions all cleave at potentials below -2 V and this is also the case with the groups we studied. Although the peptide molecules seemed to be stable under these conditions and the results we obtained suggest that we may feel confident with regard to the use of electrochemical methods for removal of the tosyl and the benzyloxycarbonyl groups in routine peptide synthesis, at such negative potentials many solvents and supporting electrolytes may decompose; this tends to narrow the choices for these auxiliary materials and thus bring difficulties during the work-up of the reaction products. Therefore, we believed that it would be appropriate to develop new groups suitable for electrochemical cleavage at less negative potentials, possibly modifications of the classical protecting groups in order to help peptide chemists to feel confident in their use. Following this line, we decided to start by studying a classical modification of a classical protecting group, i.e. the *p*-nitrobenzyloxycarbonyl group. We report here the results obtained so far with monofunctional amides.

Materials and Methods

The kinetic and mechanistic studies were carried out by cyclic voltammetry and coulometry, while preparative studies were performed by electrolysis at constant potential. A potentiostat, a function generator, an X,Y-recorder, and an integrator were used for the electrochemical techniques in conjunction with three electrode cells, being the cathode (a vitreous carbon disc), the secondary electrode (a platinum grid), and the reference electrode (a saturated calomel electrode). All experiments were carried out in *N,N*-dimethylformamide containing 0.1 mol dm^{-3} of tetrabutylammonium tetrafluoroborate and the compound under study (2.5-5 mmol dm^{-3} in cyclic voltammetry and 20 mmol dm^{-3} in electrolysis). Product analysis was carried out by gas-liquid chromatography.

Results and Discussion

The *p*-nitrobenzyloxycarbonyl derivatives of the following four amines were studied: butylamine, cyclohexylamine, diethylamine and piperidine. In cyclic voltammetry all compounds showed three reduction peaks at potentials close to -1.1, -1.8, and -2.3 V. Although the first peak showed a reversible process within the time scale of voltammetry, we were able to remove the protecting group at a potential just after this peak, i.e. -1.1 V. Coulometry results indicate that under these conditions the number of electrons involved per molecule is 2. This means that within the time scale of the electrolysis the radical anion is no longer stable and decomposes according to the following mechanism that accounts for only one of the electrons involved:

$$NO_2C_6H_4CH_2OCONR^1R^2 + e \rightleftharpoons [NO_2C_6H_4CH_2OCONR^1R^2]^{\bullet-}$$

$$[NO_2C_6H_4CH_2OCONR^1R^2]^{\bullet-} \longrightarrow NO_2C_6H_4CH_2^{\bullet} + R^1R^2NCO_2^-$$

$$R^1R^2CO_2^- + H^+ \text{(solvent)} \longrightarrow R^1R^2NH + CO_2$$

The yields in amine were as high as 80-90%, but only a small percentage of *p*-nitrotoluene was found. Although this may be unimportant regarding the aim of the present work, we are investigating the fate of the second electron and that of the *p*-nitrotoluene radical which may be further reduced and/or dimerize, the product undergoing further decomposition at the lower potentials corresponding to the two other reduction peaks. No change in the cleavage process was observed when methanol was added as a proton donnor; however, addition of acetic acid shifts the first reduction peak by as much as 0.18 V towards a less negative potential; it is clear that a change of mechanism takes place when acid is added and, thus, we are still investigating if this change will affect the cleavage process in any way.

As one would expect, the nitro group bonded to the aromatic ring allows cleavage of the *p*-nitrobenzylowycarbonyl group at a potential much less negative (-1.1 V) than that required for removal of the benzyloxycarbonyl group (-2.8 V). This result is completely satisfactory with regard to the possible difficulties often arising when cleavage has to be performed at potentials below -2 V, and offers *p*-nitrobenzyloxycarbonyl as a protecting group of election for electrochemical removal. Moreover, it offers a new strategic combination with the benzyloxycarbonyl group. In fact, both groups can be cleaved simmultaneously either by catalytic hydrogenolysis or by electrolysis, while it is well known that the former is rather stable in the presence of acid and that under the same conditions the latter may be cleaved; however, this selective behaviour is *reversed* when cleavage is performed *by electrolysis*, i.e. the *p*-nitrobenzyloxycarbonyl group can be cleaved selectively in the presence of the benzyloxycarbonyl group with a selectivity range of 1.7 V.

Aknowledgement

This work was financially supported by Junta Nacional de Investigação Científica e Tecnológica (contract nº 402.82.05).

References

1. M.I. Montenegro. 1986. Electrochim. Acta 31, 607.
2. H.L.S. Maia, M.J. Medeiros, M.I. Montenegro, D. Pletcher. 1984. J. Electroanal. Chem. 154, 347.
3. H.L.S. Maia, M.I. Montenegro, D. Pletcher. 1983. Portugaliae Electrochim. Acta 1, 201.
4. H.L.S. Maia, M.J. Medeiros, M.I. Montenegro, D. Pletcher. 1986. J. Electroanal. Chem. 200, 363.
5. *Ibid*. 1984. Portugaliae Electrochim. Acta 2, 1.

AMMONOLYSIS MEDIATED SIDE REACTIONS OF β-T-BUTYL-ASPARTYL PEPTIDES

István Schön and Attila Rill
Chemical Works of Gedeon Richter, H-1475 Budapest, POB 27, Hungary

During an attempted synthesis of a pentagastrin analog we observed that ammonolytic cleavage of the peptide-polymer bond of this peptide containing a β-tert-butyl-aspartyl residue did not at all result in the desired product /1/. Having known the base sensitivity of the half esters of aspartyl peptides /2-4/ we examined the behavior of Z-Asp/O^tBu/-Phe-NH_2 /LL/ in methanol saturated with ammonia at room temperature. A complete conversion of LL was detected within 24 h on tlc. Preliminary physico-chemical characterization of the isolated product seemed to support that it was the isoasparaginyl derivative, the melting point and optical rotation, however, did not conform to those of an authentic sample of Z-isoAsn-Phe-NH_2 synthesized directly from Z-isoAsn-OPfp and H-Phe-NH_2 /5/. Finally a successful resolution of the product by hplc gave 4 peaks, suggesting that in addition to transpeptidation a partial epimerization occurred /6/.

To confirm this hypothesis together with an assumption that racemization occurred at the C-terminal residue, at first on hplc we identified the individual peaks of the ammonolytic mixture with authentic samples of the newly expected products, i.e. with Z-Asn-Phe-NH_2 /α-LL/, Z-isoAsn-Phe-NH_2 /β-LL/, Z-Asn-D-Phe-NH_2 /α-LD/ and Z-isoAsn-D-Phe-NH_2 /β-LD/. Thus knowing the product ratio of the ammonolytic mixture and the optical rotation of each compound we calculated the sum of the partial optical rotation values of the individual products. The significant difference between the measured and calculated values led us to the conclusion that our assumption concerning the place of the chirality center partially racemized proved to be wrong. Really, taking into account of the optical rotation values of α-LL and β-LL and those of the other epimers Z-D-Asn-Phe-NH_2 /α-DL/ and Z-D-isoAsn-Phe-NH_2 /β-DL/ with an opposite sign, i.e. assuming racemization at the Asp residue, we got excellent adequacy in the measured and calculated values.

Peptides 1986

Table. Product Ratio /%/ of Ammonolyses of Z-Asp/O^{t}Bu/-Phe-NH_2 /LL/, Z-D-Asp/O^{t}Bu/-Phe-NH_2 /DL/ and Z-Asp/O^{t}Bu/-D-Phe-NH_2 /LD/[a]

Starting compound		LL				DL	LD[b]
Solvent /No. of hplc resolutions/		MeOH /5/	EtOH /3/	BuOH /1/	DMF /1/	MeOH /2/	MeOH /4/
Product /its enantiomer/[b]							
α-LL	/α-DD/	7.9-11.4	5.2-10.1	10.7	4.8	5.2-6.0	/3.8- 4.9/
β-LL	/β-DD/	50.8-61.4	66.5-75.8	76.4	76.7	23.6-23.7	/21.2-25.1/
α-DL	/α-LD/	12.6-18.3	5.2- 9.7	5.3	3.6	30.0-30.8	/30.2-32.3/
β-DL	/β-LD/	16.2-21.2	13.6-13.9	7.9	14.9	40.0-40.4	/37.9-44.8/
α-LL + α-DL	/α-LD + α-DD/[c]	21.2-28.3	10.3-19.7	15.8	8.4	35.9	/34.0-36.9/
Σepimerization[c,d]		30.7-38.9	19.0-23.4	13.2	18.5	28.8-29.6	/25.0-29.8/

[a]Because of high deviations of parallel results supposedly coming from sample inhomogeneity, minimal and maximal values are presented instead of averaging.

[b]Symbols and data parenthized and unparenthized correspont to each other, respectively.

[c]These data are taken from the actual individual resolutions, not from the summation of minimal and maximal values presented above.

[d]Data referring to the compounds epimerized are underlined.

In this case the same solution was used for determination of the optical rotation and hplc analysis.

Having reached the evidence on the correctness of our original hypothesis we were curious to know how the ratio of the ammonolytic products depends on the chirality of the starting material and on the solvent. Thus we examined the ammonolysis of Z-Asp-/O^tBu/-D-Phe-NH_2 /LD/ and Z-D-Asp/O^tBu/-Phe-NH_2 /DL/ in methanol, and that of LL in ethanol, butanol and dimethylformamide, as well. Percentages of the product ratios calculated from hplc by area normalization are shown in the Table.

Ammonolyses were complete in methanol within 24 h, in ethanol within 48 h, in butanol within 1 week and in dimethylformamide within 2 weeks. All experiments were carried out in solvents saturated with ammonia at 0 °C, then the reaction mixtures were kept at room temperature in closed vessels. In alcohols the mixture of products precipitated, in dimethylformamide it remained in solution. In each case the reaction mixture was evaporated, and filtration of the suspension in hot ethanol gave the mixture of products with 80 to 86 % yield.

In protic solvents the extent of epimerization decreases with diminution of the polarity, and it is very significant in dipolar aprotic dimethylformamide, too. A nearly similar relationship can be observed between the polarity of solvents and the overall amount of α-peptides, except in dimethylformamide where the highest ratio of β/α-peptides were detected. In the ammonolytic mixtures of both DL and LD the overall amount of α-peptides as well as of the epimerized products were identical within the experimental error, respectively. Attention should be called on two facts: a. in methanol ammonolysis of LL results in more epimerized produts than those of its epimer DL and LD; b. significantly greater amounts of α-peptides come from DL and LD than from LL. We assume that both of these differences are related to steric factors, i.e. to the solution conformation of the reactive intermediates. It is no doubt that ammonolysis of these β-tert-butyl-aspartyl peptides follows the well-known reaction mechanism, i.e. proceeds through the aminosuccinyl derivatives. The vulnerability to epimerization of both the educts and products can be excluded as well as that of the transitional state species leading to and from the aminosuccinyl derivatives, because all of them are ionized species. Thus it is reasonable to propose that it is the appropiate aminosuccinyl intermediate that epimerizes by base catalysis, presumably through the α-proton abstraction-enolization mechanism /7/.

These results are in complete accordance with observations of Clarke's group reporting a significant formation of aminosuccinyl derivatives and their epimerizations during syntheses of peptides containing an L- or D-Asp residue /8/. Biologically active proteins supposedly containing a cyclic imide residue and/or epimerized sequences were brought into connection with their biochemical repair mechanism and metabolism /8 -12/.

Acknowledgement

We thank Dr. L. Kisfaludy and M. Lőw for valuable discussions, Ms. Zs. Muck and Ms. I. Überhardt for skilful technical assistance.

References

1. Schőn, I. and L. Baláspiri: unpublished observation.
2. Schwyzer, R., B.Iselin, H.Kappeler, B.Riniker, W.Rittel and H. Zuber. 1963. Helv.Chim.Acta 46, 1975.
3. Bajusz, S., T.Lázár and Z. Paulay. 1964. Acta Chim.Acad.Sci. Hung. 41, 329.
4. Bodanszky, M. and J.Martinez. 1981. Synthesis 5, 333.
5. Schőn, I., T.Szirtes and G. Iványi. Kémiai Közl. /in press/.
6. Results were partly reported at the Workshop of the Peptide Committee of the Hung.Acad.Sci., December, 1984, Budapest and at the 8th Polish Pept. Symp., October, 1985, Jastrzebia Góra.
7. Kemp, D.S. 1979. In: The Peptides, Vol.1 /E.Gross and J. Meienhofer, eds./. Academic Press, New York, p. 317.
8. Murray, E.D.Jr. and S.Clarke. 1984. J.Biol.Chem. 259, 10722.
9. Terwilliger, T.C. and S. Clarke. 1981. J.Biol.Chem. 256, 3067.
10. Barber, J.R. and S. Clarke. 1985. Biochemistry 24, 4867.
11. Johnson, B.A. and D.W. Aswad. 1985. Biochemistry 24, 2581.
12. McFadden, P.N. and S. Clarke. 1982. Proc.Natl.Acad.Sci. /USA/ 79, 2460.

THE PHENOTHIAZINYL-(10)-CARBONYL(PC)-RESIDUE, A NOVEL AMINO-PROTECTING GROUP FOR PEPTIDE SYNTHESIS

J. Gante, W. Hechler, R. Weitzel

E. Merck Darmstadt, Präklinische Pharmaforschung, Pharmazeutische Chemie, D-6100 Darmstadt, West-Germany

Introduction

It would be desirable to have an amino-protecting group which is cleavable under mildly basic conditions but does not have the disadvantages of those already existing (1-3). Of special interest are "two-step" protecting groups which exist in a "safety form" (stable during peptide synthesis) and a labile "cleavage form" (generated directly before cleavage) (4,5).

Results

Amino acids or their esters can be reacted with the easily accessible and stable phenothiazinyl-(10)-carbonylchloride (6) in the presence of a base (1 N NaOH/water/acetone or N-methylmorpholine/THF, respectively; 0-20 C) to the corresponding, hitherto unknown, N- phenothiazinyl-(10)-carbonyl (PC)-derivatives:

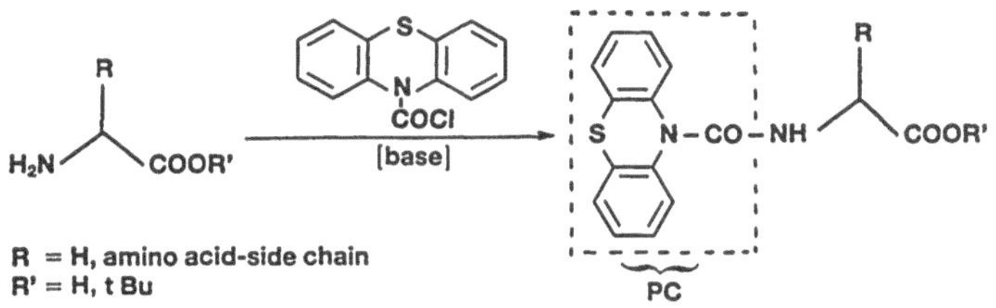

Peptides 1986

The PC-compounds are easily isolable and stable solids (see table 1).

Table 1: PC-Amino acid derivatives PC-AA-OR

AA	R	yield (%)	mp. (C)	(c=1, CH_3OH)
Gly	H	78	182-183	---
Ala	H	70	189	+ 42,4
Ala	tBu	64	97-98	+ 34,9
Val	H	59	79	+ 25,1
Leu	H	82	65	+ 14,0
Ile	H	51	75	+ 34,8
Pro	H	72	99	+ 14,4
Phe	H	78	75	- 2,9

The PC-residue represents a novel "two-step" amino protecting group of an "activatable" urea-type. Thus, using PC-amino acids and the DCC/HOBt-method (7) peptide coupling with various amino acid esters can be performed without observable side reactions and racemization.

The ester function of the PC-amino acid esters and PC-peptide esters can be cleaved as usual to the corresponding acids (e. g. t.-butylesters with TFA, ethylesters with NaOH/H_2O) without affecting the protecting group (see table 2).

Removal of the protecting group can be performed by "oxidation-induced" labilization of the PC-compounds with hydrogenperoxide/glacial acetic acid (excess of 30 % H_2O_2 in glacial acetic acid, 65-80 C, 3-5 h) to form the corresponding, isolable, phenothiazine-sulfones (PCS-compounds). Subsequent mild alkaline hydrolysis of the PCS-compounds with aqueous $Ba(OH)_2$ (10 %, molar ratio 1:1,5; 0 C, 5-10 min., yield 80-90 %) yields the unprotected products, which in all measured parameters have been shown to be identical with the known peptides.

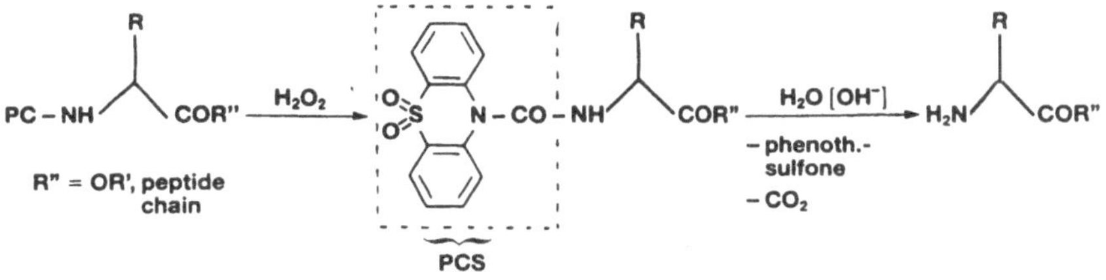

Table 2: PC-Peptide derivatives PC-X-OR

X	R	Yield (%)	mp. (C)	(c=1, CH_3OH)
Gly-Ala	H	87	169-170	- 10,1
Gly-Ala	Et	73	128-129	- 25,3
Gly-Ala	tBu	52	163	- 28,1
Gly-Pro	tBu	69	67	- 63,1
Ala-Pro	H	96	198-199	- 62,3
Ala-Pro	tBu	80	218-219	- 31,0 (c=1, $CHCl_3$)
Leu-Gly	Et	68	117-118	- 5,4
Leu-Gly	BZL	66	97-100	- 4,9
Phe-Leu	BZL	89	Oil	- 47,3

The protecting group can also be removed in a "one-pot" procedure by reacting 1 N NaOH, 30 % H_2O_2 and the PCS-compound (molar ratio 1,5:10:1; 20 C, 24 h, 50-60 %) without intermediate isolation of the sulfones.
The PCS-amino acids and their derivatives are mostly crystalline and, in the solid state, fully stable compounds (see table 3).

Because of the stability of the PCS-residue under the acidic or neutral conditions necessary to cleave the esters (t.-butylesters can be cleaved by TFA, benzylesters by hydrogenolysis), the PCS-compounds themselves can be used for peptide syntheses if strongly basic conditions are avoided.

Table 3: PCS-amino acid and PCS-peptide derivatives PCS -X-OR

X	R	prep. method	yield	mp. (C)	(c=1, CH_3OH)
Ala	H	A	72	216	+ 17,8
Ala	tBu	A	50	133-135	+ 12,9
Phe	H	A	66	95	- 18,2
Gly-Ala	H	B	94	175	- 10,8
Gly-Ala	tBu	A	82	170-173	- 9,8(c=1,THF)
Ala-Pro	H	A	72	199	- 49,0
Ala-Pro	tBu	A	98	Oil	- 69,4
Leu-Gly	H	B	95	215	- 7,8
Leu-Gly	BZL	A	87	134-138	- 10,5
Phe-Leu	H	B	75	128	- 9,6
Phe-Leu	BZL	A	95	80	- 21,6
Phe-Leu-Ala	tBu	C	44	136-140	- 22,0
Leu-Ala	tBu	C	35	143-145	- 34,9

A: by oxidation; B: by cleavage; C: by coupling

Therefore, in special cases, the new protecting group principle is also applicable to oxidation-sensitive amino acids (e. g. Cys, Met).

References

1. König, W., A. Volk. 1977. Chem. Ber. 110, 1

2. Bodanszky, M., S. S. Deshmane, J. Martinez. 1979. J. Org. Chem. 44, 1622

3. Carpino, L. A., J. R. Williams, A. J. Lopusinski. 1978. J. Chem. Soc. Chem. Comm. 450

4. Barthels, R., H. Kunz. 1982. Angew. Chem. 94, 302

5. Kunz, H. 1976. Chem. Ber. 109, 3693

6. Friederich, H. H., J. Bindig. 1960. DBP 1 072 994 (BASF)

7. König, W., R. Geiger. 1970. Chem. Ber. 103, 788

TETRACHLOROETHYL MIXED CARBONATES : THEIR USE IN THE SYNTHESIS OF PEPTIDES

G. Sennyey, G. Barcelo, J.P. Senet
SNPE, Centre de Recherche du Bouchet, BP 2 91710 VERT-le-PETIT FRANCE

J. Martinez, M. Jaouadi, B. Castro
CNRS-INSERM, Centre Pharmaco-endocrinologie, BP 5055, 34033 MONTPELLIER FRANCE

A. LOFFET
PROPEPTIDE, BP 12 91710 VERT-le-PETIT FRANCE

Phosgenation of chloral gives as product 1,2,2,2 tetrachloroethyl chloroformate (1,2) which can react with a variety of hydroxy compounds to give mixed carbonates (3,4)

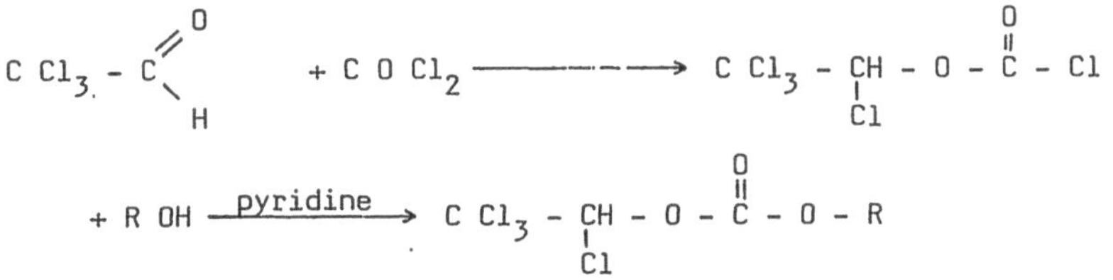

Some of these products are very useful in peptide synthesis :

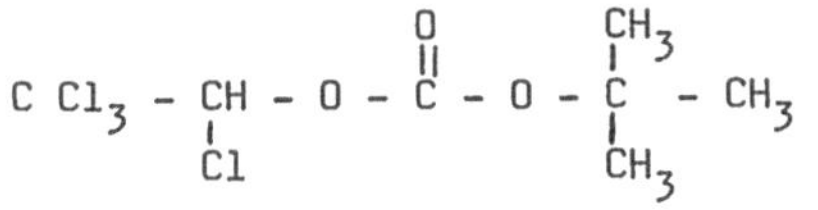

(Boc - OTCE) I

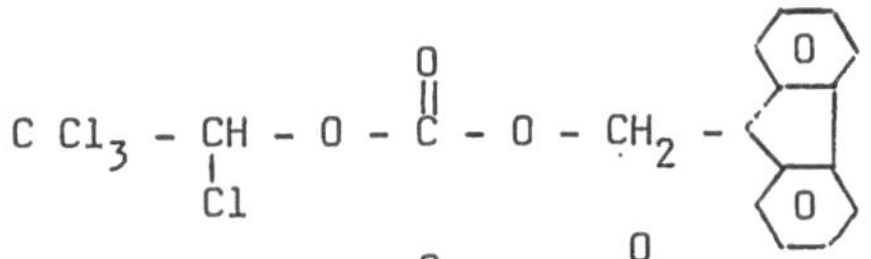

(Fmoc - OTCE) II

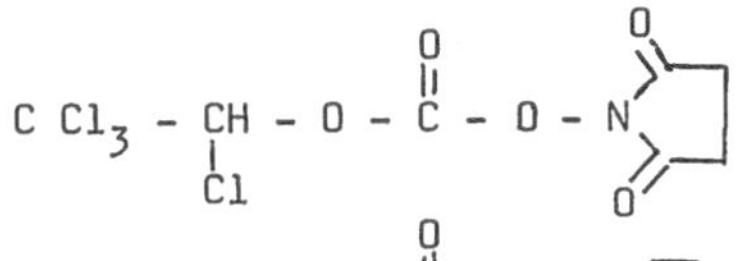

(TCE - ONSu) III

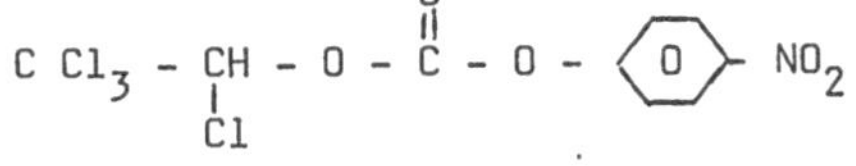

(TCE - ONp) IV

Peptides 1986

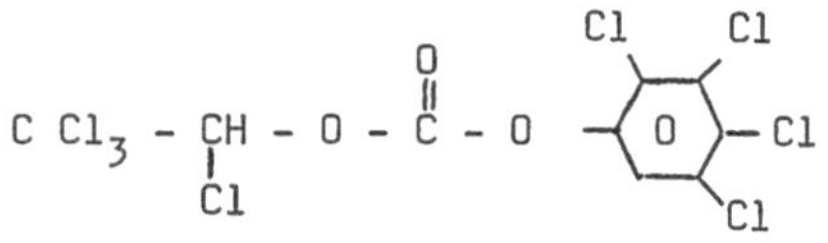

(TCE - OTcp) V

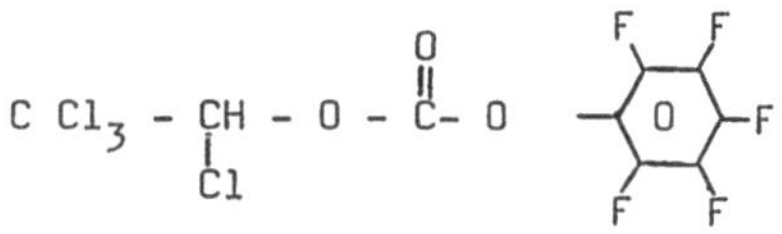

(TCE - OPfp) VI

Boc and Fmoc amino-acid derivatives can be easily prepared using respectively the reagents I and II (4,5).

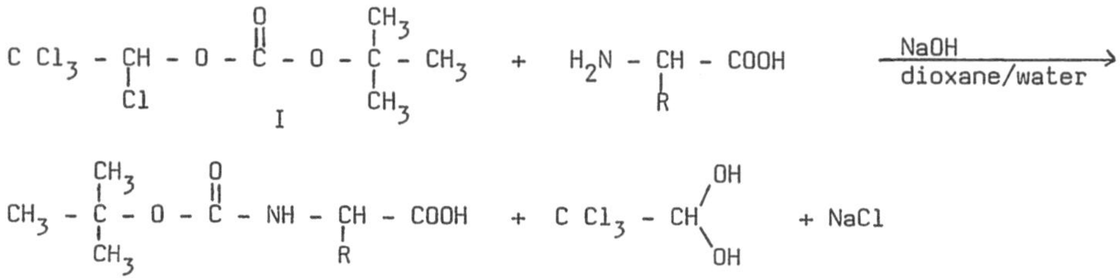

All common amino-acids have been prepared on a production scale with 50-90 % yields. Most Fmoc derivatives have also been synthetized in the laboratory.

TABLE 1

Boc	MP (ºC)		$[\alpha]_D^{20}$		
	Found	Litt.	Found	Litt.	C, Solvent
Ala	81	83-84	-26.6	-27	c=2, HOAC
Arg (NO_2)	109	110-114	-5.4	-5.6	c=1, DMF
Arg (Tos)	88	100	-3	-3.3	c=4, DMF
Asn	179	181-182	-7.6	-7.8	c=1, DMF
Asp (OBzl)	98	101	-19.9	-19.5	c=2, DMF
Gln	118	118	-3.5	-3	c=2, EtOH
Glu	61	57	-6.1	-5.2	c=1, HOAC
Gly	88	90			
Ile	59	61	+2.5	+3	c=5, HOAC
Leu H_2O	80	82	-25	-25	c=2, HOAC
Met	43	49	-22.5	-20	c=1, MeOH

TABLE 3 : N-HYDROXYSUCCINIMIDE ESTERS

Boc	Yield		M P (°C)		$[\alpha]_D^{20}$		
	Found	Litt.	Found	Litt.	Found	Litt.	c, solvent
Ala	94	71	158	161	-50.7	-49	2,Dioxane
Lys (Z)	85		110	110	-21.5	-21.1	2,Dioxane
Phe	91	81	147	152	-19.8	-20.7	2.2,Diox.
Gly	91	62	162	168			
Val	92	74	125	125	-37.4	-37	2,Dioxane
Pro	82	74	133	135	-54.7	-55.3	2,Dioxane
Tyr	74		181	180	-12.1	-12.5	1,Dioxane
Trp	89	37	140	153	-23.1	-22.4	2,Dioxane
Met	92	59	124	128	-21.6	-20.6	2,Dioxane
Cys (Bzl)	61	49	106	117	-54.0	-54.0	2,Dioxane
Tyr (Bzl)	73		144	149	-6.2	-6.3	2,Dioxane

TABLE 4 : PENTAFLUOROPHENYL ESTERS

Boc	Yield		M P (°C)		$[\alpha]_D^{20}$		
	Found	Litt.	Found	Litt.	Found	Litt.	c, solvent
Ala	90	85	85	83	-34.6	-31.2	1,Dioxane
Gly	90	70	80	80			
Phe	86	97	112	112	-16.9	-26.9	1,Dioxane
Val	94	78	64	64	-31.8	-18.4	1,Dioxane
Pro	98	88	50	51	-48.2	-53	1,Dioxane
Trp	91	88	111	111	-12.8	-28.1	1,Dioxane
Met	81	89	79	80	-20.7	-22.5	1,Dioxane

Phe	86	87	+25.3	+24.7	c=1.5, EtOH
Pro	133.5	137	-60.9	-60.2	c=1, HOAC
Ser	88	78	-3.9	-4.6	c=1, HOAC
Ser (Bzl)	60	61	+23.7	+23.5	c=2, EtOH
Thr	76	80	-8.7	-8.5	c=1, HOAC
Val	76	79	-6.3	-5.8	c=1, HOAC

Even more interesting is the fact that active esters of N protected amino-acids can be synthetized easily and in good yields without the use of dicyclohexylcarbodiimide (6).

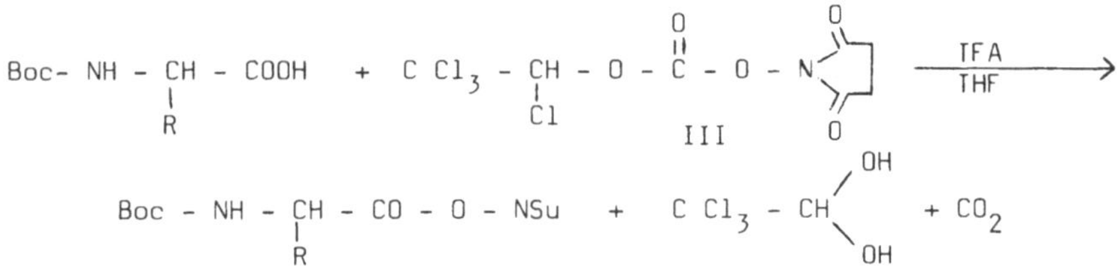

The same reaction can be done with reagents IV to VI to give the corresponding active esters.

TABLE 2 : 2,4,5 TRICHLOROPHENYL ESTERS

Boc	Yield		M P (°C)		$[\alpha]^{20}_{D}$		
	Found	Litt.	Found	Litt.	Found	Litt.	c, solvent
Ala	79	73	82	82	-43.5	-44	2, Ac.Acet
His (DNP)	58		98		+1.8		2,Dioxane
Glu (OBzl)	60	60	107	108	-33.5	-33.5	2,Methanol
Lys (Z)	85	86	99	99	-19.1		2,Dioxane
Met	79	71	90	91	-38.0	-38.5	2,DMF
Ser	22	5	105	105	-42.0	-41	2,DMF
Tyr	73	40	168	158	-31.4	-26.4	3.5, Ac.Ac

References

1. Cagnon, G., Piteau, M., Senet, J.P., Olofson, R.A., Martz, J.T., Eur. Pat. Appl. E.P. 40, 153

2. Olofson,R.A., Martz, J.T., Senet, J.P., Piteau, M. and Malfroot, T., J. Org. Chem. 49 , 2081 (1984)

3. Barcelo, G., Senet, J.P., Sennyey, G., J. Org. Chem. 50 , 3951, (1985)

4. Barcelo,G., Senet, J.P., Sennyey, G., Bensoam, J., Loffet, A., Synthesis in the press (1986)

5. Barcelo, G., Senet, J.P., Sennyey, G., Eur. Pat. Appl. 85 400 251.6

6. Barcelo, G., Senet, J.P., Sennyey, G., Jaouadi, M., Martinez, J., Castro, B., Eur. Pat. Appl. 85 402 298.5

TFMSA/TFA CLEAVAGE AND DEPROTECTION IN SPPS

B. John Bergot*, Richard F. Noble and Timothy Geiser
Applied Biosystems Inc., Foster City, California 94404

INTRODUCTION

The use of MSA/TFA and 1M TFMSA/TFA scavenger mixtures for efficient deprotection of peptides synthesized by the Boc/benzyl strategy is exemplified in the remarkable classical (solution) synthesis of bovine RNase by Yajima and Fujii (1). A 6.6% yield of crystalline synthetic RNase with full enzymatic activity was obtained from the fully protected precursor (33 of the 124 amino acids variously protected with benzylic or benzene sulfonyl groups) comprised of a large number of most of the protected amino acid combinations that have been shown to suffer serious side reactions under a variety of acidic deprotection conditions (2).

We were interested in applying the TFMSA/TFA approach to peptides sythesized by the BOC/benzyl strategy on solid phase, specifically utilizing the PAM (3) or MBHA (4) linkage to 1%-crosslinked (styrene-divinyl benzene) resin. Since few examples (5) and no systematic study of TFMSA/TFA cleavage deprotection in the solid phase context were evident in the literature, we undertook a somewhat systematic effort to determine the efficacy of this approach relative to HF as a general cleavage/deprotection technique in SPPS.

METHODS

Table 1 lists all peptides to which we have applied TFMSA cleavage methodology. Several among them have also been prepared with various combinations of protecting groups (e.g. ACTH (1–10) was synthesized with eleven different protecting group combinations on Met, His, Arg, and Trp) affording a data base of over 40 peptides exposed to the TFMSA technique. Peptides were assembled on an Applied Biosystems Model 430A Peptide Synthesizer.

Table 1

PEPTIDES SUBJECTED TO TFMSA/TFA + ADJUVANTS CLEAVAGE/DEPROTECTION
1. H-Leu-Ala-Gly-Val-OH
2. H-Leu-Ala-Gly-Val-NH_2
3. H-Glu-Asp-Gly-Thr-OH
4. H-Val Gln Ala Ala-Ile-Asp-Tyr-Ile-Asn-Gly-OH (ACP 65-74)
5. [Asp^{10}]-ACP
6. [Ser^{10}]-ACP
7. [Thr^{10}]-ACP
8. H-Ser-Tyr-Ser-Met-Glu-His-Phe-Arg-Trp-Gly-OH (ACTH 1–10)
9. H-Arg-Pro-Lys-Pro-Gln-Phe-Phe-Gly-Leu-Met-NH_2 (Substance P)
10. H-Ser-Thr-Glu-Asp-Ala-His-Lys-Tyr-Cys-Arg-Phe-Ala-OH
11. H-Leu-Tyr-Pro-Gly-Glu-Val-Cys-Pro-Gly-Met-Asp-Ile-Arg-Asp-OH
12. H-Cys-Tyr-Ile-Gln-Asn-Cys-Pro-Leu-Gly-NH_2 (Oxytocin)
13. H-Ala-Gly-Cys-Lys-Asn-Phe-Phe-Trp-Lys-Thr-Phe-Thr-Ser-Cys-NH_2 (Somatostatin amide)
14. [Pro^7]-ACP
15. [Gly^7]-ACP
16. H-His-Val-Gln-Ala-Ala-Ile-Asp-Tyr-Ile-Asn-Gly-OH
17. Somatostatin-OH
18. H-Ser-Tyr-Ser-Met-Glu-His-Phe-Arg-Oh
19. H-Ser-Tyr-Ser-Met-Glu-His-Phe-Trp-Gly-OH
20. H-Ile-Ser-Gln-Ala-Val-His-Ala-Ala-His-Ala-Glu-OH
21. H-Ser-Tyr-Ser-Met-Glu-His-Phe-NH_2
22. H-Ser-Ser-Cys-Phe-Gly-Gly-Arg-Ile-Asp-Arg-Ile-Gly-Ala-Gln-Ser-Gly-Leu-Gly-Cys-Asn-Ser-Phe-Arg-OH (ANF 23)
23. p-Glu-Gly-Pro-Trp-Leu-Glu-Glu-Glu-Glu-Glu-Ala-Tyr-Gly-Trp-Met-Asp-Phe-NH_2 (Gastrin 1)
24. H-Gly-Ile-Val-Glu-Glu-Cys-Ser-Thr-Ser-Ile-Cys-Ser-Leu-Tyr-NH_2 (porcine insulin 1–14)
25. H-Val-Asp-Cys-Ser-Glu-Tyr-Pro-Lys-Pro-Ala-Cys-Thr-Leu-OH

Peptides 1986

Table 2 summarizes the appropriate side chain protection found experimentally to be compatible with the described TFMSA/TFA cleavage/deprotection chemistry.

SIDE CHAIN PROTECTION COMPATIBLE WITH TFMSA/TFA CLEAVAGE/DEPROTECTION	
Asp(OBzl)	Met(O)
Arg(MTS)* (ref.5)	Ser(Bzl)
Cys(Mob)**	Thr(Bzl)
Glu(OBzl)	Trp(MTS or CHO) (ref.6)
His(Z or Tos)	Tyr(BrZ)
Lys(ClZ)	

*MTS = mesitylene-2-sulfonyl-
**Mob = p-methoxybenzyl

Standard TFMSA/TFA Cleavage/Deprotection Procedure:

Two variations are described for small scale (100mg peptide-resin) and large scale (1g peptide-resin) cleavage/deprotection. TFMSA was supplied in 10g ampules (Aldrich Chemical Co.), transferred to screw-top amber bottles and stored at 5°C under argon between uses. It is important to discard TFMSA stored in this fashion after 4 weeks. The N- BOC group is removed from the peptide-resin prior to treatment with the cleavage/deprotection reagents.

A) Small Scale

100 mg peptide resin are combined with 150 uL thioanisole:ethanedithiol (2:1 v/v) and equilibrated with stirring for 10 minutes, followed by addition of 1.0 mL neat TFA and stirring another 5–10 minutes. 100 uL of TFMSA is then added dropwise with vigorous stirring (heat is evolved in this step but minimized by careful addition of TFMSA). Allow the mixture to react for 15–30 minutes at room temperature (peptides anchored *via* the amide bond to BHA type resin require 1.5–2.5 hours reaction time for maximum cleavage yields).

Pipette the reaction medium to a sintered glass funnel (coarse porosity) that has been fitted with a collection flask containing 25 mL of methyl-t-butyl-ether (MBE). This will trap the resin on the filter allowing the TFA solution to pass through, whereby the peptide will precipitate and the acids and scavengers will remain in solution at greatly diluted concentration. Rinse the reaction vial three times with 0.5 mL TFA, transferring each rinse to the glass funnel and re-suspending the solid residue to ensure complete washing. The rinses are combined with the original filtrate/MBE mixture and refrigerated briefly to complete peptide deposition.

B) Large Scale

1 g peptide resin in combined with 1.5 mL thioanisole:ethanedithiol (2:1 v/v) and stirred 5–10 minutes. Then *slowly* add 15 mL of a *freshly prepared* solution of TFMSA:TFA (1:10 v/v) (note: if convenient, the reaction flask may be immersed in an ice-bath to facilitate more rapid addition of acidic reagent with minimal heat evolution). Stir 0.5 hr. (or ca. 2 hr. in the case of amide linked peptide-resin) at room temperature (20°C).

The reaction medium is then transferred to a sintered glass funnel (coarse porosity) atop a collection flask containing ca. 250 mL of MBE, and then treated in the manner described above.

To scale up or down from either of the two methods described, it is sufficient to merely adjust the amount of reagents to the weight of peptide-resin in direct proportion to the ratios described above.

To assess the degree of cleavage achieved, the recovered resin is neutralized with a dilute solution of diisopropylethylamine in CH_2CL_2, rinsed thoroughly, dried, then subjected to the Quantitative Ninhydrin Method (6) to determine residual amine content.

RESULTS AND DISCUSSION

Using ACTH(1–10), H-Ser-Tyr-Ser-Met-Glu-His-Phe-Arg-Trp-Gly-OH, as a model, a variety of TFMSA/TFA concentrations, scavenger combinations and concentrations, reaction times and temperatures were tested in conjunction with various side chain protection functionality to determine a generally efficient system and optimal side chain protection for deprotection and cleavage from PAM(SDVB) resin. The approach was gradually expanded to include the peptides listed in Table 1, many of which have amide termini derived from MBHA resin. Over 600 cleavage/deprotection reactions were executed in this study that resulted in a standard TFMSA/TFA/thioanisole/ethanedithiol (8:80:8:4 v/v) system that worked extremely well on all peptides listed in Table 1 when side chain protection conformed to that listed in Table 2. The single exception was porcine insulin (1–14) which required 5 hrs. for ►90% cleavage and whose insolubility posed problems in characterization. The overall results compared very favorably to those obtained from standard HF/anisole cleavage/deprotection of analagous peptides with side chain protection amenable to HF removal.

Our results confirm those of other workers (2a,b) that superior results are obtained with Arg(MTS) (7) relative to Arg(Tos) and with Cys(Mob) relative to Cys(4-methyl benzyl) although we have not yet examined Cys(Ad) [Ad = S-1-adamantyl (8)]. Excellent results were also obtained with Trp(N^{im}-MTS) (9) and Trp(CHO). Methionine should be incorporated as its sulfoxide, both to avoid alkylation during chain assembly and to minimize side products on cleavage. Normally, one can expect 10–40% of the reduced Met-peptide to be formed under the standard TFMSA cleavage of Met[0]-peptides, but complete reduction can usually be achieved with N-methylmercaptoacetamide (10).

The well known sequence dependant tendency to form "aspartimide" derivatives from -Asp(OBzl)-X- (X = Gly, Ser, Thr, Ala, or Asn) by base catalysis during chain assembly has been greatly minimized by the use of O-cyclopentyl (11a), O-cyclohexyl (11b), O-cycloheptyl and O-cyclooctyl (11c) for Asp side chain protection. Additionally, 1M TFMSA/TFA has been shown to minimally form (◄4%) aspartimide in Z(OME)-Ala-Asp-(OR)-Gly-OBzl (R = Bzl, Cpe, Chx, Chp, and Coc) (11c). Our standard TFMSA/TFA mixture is only 0.7M and failed to effectively cleave the cyclohexyl group from Asp (O-Chx)-peptides. However, in the only case, peptide *15*, where appreciable rearrangement of an Asp (OBzl)-peptide occurred (note that peptides *3, 4, 10, 15,* and *16,* Table 1, contain the sensitive -Asp-X- sequence), simply lowering the cleavage temperature to −10°C and increasing the cleavage/deprotection time to 60 minutes substantially suppressed this un-wanted by-product.

After performing hundreds of cleavages on over 40 differently substituted and/or protected peptides on solid support, we conclude that the type and composition of so-called scavengers used in the reaction mixture profoundly affect both quantitative and qualitative aspects of product yield. Clearly, thiol ethers and thiols function as *reaction promoters* as well as mere carbocation acceptors. Indeed, we find that thioanisole is a requirement for efficient cleavage of peptide from PAM and, especially, BHA type supports, presumably functioning as a "push-pull" catalyst as previously suggested (12). Typically we can cleave 90–99% of theoretically available peptide off PAM-resin and 70–80% off BHA-resin using thioanisole as described in the standard procedure. Substituting phenols or phenolic ethers for thioanisole reduces overall cleavage yields to 25–90%, based on quantitative ninhydrin assay of recovered resin. In addition, the presence of alkylthioethers severely retards scission of peptide from the support, despite the inclusion of thioanisole. Ehylmethyl sulfide, dimethl sulfide, tetrahydrothiophene, etc., all suppress in varying degrees (40–80%) peptide cleavage when present in amounts equal to or as low as one-half (v/v) the thioanisole in the reaction mixture.

Ethanedithiol (EDT) is an absolute requirement for the removal of the formyl group from Trp(N^{in}-CHO)-peptides as noted previously for HF mediated cleavage/deprotection reactions (13). EDT also minimizes formation of dimers and other types of -S-S- aggregates in cysteinyl peptides, apparently serving as a reducing agent under the acidic cleavage conditions.

Adjuvants such as m-(or p-)thiocresol or p-methylthiophenol are deleterious when cleaving cysteinyl peptides, as their use results in a variety of S-adducts and/or -S-S- dimers. Anisole or the cresols used in conjunction with thioanisole appear to be neutral in their effects, neither enhancing nor degrading product quality.

Table 3 provides cleavage rate data for a random selection of peptides cleaved in accordance with the Standard Procedure.

Table 3

Peptide	%Cleavage[a] (@ time in minutes)			
4 (-Gly-PAM)	100 (15m)			
5 (-Asp-PAM)	61 (10m)	97(20m)	95(30m)	100(60m)
25 (-Leu-PAM)	58 (10m)	95(20m)	90(30m)	100(60m)
22 (-Arg-PAM)	82 (10m)	90(20m)	95(30m)	100(60m)

[a]see Methods

The following HPLC profiles Fig. 1, Fig. 2, and Fig. 3 of crude, unpurified peptides are, we believe, examples of the exceptional capabilities of the standard TFMSA/TFA/thioanisole/EDT (8:80:8:4 v/v) procedure when applied to fully protected peptide-resins that have been assembled under conditions of very high average step yield efficiencies.

Fig. 2 **ATRIOPEPTIN II (rat)**

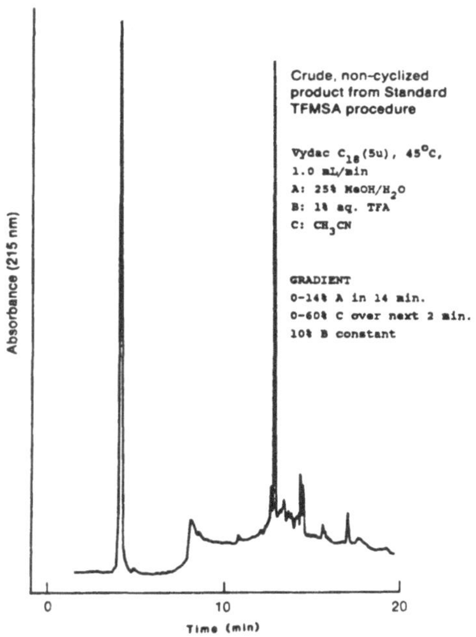

Fig. 1 **ACTH (1-10)**

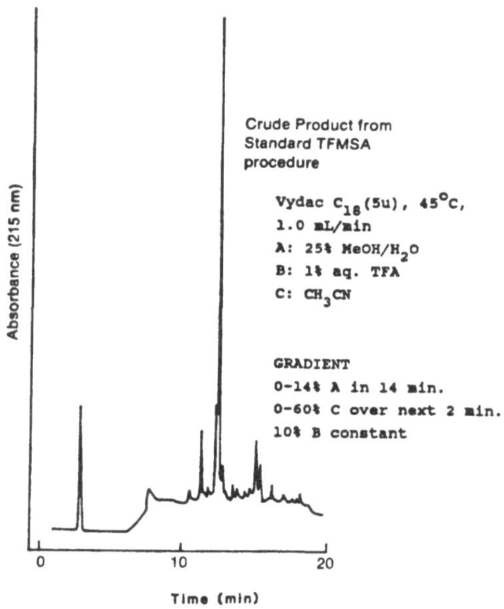

Fig. 3 **GASTRIN I, CRUDE PRODUCT**

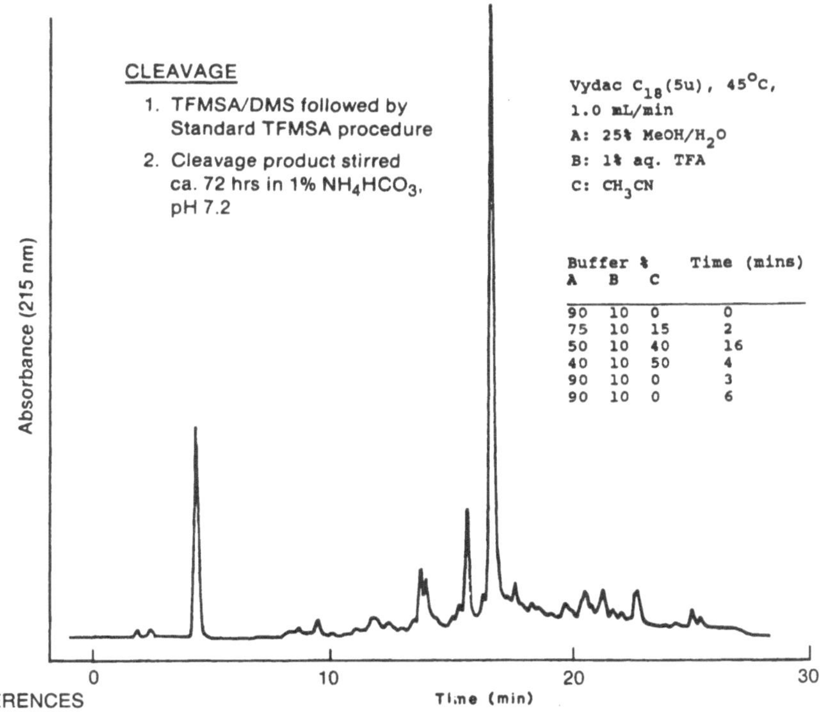

REFERENCES

1. Yajima, H. and Fujii, N. 1981. J.Am.Chem.Soc. *103,* p.5867
2. (a) Yajima, H. and Fujii, N., (1983), IN: The Peptides Vol.5 (Gross, E. and Meienhofer, J., eds.). Academic Press, New York, pp.65–109. (b) Yajima, H., Funakoshi, S. and Akaji, K. 1985. Int.J.Peptide Protein Res., *26,* pp.337–361.
3. Mitchell, A.R., Erickson, B.W., Ryabtsev, M.N., Hodges, R.S., and Merrifield, R.B. 1978. J.Am.Chem.Soc. *98,* p.7357.
4. Matsueda, G.R. and Stewart, J.M. 1981. Peptides Vol.2, pp.45–50.
5. a) Tam, J.P. and Heath W.F. (1983) in Proceedings of the 8th American Peptide Symposium (V.J. Hruby and D.H. Rich, eds.). Pierce Chemical Co., Ill., pp.83–86. (b) Bodanszky, M., Natarajan, S., Gardner, J.D., Makhlouf, G.M., and Said, S.I. 1978. J.Med.Chem. *21,* p.1171. (c) Heavner, G.A., Doyle, D.L., and Riexinger, D. 1985. Tet.Lett., *26,* pp.4583–4586.
6. a) User Bulletin NO. 1 (1985), Peptide Synthesizer, Applied Biosystems Inc. (b) Sarin, V., Kent, S.B.H., Tam, J.P. and Merrifield, R.B. 1981. Anal.Biochem. *117,* pp.147–157.
7. Yajima, H., Takeyama, M., Kanaki, J., Nishimura, O., and Fujino, M. 1978. Chem.Pharm.Bull. *26,* pp.3752–3757.
8. a) Nishimura, O., Kitada, C., and Fujino, M. 1978. Chem.Pharm.Bull. *26,* p.1576. (b) Fujii, N., Yajima, H., Otaka, A., Funakoshi, S., Nomizu, M., Akaji, K., Yamamoto, I., Torizuka, K., Kitagawa, K., Akita, T., Ando, K., Kawamoto, T., Shimonishi, Y., and Takao, T. 1985. J.Chem.Soc., Chem.Commun., pp.602–603.
9. Fujii, N., Futaki, S., Yasumura, K., and Yajima, H. 1984. Chem.Pharm.Bull. *32,* pp.2660–2665.
10. Houghten, R.A. and Li, C.H. 1979. Anal.Biochem. *98,* pp.36–40.
11. a) Blake, J. 1979. Int.J.Peptide Protein Res. *13,* p.418. (b) Tam, J.P., Wong, W.T., Riemann, M.W., Tsoeng, F.S., and Merrifield, R.B. 1979. Tet.Lett. *42,* p.4033. (c) Fujii, N., Nomizu, M., Futaki, S., Otaka, A., Funakoshi, S., Akaji, K., Watanabe, K., and Yajima, H. 1986. Chem.Pharm.Bull. *34,* pp.864–868.
12. Kiso, Y., Ukawa, K., and Akita, T. 1980. J.Chem.Soc., Chem.Commun., pp.101–102.
13. a) Matsueda, G. 1982. Int.J.Peptide Protein Res. *20,* pp.26–34. (b) Tam, J.P., Heath, W., and Merrifield, R.B. 1982. J.Am.Chem.Soc. *105,* pp.6442–6455.

IS THERE HOPE THAT FOUR COMPONENT CONDENSATIONS WILL BECOME USEFUL FOR PEPTIDE AND PROTEIN CHEMISTRY ?

I.Ugi, M.Baumeister, Ch. Fleck, R. Herrmann,
R. Obrecht, F. Siglmüller and J.-H. Youn
Institut für Organische Chemie, Lichtenbergstr. 4,
8046 Garching, FRG.

Since more than 25 years considerable efforts have been made to develop methods for the synthesis of peptide derivatives that are based on four component condensations (4CC); but up to now no generally usable 4CC peptide synthesis has evolved (1). What is the prognosis of 4CC for peptide syntheses?

The ß-lactam antibiotics and related compounds are basically peptide derivatives. The 4CC is almost ideally suited for the synthesis of the ß-lactams from ß-amino acids or their derivatives (2). In the past years some substantial progress has been made with respect to the conditions of 4CC ß-lactam syntheses: It was observed that trifluoroethanol is a magic solvent for such syntheses; in almost all studied cases the yield and purity of the ß-lactams was significantly better when the 4CC was carried out in trifluoroethanol instead of the previously used other solvents (3).

Since the amides of the ß-lactam antibiotics and related compounds are formed through the 4CC instead of the free carboxylic acids, it is necessary to convert the ß-lactam carbonamides into the respective carboxylic acids in order to obtain the target compounds of the syntheses. Improved methods for the selective cleavage of carbonamide groups in the presence of the ß-lactam moiety have been developed lately (3,4).

Coupling of peptide segments by 4CC (4CC-SC) is,in principle, possible (1). However, no aldehyde component has as yet been found that affords in good yield and without racemization N-

Peptides 1986

substituted peptide derivatives by 4CC-SC, such that the auxiliary, solubility improving N-substituent which contains, in essence, the aldehyde component and the isocyanide component, can be removed satisfactorily. The search for a well-suited aldehyde component is still on.

The synthesis of peptide segments by stereoselective 4CC (4CC-SSS) is also not yet ready for routine peptide syntheses. Although the sequential application of productively selective (5) 4CC of protected carboxypeptides with chiral α-ferrocenyl alkylamines, aldehydes and α-isocyano acid derivatives, followed by destructively stereoselective (5,6) acidolytic cleavage of the α-ferrocenyl alkyl N-substituent from an amide group proceeds remarkably well in some instances, it is not yet usable as a routine method for peptide syntheses.

Up to now the chiral ferrocenyl sec.-alkyl carbinylamines were considered to be the most promising chiral amine components, because at temperatures below -40° C these amines undergo extremely stereoselective 4CC. In favorable cases, the components react by 4CC in supersaturated solution at these low temperatures,whereas in other cases the same components crystallize out, and the 4CC occurs when the reaction mixture is warmed up. Thus, reproducible results are hard to obtain.

Furthermore, under the customary 4CC conditions, a variety of side reactions takes place to a relative extent that depends on the structure of the components and the reaction conditions. Among the by-products malonamide derivatives and N-dealkylation products predominate (1). The formation of the malonamide derivatives is minimal if the 4CC are carried out in trifluoroethanol (7).

Recently it was found that the relative amounts of the N-dealkylation products depend on the workup protocol, and the avoidance of acidic conditions and strogly polar solvents during workup minimizes N-dealkylation of the 4CC products (8).

Since the 4CC do not proceed by a uniform reaction mechanism, they are not suitable for determining the relative asymmetric inducing powers of chiral amines.

At -60°C to +60°C the stereoselective acylation of racemic primary amines by trifluoromethyl-phenylketene was found to be a pair of corresponding reactions (5,10,11) and it was used to evaluate diverse amines as candidates for chiral templates of asymmetric 4CC. The temperature dependent stereoselectivity experiments were carried out in analogy to a previous stereoselectivity study on racemic alkohols, where the relative amounts of the diastereoisomers were determined by 19 F-NMR (12).

It was found that there exist two distinct types of chiral amines with respect to this stereoselective reaction:The amines of type I (e.g. α-ferrocenyl ethylamine) react with a high degree of stereoselectivity at high temperatures, whereas the amines of type II (e.g. α-ferrocenyl isobutylamine) show more stereoselectivity at low temperatures (10).

On the basis of our previous experience, we consider the amines of type I more promising as chiral amine components of asymmetric 4CC, provided that these amines behave similarly in both stereoselective reactions. The use of Mosher's acid is a good acid component for such 4CC studies, because it yields 4CC products that are amenable to 19F-NMR stereoisomer analyses.

The results of a systematic study of yield and stereoselectivity of such 4CC would be used on a computer assisted optimization (13) of amine components and reaction conditions (14).

Thus, with regard to 4CC-SSS we are back to square one , but now we have much better orientation where to go. The 4CC still has its known advantages and attractions. We hope to be able to present useful 4CC methods for peptide chemistry within a few years.

An interesting new application of the 4CC is the enclosure-immobilisation of enzymes. The 4CC of natural and synthetic polymers that carry carboxyl groups, with difuctional prim.

amines, their isocyanides and reactive aliphatic aldehydes in a well-stirred two-phase system of an aqueous buffer and an immiscible organic solvent in the presence of enzymes yields beads of enclosure-immobilised enzymes of often remarcably high activity and stability (14).

Acknowledgement: We acknowledge gratefully the financial support of our work by Deutsche Forschungsgemeinschaft (SFB 145) and Fonds der Chemischen Industrie.

References

1. Ugi, I., D. Marquarding, R. Urban. 1982. In: Chemistry and Biochemistry of Amino Acids, Peptides and Proteins. New York (B. Weinstein, ed.). Marcel Dekker, p. 245.
2. Ugi, I.. 1982. Angew. Chem. 94, 826, Angw. Chem. Int. ed. Engl. 21, 810.
3. Obrecht, R.. 1986. Diss. Techn. Univ. München.
4. Ugi, I., J. Geller. 1983. Chem. Scripta 22, 85; Ugi, I., R. Obrecht, S. Toure. 1984. Heterocycles 21, 271.
5. Ugi, I.: Chimia (in press)
6. Brandt, J., R. Obrecht, C. Jochum, I. Ugi, P. Jochum. 1977. Tetrahedron 33, 1353.
7. Siglmüller, F.: Diss. Techn. Univ. München (in press)
8. Fleck, C. (unpubl. results).
9. Ugi, I., G. Kaufhold. 1967. Liebigs Ann. Chem. 709, 11.
10. Youn, J.-H.. 1986. Diss. Techn. Univ. München.
11. Ugi. I., H. Aigner, M.L.V. Arnaez, G. Glahsl, P. Lemmen, R. Stöcklein-Schneiderwind, M. Balla-Tamasi. 1984. In: Vorträge anläßlich der Königsteiner Chromatographie-Tage. (H. Aigner, ed.). Waters GmbH, Eschborn, p. 1.
12. Anders, E., E. Ruch, I. Ugi. 1973. Angew. Chem. 85, 16; Angew. Chem. Int. Ed. 12, 25.
13. Carlson, R., T. Lundstedt, R. Shabana. Acta che. Scand. B40, 534; see also: Wold, S., C. Albano, W.J. Dunn III, U. Edlund, K. Esbensen, P. Geladi, S. Hellberg, E. Johansson, W. Lindberg and M. Sjöström. 1984. In: Chemometrics - Mathematics and Statistics in Chemistry. (B.R. Kowalski, ed.). Reidel Publ. Co., Dordrecht, p. 17.
14. Baumeister, M.. 1986. (unpubl. results).

USE OF FMOC AMINO ACID CHLORIDES IN THE SYNTHESIS OF TACHYKININ SEQUENCES AND INVESTIGATION OF THEIR HISTAMINE RELEASING ACTIVITY

M. Beyermann, M. Bienert, H. Repke
Akademie der Wissenschaften der DDR,
Institut für Wirkstofforschung
DDR-1136 Berlin

L. A. Carpino
Department of Chemistry, University of Massachusetts,
Amherst, MA 01003

Introduction

A new method for rapid peptide synthesis has been developed (1) in which Fmoc amino acid chlorides are taken to elongate the peptide chain. N-methylpiperazine and 4-(aminomethyl)piperidine (AMP) are used to scavenge the excess acid chloride and split off the Fmoc-group, respectively. Excess reagents and by-products are removed by aqueous extractions with the growing peptide chain being always retained in the organic phase (for other "hold in solution" techniques, see (2) and references cited therein). Investigating the feasibility of this novel technique, the synthesis of H-Phe-Phe-Gly-Leu-Nle-OBzl (I) was thoroughly studied and an improved procedure has been found convenient for the rapid preparation of mmol-amounts of C-term. tachykinin- and N-term. substance P sequences. Some of the analogs described here were designed to investigate the structural requirements for substance P-induced histamine release from mast cells.

Results and Discussion

<u>Peptide synthesis:</u> The preparation of Fmoc amino acid chlorides of Gly, Phe, Glu(OBzl), Pro and Lys(Z) was carried out by treatment of the Fmoc amino acids with thionyl chloride under reflux for 60 min and crystallization from dichloromethane-hexane (1). Fmoc-δ-aminovaleric acid (Ava) was treated with

thionyl chloride for 15 min to avoid the formation of 1-Fmoc-piperidon-(2). With the exception of Fmoc-Pro-Cl and Fmoc-Ava-Cl the chlorides can be stored without decomposition for several months.

The following protocol for chain elongation has been found suitable in the syntheses of N- and C-term. substance P sequences: To 0.5 mmol amino acid ester in 5 ml $CHCl_3$ a solution of 0.55 - 0.75 mmol Fmoc amino acid chloride, dissolved in 5 ml $CHCl_3$, and 5 ml 5% Na_2CO_3 are added. After vigorously stirring for 5 min, the organic phase is separated and 5 ml AMP are added to scavenge the excess of acid chloride and to split off the Fmoc group simultaneously. After 20 min the solution is diluted by addition of 40 ml $CHCl_3$ and extracted five times with 10% phosphate buffer pH 5.5, with the peptide being retained in the organic phase. The organic phase, washed with water and reduced in vacuo to 5 ml, is ready for the addition of the next Fmoc amino acid chloride (Fig. 1).

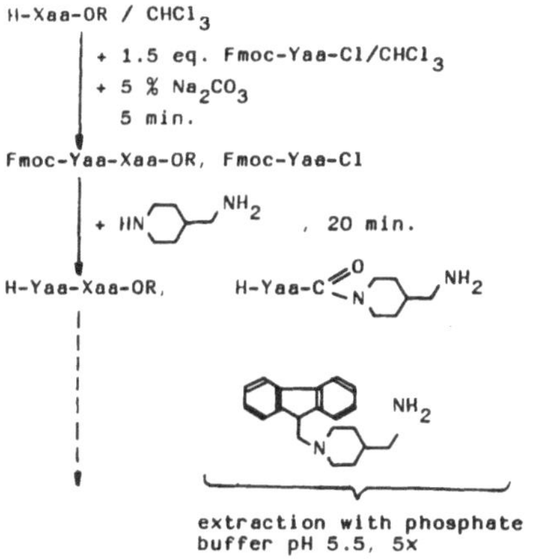

H-Phe-Phe-Gly-Leu-Nle-OBzl	I
H-Phe-Phe-Gly-Leu-Nle-NH_2	II
H-Ava-Phe-Phe-Gly-Leu-Nle-NH_2	III
H-Lys(Z)-Phe-Phe-Gly-Leu-Met-NH_2	IV
H-Phe-Phe-Val-Gly-Leu-Met-NH_2	V
Z-Arg(NO_2)-Pro-Lys(Z)-Pro-OH	VI
H-Arg-Pro-Lys-Pro-NH-$C_{12}H_{25}$	VII
H-Lys-Pro-Lys-Pro-NH-$C_{12}H_{25}$	VIII

Fig. 1. General scheme for chain elongation by use of Fmoc amino acid chlorides without isolation of intermediates (R = tert. butyl, benzyl).

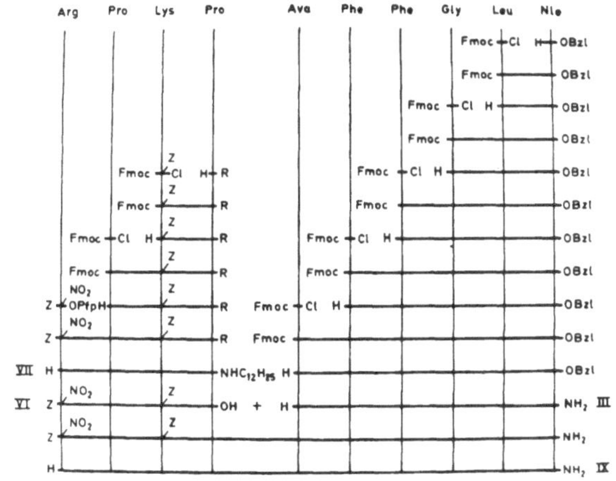

Fig. 2. Synthesis of substance P analogs and fragments by use of Fmoc amino acid chlorides. Ava = δ-aminovaleric acid, R = tert. butyl, NH-$C_{12}H_{25}$.

This general procedure was applied to the synthesis of the sequences listed in Fig. 1. In the synthesis of C-term. peptides, benzyl ester protection was used which ensures a sufficient solubility of the growing peptide chain and permits the smooth conversion into the corresponding amides. However, 10 - 20% diketopiperazines are formed during the treatment of Fmoc dipeptide benzylesters with AMP. This side reaction was diminished to less than 5% by reduction of the reaction time to 5 min. Synthesizing the N-term. fragment VI, tert. butyl ester protection was used (Fig. 2) to avoid the formation of cyclo [Lys(Z)-Pro] which was quantitatively formed in 2 min from Fmoc-Lys(Z)-Pro-OBzl and AMP. Fmoc-Glu(OBzl)-peptides form considerable amounts of the corresponding pyroglutamyl derivatives even when the AMP-treatment is reduced to 5 min, and the elongation by glutamine residues via Fmoc(Glu(OBzl)-Cl and ammonolysis has so far been unsuccessful. The time required for one cycle is about one hour (in a few cases the phase separation is delayed by the formation of emulsions or precipitates). Small amounts of remaining dibenzofulvene and other impurities were removed by chromatography on silicagel or, after ammonolysis, by reprecipitation. The peptides obtained in an over all yield of 30 - 60% had a purity of >95% (tlc, hplc, nmr). The strategy was also successfully applied to the synthesis of the methionine-containing peptides IV and V; it is noteworthy, that during the numerous washing steps no S-oxidation took place. Pentafluorophenyl esters may also be used when the synthesis of the corresponding chlorides is impossible (Fig. 2, VI, VII). VI and III were used to synthesize the substance P analog IX as demonstrated in Fig. 2.

Histamine Releasing Activity: Searching for the structural requirements of substance P induced histamine release we found that the activity on rat peritoneal mast cells of six investigated analogs increases with the lipophilicity of the C-term. heptapeptide. A positively charged cluster in connection with a hydrophobic domain (3) seem to be sufficient for mast cell degranulation with no indication for substance P specific stereochemical requirements. This view is supported by the newly synthesized analogs VII and VIII, which were found to be much more active than substance P on rat peritoneal mast cells (ED_{50}: VII 0.022 μM, VIII 0.09 μM, substance P 4.5 μM), indicating, that the existence of substance P specific mast cell receptors is unlikely.

References

1. Carpino, L.A., B.J. Cohen, K.E. Stephens, Jr., S.Y. Sadat-Aalaee, J.-H. Tien, D.C. Langridge. J.Org.Chem. (in press)

2. Dölling, R., K.-D. Kaufmann. 1984. J.Prakt.Chem. 326, 171.

3. Jasani, B., G. Kreil, B.F. Mackler, D.R. Stanworth. 1979. Biochem. J. 181, 623.

IMPROVED PREPARATION OF CRYSTALLINE N-(9-XANTHYL) AMINO ACID N-CARBOXYANHYDRIDES FOR PEPTIDE SYNTHESIS

J. Halstrøm, K. Kovács
Novo Research Institute, Novo Allé, DK-2880 Bagsvaerd, Denmark

Introduction

Strategies allowing tert.butyl based side chain protection during peptide synthesis are attractive because of the mild conditions for final deprotection of the peptide. Furthermore, the use of crystalline, activated derivatives of amino acids greatly facilitates automated peptide synthesis, since the necessity for coupling or activating reagents is dispensed with.
A recent approach, employing crystalline Fmoc amino acid pentafluorophenyl esters (1), seems promising, since the esters, in contrast to the symmetrical anhydrides, possess sufficient stability to allow storage.
Some time ago (2), we reported the preparation and use in peptide synthesis of equally stable, crystalline N-(9-xanthyl) amino acid N-carboxyanhydrides (X-NCAs). The X substituent, after coupling, affords a trityl-like protection of the amino terminus of the formed peptide, and is removed by e.g. 10% acetic acid in methanol. However, the conditions employed at that time for NCA formation, i.e. phosgene in hot THF, did not permit the use of tert.butyl based side chain protection due to the release of two equivalents of HCl during the cyclization.

Results

In recent syntheses involving succinimidyl (Su) esters, we found that the Su ester forming reagent di-(N-Su) carbonate (DSC) when applied to several N-aralkyl amino acids, brings about formation of the corresponding NCAs by carbonyl insertion (3). Figure 1 shows the cyclization of X amino acid to X-NCA.

Peptides 1986

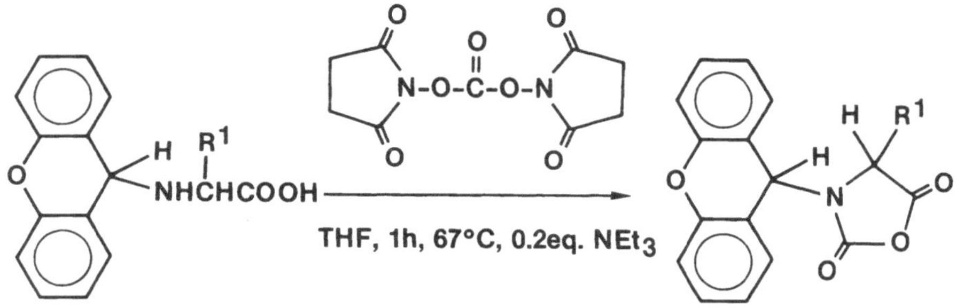

Figure 1

Unlike HCl, the by-product N-hydroxysuccinimide (HOSu) is only weakly acidic, and therefore not deleterious to Boc, tBu or X groups. An extra benefit is the substitution of the safe, commercially available solid DSC for the toxic gas phosgene.
In addition to X-Lys(Boc) NCA (3), X-Ser(tBu) NCA and X-Glu(OtBu) NCA have now been prepared, demonstrating that carbonyl insertion with DSC is indeed a general route for obtaining tert.butyl protected X-NCAs.
Yields and physical data for the X-NCAs so far synthesized with the aid of DSC are collected in Table 1. For measurement of the specific optical rotation, ethyl acetate is preferred over benzene (2,3), since the rotations are generally slightly higher.

Table 1. X-NCAs Prepared by Carbonyl Insertion with DSC

X-NCA	Yield/%	M.p./°C	$[\alpha]^{20}_{578}/^{0}$ (c=1,ethyl acetate)	ir: νC=O/cm^{-1} (chloroform)
Val	81	113-114	+76.9	1765,1845
Leu	89	129-130	+84.6	1765,1840
Phe	54	118-119	+130.0	1765,1845
Lys(Boc)	62	103-105	+92.8	1705,1770,1845
Gly	65	165-166	---	1770,1845
Ser(tBu)	64	153-154	+88.5	1765,1845
Glu(OtBu)	53	138-139	+77.0	1720,1765,1840

The purity and identity of the NCAs was confirmed by thin-layer chromatography, C,H,N analysis and ^1H nmr spectroscopy. For the Val, Phe and Leu derivatives, the absence of racemization was established by comparison of the specific optical rotation in benzene with that of previously synthesized X-NCAs (2). Formation of the title compounds (R=X) presumably proceeds via path A (Figure 2).

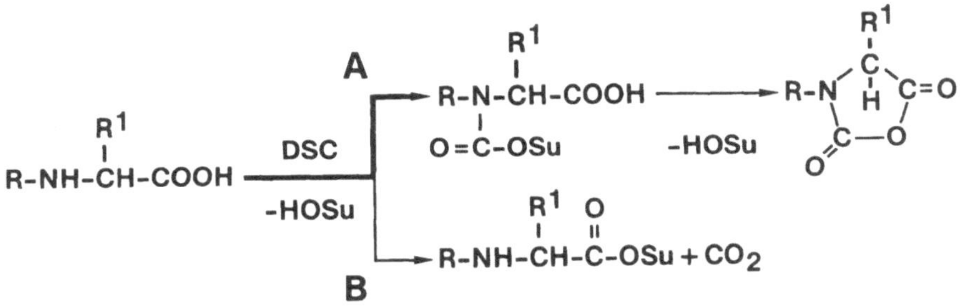

Figure 2

Trityl alanine (R=Ph_3C-) under identical conditions produces exclusively the Su ester (path B). For the NCAs reported here (R=X), formation of a less soluble by-product is observed, which may well be the Su ester also. Apparently, the degree of sterical hindrance exerted by R determines the ratio of products formed by paths A and B.

Concerning the application of X-NCAs in practical peptide synthesis, there are drawbacks as well as advantages. Apart from the fact that N-protected NCAs of proline and other imino acids cannot be made, there are certain limitations associated with the aralkyl nature of the X group. As it appears from the proposed reaction path A in Figure 2, the amino protection afforded by X is not absolute, but depends on the bulk of the acylating agent. Thus, treatment of X-LeuOMe at 20°C for 2 days with excess Ac_2O results in partial acetylation, whereas acylation is not observed with excess Boc_2O even after prolonged treatment. For solid phase peptide synthesis this implies that acetylation of residual amino

groups may not be applicable. Furthermore, since N-aralkyl peptides react with ninhydrin, the ninhydrin test cannot be employed to control coupling completion.
On the other hand, X-NCAs possess a number of features, which render them attractive for use in automated peptide synthesis. First of all, the unique release of an equivalent quantity of carbon dioxide during coupling permits a direct monitoring of the extent of coupling by titration of the amount of carbon dioxide formed (4). Long-term stability at -20°C, a high solubility in common solvents and the magnitude of specific optical rotation are practical details of importance in the choice of crystalline, activated derivatives. The strong UV absorption of X groups at 292 nm, as well as the intense, yellow colour produced in the presence of trifluoroacetic acid, may be exploited for controlling deprotection during solid phase synthesis, as well as in chromatographic procedures. Last but not least, the low level of racemization associated with the use of X-NCAs (5) in aqueous media hopefully will also apply to conditions of solid phase synthesis. Apart from the possibilities offered by X-NCAs in their own right, their application could be complementary to the use of Fmoc derivatives, particularly where piperidine treatment for some reason should prove unsuitable.

References

1. Atherton, E., A. Dryland, P. Goddard, L. Cameron, J.D. Richards, R.C. Sheppard. 1985. In: Peptides, Structure and Function (C.M. Deber, V.J. Hruby and K.D. Kopple, eds.). Pierce Chemical Company, p. 249.

2. Halstrøm, J., K. Brunfeldt, K. Kovács. 1979. Acta Chem.Scand. B33, 685.

3. Halstrøm, J., K. Kovács. 1986. Acta Chem.Scand. B40, 464.

4. Patchornik, A., Y. Shalitin. 1961. Anal.Chem. 33, 1887.

5. Halstrøm, J., K. Kovács. 1979. In: Peptides 1978 (I.Z. Siemion and G. Kupryszewski, eds.). Wroclaw University Press, p. 119.

THE OPTIMIZATION OF REACTION CONDITIONS FOR PEPTIDE SYNTHESIS BY MEANS OF N,N-BIS/2-OXO-3-OXAZOLIDINYL/PHOSPHORODIAMIDIC CHLORIDE

A.M. Kołodziejczyk, Z.J. Wodecki
Dept. of Organic Chemistry, Technical University of Gdańsk, 80-952 Gdańsk, Poland

Introduction

N,N-Bis/2-oxo-3-oxazolidinyl/phosphorodiamidic chloride - BOP-Cl, a new very promising coupling reagent was successfully applied for synthesis of amides and esters in 1980 /1/ and later for peptide synthesis /2-4/. The reagent, commercially available and easy to prepare in a chemical laboratory /1,4/, is stable under normal conditions and according to the authors smoothly produces peptide bond with high yield and low racemization. Up till now, however, BOP-Cl has been evaluated mainly for activation of N-alkylamino acids masked only with urethane protection groups, so we decided to check up its properties in coupling of N-acylamino acids.

Results

Synthesis of Bz-L-Phe-GlyOMe by means of BOP-Cl appeared unexpectedly less efficient as compared to the synthesis by DCC method /Table 1/. Hence, we tried to find reaction conditions leading to the increase of the yield and purity of the reaction products by varying solvents, bases and applying additives. Among the checked solvents, dioxane appeared better than CH_2Cl_2, recommended for this reagent; whereas DMF was completely useless /Table 2/. Triethylamine used as a base gave better results than N-methylmorpholine /Table 3/, which was in an agreement with the data reported by Tung and Rich /3/. The action of the additives such as HONSu or HOBt was surprisingly different than in case of reactions carried by DCC. HONSu was completely inefficient as a racemiza-

Peptides 1986

tion suppresor whereas HOBt increased tangibly the peptide yield only, not influencing the racemization degree /Table 4/. The failure of the additives in suppression of the racemization was probably due to a high rate of the oxazolone formation since the very beginning of the reaction. Kinetic investigation showed that yield of the oxazolone in the reaction: Bz-L-Phe + TosOH·L-PheOMe exceeded 30% within five minutes and practically all the formed dipeptide came from aminolysis of the oxazolone, which within one hour was completely racemized.

Summarizing our observations and hitherto information reported by the cited authors it locks that BOP-Cl has advantages over DCC in peptide synthesis in the case of activation of N-alkylamino acids protected with urethane-type groups. According to the authors the reactions proceeded fast, the side products and unreacted reagents were easily removed by acid-base washing and the main products were obtained very pure and with high yield. BOP-Cl however, appeared less useful for coupling of N-acylamino acids due to high racemization and a low rate of the peptide formation as a result of a very fast formation of the oxazolone.

Table 1. Comparison of Coupling Properties of BOP-Cl and DCC in the Reaction: Bz-L-Phe + X + n·NEt_3 carried in CH_2Cl_2.

Coupling reagent	X	n	Y_L %	Y_D %	$Y_L + Y_D$ %	Y_N %	R %
BOP-Cl	HCl·GlyOMe	3	26.8	19.7	46.5	-	42.4
"	GlyOMe	2	28.6	13.1	41.7	-	31.4
DCC	HCl·GlyOMe	1	33.5	20.1	53.6	8.5	37.5
"	GlyOMe	0	25.9	8.8	34.7	4.8	25.4

Y_L, Y_D, Y_N and R stand for the yield of L-, D-dipeptide, N-acylurea and racemization degree.

Table 2. Influence of Solvents on Yield and Purity of the Product Obtained in the Reaction: Bz-L-Phe + TosOH·L-PheOMe + $3NEt_3$ + BOP-Cl.

Solvent	Y_{Ox} %	Y_{LL} %	Y_{DL} %	Y_{LL} + Y_{DL} %	E %
CH_2Cl_2	25.7	29.7	24.0	53.7	44.7
$CHCl_3$	24.9	25.4	16.7	42.1	39.7
THF	24.5	30.4	17.8	48.2	36.9
Dioxane	9.4	33.1	26.4	60.0	44.0
DMF	3.2	21.0	11.7	32.7	35.8

Table 3. Influence of Tertiary Amines on the Above Reaction Carried in Dioxane.

Amine	Y_{Ox} %	Y_{LL} %	Y_{DL} %	Y_{LL} + Y_{DL} %	E %
Bzl_3N	14.4	18.6	6.3	24.9	25.3
Bu_3N	6.9	32.5	23.1	55.6	41.5
Et_3N	9.2	32.8	26.3	59.1	44.5
NMM	15.8	24.8	28.0	52.8	53.0

Table 4. Influence of Additives on the Above Reaction Carried in Dioxane in the Presence of Triethylamine.

Suppressor 1 equiv.	Y_{Ox} %	Y_{LL} %	Y_{DL} %	Y_{LL} + Y_{DL} %	E %
HONSu	-	27.8	13.3	46.2	32.7
HOBt	-	48.2	28.1	96.3	32.1

Y_{Ox}, Y_{LL}, Y_{DL}, Y_N and E stand for the yield of oxazolone, LL-, DL-dipeptide, N-acylurea and racemization degree respectively.

Plot of the Product Rate Formation vs.Time in the Reaction: Bz-L-Phe + TosOH•L-PheOMe + $3NEt_3$ + BOP-Cl Carried in Dioxane.

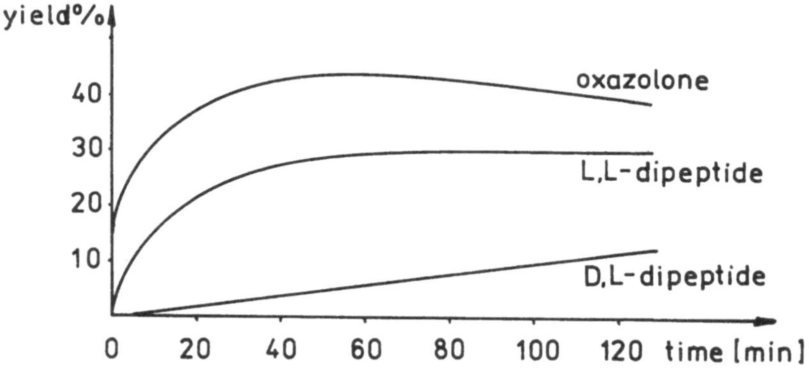

BOP-Cl was prepared according to Diago-Meseguer at al. /1/. The described reactions were carried out at 25.0°C in thermostatic vessels with magnetic stirring. Amino acid components, BOP-Cl and additives were added as solid and amines as titrated solutions. The initial concentration of the reagents was maintained at 0.050 M. The chromatographic analyses /5/ were carried out after 24 hrs of the reaction with two HPLC systems: a Kabid 5101 equipped with a 254 nm UV detector and a Hewlett-Packard HP1090A equipped with an auto-injector, a diode-array detector and controlled by HP-85B workstation. The work was financially supported in part by The Central Programme of Basic Research and by The Commitee of Chemical Sciences of the Polish Academy of Sciences.

References

1. J. Diago-Meseguer, A.L. Palomo-Coll, L.R. Fernandez-Lizarbe, A. Zugaza-Bilbao. 1980. Synthesis, 547.
2. A. Omodei-Sale, G. Sindona, D. Sola, N. Uccella. 1984. J.Chem. Res./S/, 50.
3. R.D. Tung, D.H. Rich. 1985. J.Am.Chem.Soc. 107, 4342.
4. C. Van der Auwera, M.J.O. Anteunis. 1986. Bull.Soc.Belg. 95, 203.
5. M. Ślebioda, A.M. Kołodziejczyk. 1985. Analyst, 110, 27.

APPLICATION OF MIXED PHOSPHINIC ANHYDRIDES IN FRAGMENT CONDENSATION OF PEPTIDES

C.Poulos*, Th.Tsegenidis*, and R.Ramage**

*Department of Chemistry, University of Patras, Patras, Greece.
**Department of Chemistry, University of Edinburgh, West Mains Road, Edinburgh EH9 3JJ, U.K.

Introduction

Carboxylic phosphinic mixed anhydrides have been applied succesfully in the stepwise synthesis of peptides either in liquid(1) or solid phase(2). The results from the above investigations and the advantages of the phosphinic mixed anhydrides(3) R.CO.O(O)P (I) compared to carboxylic-carbonic mixed anhydrides prompted us to investigate the efficiency of the method in the fragment coupling of peptides. For our investigation we employed the phosphinic acid P(O)OH (II) to activate the carboxyl component of peptides as the corresponding mixed anhydrides I and these were applied in the fragment peptide synthesis of Leu^5-Enkephalin, H-Leu-Pro-Leu-Arg-Phe-NH_2 and H-Phe-Met-Arg-Phe-NH_2. The performance of several mixed anhydrides of type I was compared, in terms of yield and purity of the products obtained, with other acylating agents used in fragment peptide synthesis(3).

Synthesis

The synthesis of the peptide fragments was performed in solution, in a stepwise manner, using the t-butyloxycarbonyl(Boc) group for Nα-amino protection and the mixed phosphinic anhydride method for the couplings, the exception being Boc-Leu-OH.H_2O which was introduced to the peptide chain by the DCC/HOBt method yielding the tripeptide Boc-Leu-Pro-Leu-OBzl(VI) as an oil. This peptide failed to solidify, it was purified by filtration on Sephadex LH-20(methanol) and it was then converted to the corresponding acid VII by hydrogenolysis over 10% Pd/C. The other acids (Table 1), which were used for condensation, were obtained from the corresponding methyl ester

Peptides 1986

Table 1. Physical constants of the synthesized peptide fragments

No.	Compound	M.p °C	$[\alpha]_D^{22}$,deg. c 1 DMF	R_f^a A	B	C
III	Boc-Phe-Met-OH	135-137	-11.3	0.00	0.66	0.60
IV	Boc-Tyr(OBzl)-Gly-OH	148-149	- 4.4	0.00	0.68	0.57
V	Boc-Leu-Arg(NO_2)-OH	117-120	-11.8	0.00	0.64	0.62
VII	Boc-Leu-Pro-Leu-OH	70- 72	-69.3	0.00	0.79	0.68
VIII	Boc-Tyr(OBzl)-Gly-Gly-OH	150-152	- 6.9	0.00	0.60	0.62
IX	Boc-Phe-Leu-OCH_3	105-106	-16.7	0.69	0.77	0.78
X	Boc-Arg(NO_2)-Phe-NH_2	124-125	-13.7	0.45	0.66	0.64

[a]A,chloroform/methanol 6:1; B,n-butanol/acetic acid/water 4:1:1; C, n-butanol/acetic acid/water/pyridine 30:6:24:20.

by alkaline hydrolysis with 2N NaOH in MeOH:H_2O (4:1 v/v).The mixed phosphinic anhydrides were formed in CH_2Cl_2 and/or DMF at 0°C by reacting the peptide carboxyl component with P(O)Cl in the presence of one equivalent of N-methyl morpholine. After an activation time of 15-30 min. the amino component was introduced in the reaction mixture as an hydrochloride salt and was neutralized in situ with N-methyl morpholine to yield the desired products (Table 2).

Table 2. Physical constants of peptides synthesized by the fragment condensation procedure

No.	Compound	M.p °C	$[\alpha]_{D}^{22}$,deg. c 1 DMF	Yield %	t_R^a min.
XI	Boc-Tyr(OBzl)-Gly ↑ Gly-OCH_3	104-106	- 2.9	90	7.5[b]
XII	Boc-Leu-Arg(NO_2) ↑ Phe-NH_2	130-134	-23.5	71	19.5[c]
XIII	Boc-Tyr(OBzl)-Gly-Gly ↑ Phe-Leu-OCH_3	145-147	-18.7	86	19.0[b]
XIV	Boc-Leu-Pro-Leu ↑ Arg(NO_2)-Phe-NH_2	135-138	-48.4	69	17.0[d]
XV	Boc-Phe-Met ↑ Arg(NO_2)-Phe-NH_2	157-160	-14.6	67	15.0[d]

↑ Indicates position of coupling; [a]Reverse phase HPLC,Column 250X5 mm ODS-Hypersil(5μm),Solvent MeOH-H_2O,Isocratic elution,Flow rate 1ml/min.,Detector UV; 75% MeOH, 244nm; 60% MeOH, 264nm; 63% MeOH, 264nm.

The mixed phosphinic anhydride methodology described above was applied to the solid phase synthesis of the peptide XIII. Thus the mixed phosphinic anhydride of the acid VIII was formed in DMF and was coupled to the H-Phe-Leu-OCH_2-C_6H_4-Polymer. This resin and the protocol according to which the synthesis was performed in a

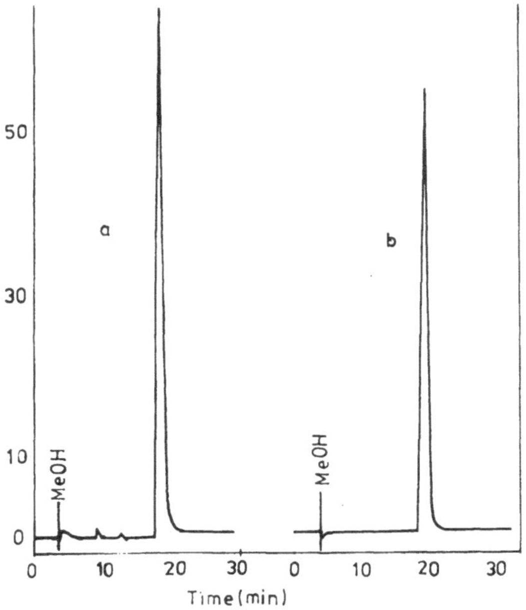

Figure 1. Reverse Phase HPLC profile of Boc-Tyr(OBzl)-Gly-Gly-Phe-Leu-OCH_3. Column 250X5mm ODS-Hypersil(5μm),Isocratic elution,MeOH-H_2O 75:25,Flow rate 1ml/min.,Detector UV_{244}.(a) solid phase (b) liquid phase.

Schwartz-Mann peptide synthesizer have been described previously(2). The peptide XIII was removed from the resin by transesterification with Et_3N/MeOH. The purity was checked by HPLC and was compared with that of the peptide obtained by the liquid phase method(Fig.1). All peptides were homogeneous on tlc and gave satisfactory amino acid analysis. The purity of the final products was also checked by HPLC, ^1H nmr and FAB-mass spectroscopy (XIII, M^+ 760; XIV, M^+ 789;XV, M^+ 692).

Results and Discussion

Peptides with a free C-terminal carboxyl group can form mixed phosphinic anhydrides which undergo aminolysis with comparable rates and yields to those formed from Nα-protected amino acids. The crude products obtained were pure as indicated by HPLC, tlc, ^1H nmr and FAB-mass spectroscopy and the yields are high compared with other methods. Improved yields(over 90%) were obtained for peptides XI and XIII when a 20% excess of the mixed phosphinic anhydride was used while the yields of the peptides containing nitro-arginine remained unchanged although completion of the reaction was indicated by tlc. The behaviour of the nitro-arginine peptides may be explained in terms of solubility in acidic solutions, as we found that these peptides have some solubility in 10% citric acid which was used in the work up procedure.

Application of the above methodology in the solid phase synthesis of the peptide XIII using 50% excess of the mixed anhydride showed that completion of the reaction occurred in two hours and the recovery of the peptide from the resin was 80% and of high purity as shown by HPLC. The above coupling was performed in the presence of one equivalent of 2,6-lutidine which facilitates the reaction because it traps the P(O)OH formed after the aminolysis, thus avoiding protonation of the attacking amino component.

Studies with ^{31}P nmr on the rates of formation, stability of the mixed phosphinic anhydrides and their tendency for oxazolone formation are now in progress. Preliminary results showed that oxazolone formation depends on the nature of the phosphinic acid. The phosphinic acid 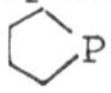P(O)OH seems not to have such a tendency and is the phosphinic acid of choice for the moment.

The above results show that mixed phosphinic anhydrides can be employed for fragment condensation of peptides either in the liquid or in the solid phase yielding products of high purity and in very good yields while in the solid phase the coupling is fast and the mixed phosphinic anhydride is used at much lower excesses compared with other acylating agents.

References

1. R.Ramage, D.Hopton, M.J.Parrott and R.S.Richardson, J.Chem.Soc. Perkin I, 1985,461 and references therein.

2. C.Poulos and R.Ramage 1984 in: Peptides 1984, Proceedings of the 18th European Peptide Symposium (Ulf Ragnarsson, ed.),Almqvist and Wiksell Trycheri-Uppsala, p 161.

3. J.Meienhofer, 1979 in: The Peptides-Analysis, Synthesis, Biology (E.Gross and J.Meienhofer, eds.), Academic Press, Vol. 1, p 299 and references therein.

A NEW PRINCIPLE FOR FRAGMENT CONDENSATION OF PEPTIDES

K. Gunnarsson, L. Grehn, U. Ragnarsson
Institute of Biochemistry, University of Uppsala, Biomedical Center, Box 576, S-751 23 Uppsala, Sweden

Introduction

Peptide oxazolones have a strong tendency to undergo racemization, and control of the peptide oxazolone formation is therefore required in fragment condensation of peptides using various techniques (1). One way of accomplishing such a control is to apply an additive, acting as an oxazolone scavenger. The effect of additives on racemization was first demonstrated with dicyclohexylcarbodiimide (DCC) as coupling agent (2,3).
In this communication we wish to report for the first time on our experiments to reduce the racemization in fragment condensation of peptides by excluding the opportunity for oxazolone formation. This can be done by protecting the ultimate peptide group of the carboxyl component with a Boc-group. Conditions favourable to the protection of amide and peptide groups were recently developed in our laboratory using Boc_2O under anhydrous conditions and 4-dimethylaminopyridine (DMAP) as catalyst (4,5).

Materials and Methods

Our experiments have been performed in a tripeptide model system, Gly-Phe-Phe, with properly protected precursors. After acidic deprotection the two diastereoisomers, Gly-Phe-Phe and Gly-D-Phe-Phe, were separated and quantitated by HPLC. For details on our HPLC system, see legend to Figure 1.
Boc_2-Gly-ONp (mp 82.5-83 ^{o}C; 87 % yield) was reacted with Phe-ONb to give Boc_2-Gly-Phe-ONb [mp 165-166 ^{o}C; $[\alpha]_D^{25}$ -9.9^{o} (c 1, DMF); 90 % yield], which could be converted to Boc_2-Gly-(Boc)Phe-ONb [mp 126-127 ^{o}C; $[\alpha]_D^{25}$ -77.1^{o} (c 1, DMF); 90 % yield] (4) and further, by catalytic hydrogenolysis, to Boc_2-Gly-(Boc)Phe [amorphous;

Peptides 1986

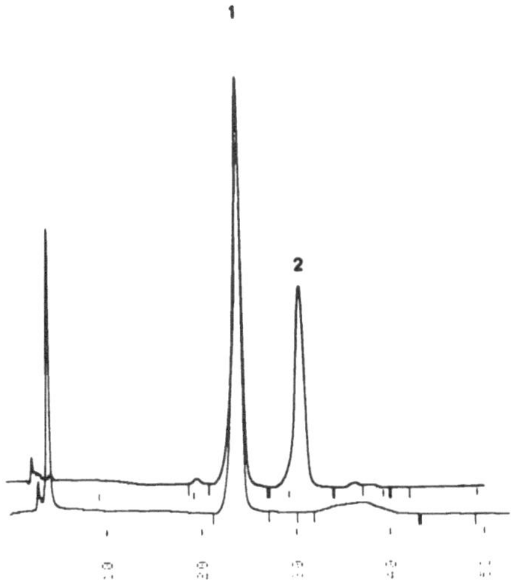

Figure 1. Lower curve: Separation of products from coupling experiment with Boc_2-Gly-(Boc)Phe and Phe-OBu^t with EDC in CH_2Cl_2 after treatment with TFA. Upper curve: Control experiment with Boc-Gly-Phe as carboxyl component. Peak 1 refers to Gly-Phe-Phe and peak 2 to Gly--D-Phe-Phe. Support: Spherisorb C_{18}, 4.6x250 mm. Mobile phase: 0.01 M ammonium acetate/ethanol; 10.5 % ethanol 10 min., gradient 10.5-21 % ethanol 5 min., then isocratically. Detection: UV at 220 nm with integrator.

$[\alpha]_D^{25}$ -87.7° (c 1.02, DMF); 80 % yield]. The last substance was purified by gelfiltration on Sephadex LH-20.- All substances in this section gave satisfactory C,H,N analyses.

A typical racemization experiment was performed as follows: Boc_2-Gly-(Boc)Phe (50 μmol) and Phe-OBu^t (55 μmol) were dissolved in 600 μL of the solvent and reacted with coupling agent (50 μmol in 200 μL of the solvent) for 5 h. After evaporation of the solvent in vacuo, the residue was dissolved in EtOAc (25 mL) and extracted with aqueous solutions of 1 M $KHSO_4$, 1 M $NaHCO_3$ and NaCl (3x5 mL each) and dried ($MgSO_4$). After filtering and evaporation, the residue was treated with TFA (2 mL) for 1 h. The free peptide, obtained after evaporation, was dissolved in CH_3OH and aliquots were applied directly to the chromatograph for determination of the percentage of Gly-D-Phe-Phe formed. In the control experiments the carboxyl component was replaced by Boc-Gly-Phe [mp 139.5-140.5 °C; $[\alpha]_D^{25}$ 14.5° (c 1, DMF)].

Synthesis of Boc_2-Gly-(Boc)Phe-Phe-OBu^t. Preparative experiment: To a solution of Boc_2-Gly-(Boc)Phe (561 mg, 1.07 mmol) and Phe-OBu^t (263 mg, 1.19 mmol) in CH_2Cl_2 (10 mL) at RT, solid EDC (215 mg,

1.12 mmol) was added. After 2 h, TLC (on silica in CH_2Cl_2-CH_3OH, 9:1) indicated that all starting material had been consumed. After evaporation, the residue was extracted and dried as described above. ^1H NMR on the product (in $CDCl_3$) indicated a mixture of the tripeptide and N-acylurea in the approximate proportion 1:1. Therefore, the oil was chromatographed in four portions on an LH-20 column in 95 % ethanol to give, after evaporation and treatment with ether, 256 mg (33 %) of the product as a white foam, pure by ^1H NMR. $[\alpha]_D^{25}$ -37.4^o (c 1.02, CH_3OH). Amino acid analysis after acid hydrolysis: Gly 1.02; Phe 1.98. Elemental analysis: C,H,N within ± 0.3 % of the theoretical values.

Results and Discussion

Under a set of experimental conditions tested, Boc_2-Gly-(Boc)Phe was shown to couple with Phe-OBu^t at RT with essentially complete retention of the configuration of the Phe-residue involved. The conditions and results are given in Table 1.

If thus this new approach to fragment condensation of peptides proved sound from racemization point of view, already the chromatographic experiments indicated that, particularly in DMF, side-product formation took place. A preparative experiment in this solvent, performed essentially as the one in CH_2Cl_2 above, proved that the product in this case was mainly the corresponding N-acyl-

Table 1. Experimental Conditions Used and Percentage of D-epimer Found in Racemization Experiments

Exp. no.	Coupling agent	Solvent	% D-epimer found sample	control	Remarks
1	DCC	DMF	0.9[a]	18	Low yield
2	DCC	DMF	0.8[a]	22	Phe-OBu^t·HCl + 1 equiv. NEt_3
3	DCC	EtOAc	0.2	16	
4	DCC	CH_2Cl_2	0.3	27	
5	EDC	$CHCl_3$	0.5	38	
6	EDC	CH_2Cl_2	0.35	37	
7	EDC	DMF	0.4	24	Low yield

[a] A different batch of Phe-OBu^t was used, which, after deprotection, contained a trace of D-Phe.

urea (^1H NMR, amino acid analysis after acid hydrolysis). Although the amount of N-acylurea formed also in this case could be reduced by a change of solvent from DMF to CH_2Cl_2 (6), so much of it was still formed that chromatography was required for the isolation of the product and the isolated yield was only 33 %. Further work is therefore required before this approach can be applied to preparative work.

Acknowledgement

This research was supported by grants from the National Swedish Board for Technical Development and the Swedish Natural Science Research Council as well as by a scholarship (to K.G.) from the Lennander Fund.

References

1. Kemp, D.S. 1979. In: The Peptides (E. Gross and J. Meienhofer, eds.). Academic Press, Vol. 1, p. 315.
2. Weygand,F., Hoffmann, D., Wünsch, E. 1966. Z. Naturforsch. 21b, 426.
3. König, W., Geiger, R. 1970. Chem. Ber. 103, 788.
4. Grehn, L., Ragnarsson, U. 1985. Angew. Chem. 97, 519; Angew. Chem. Int. Ed. Engl. 24, 510.
5. Grehn, L., Gunnarsson, K., Ragnarsson, U. 1986. Acta Chem. Scand. B40 (in press)
6. Rich, D.H., Singh, J. 1979. In: The Peptides (E. Gross and J. Meienhofer, eds.). Academic Press, Vol. 1, p. 241.

SYMMETRICAL ANHYDRIDE REARRANGEMENT LEADS TO THREE DIFFERENT DIPEPTIDE PRODUCTS.

N.L. Benoiton, F.M.F. Chen
Department of Biochemistry, University of Ottawa, Ottawa, Canada K1H 8M5

Introduction

Peptide synthesis by the solid phase method involves coupling an N-alkoxycarbonylamino acid (1) with the amino group of the polymer-supported peptide chain. Activation is achieved by the use of a carbodiimide which converts (1) into symmetrical anhydride (3) which may be prepared separately or in the presence of the amine nucleophile. In order to obtain maximum efficiency with this or any methodology, it is essential that any side-reactions which this intermediate might undergo be delineated. We have previously reported that anhydride (3) generates 2-alkoxy-5(4H)-oxazolone (2) in the presence of tertiary amines and carbodiimides (1), and that some racemization occurs when (3) is aminolyzed in the presence of a tertiary amine (2). Then in the course of developing a method for determining enantiomers H-D/L-Xxx-OH by their conversion into epimeric N-protected dipeptide acids EtOCO-L-Phe-D/L-Xaa-OH by reaction of $(EtOCO\text{-}Phe)_2O$ with H-D/L-Xaa-O^- in aqueous DMF (3), we encountered unexpected peaks in the HPLC profiles. The amounts of the pertinent products varied depending on whether $NaHCO_3$ or Na_2CO_3 had been used to convert the enantiomers to their anions. This prompted us to investigate in depth the behaviour of anhydrides (3) in the presence and absence of base in anhydrous and partially aqueous solution.

Experimental

A detailed investigation of the fate of $(MeOCO\text{-}Val)_2O$ (3a) was

Peptides 1986

carried out. (3a) (4) and (2a) (5) were obtained as described. The identity of the products (4a), (5a), and (6a) emanating from this compound were established after their isolation. The preparation of (4a) and its characterization by X-ray analysis has been published (6). Its dealkoxylated product (6a), mp 118-119°C, was obtained from (4a) in aqueous $NaHCO_3$ followed by addition of acid and its extraction into CH_2Cl_2. Its identity was established by nmr spectroscopy and mass spectrometry, and this is being confirmed by X-ray analysis. Ester (5a) was obtained conventionally. Epimeric pairs for (4a) and (7a), the latter obtained from (6a) as indicated, were prepared from racemic (3a) and separated by HPLC. The chiral integrity of rearrangement products was determined by HPLC. The amounts of oxazolone (2a), dipeptide acid (4a) and ester (5a) in solution at various times after adding N-methylpiperidine (NMP) to (3a) in anhydrous solvents were determined by nmr spectroscopy. Other information was obtained in the same manner.

Results and Discussion

Anhydrides (3) undergo decomposition by fragmentation to the parent acid anion (1) and oxazolone (2) which are in equilibrium with (3), path (d), and by rearrangement probably through the anion of (3) to the N,N'-bisalkoxycarbonyl-dipeptide acid (4), path (a). The above obtains both in anhydrous and partially aqueous solvents in the presence of tertiary amines or $NaHCO_3$, but it also occurs in the absence of added base albeit at a slower rate. Product (4) then undergoes decomposition by two different mechanisms, depending on whether the solvent is anhydrous or not. In anhydrous solution, there is migration of the alkoxy group with loss of CO_2 to give the dipeptide ester (5), path (c). No ester is formed in the presence of water. Instead, there occurs a cyclization to hydantoin (6) with loss of the alkoxy group, path (b). Representative amounts of products from (3a) after 15 min in the presence of NMP: 33% of (4a) and 0% of (2a) in DMF; 3% of (4a) and 28% of (2a) in $CDCl_3$. Maximum amounts of (4a) were 45% at 24 h in $CDCl_3$, 70% at 6-8 h in THF, and 72% at 1.5 h in DMF. A mixture of (1a) and (2a) in $CDCl_3$ gave (4a) immediately. Reaction

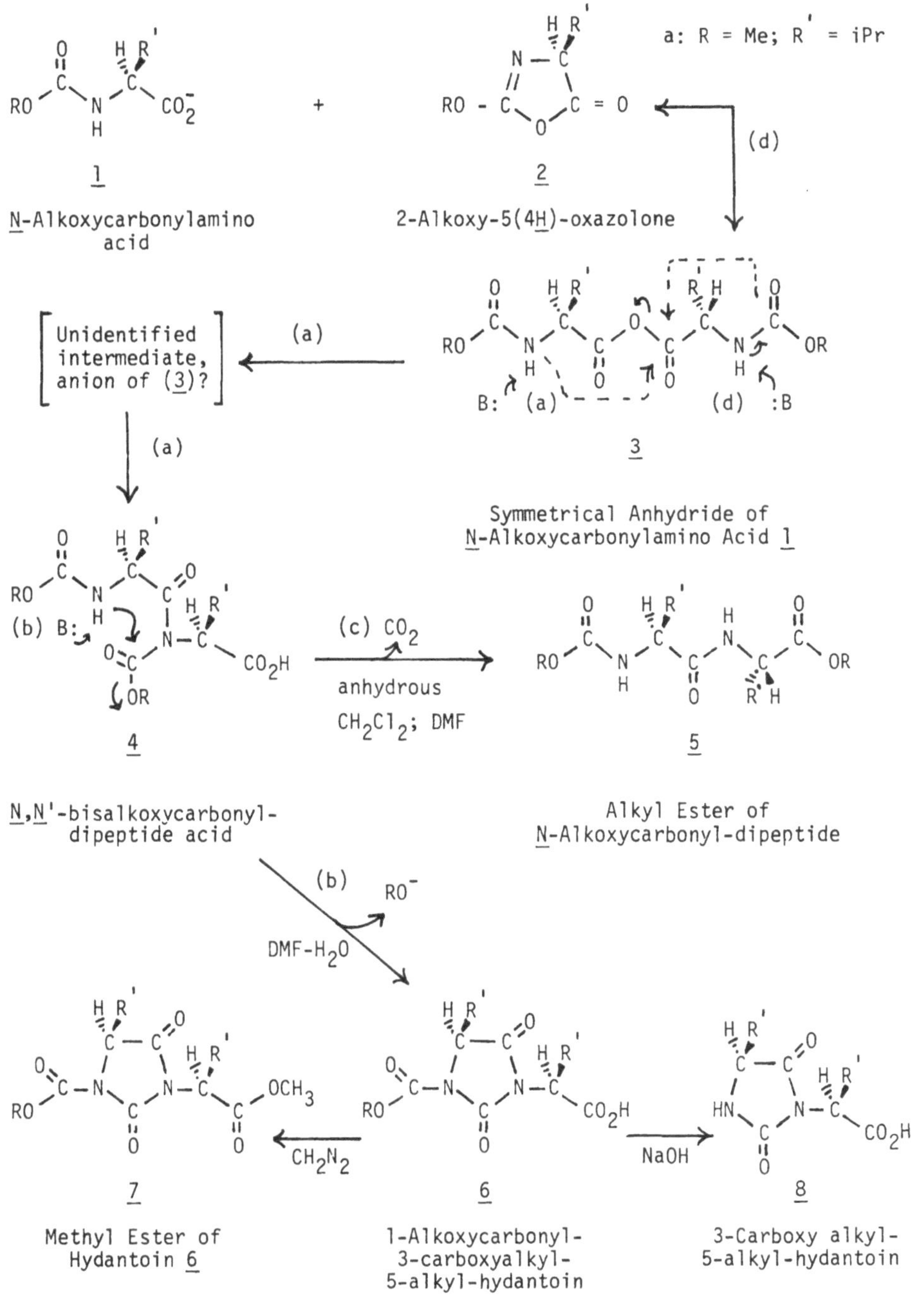
a: R = Me; R' = iPr
1
N-Alkoxycarbonylamino acid
2
2-Alkoxy-5(4H)-oxazolone
(d)
Unidentified intermediate, anion of (3)?
(a)
3
Symmetrical Anhydride of N-Alkoxycarbonylamino Acid 1
4
N,N'-bisalkoxycarbonyl-dipeptide acid
(c) CO_2
anhydrous
CH_2Cl_2; DMF
5
Alkyl Ester of N-Alkoxycarbonyl-dipeptide
(b)
RO^-
DMF-H_2O
CH_2N_2
NaOH
7
Methyl Ester of Hydantoin 6
6
1-Alkoxycarbonyl-3-carboxyalkyl-5-alkyl-hydantoin
8
3-Carboxy alkyl-5-alkyl-hydantoin

of (1a) with (2, R = $PhCH_2$) eventually produced $PhCH_2OCO$-Val-Val-OMe. After 24 h in DMF, $(Boc\text{-}Val)_2O$ had been converted primarily into (4, R = tBu). Path (b) occurs without change in chirality as evidenced by analysis of (7a) produced from optically pure (4a). Crude (4a) obtained from (3a) contained 1-2% of epimerized product, which we attribute to equilibration of (3a) with (1a) and (2a), path (d), the latter undergoing the racemization promoted by the base. Acid (8a) could be obtained in good yield by deprotection of acid (6a). (4a) and (6a) coupled smoothly with H-Ala-OCH_2Ph to give the expected N-substituted peptides. Examination of $(MeOCO\text{-}Phe)_2O$ revealed that rearrangement and decomposition occured too quickly to detect the corresponding (4, R' = $PhCH_2$). It is our belief that the reactions described are general and apply to the derivatives commonly used in peptide synthesis. The results have implications relating to the possible formation of side-products during peptide synthesis. Formation of (4) could explain the occasional report of the incorporation of two residues instead of a single residue in a peptide, in particular of glycyl, and from activation via the mixed anhydride (see ref. 6). A mixed anhydride decomposes by cyclization to oxazolone (2), which then generates anhydride (3) by reaction with parent acid (1) (7). (Financially supported by the Medical Research Council of Canada).

References

1. Benoiton, N.L., F.M.F. Chen. 1981. J.Chem.Soc.Chem.Commun., 1225-1227.

2. Atherton, E., N.L. Benoiton, E. Brown, R.C. Sheppard, B.J. Williams. 1981. J.Chem.Soc.Chem.Commun., 336-337.

3. Benoiton, N.L., R. Steinauer, Y. Lee, F.M.F. Chen. 13th Int. Cong.Biochem. Amsterdam, Abstract TU-195.

4. Chen, F.M.F., K. Kuroda, N.L. Benoiton. 1978. Synthesis 928-929.

5. Benoiton, N.L., F.M.F. Chen. 1981. Can.J.Chem. 59, 384-389.

6. Ahmed, F.R., F.M.F. Chen, N.L. Benoiton. 1986. Can.J.Chem. 64, 1396-1399.

7. Chen, F.M.F., N.L. Benoiton: Can.J.Chem. (in press).

ESTERS OF FMOC-AMINO ACIDS WITH 3,4,-DIHYDRO-3-HYDROXY-4-OXO-1,2,3-BENZOTRIAZINE. A NEW CLASS OF SELF-INDICATING, ACTIVATED INTERMEDIATES FOR SOLID PHASE SYNTHESIS.

By Morten Meldal* and Robert C. Sheppard

Medical Research Council Laboratory of Molecular Biology, Hills Road, Cambridge, CB2 2QH, England.

Since the introduction of polyamide based solid phase synthesis using Fmoc-amino acid anhydrides,[1] considerable effort has been devoted to development of new chemical methods and reagents which could facilitate the conduct and particularly the automation of such peptide synthesis. The two major requirements of (i) reagent stability, and (ii) mild, fast, and complete reactions were largely met by use of the Fmoc-amino acid pentafluorophenyl (Pfp) esters originally introduced by Kisfaludy and Schön.[2] In the absence of catalyst 1-hydroxybenzotriazole (HOBt), however, acylation reactions are appreciably slower than with symmetrical anhydrides, and spectrophotometric monitoring is more difficult. We describe now a new class of self indicating, stable, and fast reacting acylating reagents which may overcome these problems.

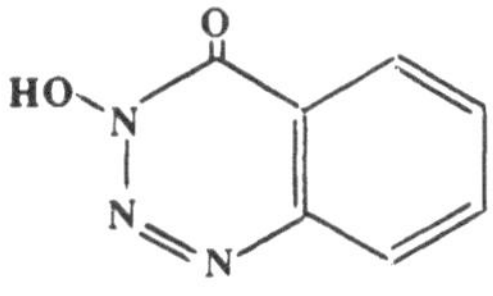

I

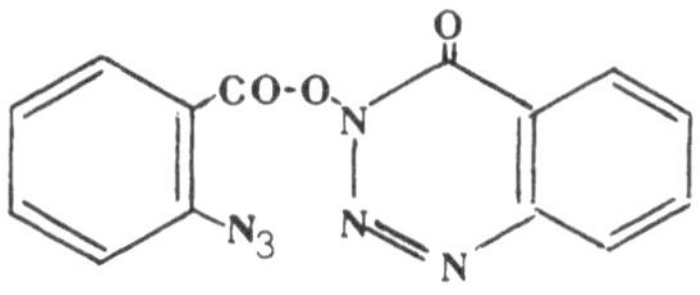

II

H-Val-Gln-Ala-Ala-Ile-Asp-Tyr-Ile-Asn-Gly-OH (III)

H-Val-Leu-Arg-Asn-Pro-Asp-Gly-Glu-Ile-Glu-Lys-Gly-OH (IV)

H-Ile-Ala-Glu-Ile-Gly-Ala-Ser-Leu-Ile-Lys-His-Trp-OH (V)

H-Gly-Lys-Lys-Lys-Cys(Acm)-Ser-Glu-Ser-Ser-Asp-Ser-Gly-Ser-Tyr-Gly-OH (VI)

Esters of Fmoc-amino acids with 3,4-dihydro-3-hydroxy-4-oxo-1,2,3-benzotriazine (I, HODhbt) are easily prepared using the DCCI-procedure. Similar

Peptides 1986

esters of benzyloxycarbonyl-amino acids were described in 1970 by König and Geiger[3] at the time of introduction of HOBt into peptide synthesis. Although HODhbt suppresses racemisation as well or better than HOBt, esters of (I) apparently did not find application in solid phase synthesis or substantially in solution. Possibly this has been due to the reported[3] formation of the o-azidobenzoate (II) as an acylating byproduct in their preparation.

We have now developed procedures to minimise the formation and detect the presence of this byproduct, and HODhbt esters of all the common Fmoc-amino acids except arginine have been prepared in crystalline and seemingly pure (hplc) form. Generally activation is performed in cold tetrahydrofuran. Nitrile formation in the case of asparagine and glutamine is suppressed by activation in cold dimethylformamide (DMF); lactam formation from Mtr-protected Fmoc-arginine during DCCI activation is minimised in cold tetrahydrofuran. In a model isoleucine to glycine coupling reaction, the Dhbt ester was found to react 5 times faster than the uncatalysed Pfp ester and at a rate comparable to that of the symmetrical anhydride. The optical rotation of Fmoc-Asp(OBu^t)-ODhbt ester was stable in neutral DMF solution but not over a long period in the presence of tertiary base. When Fmoc-Ser(Bu^t)-ODhbt was reacted with H-Leu-Gly-resin and the tripeptide product cleaved with trifluoroacetic acid (TFA), hplc separation of diastereoisomers showed negligible racemisation when the coupling was carried out in the presence or absence of tertiary base. Curiously, Dhbt esters were found to react with hydroxyl groups in the presence of dimethylaminopyridine (DMAP) approximately 5 times more slowly than Pfp esters.

When we first applied Dhbt esters to solid phase synthesis under continuous flow conditions[4] we observed an immediate bright yellow colouration of the resin which faded as acylation progressed. The solution remained colourless throughout. The colour was attributed to ionisation of the hydroxyl component by resin bound amino groups, and offered a unique possibility to monitor the coupling reaction directly on the resin beads.[5]

The efficiency of these new derivatives in the continuous flow version of the polyamide solid phase method was tested by preparation of the difficult acyl carrier protein decapeptide 65-74 sequence (III). In this method the polyamide gel resin is supported in macroporous kieselguhr particles which allow low pressure solvent flow through packed columns. The resin was functionalised with a norleucine internal reference amino acid and an acid labile linkage agent as usual,[1,4] and the terminal ester bond formed using Fmoc-Gly-OPfp in the presence of DMAP. All the remaining residues were added as Dhbt esters in DMF and intermediate Fmoc-peptides deprotected with 20% piperidine in DMF. Individual reactions were followed visually by observation of the colour of the column; reactions were generally allowed to continue 2-3 times longer than indicated by the fading colour. In agreement with previous experience, exceptionally long reaction times were observed for addition of the final valine residue.** This last reaction was investigated in more detail using both Pfp and Dhbt esters and with addition of urea.[6] Amino acid analyses showed all reactions to have an initial fast phase followed by a slow phase which could only be forced to completion by the combined use of urea and Fmoc-Val-ODhbt. The chromatogram of the TFA cleaved crude product showed this to be approximately 97% pure ACP decapeptide, a substantially better result than earlier synthesis of the same peptide using symmetrical anhydrides.

We conclude that Fmoc-amino acid-Dhbt esters are valuable alternatives to other acylating species currently employed in solid phase synthesis. They are easily prepared, usually crystalline, and thus far have proved stable to storage. They react rapidly and selectively with amino groups and offer a unique possibility to monitor acylation reactions both qualitatively and quantitatively.*** Substantiating this conclusion we have recently synthesised three further peptides using Dhbt esters, viz. an arginine-containing dodecapeptide analogue from E.coli K88 ad fimbrial protein (VI), the penicillinase C-terminal dodecapeptide sequence which includes several sterically hindered residues (V), and a pentadecapeptide sequence from protein P16 (VI). All the products gave satisfactory analytical data.

We thank Miss Linda Cameron for excellent assistance in the preparative work.

*Present address: Department of Chemistry, The University of Copenhagen, Universitets, Parken 15, Copenhagen, Denmark.

**Interestingly, this very slow reaction was not observed in a similar assembly of the same sequence on an aminopropyl-functionalised controlled pore glass support. Presumably, aggregation of the peptide chains or interaction with the polymer matrix[7] is prevented when a rigid, surface functionalised support is used.

***A photometric system for accurate measurement of resin colour has now been contructed and used in a fully automated synthesis of the acyl carrier protein test decapeptide (Meldal and Sheppard, to be published).

References:

(1) E. Atherton, C.J. Logan and R.C. Sheppard. 1981. J. Chem. Soc. Perkin I 538.

(2) L. Kisfaludy and I. Schön. 1983. Synthesis 325.

(3) W. König and R. Geiger. 1970. Chem. Ber. 103, 2034.

(4) A. Dryland and R.C. Sheppard. 1986. J. Chem. Soc. Perkin I 125.

(5) Patent applied for.

(6) F. Westall and A.B. Robinson. 1970. J. Org. Chem. 35, 2842.

(7) E. Atherton and R.C. Sheppard, Proc. 9th Amer. Peptide Symp., Toronoto, 1985, Eds. C.M. Deber, V.J. Hruby, and K.D. Kopple, Pierce Chemical Company, Rockford, Ill., 1985, p415.

DESIGN OF REAGENTS FOR SOLID PHASE SYNTHESIS

R. Ramage, S. Bielecki and M.J. Florence

Department of Chemistry, University of Edinburgh, West Mains Road, Edinburgh EH9 3JJ, Scotland

Since the advent of solid phase peptide synthetic methodology due to the pioneering researches of R.B. Merrifield,[1] there have been many investigations into the design of reagents and polymer supports with the aim of optimising the polypeptide forming stages and the final release of the product polypeptide from the polymer. In the general Scheme 1 for solid phase synthesis shown below we have investigated the design of the protecting group X, the activating group Y and the linking ester function which must finally be cleaved. We have previously reported[2] on the choice of phosphinic-carboxylic acids for solid phase synthesis in order to lower the stoiochiometric requirements of the protected amino acids and also enhance the rate of amide formation.

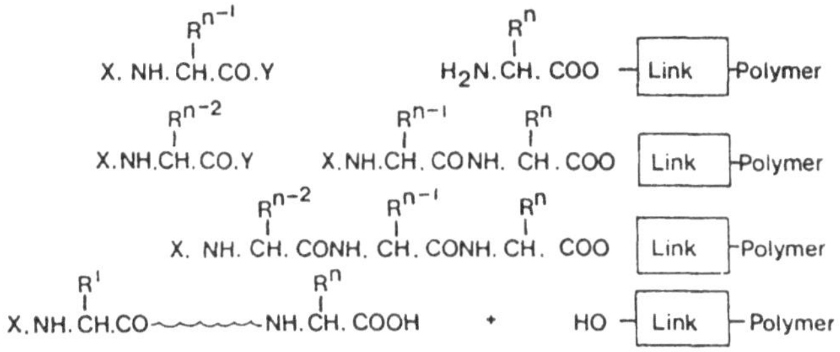

Scheme 1

NATURE OF X

The Fmoc[3] strategy chronicled by Sheppard,[4] and also Meienhofer,[5] offers the exciting possibility of constructing side chain protecting strategies which are orthogonal to both the Nα amino protection and the ester linkage to the polymer, thus producing fully protected fragments for protein

Peptides 1986

synthesis. We sought to emulate the Fmoc group by devising a protecting group which was cleaved by near stoichiometric amounts of base in a fast β-elimination reaction. The system chosen for study was that derived from bis[4'-nitrophenyl]-ethan-2-ol (**1**) in which both para-nitrophenyl groups served to activate the β-H towards base abstraction. Synthesis of (**1**) was achieved by nitration of the acetate of 2:2-diphenyl-ethanol followed by hydrolysis of the protecting acetate group. This alcohol could be converted readily into the series of useful synthetic intermediates (**2**), (**3**), (**4**), (**5**) and (**6**).[6] It was found that the ester (**7**) was cleaved by β-elimination to the olefin (**8**) and the corresponding acid by a mechanism which could be classified as E_{1cb}. The base used was DBN or DBU in stoichiometric amount and the reaction was found to proceed faster in DMF than DCM. Although the deprotection reaction is relatively rapid, the strong base DBN may be buffered by one equivalent of HOAc without serious reduction in rate in order to prevent any untoward reactions which may result from the presence of DBN.

The preparation of bis[4'-nitrophenyl]-2-ethoxycarbonyl amino acids proceeded smoothly by the method of Lapatsanis[7] using the reagent (**5**). These Bnpoc amino acids form peptide links in the normal manner either in solution or solid phase. The group is stable towards amines such as pyridine, NMM, triethylamine and cleaved to a limited extent by piperidine. It is stable and very slowly cleaved by alkali allowing preferential hydrolysis of the carboxymethyl function.

$$Ph_2CH.CH_2OH \longrightarrow (p\text{-}O_2NC_6H_4)_2CH.CH_2OH$$

1

(i), Ac_2O; (ii), HNO_3/H_2SO_4; (iii), H_3O^+

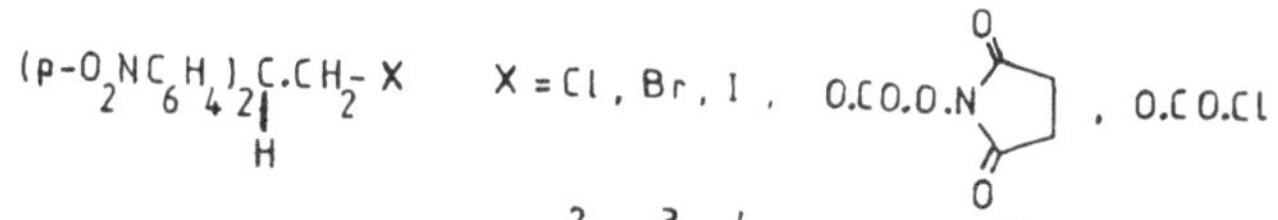

2 3 4 5 6

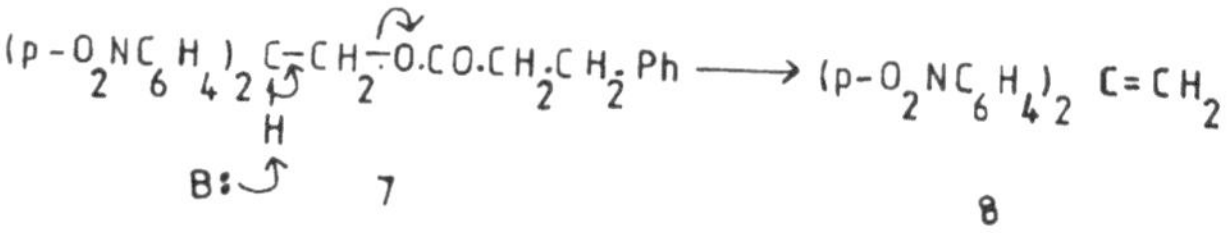

7 8

NATURE OF LINKAGE TO POLYMER

Consistent with the orthogonal protecting group strategy outlined above we designed a terminal ester link to the polymer based on the well established tendency of β-silyl substituted alcohols, or esters thereof, to suffer fragmentation by fluoride ion attack at silicon. The linking agent (**9**) selected may be synthesised by the route shown in Scheme 2.

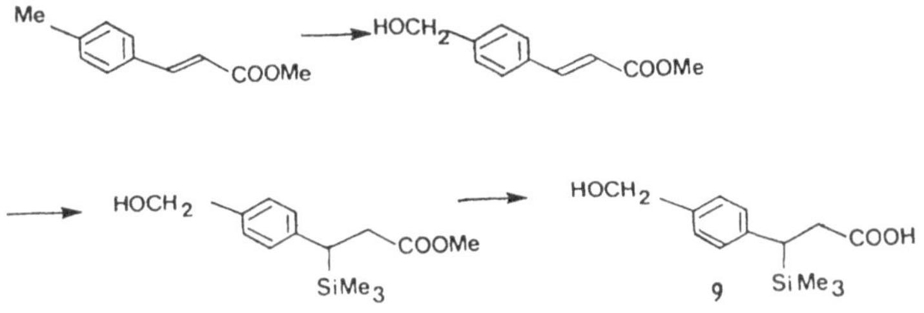

Scheme 2

The linking agent (**9**) was first converted to the amide (**10**) by reaction with Gly-OPh (DCCI) then reacted with the first Fmoc or Bnpoc amino acid (DCCI) to give the phenyl ester (**11**). Selective hydrolysis of the phenyl ester function and subsequent attachment to aminopolystyrene resin afforded the desired starting point (**12**) for solid phase peptide synthesis. Base induced cleavage of the Fmoc protecting group and subsequent elongation using 1.5 equivalents of Fmoc amino acids by the phosphinic-carboxylic mixed anhydride method was used to assemble the hexapeptide Dpp.Leu.Val.Gly.Phe.Ala.Leu bound to the polymer. It was found that $Bu_4N^+F^-$ liberated the peptide within 5 minutes by the mechanism shown using 1-2 equivalents of fluoride ion. Deleterious effects of trace amounts of water, presumably leading to the presence of HF, caused deprotection of t-butyl protection in the synthesis of Boc.Tyr.Gly.Trp.Met.Asp(OBut).Phe.Gly. However this could be avoided by using $Bu_4N^+F^-$ in the presence of Et_3N.

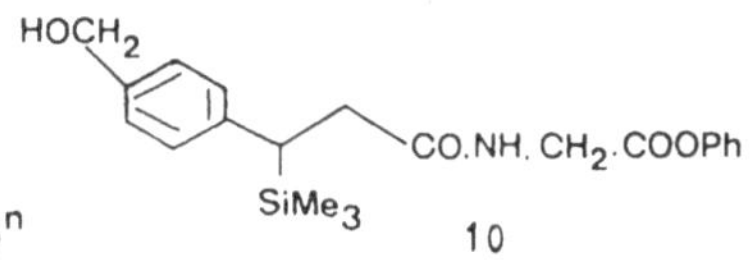

10

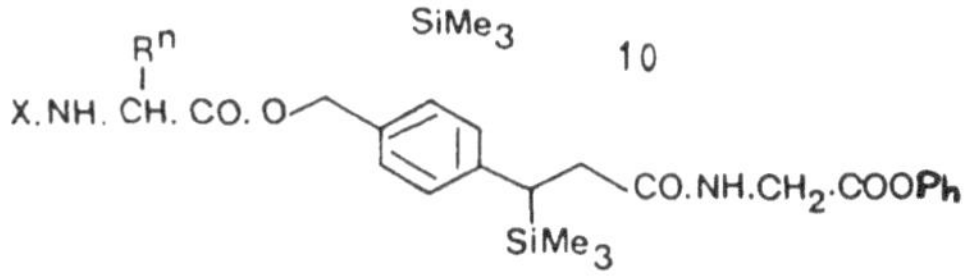

11

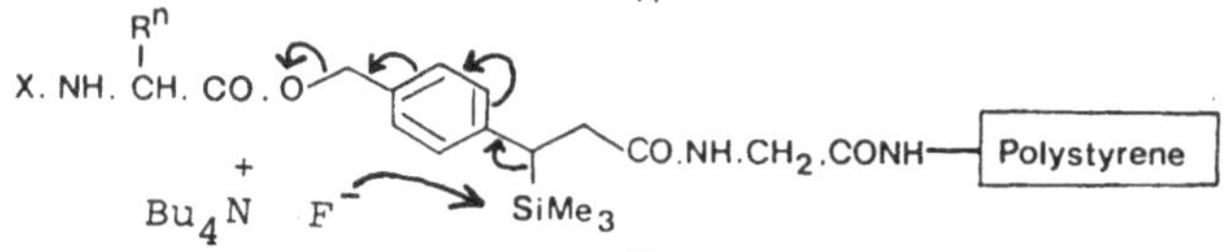

12

This work was supported by Wendstone Chemicals plc, Glaxo Research Ltd and SERC.

REFERENCES

1. R.B. Merrifield, *J. Amer. Chem. Soc.*, **85**, 2149 (1963); R.B. Merrifield, *Angew. Chem.*, **24**, 799 (1985).

2. C. Poulos and R. Ramage, 18th EPS Peptides 1984, Ed. Ulf Ragnarsson Almqvist and Wiksell Trycheri - Uppsala 161 (1984).

3. L.A. Carpino and G.Y. Han, *J. Org. Chem.*, **37**, 3404 (1972).

4. E. Atherton, H. Fox, D. Harkiss, C.J. Logan, R.C. Sheppard and B.J. Willliams, *J. Chem. Soc., Chem. Commun.*, 537 (1978); E. Atherton, H. Fox, D. Harkiss and R.C. Sheppard, *ibid*, 539 (1978); E. Atherton, M.J. Gait, R.C. Sheppard and B.J. Williams, *Bioorganic Chemistry*, **8**, 351 (1979).

5. C.D. Chang and J. Meienhofer, *Int. J. Pept. Protein Res.*, **11**, 246 (1978).

6. Wendstone Chemicals plc, patent applied for.

7. L. Lapatsanis, G. Milias, K. Froussion and M. Kolovos, *Synthesis*, 671 (1983).

SYNTHESIS OF THE CARBOHYDRATE-FREE VESPULAKININ 1 BY THE CONTINUOUS FLOW SOLID PHASE PROCEDURE AND BY A COMBINATION OF SOLID PHASE AND SOLUTION SYNTHESIS

R. Rocchi, L. Biondi, F. Filira, and B. Scolaro

Centro di Studio sui Biopolimeri del C.N.R., Dipartimento di Chimica Organica dell'Universita' di Padova, 35131 Padova, Italia

Introduction

Two structurally similar, highly basic, vasoactive glycopeptides, vespulakinin 1 (VSK 1) and vespulakinin 2 (VSK 2), have been isolated from extracts of venom sacs of the yellow jacket Vespula Maculifrons (1). The structure deduced for VSK 1 (heptadecapeptide) is:

```
          CHO CHO                              bradykinin
           |   |                  ________________________________________
Thr-Ala-Thr-Thr-Arg-Arg-Arg-Gly-Arg-Pro-Pro-Gly-Phe-Ser-Pro-Phe-Arg
 1                                                                  17
```

The precise composition of carbohydrates (CHO) O-glycosidically linked to the hydroxyl side chain function of Thr 3 and Thr 4 is not known but some evidence suggests that they are tri- or tetrasaccharides containing N-acetyl galactosamine and galactose. Vespulakinin 2 lacks the amino terminal Thr-Ala and has the same carbohydrate composition as vespulakinin 1. In order to investigate the extent to which the carbohydrate contributes to the properties of biologically active glycopeptides the synthesis was undertaken of vespulakinin analogues either devoid of carbohydrates or differing from the native molecule by the number and the composition of the carbohydrate side chains. Total synthesis of the carbohydrate-free vespulakinin 1 was achieved by the continuous flow solid phase technique as well as by a combination of solid phase and solution synthesis.

Results and Discussion

Assembly of the heptadecapeptide sequence of vespulakinin 1 by the solid phase procedure was achieved on 4-hydroxymethylphenoxyacetyl-norleucyl derivatized polydimethylacrylamide-kieselguhr sup-

Peptides 1986

ported resin, according to the standard procedure, by utilizing a manually controlled pepsynthesiser apparatus (2). The resin was esterified with pre-formed N^{α}-fluoren-9-ylmethoxycarbonyl-N^{G}-4-methoxy-2,3,6-trimethylbenzenesulphonyl-L-arginine symmetrical anhydride, [Fmoc-Arg(Mtr)$_2$O](3), in DMF, in the presence of 4-dimethylaminopyridine catalyst. The Fmoc group was cleaved with 20% piperidine-DMF solution and the synthesis was continued by using pre-formed Fmoc-amino acid symmetrical anhydrides (2) as acylating agents. The N-terminal threonine residue was introduced as the N^{α}-Boc derivative. Threonine and serine were protected in the side chain by <u>tert</u>-butyl and the Mtr group was used for masking the guanidino function of arginine residues (3). The ninhydrin colour test (4) for residual amine was negative after 60 min but the coupling of arginine residues was usually complete after a longer reaction time (3). The final peptide was cleaved from the resin with trifluoroacetic acid containing 5% thioanisole (96% cleavage). Purification by ion exchange chromatography on a CM 52 column eluted with a linear gradient (0.05 M - 1.0 M) of ammonium acetate afforded the HPLC-pure heptadecapeptide, $[\alpha]_D^{20°}$ -78.6° (c 1.0, water), single spot by high voltage electrophoresis at pH 3.6 and 9.6. The amino acid composition of the acid hydrolysate is shown in Table 1.

Table 1 - Amino Acid Ratios in Acid Hydrolysate of VSK 1

Amino acid residue	VSK 1 (a)	VSK 1 (b)	Theory
Ala	0.98	0.96	1
Gly	2.01	1.90	2
Phe	1.96	1.80	2
Pro	3.12	3.00	3
Ser	1.01	1.10	1
Thr	2.88	3.00	3
Arg	5.04	5.20	5

(a) by the continuous flow solid phase procedure
(b) by a combination of solid phase and solution synthesis

Alternatively the nonapeptide corresponding to the 9-17 amino acid sequence of vespulakinin 1 has been synthesized in solution by the fragment condensation procedure as Boc-Arg(NO_2)-Pro-Pro-Gly-Phe-Ser-Pro-Phe-Arg(NO_2)-OBzl. The Boc protecting group was removed by trifluoroacetic acid treatment and the partially protected nonapeptide was acylated, by the DCC-HOBt procedure, with the octapeptide Fmoc-Thr(Bu^t)-Ala-Thr(Bu^t)-Thr(Bu^t)-Arg(Mtr)-Arg(Mtr)-Arg(Mtr)-Gly-OH prepared by the "Fmoc polyamide" method on 4-hydroxymethyl-3-methoxyphenoxyacetyl-norleucyl derivatized polydimethylacrylamide-kieselguhr supported resin (2) followed by treatment with 1% trifluoroacetic acid in dichloromethane. Final

deblocking of the resulting heptadecapeptide was achieved by catalytic hydrogenation followed by trifluoroacetic acid treatment. The isolated product was purified by ion exchange chromatography on a CM 52 column and the peptide, $[\alpha]_D^{20°}$ -74.6° (c 1.0, water), was analytically indistiguishable from the carbohydrate-free vespulakinin 1 prepared by the continuous flow solid phase procedure.
The conformational properties of the synthetic heptadecapeptide have been investigated by circular dichroism in aqueous solution and in trifluoroethanol, which is known to cause a transition from the random to the helical form in polypeptide systems. The general features of the spectrum in aqueous solution are those typical of largely disordered polypeptides but, in trifluoroethanol, the spectrum clearly shows that vespulakinin 1 is able to undergo a solvent induced conformational transition and assumes a partially ordered conformation.
The effect of the synthetic heptadecapeptide on the arterial blood pressure of anaesthetized rats has been also investigated. Vespulakinin 1 has been reported (1) to be at least twice as potent as bradykinin (on a weight basis) in lowering rat blood pressure. The results of preliminary pharmacological experiments indicated that carbohydrate-free vespulakinin 1 is even more potent and that the duration of the response is apparently longer. This could mean that the contribution of the carbohydrate moieties present in the native molecule is not directly related to the binding to the receptor. In some cases the severe hypotensive effect was accompanied by stimulation of blood coagulation and death of the animal occurred within 90 min. Detailed pharmacological experiments and the synthesis of vespulakinin analogues O-glycosilated at threonine residues in position 3 and/or 4 are in progress.

Acknowledgement

This research was supported by the "Progetto Finalizzato Chimica Fine e Secondaria" of the National Research Council of Italy.
The authors wish to thank Polifarma S.p.A., Roma, Italy, for the pharmacological assays, mr. F. Cavaggion for skilful technical support and prof. G. Young for reading the manuscript.

References

1. Yoshida, H., R.G. Geller, J.J. Pisano. 1976. Biochemistry 15, 61.

2. for references see: Dryland, A., R.C. Sheppard. 1986. J. Chem. Soc. Perkin Trans. I, 125.

3. Atherton, E., R.C. Sheppard, J.D. Wade. 1983. J. Chem. Soc. Chem. Commun. 1060.

4. Kaiser, E., R.C. Colescott, C.D. Bossinger, P.I. Cook. 1970. Analyt. Biochem. 34, 595.

FACILE SOLID-PHASE PREPARATION OF PEPTIDES CONTAINING THE CH_2NH PEPTIDE BOND ISOSTERE AND APPLICATION TO THE SYNTHESIS OF SOMATOSTATIN (SRIF) OCTAPEPTIDE ANALOGUES

David H. Coy, Yusuke Sasaki, William A. Murphy and Mark Heiman

Peptide Research Laboratories, Department of Medicine, Tulane University School of Medicine, New Orleans, LA 70112, USA

Introduction

Isosteric replacement groups for the CO-NH peptide bond often produce interesting biological properties in peptide analogues due to effects on conformation and/or increased resistance to peptidases. Unfortunately, the chemistry required for the introduction of these groups is usually quite time consuming. Perhaps the simplest route to CH_2NH peptides involves the reaction (1) of a protected amino acid aldehyde with the α-amino group of another amino acid or peptide chain in the presence of $NaBH_3CN$ in acidified MeOH solution. This reaction can be accompanied by side reactions such as reduction of the aldehyde to the alcohol and work-up of the resulting basic pseudopeptide is problematic. It occurred to us that the reductive alkylation reaction might be adaptable to solid-phase systems and, if successful, would be interesting to utilize for peptide bond replacements in potent SRIF octapeptides (2,3).

Results and Discussion

Boc α-amino acid aldehydes were synthesized in excellent yields by the method of Fehrentz and Castro (4) (Fig. 1). In an initial experiment, the $NaBH_3CN$-mediated coupling of Boc-D-Trp-CHO (2.5 M excess) to TFA.Ala-polystyrene-resin was examined. Use of the regular solution synthesis solvent system (MeOH:AcOH/99:1), resulted in an alkylation reaction which was unacceptably slow (very positive Kaiser test at 3 h). However, when the same reaction was tried in DMF containing 1% AcOH,

Peptides 1986

complete coupling (Kaiser test although weak does not become completely negative due to secondary amine) took place within 1

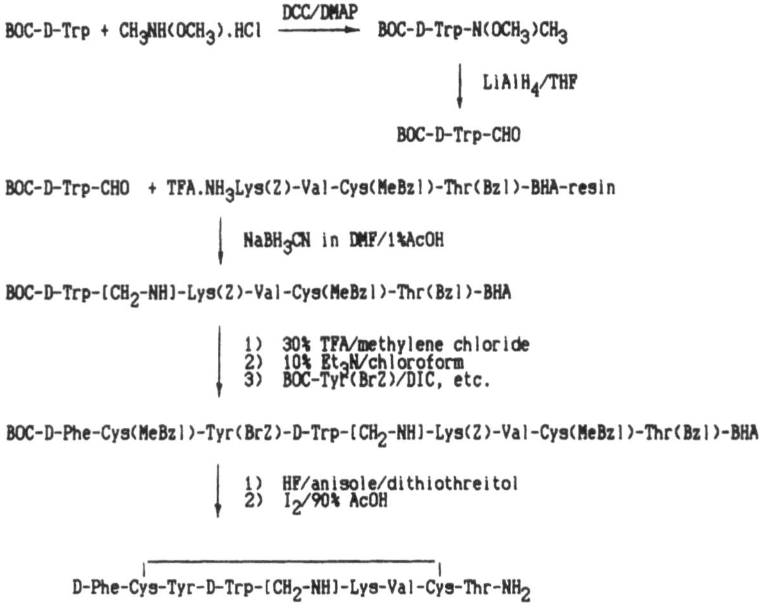

Fig 1. Reaction scheme for the solid-phase introduction of CH_2-NH groups into SRIF octapeptides.

hour. Studies (5) on the the reactivity of the psiNH group revealed very low susceptibility to subsequent DIC coupling of Boc-amino acids with the exception of Gly which could, however, be coupled as the activated ester with no discernible side-reaction. The method was then used to synthesize a series of psi[CH_2NH] octapeptide analogues of SRIF (Fig. 1) based on the potent analog, D-Phe-Cys-Tyr-D-Trp-Lys-Val-Cys-Thr-NH_2 (3), in yields similar to normal peptide bond analogues. Peptides were purified by gel filtration and preparative reverse phase hplc. The reductive alkylation procedure was compatible with the side-chain protection used. The analogues were tested for their inhibitory effects on GH release in the rat (6) and their potencies are shown in Table 1. The parent peptide was 80 times more active than somatostatin. The two most active psi-analogues were the Lys-[CH_2NH]-Val (434% potent relative to SRIF) and Cys-[CH_2NH]-Tyr- (1960%) containing octapeptides. The remaining analogues were all considerably less potent. These data appear to correlate extremely well with a recent conformation proposed for a similar octapeptide by Wynants et al. (7) based on 500 mhz NMR spectra in DMSO solution (Fig. 2). This favours a Type II′ B-bend

Table 1. Potency of SRIF Analogs on Inhibition of GH Release in the Anesthetized Rat

Bond Replaced	% Potency (Somatostatin = 100%)
None	8000
1-2	160
2-3	1960
3-4	20
4-5	2
5-6	434
6-7	2
7-8	73
2-3 and 5-6	65

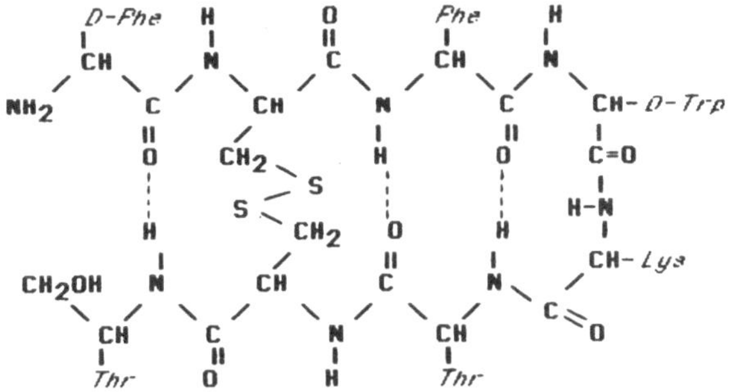

Fig. 2. Proposed solution conformation (7) for SRIF octapeptides

around residues 3-6 and hydrogen bond formation between C=O 1 to NH 8, NH 3 to C=O 6, and C=O 3 to NH 6. Replacement of these carbonyl groups by CH_2 results in the lowest biological activity, particularly within the ring. Replacement of the D-Trp-Lys (6 to 7) bond, critical to the B-bend, also drastically lowers potency. It is not presently known if the two highly active analogues exhibit improved pharmacokinetics due to increased enzymatic resistance. It seems likely that this convenient method for CH_2NH bond introduction can be extended to numerous other peptide systems within certain chain length limitations and to other alkylation reactions (eg. solid-phase N-alkyl amino acid peptide formation using a variety of alkyl and aryl aldehydes and ketones).

Acknowledgement

This research was supported by NIH grant AM-18370.

References

1. Jacobson, K.A., Marr-Leisy, D., Rosenkranz, R.P., Verlander, M., Melmon, K.L., and Goodman, M., J. Med. Chem., 26 (1983) 492.

2. Bauer, W., Briner, U., Doepfner, W., Haller, R., Huguenin, R., Marbach, P., Pletcher, T.J., and Pless, J., Life Sci. 31 (1982) 1133.

3. Murphy, W.A., Heiman, M.L., Lance, V.A., Mezo, I, and Coy, D.H., Biochem. Biophys. Res. Commun., 132 (1985) 922.

4. Fehrentz, J.A. and Castro, B., Synthesis (1983) 676.

5. Sasaki, Y. and Coy, D.H., Peptides (1986) in press.

6. Murphy, W.A., Meyers, C., and Coy, D.H., Endocrinology, 109 (1981) 491.

7. Wynants, C., Van Binst, G., and Loosli, H.R., Int. J. Pept. Prot. Res., 25 (1985) 615.

SYNTHESIS OF A KARYOPHILIC PEPTIDE COVALENTLY BOUND TO PHOSPHATIDYLETHANOLAMINE

D. Lelièvre, A. Brack, Y. Trudelle, P.F. Tosi and C. Nicolau
Centre de Biophysique Moléculaire, C.N.R.S.
45071 Orléans cedex 2, France

A segment as small as the 13 amino acid N-terminus of S. cerevisiae wild-type α2 protein (a yeast nuclear protein) covalently linked to the N-terminus of cytoplasmic β-galactosidase is sufficient to direct localization of β-galactosidase to the nucleus in yeast (1). We have prepared the compound H-Met^1-Asn^2-Lys^3-Ile^4-Pro^5-Ile^6-Lys^7-Asp^8-Leu^9-Leu^{10}-Asn^{11}-Pro^{12}-Gln^{13}-PE (with PE for phosphatidylethanolamine) combination of the karyophilic 13-peptide and a phospholipid, in order to increase the nucleotropism of liposomes used as potent carriers.

Synthesis

The synthesis was carried out by combining SPPS, liquid phase synthesis and fragment coupling. The strategy, summarized in scheme 1, was chosen in order to minimize the risk of racemization of Gln^{13}.

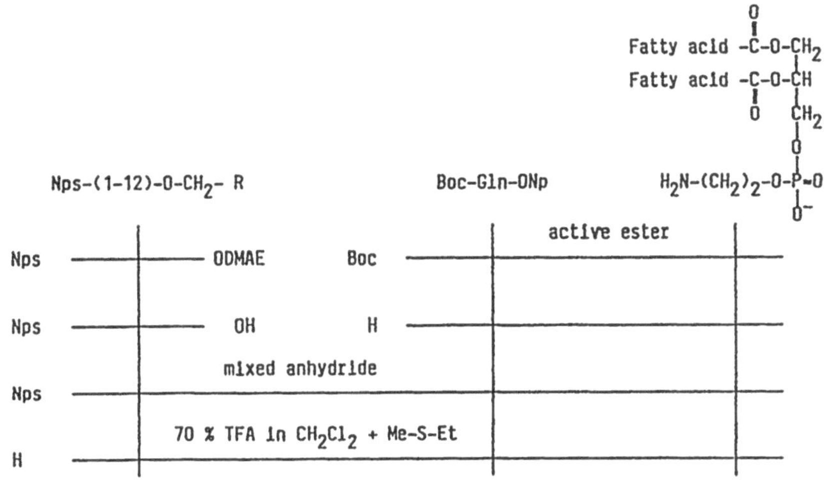

Scheme 1

Peptides 1986

Coupling Boc-Gln^{13} to the phospholipid was achieved <u>via</u> the p-nitrophenyl ester in a 1:1 (v/v) mixture of chloroform and dimethylformamid, at room temperature. Boc-Gln-PE was purified on a silica gel column (Biosil HA from Bio-Rad) eluted with a 85:10:5 (v/v) mixture of chloroform, methanol and acetic acid. The purity was checked by TLC in the same solvent : R_f 0.37 (0.11 for PE and 0.76 for Boc-Gln-ONp). The Boc-group was cleaved by HCl 1.1 N in acetic acid. After evaporation of the solvent, the product was washed with cold methanol and recovered by centrifugation. Yield = 53 % based on the PE used.

The N-terminal 12-peptide was synthesized by the solid phase procedure using the classical chloromethylated polystyrene resin. Protecting groups and coupling methods are reported in scheme 2.

FMoc (Nps-Met … Asp(OtBu)) | **Boc** (Leu … Pro)

Nps - Met - Asn - Lys(Boc) - Ile - Pro - Ile - Lys(Boc) - Asp(OtBu) - Leu - Leu - Asn - Pro- R

BOP (Met) | **DCCI/HOBt** (Asn … Asp(OtBu)) | **BOP** (Leu) **BOP** (Leu) **ONp** (Asn) **OCs** (Pro)

Scheme 2

Leucine^{10} and leucine^{9} were introduced with the BOP reagent to minimize the formation of cyclo (Asn-Pro) and to a lesser extent that of cyclo(Leu-Asn-Pro) (2). We found also that Fmoc-Asn^{2}-OH could be safely introduced with DCCI/HOBt. The Nps-group was chosen to N-protect Met because of its chromophoric properties in the visible range. The Nps-12-peptide was removed from the resin by transesterification with dimethylamino 2-ethanol in DMF (1:1) in the presence of thallium ethoxide (20 µl for 945 mg of peptide) (3) for 24 hrs at 45°C, the treatment being repeated five times. A mild saponification with 0.1 N NaOH (3 eq. in DMSO) yielded the carboxyl-free protected 12-peptide (72 % overall yield based on Pro-resin) which was purified by silica gel column chromatography using a mixture of chloroform-methanol-acetic acid-water (80:15:2:2) as eluent. Amino acid analysis after 48 hrs at 105°C in 5.6 N HCl gives the following values : Asp 2.86 (3.00), Pro 2.14 (2.00), Ile 1.90 (2.00), Leu 2.00 (2.00), Lys 1.92 (2.00).

Coupling of the protected 12-peptide to H-Gln-PE presented serious difficulties. Active esters (OSu) and condensing agents (DCCI/HOBt,BOP) were inoperent. Finally, the coupling could be achieved <u>via</u> isobutyl chloroformate using 1.5 eq. of protected-12-peptide and 1 eq. of H-Gln-PE in chloroform-dimethylsulfoxide 7:3. Column chromatography on silica gel (chloroform/acetic acid/water, 80:40:5) afforded the pure protected peptidophospholipid in 30 % yield. Final deprotection was achieved with 75 % trifluoroacetic acid-dichloromethan and 5 % methylethylsulfide for 2.5 hrs. The free peptidophospholipid was purified by gel chromatography on Sephadex G25. The product was visualized in blue on TLC silica gel plates with a reagent specific for organic phosphates (4). Its purity was also checked by ^{32}P NMR (figure 1) and gel permeation on Sepharose 4B (figure 2).

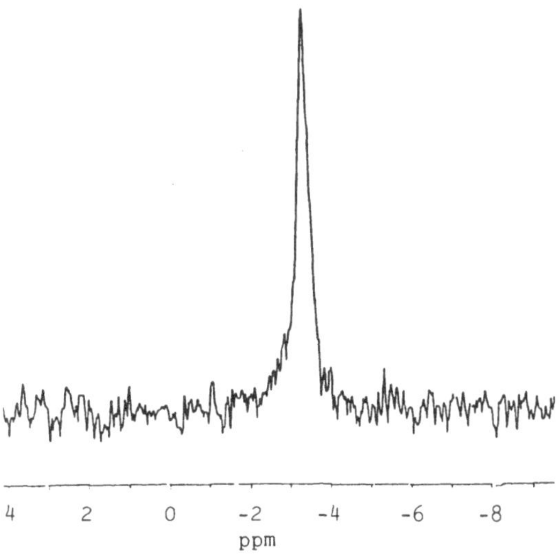

Fig. 1 - ^{32}P NMR spectrum of the free 13-peptidophospholipid in D_2O.

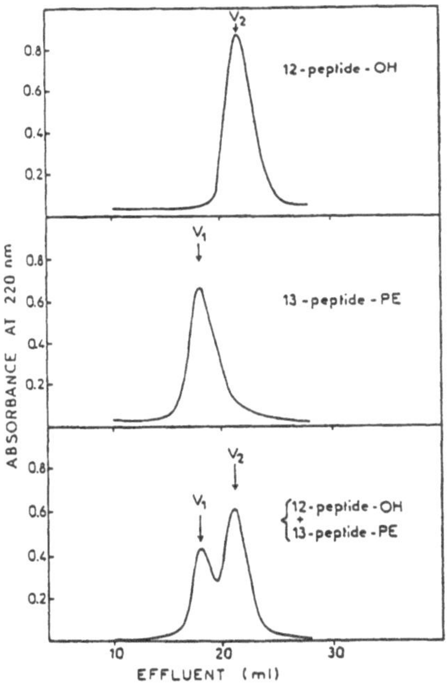

Fig. 2 - Elution profile on Sepharose 4B in 0.1 M NH_4HCO_3 of the free 12-peptide (top), of the desired compound (middle) and of a mixture of both (bottom).

Our attends to characterize the product by reversed phase HPLC on C_{18}-silica gel failed, due to a strong adsorption on the support.

Incorporation of the 13-peptide-PE into liposomes and interaction with cells

The amino groups of the peptide were labeled with a fluorescent probe, i.e., Texas Red (TR). The labeled peptide-PE (here after designed as $PEP\text{-}PE^{TR}$) was mixed to phosphatidylcholine ($PC:PEP-PE^{TR}$, 97:3), hydrated in PBS buffer and stirred on a Vortex in the presence of glass beads. The multilamellar vesicles (MLV) were then sonicated until a clear solution of small unilamellar vesicles (SUV) was obtained. Fluorescence microscopy on both MLV and SUV confirmed the incorporation of fluorescent molecules in the bilayers.

Fluorescence observation after incubation of the peptide-doped liposomes with isolated mouse liver nuclei in TBS buffer, indicated that both MLV PC^{TR} and MLV $PC:PEP-PE^{TR}$ bound to the nuclear envelopes. In the case of MLV PC^{TR}, big liposomes interact at the periphery of the nucleus. In the presence of $PEP-PE^{TR}$, the liposomes are much smaller.

When SUV $(PC:PS:PEP-PE)^{TR}$ and SUV $(PC:PS)^{TR}$ are incubated with fibroblasts, endocytosis is observed but the endosomes are very stable, even after cellular division.

Conclusion

Nuclear targetting of liposomes might be achieved by karyophilic peptides. A yeast leader sequence covalently bound to a phospholipid has been synthesized and a short sequence of SV40 large T antigen (5)(6) bound to PE is being synthesized. Fluorescence and molecular biology assays are currently used to study the interactions of liposomes bearing these peptides with isolated nuclei and with nuclei in culture cells after endocytosis of the liposomes.

References

1. Hall, M.N., L. Hereford, I. Herskowitz. 1984. Cell 36, 1057.
2. Suzuki, K., K. Nitta, N. Endo. 1975. Chem. Pharm. Bull. 23, 222.
3. Barton, M.A., R.U. Lemieux, J.Y. Savoie. 1973. J. Amer. Chem. Soc. 95, 4501.
4. Vaskovsky, V.E., E.Y. Kostetsky, I.M. Vasendin. 1975. J. Chromatogr. 114, 129.
5. Kalderon, D., B.L. Roberts, W.D. Richardson, A.F. Smith. 1984. Cell 39, 499.
6. Goldfarb, D.S., J. Gariepy, G. Schoolnik, R.D. Kornberg. 1986. Nature 322, 641.

GEL PHASE ^{13}C NMR INVESTIGATION OF 'HYPER-HIGH LOAD' SOLID (GEL) PHASE SYNTHESIS OF AN LH-RH (1-5) ASSEMBLY

R. Epton and D.A. Wellings
Wolverhampton Polytechnic, Wolverhampton, England, WV1 1LY, U.K.

Introduction

We have previously described a scheme for the ultra-high load solid (gel) phase synthesis of peptides in which a phenolic bead-form core polymer, crosslinked poly N-[2-(4-hydroxyphenyl)ethyl]-acrylamide, was used as support matrix at maximum loading (1,2). In this approach, consecutive repeating units of the core matrix each bear a growing peptide chain. The peptide chains comprise the major component of the matrix-peptide assembly throughout the synthesis. This is deemed to proceed via the stepwise elaboration of a series of quasi-homogeneous peptide gel networks.

Recently, we have been involved in the synthesis of LH-RH and LH-RH analogues modified at amino acid residues 6 and 10. Each of these molecules has LH-RH (1-5) as a common segment. This sequence has a C-terminal tyrosine residue. It occurred to us that direct condensation of Boc-Tyr-OH with the phenolic core matrix would, in the absence of side-chain protection, lead to side-chain branching, thus effecting initial amino acid loadings in excess of the apparent theoretical maximum. Subsequent elaboration of the LH-RH (1-5) sequence by solid (gel) phase methods would result in an assembly, hydrazinolysis of which would yield LH-RH (1-5) as an acylpeptide hydrazide. This 'hyper-high load' approach would be particularly amenable to investigation by gel phase ^{13}C NMR spectroscopy (3).

Solid (Gel) Phase Synthesis at 'Hyper-high Load'

The reaction and washing procedures used for solid (gel) phase synthesis were those of the ultra-high load strategy described

Peptides 1986

previously by us (1). Core polymer (1g, phenol content 5.0 mmol) was subjected to DIC/DMAP-mediated loading with Boc-Tyr-OH (4x excess) and the product divided into two equal parts. One part was subjected to a chain extension cycle consisting of N-terminal deprotection (Boc removal) by 10% $BF_3.Et_2O$ in 3-cresol followed by reaction with Boc-Ser-OBt (0.5x excess). The other part was treated similarly, except that Boc-Ser(OBzl)-OBt was used. Both syntheses were then progressed by chain extension cycles utilising Boc-Trp-OBt (0.5x excess), Boc-His(Tos)-OBt (0.5x excess) and Glp-OH/DIC (4x excess) to give LH-RH (1-5) assemblies.

Gel Phase ^{13}C NMR Spectroscopy

Network samples (50 mg) were swelled in $(CD_3)_2SO$ (2 cm^3) and subjected to 50,000 pulses at 0.5 s intervals in a Bruker WP805Y instrument operating at 20 MHz. The spectrum taken after core loading (Figure 1) showed two peaks Y_3' and Y_3, due to aromatic

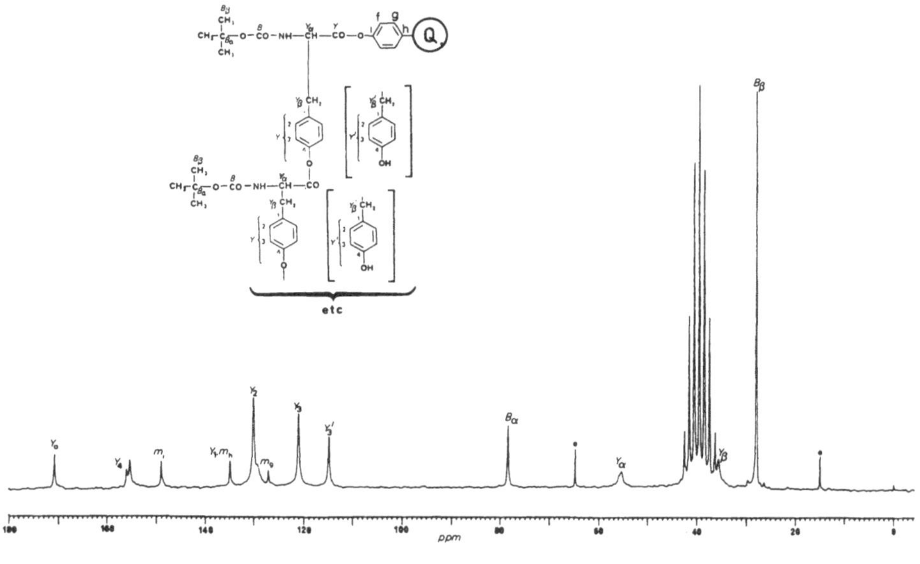

Figure 1 ^{13}C NMR spectrum of tyrosine-loaded core assembly (e is a solvent peak due to ether used in work-up)

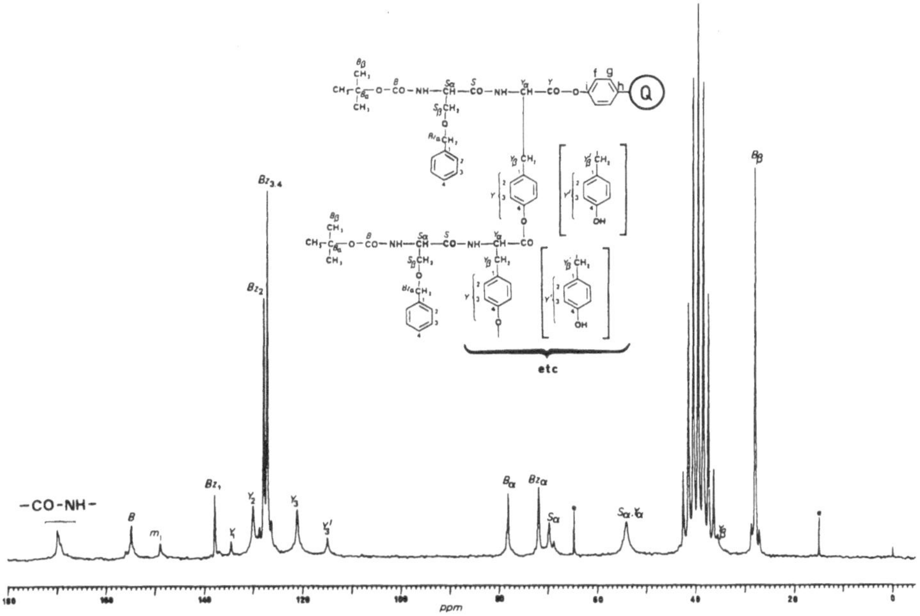

Figure 2 ^{13}C NMR spectrum of dipeptide gel network assembly (e is a solvent peak)

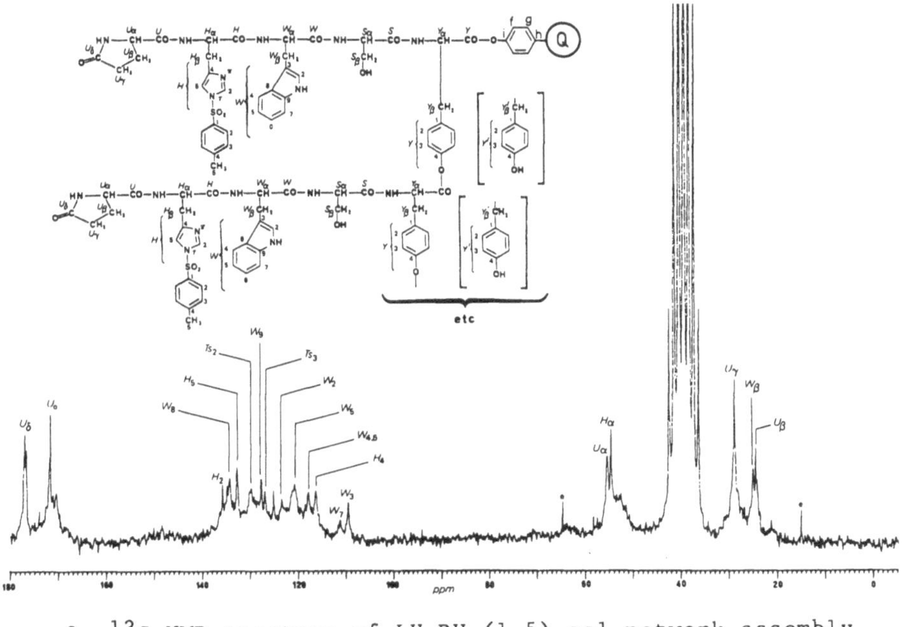

Figure 3 ^{13}C NMR spectrum of LH-RH (1-5) gel network assembly (e is a solvent peak)

carbons ortho to free and acylated -OH respectively, the relative intensities of which were consistent with a tyrosine loading of 8.0 mmol per g. On subsequent BF_3 in cresol treatment, disappearance of the intense peak B_β, due to the methyl carbons of the Boc group, was used to find the mildest N-terminal deprotection conditions consistent with total removal. The spectra obtained after chain extension cycles to add either Boc Ser or Boc Ser(OBzl) were similar except that the latter (Figure 2) exhibited peaks due to the carbon atoms of the benzyl group. Surprisingly, application of BF_3 in cresol, for just sufficient time to effect N-terminal deprotection, also effected almost total de-O-benzylation. In contrast, N-terminal deprotection with CF_3CO_2H in dichloromethane gave the usual selectivity. Unfortunately, poor swelling of the assemblies in this reagent precluded its subsequent use. Progression of the two syntheses to completion gave final quasi-homogeneous peptide gel networks with identical spectra (Figure 3) and in similar yield. In a repeat synthesis, using serine without side chain protection, 0.5 g of core polymer gave 3.9 g of final assembly. This was treated with 2% $NH_2NH_2.H_2O$ in $MeCONMe_2$ for 5 min at 25°C, after which the liquors were drawn into Et_2O at -78°C to effect precipitation of crude LH-RH (1-5) hydrazide. This, after drying, weighed 3.1 g. Purification was effected by reverse phase MPLC in which 500 mg of crude product was applied to 2.5x100 cm Whatman C18 LRP1 column and eluted with $MeCN/H_2O$ 1/9 containing 0.04% CF_3CO_2H to give 280 mg LH-RH (1-5) hydrazide, purity 97%.

References

1. Epton, R., G. Marr, B.J. McGinn, P.W. Small, D.A. Wellings, A. Williams. 1985. Int.J.Biol.Macromol. 7, 289.

2. Butwell, F.G.W., R.Epton, P.W. Small and D.A. Wellings. 1985 In Peptides-Structure and Function (C.M. Deber, V.J. Hruby and K.D. Kopple, Eds.), Pierce Chemical Company, p 273.

3. Epton, R., P. Goddard and K.J. Ivin. 1980. Polymer 21, 1367

LIQUID PHASE PEPTIDE SYNTHESIS EMPLOYING A PHENOLIC MODIFICATION OF THE POLYETHYLENE GLYCOL SUPPORT

Ian J. Galpin and Mark J. Wright
University of Liverpool, Robert Robinson Laboratories, P.O.Box 147, Liverpool, L69 3BX, England.

The use of phenyl esters in solution peptide synthesis has been clearly illustrated.[1,2] Also a phenolic linkage has been used in the traditional solid phase approach[3] and recently has been employed as a linkage in synthesis on an ultra high loading polyacrylamide.[4]

The support prepared for use with the liquid phase method is indicated in structure (1). The support was prepared using the

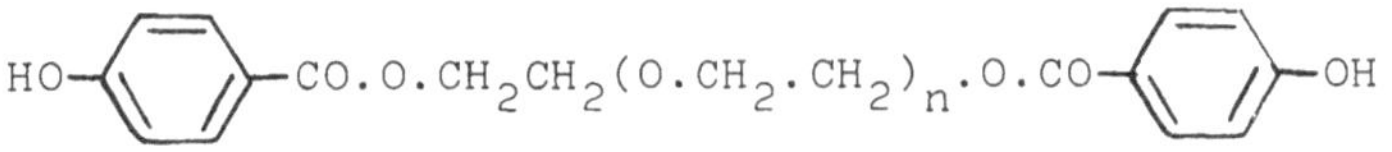

(1)

acetoxy derivative of the corresponding phenolic acid, being prepared by reaction of acetic anhydride with the parent phenolic acid. Coupling of the acetoxy derivative with the polyethylene glycol (Mw 6,000), could be achieved using DCC, DCC/HOBt, or by using the corresponding acid chloride. This coupling led to the bis(acetoxy) derivative of the polyethylene glycol support (1), which was quantitatively deacylated using the standard conditions for phenyl ester cleavage, with dioxan/water as solvent.

Attachment of the first residue was investigated using Z.Ala.OH, employing DCC as the coupling agent and dichloromethane as the solvent. Work-up involved precipitation by addition of ether, the product had infra red and proton n.m.r. spectroscopic properties which were consistent with the required bis-derivatised benzyloxy-carbonyl alanyl support (1). The alanine incorporation was determined by amino acid analysis and was found to be 0.26 mM Ala gm^{-1} which was in good agreement with the previously determined hydroxyl group content of the polyethylene glycol of 0.28 mM gm^{-1}. The FT n.m.r. gave some rise for concern, as satelite resonances

Peptides 1986

were observed for the benzyloxycarbonyl methylene protons. Eventually it became clear that a small proportion of the polyethylene glycol had not been converted to the derivative (1), so that both the required aryl ester of (1) and the alkyl ester of the polyethylene glycol were present. This was somewhat disconcerting, as prior to deacylation of the phenolic polymer the support had been treated with acetic anhydride in the presence of triethylamine in order to block any underivatised polyethylene glycol hydroxyl groups. Subsequently it was found that treatment with acetic anhydride/triethylamine/dimethylaminopyridine gave complete blocking of any unreacted polyethylene glycol hydroxyl groups. Amino acid analysis indicated that the contribution of the alkyl ester was 0.07 mM Ala gm^{-1} .

The benzyloxycarbonyl alanyl support was successfully subjected to phenyl ester cleavage and treatment with hydrazine in order to confirm cleavage from the resin. As peptide synthesis was to be achieved by elaboration of peptide chain following hydrogenolysis of the benzyloxycarbonyl group, cleavage of this group was confirmed using hydrogenolysis in DMF in the presence of 5% Pd/C. One equivalent of p-toluene sulphonic acid was used in order to trap the liberated amino group as the corresponding salt, as the phenyl ester is sufficiently activated to allow competing side reactions if the amino group remains unprotonated.

Synthesis of the peptide (2) was then investigated using the benzyloxycarbonyl group for amino protection throughout, and tert-

$$Z.Tyr(Bu^t).Thr(Bu^t).Asp(OBu^t).Ala.OH \qquad (2)$$

butyl based side chain protection. The hydrogenolysis and coupling reactions were carried out in solution, and after the addition of each residue the peptide was precipitated by the method normally employed in liquid phase synthesis.

Diphenyl phosphinic mixed anhydrides have been successfully used in solid phase synthesis,[3] and we have therefore again used this method of activation in the present liquid phase synthesis. 1.3 Equivalents of the diphenylphosphinic mixed anhydride of the benzyloxycarbonyl protected amino acid were used, and analytical control was provided by the use of fluorescamine. Qualitative amino acid analysis was carried out at each step and at the end of the synthesis the amino acid analysis was in agreement with anticipated values.

Cleavage of the peptide from the support was carried out by treatment with sodium hydroxide at pH 10.5 in the presence of hydrogen peroxide. In this case trifluoroethanol/water (4:1) was used as the solvent and a final yield of 37% of homogeneous material isolated. This low yield was similar to that which has been observed with other phenolic linked polymers, a virtually quantitative yield was, however, achieved by hydrazinolysis. As with other diphenylphosphinic mixed anhydrides, no tendency to disproportionation, or wrong-way opening, was observed with any of the amino acids in this sequence, or with the more hindered Z.Val.ODpp. This confirms the regiospecificity found previously and reiterates the results of other workers.[5]

Synthesis of two fragments of proglucagon were then investigated. The first (3), was straightforwardly assembled using the diphenyl-

$$Z.Phe.Ser(Bu^{t}).Asp(OBu^{t}).Glu(OBu^{t}).OH \qquad (3)$$

phosphinic mixed anhydrides with an initial loading of 0.13 mM Glu g^{-1}. A 73% yield of the homogeneous tetrapeptide hydrazide was obtained in this case.

A second sequence from proglucagon was investigated in which glutamine was the C-terminal residue. A number of methods for the incorporation of the benzyloxycarbonyl glutamine were examined, including the trichlorophenyl ester in the presence of HOBt, DCC/HOBt and the diphenylphosphinic mixed anhydride. In all cases very low incorporation was observed, the highest figure being 0.09 mM Gln g^{-1} (cf. 0.19 mM Ala gm^{-1}).

The only way in which glutamine could be quantitatively incorporated to a related polymer was through the initial synthesis of compound (4). This compound was prepared by reaction

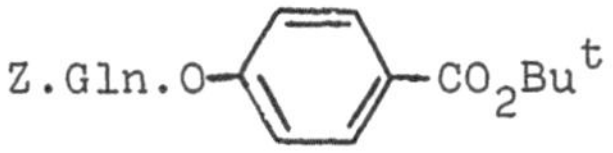

(4)

of Z.Gln.ODpp with butyl-4-hydroxybenzoate. After removal of the ^{t}butyl ester function with trifluoroacetic acid the corresponding phenolic linked glutamine was coupled to bis-glycine polyethylene glycol p-toluene sulphonate using the diphenylphosphinic mixed anhydride. On this occasion infra red spectroscopy confirmed the linkage of the phenyl ester containing moiety to the polyethylene glycol, and proton n.m.r. was consistent with a glutamine to glycine ratio of 1:1, amino acid analysis confirmed this {$Glu_{0.98}$(1), $Gly_{1.00}$(1)}. It therefore appears that coupling the linker to glutamine prior to incorporation onto the liquid phase support is a superior method.

This work thus demonstrates that a phenolic linked liquid phase support of the type (1) is suitable for peptide synthesis using diphenylphosphinic mixed anhydride activated N-protected amino acids.

References

1. Galpin, I.J., Hardy, P.M., Kenner, G.W., McDermott, J.R., Ramage, R., Seely, J.H., and Tyson, R.G, *Tetrahedron,* 1979, 35, 2577.

2. Choudhury, A.M., Kenner, G.W., Moore, S., Ramachandran, K.L., Thorpe, W.D., Ramage, R., Dockray, G.J., Gregory, R.A., Hood, L., and Hunkapiller, M., *Hoppe-Seylers.Z.Physiol.Chem.,* 1980, 361, 1719.

3. Galpin, I.J., and Robinson, A.E., *Tetrahedron,* 1984, 40, 627.

4. Epton, R., Marr, G., McGinn, B.J., Small, P.W., Wellings, D.A., and Williams,A ., *Int.J.Biol.Macromol.,* 1985, 7, 289.

5. Ramage, R., Hopton, D., Richardson, R.S., Kenner, G.W., and Moore, G.A., *J. Chem.Soc., Perkin Trans. 1,* 1985, 461.

A COMPARISON OF DIFFERENT METHODS FOR THE PREPARATION OF BENZHYDRYLAMINE RESINS

E. Pedroso, J. Josa, J-R. Peñalver and E. Giralt
Departament de Química Orgànica, Facultat de Química, Universitat de Barcelona, 08028 Barcelona, Spain

Benzhydrylamine (BHA) resin is the support of choice for the solid phase synthesis of C-terminal carboxamide peptides (e.g. hormones, toxines), as well as the starting resin to obtain more elaborate supports such as $ClCH_2$-Pab-resin (1,2) or $BrCH_2$-Nbb-resin (3,4). Homogeneity of the starting polymer is one of the key factors to achieve the synthesis of long peptides in high purity and without side-reactions. We have recently shown by gel-phase ^{13}C-NMR that the synthesis of BHA resin *via* the reduction of benzoylpolystyrene oxime gives high amounts of secondary amine derivatives (5). On the other hand the most commonly used procedure, *i. e.* Leuckart reductive amination of benzoylpolystyrene (6), is far from quantitative as can be easily demonstrated both by ^{13}C-NMR and by infrared spectroscopy. The unreacted carbonyl groups could in some circumstances react with the amino groups of the growing peptide chain in a similar way to that described by Kent (7) for the case of aldehyde groups linked to the polystyrene.

Among the different alternative ways to the synthesis of BHA resin recently reported (8,9) little atention has been paid to the procedure of Hirao *et al.* based on the reaction of lithiated polystyrene with N-(trimethylsilyl)-phenylmethan-imine. To the best of our knowledge, there have been no reports of peptides synthesized on a BHA resin prepared by this procedure, in spite of the fact that, in contrast with other described procedures, this BHA resin is obtained in one

Peptides 1986

single reaction step on the polystyrene, thus resulting in a more homogeneous polymer.

In our laboratory we have optimized the procedure of Hirao in the sense of facilitate the synthesis of the intermediate aldimines and easily control the amine content of final BHA and p-methyl-BHA resins. In our hands, preparation of N-(trimethylsilyl)-phenylmetanimine and N-(trimethylsilyl)-4-methylphenylmetanimine by the Hirao procedure gave erratic yields complicated by the lack of stability of the intermediates. We have found that lithium bis(trimethylsilyl)amide, in Ar atmosphere, substitutes advantegously for sodium bis(trimethylsilyl)amide in this synthetic step (10). ^{13}C-nmr is the method of choice to control the purity of aldimines (N-(trimethylsilyl)-phenylmetanimine: δ -0.9, 128.7, 128.8, 131.5, 139.1, 168.7; N-(trimethylsilyl)-4-methylphenylmetanimine: δ -1.1, 21.5, 128.5, 129.2, 136.5, 141.6, 168.4). The reaction with lithiated polystyrene can be effected at room temperature in a shaken vessel under Ar atmosphere and the desired final amine substitution of the resin can be effectively controled by appropriate selection of the lithiation time (see table). We regularly use a reaction time of 5hr at room temperature for the subsequent reaction with aldimines . In our experiments we use at least a two-fold aldimine excess and it appears that the final substitution of the resin is the same if larger excesses are used. Both i.r. and gel-phase ^{13}C-nmr spectroscopy, when applied to the resin after the aldimine condensation, show the presence of some weak signals (1240, 830 cm^{-1} and δ -0.5 respectively) that can be assigned to trimethylsilyl groups attached to polystyrene. These signals dissapear after a final washing with trifuoroacetic acid. The final resins can be caractherized as pure <u>carbonyl-free</u> BHA and p-Me-BHA resins by both i.r. and gel phase ^{13}C-NMR spectroscopy.

Leaving aside its probable irrelevance from the point of view

Table:	amount (g) of starting resin	lithiation time (hr)	final amine content (mmol/g)
	0.1	0.5	0.10
	1.25	1	0.16
	0.1	2	0.30
	2.5	4.5	0.37
	2.5	7	0.55
	0.1	16	0.84
	2.5	17	0.87

of their use in solid phase peptide synthesis, it must be signaled that in the gel-phase ^{13}C-nmr spectra of BHA and p-Me-BHA resins obtained by our optimized Hirao's procedure the methine carbon resonances appear as rather sharp singlets with line width of the same order that those observed for benzhydrylamine obtained via Leuckart reaction. This observation strongly suggests that in our experimental conditions reaction with aldimine takes place exclusively on the para position of polystyrene phenyl rings.

Synthesis of luliberin: The decapeptide luliberin has been chosen to test the performance of our carbonyl-free p-Me-BHA. The synthesis was carried with Boc-amino acids following a double coupling protocole: i) Standard DCC-mediated coupling and ii) symmetric anhydride. HF cleavage gave a 89% yield (determined from aminoacid analysis of the resin before and after the cleavage). HPLC of the crude after filtration on a G-15 Sephadex column showed a single major peak (95%). After reversed-phase chromatography the purified decapeptide gave a correct amino acid analysis, was shown to be homogeneous by HPLC and coeluted with an authentic luliberin sample.

In summary, reaction of aldimines with lithiated polystyrene following the modified procedure described in this communication provides an extremely convenient procedure for the preparation of carbonyl-free BHA and p-Me-BHA resins. The desired final amine substitution can be easily controled and the resins look homogeneous both by i.r. and gel-phase ^{13}C-nmr spectroscopy. Finally, our synthesis of luliberin shows that these resins are suitable for the synthesis of highly pure peptides.

Acknowledgement

This work was supported by funds from CAICYT (grant 2899/83) and US-Spain Joint Commitee for Scientific and Technological Cooperation. One of us (J. J.) thanks ERT for a doctoral fellowship.

REFERENCES

1. Giralt, E., D. Andreu, M. Pons, E. Pedroso. 1981. Tetrahedron 37, 2007.

2. Giralt, E., D. Andreu, P. Miró, E. Pedroso. 1983. Tetrahedron 19, 3185.

3. Giralt, E., F. Albericio, E. Pedroso, C. Granier, J. Van Rietschoten. 1982. Tetrahedron 38, 1193.

4. Giralt, E., R. Eritja, E. Pedroso, C. Granier, J. Van Rietschoten. 1986. Tetrahedron 691, 691.

5. Giralt, E., J. Rizo, E. Pedroso. 1984. Tetrahedron 40, 4141.

6. Orlowsky, R.C., R. Walter, D. Winkler. 1976. J. Org. Chem. 41, 3701.

7. Kent, S.B.H.. 1983. In: Peptides: Structure and Function (V.J., Hruby and D. H. Rich, eds.). Pierce Chem. Co., Rockford, p. 99.

8. Hirao, A., S. Itsuno, I. Hattori, K. Yamaguchi, S. Nakahama, N. Yamazaki. 1983. J. Chem. Soc. Chem. Commun. 25.

9. Bryan, W. M.. 1985. In: Peptides: Structure and Function (C.M. Deber, V.J. Hruby and K.D. Kopple, eds.). Pierce Chem. Co., Rockford, p. 261.

10. Hart, D.J., K. Kanai, D.G. Thomas, T. Yang. 1983. J. Org. Chem. 48, 289.

THIOETHER LINKAGES BETWEEN PEPTIDE SYNTHETIC INTERMEDIATES AND SOLID SUPPORTS

John D. Glass and Angeliki Buku

Department of Physiology and Biophysics and the Center for Polypeptide and Membrane Research, Mount Sinai School of Medicine, New York, New York, USA

Chandra S. Pande

Department of Chemistry, H. P. University, Summerhill, Simla, India

Samples (5 g each) of chloromethylated copolystyrene-1%-divinylbenzene (Merrifield resin, 1-2 meq active chlorine per gram of polymer) were shaken in 50 ml each of dimethylformamide containing 150 mmoles of triethylamine or diisopropylethylamine, and 50 mmoles of β-mercaptopropionic acid or cysteine hydrochloride or cysteamine hydrochloride. The resins which had been treated with cysteamine and with cysteine methyl ester gave strong reactions with the Kaiser ninhydrin reagent (1). Subsequent experiments indicated that at least 0.08 meq/g of β-mercaptopropionic acid, 0.1 meq/g of cysteine methyl ester, and 0.2 meq/g of cysteamine had been covalently attached to the respective polymer derivatives through alkylation of their thiol functions. These polymer derivatives are substituted S-benzyl derivatives and subject to cleavage by HF with regeneration of the thiol function.

Several peptide-resin derivatives were synthesized by sequential acylations of the cysteamine-resin derivative (H-Cya-Resin) with amino acid and peptide segment derivatives. Both Boc and Z groups were used for temporary protection of alpha amino groups during solid phase assembly of the peptide chains. Peptide-resin

Peptides 1986

derivatives of the form β-Mp(Meb)-X-Gly-Arg(Tos)-Val-Val-Gly-Gly-Cya-Resin, where X = Pro or Gly, were treated with anhydrous HF for an hour at 0° . The peptides cleaved from the polymers by this treatment gave a positive Ellman test for thiols (2) and they were oxidized with potassium ferricyanide (3) to yield the corresponding intramolecular disulfide loop peptides. These were isolated by HPLC on a C-18 reversed phase column and characterized by amino acid analysis of acid hydrolysates. Hydrolysis of a sample of residual resin and amino acid analysis of the hydrolysate indicated that some peptide was left on the resin.

NO_2-Z-Arg(Tos)-Val-Gly-Gly-Cya-Resin was cleaved with anhydrous HF. The thiol-containing peptide (Ellman test) cleaved from the resin was taken up in aqueous solution. Part of the aqueous solution was aerated at pH 8.5. The symmetrical disulfide peptide precipitated from solution as it formed. The remaining, unaerated portion of the thiol-containing peptide solution was treated with chloracetylglycylglycine at about pH 9. NO_2-Z-Arg-Val-Gly-Gly-NH-CH_2-CH_2-S-CH_2-CO-Gly-Gly-OH was desalted on a cation exchanger and characterized by amino acid analysis of an acid hydrolysate.

Discussion

The few syntheses which have been carried out using a substituted S-benzyl linkage between peptide and polymer have pointed up several attractive features of the method and some aspects which require further modification and optimization. The synthetic mode in which cysteamine is attached to the resin through its thiol group provides a convenient solid phase route to peptides containing C-terminal cysteamine. Such peptides may be valuable for formation of cyclized pharmaceutical peptides (4) with loop closure achieved either through a disulfide bond or a thioether linkage. Peptide antigen sequences with C-terminal cysteamine residues should be very convenient molecules for coupling with carrier protein derivatives (e.g., maleimide-substituted keyhole limpet hemocyanin) in immunochemical screening work.

Solid phase syntheses on resins with cysteine residues anchored to the support through an S-benzyl linkage may have some interesting applications, especially for the bidirectional (5) solid phase synthesis of peptides.

Neither the conditions for attachment of thiols to the polymer nor the conditions for cleavage of the thioether linkage have been optimized. Although the substitution levels of polymer-bound cysteamine and cysteine methyl ester are adequate for solid phase synthesis of peptides, only a fraction of the chloromethyl groups alkylated a thiol group. The cleavage of the peptide-resin linkage is apparently not complete in one hour at 0° in anhydrous HF. This is a matter of little consequence when small amounts of simple peptides are being prepared, but the cleavage efficiency will have to be optimized for situations where yields of product are of the essence.

Acknowledgements

This work was supported by NIH Grants GM 18652 and AM 10080.

References

1. Kaiser, E., R.L. Colescott, C.D. Bossinger, and P.I. Cook. 1970. Anal. Biochem., 34, 595.

2. Ellman, G.L. 1959. Arch. Biochem. Biophys. 82, 70.

3. Hope, D.B., V.V.S. Murti, and V. du Vigneaud. 1962. J. Biol. Chem. 237, 1563.

4. Schiller, P.W. and J. DiMaio (1983) Proc. 8th Amer. Pept. Symp. (V.J. Hruby and D. H. Rich, Eds.) Pierce Chemical Company, Rockford, Illinois, p 269.

5. Glass, J.D., I.L. Schwartz, and R. Walter (1972) J. Amer. Chem. Soc. 94, 6209.

SOLID PHASE COUPLING OF PROTECTED PEPTIDE SEGMENTS. INFLUENCE OF THE POLYMER SUPPORT.

Fernando Albericio, Anna Porta, Enrique Pedroso, and Ernest Giralt

Departament de Química Orgànica, Facultat de Química, Universitat de Barcelona. 08028 Barcelona, Spain.

INTRODUCTION

Current reports from this laboratory provide a useful methodology for the preparation of peptides by a "convergent approach" (1,2). Convergent solid phase peptide synthesis involves: i) solid phase synthesis and purification of protected peptides on a solid support, ii) assembly of different protected peptides on a new solid support, and iii) cleavage of the full unprotected peptide and final purification.

In this paper, we describe some results concerning the influence of the polymer support on the coupling efficiencies of protected peptides. For this purpose, we have chosen five polymers which differ in chemical and physical constitution. Thus, in addition to common crosslinked polystyrene-1%-divinylbenzene (I), we have tested a more porous support (polystyrene-macroporous) (II), a non-crosslinked one (Kel-F-g-styrene) (III), a more polar one (polyacrylamide resin) (IV), and a rather rigid matrix (controled pore glass) (V) (3).

The Fmoc-(60-67) uteroglobin-OH protected segment was chosen as model compound for these studies. Kinetics of coupling of this peptide on H-(68-70)-uteroglobin previously assembled on each one of the polymeric supports were studied.

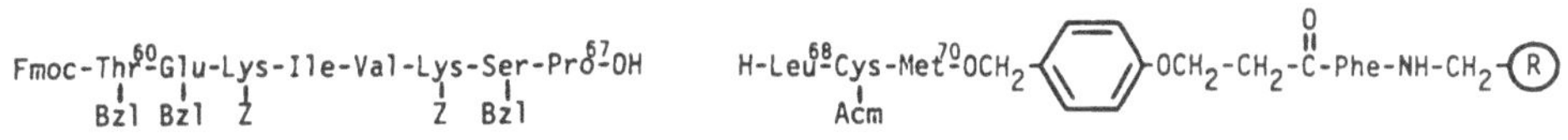

First at all, Merrifield's model tetrapeptide (leucylalanylglycylvaline) was synthesized on each support in order to check their feasibility for solid phase peptide synthesis.

RESULTS

Syntheses of model tetrapeptide

The approach for the synthesis of Leu-Ala-Gly-Val is based on the use of base-labile fluorenylmethoxycarbonyl group for α-amino protection combined with trifluoroacetic acid-labile p-alcoxybenzyl anchoring linkage. The first amino acid was incorporated to the support via preformed handle (4). The remaining amino acids (2.5 equiv.) were attached by single dicyclohexylcarbodiimide-coupling. Cleavage was effected by trifluoroacetic acid-methylene chloride (1:1) for 1 h at 25°C. In all cases, the ratio upon acid hydrolysis of amino acids remaining on the cleaved resin to Phe "internal reference" indicated that the yields of the cleavage were < 85% (table I). The purity of these tetrapeptides was judged by ion-exchange chromatography on an amino acid analyzer and shown to be 97-99%, except for the synthesis carried out on macroporous resin (table I).

Support	cleavage yield	LAGV	AGV	GV	LAV	LGV	LA(5)	LV
I	99%	99.07	0.08	0.12	0.32	0.10	0.31	-
II	98%	90.54	0.17	0.09	4.23	1.22	1.13	2.48
III	97%	97.98	0.03	0.22	0.71	0.17	0.85	0.04
IV	85%	98.94	0.14	0.13	0.49	0.08	0.22	-
V	95%	99.07	0.02	0.15	0.18	0.25	0.31	0.02

TABLA I

Synthesis and purification of Fmoc-(60-67) uteroglobin protected segment

The title peptide was synthesized on a photo-labile resin (Nbb-resin) (5), using HF-labile side chain protection and t-butoxy-carbonyl group for α-amino protection, except for the last amino acid where Fmoc group was used. The presence of this cromophore on the protected peptide allowed to monitor the purification by UV absorbance (280-300 nm). The crude product was shown to be about 85% pure on analytical reversed-phase

HPLC chromatography. The product was purified on a Merck Lobar Size A RP-8 column eluted with a convex gradient (dimethylformamide/acetonitrile/water/propionic acid mixtures) to provide pure protected peptide which was characterized by its amino acid analysis (Thr,0.92; Ser,0.95; Glu,0.98; Pro,1.02;Val,1.03; Ile,1.00; Lys,2.06.) , analytical HPLC and high field ^{1}H-NMR.

Syntheses of H-(68-70) uteroglobin resins

The different peptide-resins were synthesized using the approach described previously for the preparation of Merrifield's model tetrapeptide. Samples of peptidyl resins were subjected to trifluoroacetic acid/dichloromethane (1:1) treatments and the purity of protected peptides was <95% as judged by HPLC.

Coupling Experiments

H-(68-70)-uteroglobin resins (5 μmol) were stirred with Fmoc-(60-67)-uteroglobin-OH (1.25/2.5 equiv.) in 300 μl of dimethylformamide, at 0°C; hydroxybenzotriazole (1.25/2.5 equiv.) and dicyclohexylcarbodiimide (1.25/2.5 equiv.) in 50 μl of dimethylformamide each one were added and the mixture was stirred for 2h at 0°C and then at 25°C. The extension of the coupling was checked by amino acid analysis of the acid hydrolysate of samples of peptide resins removed at different times (table II).

Support	equiv. of peptide	2h	7h	24h	48h	72h
I	2.5	31%	100%			
I	1.25	8%	24%	56%	-	80%
II	2.5	6%	12%	22%	34%	37%
II	1.25	2%	4%	-	8%	17%
III	2.5	24%	49%	92%	100%	
III	1.25	3%	9%	15%	23%	35%
IV	2.5	24%	87%	100%		
IV	1.25	4%	13%	22%	32%	44%
V	2.5	18%	35%	41%	46%	66%
V	1.25	7%	21%	31%	53%	64%

TABLA II

DISCUSSION

Coupling of Fmoc-(60-67)-uteroglobin-OH protected segment on H-(68-70)-uteroglobin resins is always a rather slow reaction but in spite of the hydrophobicity and high density of protecting groups, quantitative coupling yields can be obtained when a sufficient excess, between 1.25 and 2.5 equiv., is used. Conventional microporous polystyrene-1%-divinylbenzene is the best choice for this model. Both polystyrene grafted on Kel-F and polyacrylamide allows the assembling of (60-70)-uteroglobin with quantitative yield but with lower rate. Controled pore glass and macroporous polystyrene are, in our hands and for this model less, suitable polymers for the coupling of protected peptide segments.

Acknowledgment

This work was supported by funds from CIRIT, CAICYT and US-Spain Joint Commitee for Scientific and Technological Cooperation. We thank ICI (Australia) and Rohm & Haas (Italy) for the gift of Kel-F-g-styrene and polystyrene-macroporous respectively. The excellent technical assistance of F. Cárdenas in running the amino-acid analyzer is gratefully acknowledged.

REFERENCES

1. Pedroso, E., A. Grandas, M.A. Saralegui, E. Giralt, C. Granier, and J. van Rietschoten. 1982. Tetrahedron 38, 1183.

2. Giralt, E.,R. Eritja, E. Pedroso, C. Granier, and J. van Rietschoten. 1986. Tetrahedron 42, 691.

3. Polystyrene-1%-divinylbenzene (Bio-Rad, USA), polystyrene-macroporous (Rohm & Hass, Italy), and Kel-F-g-styrene (ICI, Australia) were functionalyzed as aminomethyl according to Mitchell, A.R., S.B. Kent, M. Engelhard, and R.B. Merrifield. 1978. J. Org. Chem. 43, 2845. Polyacrylamide resin (Expansia, France) and controled pore glass (Pierce, USA) were used without further modification.

4. Albericio, F and G. Barany. 1985. Int. J. Peptide Protein Res. 26, 92.

5. Due to diketopiperazine formation during deprotection step of second residue. Pedroso, E., A. Grandas, X. de las Heras, R. Eritja, and E. Giralt. 1986.Tetrahedron Lett.27, 743.

6. Giralt, E., F. Albericio, E. Pedroso, C. Granier, and J. van Rietschoten. 1982. Tetrahedron 38, 1193.

ATTEMPTS AT THE SYNTHESIS OF FOLLICLE STIMULATING HORMONE RELEASING PROTEIN (FRP)

J. Rivier, C. Hoeger, R. Galyean

The Clayton Foundation Laboratories for Peptide Biology, The Salk Institute, 10010 N. Torrey Pines Road, La Jolla, California 92037

B. Penke

Institute of Medical Chemistry, Medical University of Szeged, H-6720 Szeged, Dom Ter 8, Hungary

INTRODUCTION

In the process of isolating inhibin from porcine follicular fluid (1), we discovered a FSH releasing activity which we isolated and characterized (2). This FSH releasing protein (FRP) was found to be the dimer of the β_A chain of inhibin (3) the monomeric sequence of which is given in Table 1. In the absence of an abundant natural source for FRP, a synthetic preparation with the biological properties of FRP would be extremely useful for our understanding of the biological role of FRP and for the development of RIA's which would be specific for FRP and not read inhibin. Finally, if pure enough, one could use such a preparation for the determination of FRP's secondary structure. We report here on the strategy which we followed to build the 116 protected peptide on a Pam-resin.

RESULTS AND DISCUSSION

The Pam resin was used because of its greater stability toward acids (4). In order to minimize the number of chemical steps to which the growing polymer would be exposed, we generated a series of dipeptides (with the general sequence X-Pro or X-Gly) by coupling of the respective Boc amino acid hydroxysuccinimide esters onto Pro or Gly according to the general procedure of Anderson et al. (5) (Tables 2, 3). Boc-(OBzl)Ser-OCH$_2$-Pam resin was obtained by a modification of the procedure of Mitchell et al. (4). We prepared an aminomethyl resin with medium substitution (0.38 mmol NH$_2$/g) to which Boc(OBenzyl)-seryl-(4-oxymethyl)phenylacetic acid was coupled (DCCI);

Peptides 1986

the latter was prepared by condensation of Boc-(O-benzyl)serine with 4-(Bromomethyl)phenylacetic acid phenacyl ester in the presence of KF (6) followed by zinc/acetic acid reduction (4). Because of our earlier experience with non-predictable, sequence dependent rate of condensation, we chose manual synthesis and monitoring by the Kaiser test (7) to gain an appreciation of the extent of coupling. This sequence was assembled using α-Boc protection and TFA (1% ethanedithiol) for deprotection; the following protecting groups were used: Cys(SpOMeBzl), Lys(2ClZ), Ser(OBzl), Thr(OBzl), Tyr(2BrZ), Asp(β-cyclohexyl), His(Tosyl), Arg(Tosyl), Glu(Bzl). While the individual amino acids were coupled (average of 1 h) using equimolar amounts of the amino acid and diisopropylcarbodiimide (DIPC), the dipeptides (see Table 3), either as the free acids or their corresponding dicyclohexylamine salts, were coupled using Benzotriazoyloxytri(dimethylamino)phosphonium hexafluorophosphate (Bop) in the presence of three equivalents of diisopropylethylamine for 10 to 15 min (8). This coupling procedure was also used for residues after regular DIPC coupling had either failed or partially failed. Synthesis was carried out around the clock. The coupling of dipeptides of the type X-Pro or X-Gly had advantages: it reduced the number of coupling steps while assuring the introduction of equivalent amounts of both amino acids on the growing chains (9-10); in addition dipeptides usually couple 6 to 15 times faster than the corresponding monomers (11). Finally, using the X-Pro dipeptides eliminated the need to assess the extent of coupling onto proline which is notorious for its weak positive ninhydrin test. The known limitation for the wider use of coupling of small peptides in SPPS is the possibility of racemization of the C-terminal amino acid, hence our decision to only use X-Pro or X-Gly dipeptides. In this manner we were able to assemble the protected peptide resin in less than two weeks. The speed at which the peptide resin was assembled reflected our concern for side reaction or structural rearrangements that may occur by prolonged sitting of the resin in either solvents or reactive media (12). The final peptide resin obtained (27.1 g) had a satisfactory amino substitution level when monitored by the Kaiser test (7). Peptide hydrolyses and subsequent amino acid analyses were performed on several preparations of crude, HF cleaved material. Based on the average values obtained for Gly(10.0/10.0), Ala(3.6/4.0), Phe(4.7/5.0), and Lys(8.5/8.0), which in our experience are relatively stable to the hydrolysis conditions, we feel that the crude peptide preparation may contain monomeric FRP. An oxidation procedure which would avoid problems associated with aggregation is being sought.

TABLE 1. Sequence of the βA Chain of Porcine/Human Inhibin or Monomeric Form of FRP.

Gly-Leu-Glu-Cys-Asp-Gly-Lys-Val-Asn-Ile-Cys-Cys-Lys-Lys-Gln-Phe-Phe-Val-Ser
-Phe-Lys-Asp-Ile-Gly-Trp-Asn-Asp-Trp-Ile-Ile-Ala-Pro-Ser-Gly-Tyr-His-Ala-Asn
-Tyr-Cys-Glu-Gly-Glu-Cys-Pro-Ser-His-Ile-Ala-Gly-Thr-Ser-Gly-Ser-Ser-Leu-Ser
-Phe-His-Ser-Thr-Val-Ile-Asn-His-Tyr-Arg-Met-Arg-Gly-His-Ser-Pro-Phe-Ala-Asn
-Leu-Lys-Ser-Cys-Cys-Val-Pro-Thr-Lys-Leu-Arg-Pro-Met-Ser-Met-Leu-Tyr-Tyr-Asp
-Asp-Gly-Gln-Asn-Ile-Ile-Lys-Lys-Asp-Ile-Gln-Asn-Met-Ile-Val-Glu-Glu-Cys-Gly
-Cys-Ser-OH

Residues that are underlined were introduced as dipeptides. Those residues with a dotted underline had to be recoupled.

TABLE 2. Boc-amino Acid N-hydroxysuccinimide Esters.

Amino Acid	Mp,°C	[α]D	C[a]	Solvent for Crystallization	Yield %
Boc-Asp(OcHex)-OSu	110-112	-21.7	1.0	ether	85
Boc-Arg(Tos)-OSu	120-133	-14.9	1.1	ether-hexane	95
Boc-Cys(pOCH$_3$-Bzl)-OSu	48-51	-24.4	1.1	ether	93
Boc-His(Tos)-OSu	118-120	- 9.0	1.0	ether-hexane	88
Boc-Tyr(2BrZ)-OSu	114-115	-31	1.1	ether	82

a) Specific rotation was measured in DMF.

TABLE 3. Characterization of Dipeptides Used in Solid Phase Coupling.

Protected Dipeptide	Mp,°C	[α]D	c	Solvent	Cryst. Solvent[e]	Yield %
Boc-Ala-Pro-OH (a)	152-55	-61.5	1.1	DMF	E	79
Boc-Asp(OcHex)-Gly-OH	148-49	- 5.0	1.0	50% HOAc	E-PE	82
Boc-Glu(OBzl)-Gly-OH (b)	111-13	- 5.0	1.0	"	PE	84
Boc-Ile-Gly-OH (c)	137-38	- 8.3	1.1	"	E	65
Boc-Ser(Bzl)-Gly-OH.DCHA	142-43	- 0.5	1.1	"	E	89
Boc-Val-Pro-OH (d)	145-47	-62	1.0	"	E-PE	83
Boc-Leu-Gly-OH	118-21	-19	1.0	"	E	84
Boc-Ser(Bzl)-Pro-OH.DCHA	108-114	-21	1.1	DMF	E-PE	92
Boc-Cys(pOCH$_3$-Bzl)-Gly-OH	93-96	-31	1.0	"	E-PE	85

a) Mp 155-56° ref 14; b) Mp 116°, [α]D = -8.4, c = 1 (EtOH) ref 13; c) Mp = 127-29° [α]D = -11.4, c = 1 (DMF) ref 15; d) Mp 143-44°, [α]D = -86.8°, c = 1 EtOH ref 16; e) E = ether; PE = petroleum ether; EtOAc = ethyl acetate

All dipeptides were checked by HPLC and were shown to be $\geq$ 97% pure. Amino acid analyses showed the correct ratio for each dipeptide.

Acknowledgement

Research was supported by NIH grants HD13527, AM26741 and contract N01-HD-2-2824. We thank J. Dykert for amino acid analysis and R. Hensley for manuscript preparation.

REFERENCES

1. Rivier, J., Spiess, J., McClintock,R., Vaughan, J., Vale, W. (1986) Biochem. Biophys. Res. Commun. 133, 120-127.

2. Vale, W., Rivier, J., Vaughan, J., McClintock, R., Corrigan, A., Woo, W., Karr, D., & Spiess, J. (1986) Nature 321, 776-779.

3. Mason, A.J., Hayflick, J.S., Ling, N., Esch, F., Ueno, N., Ying, S.-Y., Guillemin, R., Niall, H. & Seeburg, P.H. (1985) Nature 318, 659.

4. Mitchell, A.R., Kent, S.B.H., Engelhard, M., & Merrifield, R.B. (1978) J. Org. Chem. 43, 2845.

5. Anderson, G.W., Zimmerman, J.E. & Callahan, F.M. (1964) J. Am. Chem. Soc. 86, 1839.

6. Horiki, K., Igano, K., & Inouye, K. (1978) Chem. Lett. 165-168.

7. Kaiser, E., Colescott, R., Bossinger, C., & Cook, P. (1970) Anal. Biochem. 34, 595.

8. Le-Nguyen, D., & Rivier, J. (1986) Int'l. Journal of Peptides and Proteins 27, 285-292.

9. Sakakibara, S., Kishida, Y. Kikuchi, Y., Sakai, R. & Kakiuchi, K., (1968) Bull. Chem. Soc. Japan 41, 1273.

10. Gisin, B.F., Merrifield, R.B. & Tosteson, D.C. (1969) J. Am. Chem. Soc. 91, 2691.

11. Kovacs, J., in "The Peptides: Analysis, Synthesis, Biology." (E. Gross and J. Meienhofer, eds.) Acad. Press, New York 1980, Vol. 2 p 519.

12 Clark-Lewis, I., Aebersold, R. Ziltener, H., Schrader, J.W , Hood, L.E., & Kent, S.B.H. (1986) Science 231, 134-139.

13. Sugihara, T., Imanishi, Y., Higashimura, T. & Shimonishi, Y. (1976) J. Org. Chem. 41, 2584.

14. Wieland, T., Lapatsanis, L., Faesel, J. & Konz (1971) W. Annalen. Chem. 747, 194.

15. Neubert, V.K., Mansfeld, H.-W., Mansfeld, J. & Jakubke, H.D. (1977) J. Prakt. Chem. 319, 34.

16. Wuensch, E., Wendlberger, A. & Stocker, H. (1976) Hoppe Seyler's J. Physiol. Chem. 357, 459.

PEPTIDE CYCLIZATIONS ON POLYMERIC SUPPORTS -
SITE ISOLATION VERSUS SITE - SITE INTERACTION

M. Rothe, W. Fischer, K. Hornung, W. Taiber, G. Schmidtberg
Lehrstuhl Organische Chemie II, University of Ulm, 7900 Ulm/FRG

One of the most interesting problems of the chemistry on polymeric supports is the question how to achieve spatial separation of functional groups within the polymers. Accordingly, a steric isolation of reactive groups should be obtained using insoluble, cross-linked polymers with low substitution. Reactive polymer-bound molecules should then behave like at high dilution eliminating the possibility of mutual interaction within the rigid support.

In peptide chemistry, this principle has been applied to ring-closure reactions as an alternative to the cyclization in highly diluted solutions in order to avoid oligomerizations leading to large ring oligomers, but mainly with little success. The first cyclizations on polymeric supports (1) were supposed to yield only monomeric rings by an apparently complete site isolation of the activated peptide chains. Subsequent studies (2), however, showed extensive interaction of the sites leading to dimeric and oligomeric cyclopeptides.

Therefore we took up again the problem and investigated cyclization reactions as a function of the degree of cross-linking and structure of the polymeric support, the loading of the resin, and the swelling properties of the solvents as well as of the reaction time and temperature.

Polystyrenes with different cross-linking (1-20% DVB) and morphology (gel-type, macroporous, popcorn polymers) were used as supports. For anchoring the peptide chains 4-hydroxy-(phenylthiomethyl) groups (1) were introduced into the polymers. Esterification with N-protected peptides and subsequent oxidation of the thioether group with m-chloroperbenzoic acid (1,2) yielded polymeric active sulfonylphenyl esters. After cleavage of the protecting group by TFA and subsequent neutralisation by excess tert. amine of the protonated amino group formed, the free polymeric peptide active

Peptides 1986

ester cyclized spontaneously in suspension in swelling solvents. The rings formed were released into solution, whereas linear peptide oligomers remained fixed to the support.
Surprisingly, all cyclizations yielded mainly large rings with about 20 up to more than 80 ring atoms which were formed by cyclooligomerization in the course of multiple intrapolymeric reactions. They were separated by GPC and identified by MS. On the other hand, steric isolation of the grafted peptide chains leading to monomeric rings generally occurs only to a minor degree.
Among many investigated examples ring-closure of resin-bound trisarcosine yielded cyclic oligomers up to the nonameric tripeptide (27 Sar residues). A similar behaviour was found with Pro-Pro-Gly and Gly-Pro-Pro which form a homologous series of ring peptides containing up to 8 tripeptide residues. Even resin-bound amino acids (Gly, Ala) oligomerize on the resin. Apart from a large quantity of diketopiperazines, higher cyclopeptides containing 6-9 amino acid residues could be detected. In the case of the tetrapeptide partial sequence of valinomycin and false-retro-valinomycin (Ala-D-Hyv-D-Val-Hyv) complete series of cyclopeptides with up to 7 tetrapeptide residues are obtained. As expected, the yield of the rings always decreases with increasing chain length. The distribution of the ring sizes, however, also depends on the reaction conditions. Depending on solvent and temperature, the main product can be valinomycin (39%, toluene), or either the dimeric (15%, DMF) or tetrameric cyclotetrapeptide (23%, pyridine). Even the monomeric cyclotetrapeptide may be obtained in high yields (75%, methyl ethyl ketone). Such a preferred formation of specific ring sizes under different conditions is not possible in the case of cyclizations in solution.
Cyclooligomerization even occurs at low peptide loadings, provided that the polymer has sufficient flexibility to allow peptide residues from different polymeric chains to approach each other. In fact, this has been found even with highly cross-linked polymers, although the effect is surprisingly limited. Only the yield of the very large rings decreases with increasing cross-linking. Extensive intersite reactions even occur on macroporous resins, although to a somewhat smaller extent than on gel-type polymers. More important are the swelling properties of the solvent.
Generally, intrapolymeric reactions are favoured by good swelling

solvents and by decreasing cross-linking, especially at low temperatures and long reaction times, although dimeric to tetrameric rings are already formed during the first minutes of cyclization. On the other hand, steric isolation of reactive sites is favoured only to a low degree by poor swelling solvents and high cross-linking, but above all by short reaction times due to high temperatures and in particular to base catalysis.

The formation of macrocyclic rings by cyclooligomerization on polymeric supports corresponds in principle entirely to the results of ring-closure in solution. When we compared cyclizations on the resin and in solution carried out at approximately the same dilution, cyclopeptides with even larger ring sizes were obtained on the resin. Obviously, the specific microenvironment by the polymer particularly favours intermolecular reactions. Therefore it was suggested that cyclization and also oligomerization do actually not occur on the polymer itself, but in solution, after the pendant peptide chains have been released from the support by the added tert. amine. As is well known, peptide active esters are easily converted into oxazolones in the presence of tert. base. Oxazolones containing free amino groups are active peptide derivatives and can oligomerize as well as cyclize. Such a reaction could thus explain the formation of large peptide rings in solution (3).

To clarify this problem of cyclooligomerization by true intersite reactions on polymers we proceeded as follows. If two different supports each of which is loaded with a different amino acid are mixed and then subjected to cyclization, peptide formation between the amino acids on the two supports is not possible, because a mutual penetration of the polymer chains of different beads does not occur. Therefore it could be expected that exclusively rings containing a single amino acid would be formed by intrapolymeric reactions within one bead. On the contrary, the formation of soluble peptide oxazolones from both supports would lead to co-cyclopeptides containing both amino acids.

In fact, the cyclization of mixtures of a support no. I loaded with alanine and another support no. II loaded with proline led to more than 99% of homo-cyclopeptides, e.g. the cyclic di-, hexa-, and heptapeptide of alanine and the cyclic dipeptide of proline. Only less than 0.5% of the rings were co-cyclopeptides obtained

from alanine plus proline which could have been formed via soluble intermediates.
When the same cyclization was performed in the presence of imidazole as a strong nucleophile, corresponding to a three-phase test according to Rebek (4), only about 60% of the rings were obtained as homo-cyclopeptides, and as much as 40% as co-cyclopeptides containing alanine plus proline. In this case their formation can be explained by release of the peptide chain from the support in form of the imidazolide. After this active peptide derivative has diffused through the solution into the second support, it can be trapped by its amino acid. Subsequent cyclization can then lead to co-cyclopeptides.
Thus it has been proved without any doubt that the normal cyclo-oligomerization on the support (without additional imidazole) does really proceed as a true intrapolymeric reaction by site-site interaction and not after release of the peptide chain into solution.
Our results show that cyclizations on polymeric supports are generally not suitable for producing small monomeric rings. On the contrary, the polymeric supports known up to now are not rigid enough, even in case of high cross-linking, to allow a true site separation of functional groups, unless the reactions concerned proceed at a higher speed than that corresponding to the chain mobility of the polymer.

This paper is dedicated to Prof. L. Horner on the occasion of his 75th birthday. We gratefully acknowledge financial support for this work by a grant from the Fonds der Chemischen Industrie.

References

1. Fridkin, M., A. Patchornik, E. Katchalski. 1965. J. Amer. Chem. Soc. 87, 4646; Flannigan, E., G.R. Marshall. 1970. Tetrahedron Lett. 506.

2. Rothe, M., A. Sander, E. Fischer, W. Mästle, B. Nelson. 1977. In: Peptides - Proc. 5th Amer. Peptide Symposium (M. Goodman, J. Meienhofer, eds.). Wiley, New York, p. 506.

3. Mitchell, A.R., R.B. Merrifield. 1985. In: Peptides, Proc. 9th Amer. Peptide Symposium (C.M. Deber, V.J. Hruby, K.D. Kopple, eds.). Pierce Chem. Co., Rockford, Ill., p. 289.

4. Rebek, J., 1979. Tetrahedron 35, 723.

SOLID PHASE PEPTIDE SYNTHESIS WITH FMOC-AMINO ACIDS : AN ALTERNATIVE PROTOCOL.

J. Van Rietschoten, P. Fourquet, J.M. Sabatier and C. Granier.

CNRS UA 1179, INSERM U 172, Laboratoire de Biochimie, Faculté de Médecine Secteur Nord, 13326 Marseille Cedex 15, France.

Introduction

For the solid phase synthesis of protected fragments, use of the 9-fluorenylmethyloxycarbonyl (Fmoc) group (1) seemed a promising strategy when combined with benzyl type (HF labile) side chain protection and p-alkoxybenzyl ester peptide-resin (Wang type (2)) linkage (3). It has become true, several protected segments (up to the size of fourteen residues) have thus been obtained, but, going back a few years, the Fmoc methodology was not so easy to handle as it seemed from the first successes (4-6). Our early trials with this strategy (3) pointed to several difficulties, some of them linked to the use of the Fmoc group itself and some others due to the combination of the Fmoc group with benzyl type side chain protection. Enumeration of them may help to understand why the syntheses with the Fmoc group were not always perfect and will explain why our synthesis protocol has been adapted to the one that is described furtherway.

Results and discussion

The first side reactions were shown to occur during the synthesis of the Fmoc-amino acids with Fmoc-chloride, ie, mainly a polycondensation of the free amino acids in presence of this strong activating agent (7). The dipeptides formed are difficult to remove. Recommendation has been made (7) to use Fmoc-azide as acylating agent. The very long reaction time described in that paper is not really necessary. It can be shortened to two hours without noticable loss of yield (personnal communication by F. Albericio, Departament de Quimica Organica, Facultat de Quimica, Barcelona, Spain). Fmoc-succinimidyl ester has also be recommended (8, 9) and is now commonly used.

Peptides 1986

Another difficulty was the incorporation of the first Fmoc-amino acid on the Wang resin which, in the presence of dimethylaminopyridine yields up to several per cent of dipeptide on the resin (5, 6, 3). Pedroso et al. (10) adjusted the conditions to obtain a good compromise between yield of incorporation (30-50 %) and minimal amount of dipeptide formed on the resin (<0.3%). Van Nispen et al. (11) found that addition of HOBt improved quality of this reaction by suppression of racemisation. Atherton et al. use biphenyloxycarbonyl N-α protection for this first C- terminal amino acid (6) and Albericio and Barany propose use of N,N-dimethylformamide dineopentyl acetal for coupling of the Fmoc-amino acid to an alkoxybenzoyl handle (12). Colombo et al. (13) propose use of a chloromethylphenoxy resin.

Having encountered an important loss of peptide after coupling a third amino acid on a Pro-Gly-O-Resin sequence (3) it was useful to investigate how much the Fmoc deprotection conditions promote diketopiperazine formation. Pedroso et al. (14) found that use of 50 % piperidine in DMF favours cyclisation almost as much as does 20 % piperidine in DMF but deprotection rate is greater. Recommendation would then be rather a short deprotection reaction time with higher concentration of piperidine than reverse.

Use of benzyl type side chain protection in combination with Fmoc N-α protection may lead also to difficulties, especially with aspartic acid and glutamic acid which, in the usual deprotection conditions, was found to cyclise into pyroglutamic acid up to an extent of 30 %. All these nucleophilic side reactions that occur during the deprotection step prompted us to select the most efficient protocol for complete removal of the Fmoc group. As the time for 95 % deprotection was found to be close to 30 seconds with 50 % piperidine in DMF (14), a three times one minute reaction seemed sufficient to achieve completeness (>99,99%) and thus minimize the side reactions. In this way benzyl-glutamic acid did not cyclise in a measurable way and deprotection of benzyl-aspartic was limited to an acceptable minimum after five cycles.

Use of symmetrical anhydrides in DMF of the Fmoc-amino acids requires that DMF be very clean to avoid Fmoc deprotection and autocondensation reactions. Atherton et al. for some time recommended use of dimethylacetamide instead of DMF (6). DMF can be cleaned up to a great extent by 48 hrs storage over molecular sieve and by overnight nitrogen bubbling. But we felt it worth trying coupling reaction devoid of DMF and in weaker acylating conditions than symmetrical anhydride.

All these considerations prompted us to try a modified protocol on the "Wang type" polystyrene resin (P/S) for the synthesis of both free and protected peptides, the sequences of some of them being shown below.

```
               Boc tBu       tBu         tBu
                |   |         |           |
I      Fmoc-Val-Lys-Asp-Gly-Tyr-Ile-Val-Asp-O—[ Wang P/S ]

                 +  tBu       tBu         tBu
                     |         |           |
II     Fmoc-Val-Arg-Asp-Ala-Tyr-Ile-Ala-Asp-O—[ Wang P/S ]

            Chex          Bzl Bzl         Acm Bzl Chex
             |             |   |           |   |   |
III    Fmoc-Tyr-Ile-Val-Asp-Asp-Val-Asn-Cys-Thr-Tyr-
                                    Acm
                                     |
                                -Phe-Cys-Gly-O—[ Wang P/S ]

            Tos          Chex Acm        Bzl Bzl Acm Bzl
             |             |   |          |   |   |   |
IV     Fmoc-Arg-Asn-Ala-Tyr-Cys-Asn-Glu-Glu-Cys-Thr-
                             Z       Z
                             |       |
                            -Lys-Leu-Lys-Gly-O—[ Wang P/S ]
```

The protocol was as follows : DMF (4X1') ; 50 % piperidine in DMF (3x1') ; DMF (4x1') ; if arginine is in the peptide and has been introduced protected just by ionisation : 1 Eq HOBt/DMF (2') ; CH_2Cl_2 (4x1') ; addition of Fmoc-amino acid (2.5 eq) as a solid followed by mixing with a minimum of CH_2Cl_2 (5-10') ; DCC (2.5eq) in CH_2Cl_2 ; CH_2Cl_2 (4x1').

Controls have shown that the Fmoc group is practically quantitatively removed in the first piperidine/DMF treatment. By this protocol it is not always possible to accurately measure the amount of fluorenyl-piperidine formed during deprotection of the peptide because some insoluble Fmoc-amino acids may remain after CH_2Cl_2 and DMF washes. In that case, the dosage show a high value. For sure the peptidyl resin is then cleaned up by the piperidine/DMF treatments before next coupling. Coupling in just methylene chloride occured very rapidly and often completely as juged by the ninhydrin test after 5 to 15 minutes of reaction eventhough the starting Fmoc-amino acid is very insoluble in CH_2Cl_2. In fact we found that for peptide IV it was the only way to couple Fmoc-Arg(Tos)-OH. For synthesis of free peptides, arginine was protected by ionisation with HOBt just before coupling.

After cleavage with TFA the two peptides I and II were found very clean by HPLC of the crude material (>90%). Protected peptides III and IV were purified by semi preparative RPC_{18} HPLC and showed a single peak on analytical HPLC.

This protocole is very much shortened (approximately 60 min per cyle), especially, compared to an original one that was set up in the early days of use of Fmoc-amino acids for solid phase peptide synthesis (4), and is cheap in Fmoc-amino acids.

Acknowledgements : especially to the team of Drs E. Giralt and E. Pedroso from the Facultat de Quimica, Barcelona, for the many discussions and exchanges of information.

References

1. Carpino, L.A., and Han, G.Y. (1972) J. Org. Chem., 37, 3404-3409
2. Wang, S.S. (1973) J. Am. Chem. Soc., 95, 1328-1333.
3. Pedroso, E., A. Grandas, M.A. Saralegui, E. Giralt, C. Granier and J. Van Rietschoten (1982) Tetrahedron, 38, 1183-1192.
4. Chang, C.D. and J. Meienhofer, (1978) Int. J. Peptide Protein. Res., 11, 246-249.
5. Chang, C.D., A.M. Felix, M.H. Jimenez and J. Meienhofer (1980) Int. J. Peptide Protein Res., 15, 485-494.
6. Atherton, E., C.J. Logan and R.C. Sheppard (1981) J. Chem. Soc. Perkin I, 538-546.
7. Tessier, M., F. Albericio, E. Pedroso, A. Grandas, R. Eritja, E. Giralt, C. Granier and J. Van Rietschoten (1983) Int. J. Peptide Protein Res., 22, 125-128.
8. Sigler, G.F., W.D. Fuller, R.C. Chatuverdi, M. Goodman and M. Verlander (1983) Biopolymers, 22, 2157-2162.
9. Ten Kortenaar, P.B.W., B.G. Van Dijk, J. Marjolijn Peeters, B.J. Raaben, P.J. Hans, M. Adams and G.I. Tesser (1986) Int. J. Peptide Protein Res., 27, 398-400.
10. Pedroso, E., A. Grandas, R. Eritja and E. Giralt (1983) In: Peptides 1982. Proceedings of seventeeth European Peptide Symposium (Blaha, K. and P. Malon, eds) pp. 237-240, de Gruyter, Berlin.
11. Van Nispen, J.W., J.P. Polderdijk and H.M. Greven (1985) Rec. Trav. Chim. Pays-Bas, 104, 99-100.
12. Albericio, F. and G. Barany (1984) Int. J. Peptide Protein Res., 23, 342-349.
13. Colombo, R., E. Atherton, R.C. Sheppard and V. Wooley (1983) Int. J. Peptide Protein Res., 21, 118-126.
14. Pedroso, E., A. Grandas, X. de las Heras, R. Eritja and E. Giralt (1986) Tetrahedron Lett., 27, 743-746.

USE OF PROTEASES FOR THE SYNTHESIS OF LH-RH

A.J. Andersen, F. Widmer and J.T. Johansen
Carlsberg Biotechnology Ltd., Tagensvej 16, DK-2200 Copenhagen N

Introduction

The practical, preparative value of using proteases as reagents in peptide synthesis has been examined by several groups in recent years (reviewed in ref. 1-3) and has so far resulted in the synthesis for aspartame (4), enkephalin (5), fragments of epidermal growth factor (6,7) and oxytocin (8,9).
We have been working on a synthesis for LH-RH for some time. This decapeptide (pGluHisTrpSerTyrGlyLeuArgProGlyNH_2) and its analogs are increasingly used for human and veterinary purposes (10). It was thus an interesting peptide on which to test the enzymatic methodology. We report here some of our results, emphasizing concept and strategy, as well as scope and limitations, rather than experimental detail.

Materials and Methods

The enzymes were commercial preparations from the following sources: Sigma Chemicals, St. Louis, USA *(CT;T;P;Cl.P); Carlsberg Biotechnology Ltd., Copenhagen, Denmark (CPD-Y); Seikagaku Kogyo Ltd., Tokyo, Japan (PPSE, from Flavobact. meningosepticum).
The small-scale optimization (in terms of pH, enzyme and reactant concentrations)of the reactions, and the actual preparative synthesis were monitored by HPLC as described earlier (6,11). Products were separated by reverse-phase- and ion-exchange chromatography.

---

*<u>Abbreviations</u>: LH-RH, luteinizing hormone releasing hormone, CT, chymotrypsin; T, trypsin; P, papain; CPD-Y, carboxypeptidase Y; PPSE,post-proline specific endoprotease; Cl.P, chlostripain.

Peptides 1986

They were isolated as solids by either solvent precipitation or lyophilization, and analyzed and characterized by standard methods.

Results and Discussion

The coupling strategy we report here involves the synthesis of the N- and C-th pentapeptide ester and amide, which were then joined with chymotrysin to give the final decapeptide amide.

The N-th pentapeptide ester

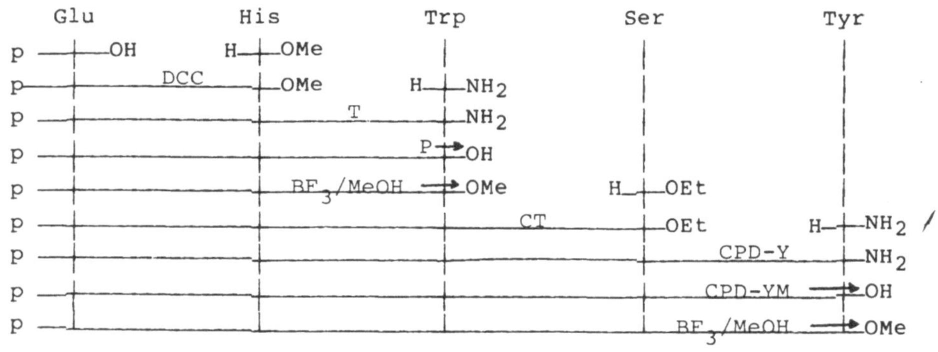

The N-th pGluHisOMe is best made chemically as shown. $HisNH_2$ can be coupled to e.g. pGluOMe, but the ester is such a poor substrate that the product, $pGluHisNH_2$, is partially hydrolyzed (and a second $HisHN_2$ coupled in low yield by transpeptidation) before all pGluOMe is converted. This situation ("overreaction") occasionally occurs when using proteases, and chemical coupling may then be preferred over separation of the complex reaction mixture resulting from an uncontrolled enzymatic reaction. The next step, tryptic coupling of $TrpNH_2$ at His(12) proceeds well. It was not possible to couple the TrpOMe - which would be the ideal strategy (6,13) - since the yield was low and the product ester, pGluHisTrpOMe, was partially hydrolyzed.

After deamidation with P, followed by BF_3/MeOH esterfication, Ser-

OEt is directly coupled with CT, a reaction that proceeds well with most typical CT substrates (12). The final coupling of $TyrNH_2$ with CPD-Y proceeds at very high yield (>95%) and is thus preferred over a direct ester coupling at lower yield, particularly because subsequent deamidation using a modified CPD-Y (CPD-YM, cf.ref.14) is possible without significant loss of the tyrosine residue from the peptide.

It is also possible to couple SerOEt and $TyrNH_2$ in a "one-pot" double reaction without isolation of the intermediates, i.e., after coupling of SerOEt with CT, $TyrNH_2$ and CPD-Y are added directly to the reaction mixture. Little or no coupling of excess SerOEt by CPD-Y will occur, because $TyrNH_2$ is a much better nucleophile for this enzyme than SerOEt (15).

The C-th Pentapeptide Amide

The reaction scheme shows the "ideal" strategy (6,13) for this peptide in terms of the number of reaction steps and cost of starting materials.

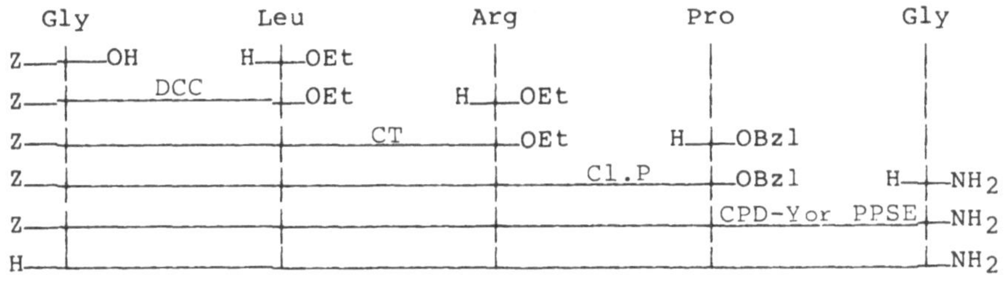

However, this approach is not (yet) practically feasable for large scale synthesis, as discussed below.

The N-th dipeptide of this fragment can of course be made chemically as shown. A clean papain coupling between ZGlyOBzl and LeuOEt is also possible (13,16) because papain has a primary specificity for hydrophobic amino acids in S_2(17). In a substrate such as ZGlyOBzl, the hydrophobic Z-group presumably occupies the S_2-subsite, resulting in acylation, and subsequent aminolysis, at the glycine carbonyl group. This is a very useful reaction (13,16), but here not necessary, because glycine is achiral, and leucine poses no chemical difficulties.

Direct coupling of ArgOEt with CT is possible at an acceptable yield (70%). However, probably due to residual T in CT, some over-reaction, i.e. product hydrolysis and partial coupling of a second ArgOEt, is difficult to avoid. To obtain a clean reaction here, a highly purified CT may be needed. A high-yield (80%) CPD-Y coupling of free Arg to get ZGlyLeuArgOH, followed by HCl/EtOH esterfication, is therefore preferred.

Andersen described recently (18) that Pro-derivatives and -peptides can be coupled to Nα-acyl-arginine esters by Cl.P. This was confirmed here with ProOBzl and $ProGlyNH_2$. However, the yields were low and the enzyme is expensive and its supply uncertain. Clearly, a better supply situation is required before this coupling step can become a routine preparative method.

The final coupling of $GlyNH_2$ to ZGlyLeuArgProOBzl by either CPD-Y or PPSE proceeds well. The yield is better with CPD-Y, which generally couples amino acid amides of high yields (11), but faster with PPSE.An alternative strategy for this peptide would be to make R_1-ArgProOR_2 chemically, add $GlyNH_2$ enzymatically, deprotect and couple $ArgProGlyNH_2$ to ZGlyLeuOEt with CT; in this step, any diastereomeric D-$ArgProGlyNH_2$ would be eliminated as was described for other peptides (19).

Fragment Coupling

This reaction repesents one of the best enzymatic fragment couplings we have observed to date, and proceeds in 95% yield at the conditions shown below:

pGluHisTrpSerTyrOMe + H-$GlyLeuArgProGlyNH_2$
(0,10 M) (0,12 M)
↓ CT(2μM);pH8,5;40%DMF
pGluHisTrpSerTyr$GlyLeuArgProGlyNH_2$

The high yield is due to the high affinity of CT in its S' binding sites for this particular N-component; CT apparently has a site for hydrophobic amino acids in S_2'. Information to that effect was also obtained from model studies with $GlyLeuNH_2$ and $LeuGlyNH_2$ with a number of different substrates. The yield with $GlyLeuNH_2$ was consistently higher.

Summary and Conclusions

The various enzyme reactions described here clearly show some attractive features of enzymatic coupling, e.g. racemization-free coupling of $TrpNH_2$ to unprotected pGluHisOMe, and the fine, high-yield coupling of the two fragments.Another great advantage of enzymic coupling, not mentioned above, is the fact that one does not always need absolutely pure intermediates. For example: some residual pGluHisTrpOH in pGluHisTrpSerOEt will not interfere in CPD-Y coupling of $TyrNH_2$, because CPD-Y will not react - on the time scale of the ester aminolysis reaction, anyhow - at pGluHis-TrpOH. In other words, this impurity is reaction-inert, and can be purified off at a later, more convenient, i.e. cost-effective, stage in the synthesis.
However, the examples shown also clearly point out some current weaknesses of enzymatic peptide synthesis, i.e. the fact that the preferred strategy of simple C-th extension of the peptide chain by amine acid esters (13) is often not possible in praxis, because the appropriate enzymes - although often known, and available for sequencing purposes - are too expensive at the moment for preparative synthetic purposes. Further, available and "cheap" enzymes - while useful for many purposes - are not always pure enough for coupling of certain sequences, the result being overreaction, i.e. product hydrolysis, or worse, intrachain peptide bond cleavage. Nevertheless, as the demand for peptides increases in general, and as it is realized more widely that the enzymatic methodology makes possible the synthesis of certain peptides at a purity, quantity and price not previously expected, the demand for suitable enzymes will increase and hopefully result in a better supply situation.

References

1. Chaiken, I.M., Komoriya, A., Ohno, M. and Widmer, F. Appl. Biochem.Biotechnol.7 (1982) 385.

2. Jakubke,H.D.,Kuhl,P.andKönnecke,A.,Angew.Chem.Int.Ed.24,85(1985)

3. Kullman, W., J. Prot. Chem. 2, 289 (1983)

4. Oyama, K. and Kihara, K., Chemtech. p.100 (1984)

5. Widmer, F., Breddam, K. and Johansen, J.T., Peptides 1980 (ed. K. Brunfeldt), p.46. Scriptor, Copenhagen (1981).

6. Widmer,F., Bayne,S., Houen,G., Moss,B., Rigby,R., Whittaker,R. and Johansen,J.T., in: Peptides 1984 (ed. U.Ragnarsson) p.193. Almqvist, Stokholm (1984)

7. Widmer,F., Bayne,S., Houen,G., Moss,B., Rigby,R., Whittaker,R.- and Johansen,J.T. in: Forum Peptides Le Cap d'agde 1984,(ed.B. Castro, J.Martinez)p.439. Montpellier (1986)

8. Cerovsky, V. and Jost, K., Coll. Czech. Commun. 50 2775 (1985)

9. Thorbek, P., Lauridsen, J., Johansen, J.T. and Widmer, F., in preparation.

10. Labrie, F., Belanger, A., and Dupont, A.: LH-RH and its Analogues - Basic and Clinical Aspects. Elsevier, Amsterdam(1985).

11. Widmer,F.and Johansen,J.T., Carlsberg Res. Commun. 44,37 (1979)

12. Widmer, F., Ohno, M., Smith, M., Nelson, N., Anfinsen, C.B., in: Peptides 1982 (ed. K. Bláha, P. Malon), p.375. Walter de Gruyter, Berlin (1983)

13. Widmer, F. and Johansen, J.T. in: Synthetic Peptides in Biology and Medicine (ed. K. Alitalo, P. Partanen, A.Vaheri) p. 79. Elsevies, Science Publisher, Amsterdam (1985)

14. Breddam, K., Carlsberg Res. Commun. 49, 535 (1984)

15. Widmer, F., Breddam, K. and Johansen, J.T., Carlsberg Res. Commun 45, 453 (1980)

16. Mitin, Yu.V., Zapevalova, N.P. and Gorbunova E.Yu., Int. J. Peptide Protein Res., 23, 528 (1984)

17. Johansen, J.T. and Ottesen, M., Compt. Rend. Trav. Lab. Carlsberg 36, 265 (1968)

18. Andersen,A.J.in: Peptides: Structure and Function(ed.C.M.Deber, V.J.Hruby,K.D.Kopple) p. 355. Pierce Chemical Company (1985)

19. Thorbek, P. and Widmer, F. in: Peptides: Structure and Function (ed. C.M. Deber, V.J. Hruby, K.D. Kopple) p. 359. Pierce (1985)

BIOSYNTHESIS OF HUMAN INSULIN IN YEAST VIA SINGLE-CHAIN PRECURSORS.

Jan Markussen, Ulrik Damgaard, Ivan Diers, Niels Fiil, Mogens Trier Hansen, Per Larsen, Fanny Norris, Kjeld Norris, Ole Schou, Leo Snel, Lars Thim and Hans Ole Voigt.

Novo Research Institute, Novo Alle, DK-2880 Bagsvaerd, Denmark.

Introduction

Biosynthetic insulin has hitherto been synthesized in E. coli either via separate A- and B-chains (1) or via scrambled proinsulin B-C-A-chain (2), fused with leader sequences as chimeric proteins and precipitated in the cytoplasm. We wish to report an alternative method of manufacture of human insulin in which post fermentation chemistry has been minimized. Single-chain insulin precursors without pairs of basic amino acid residues in the chain are exported to the broth with free B-chain N-terminal phenylalanine and with correct disulfide bridges. A tryptic transpeptidation in organic-aquoeus medium and in the presence of a threonine ester converts the precursor to the double-chain Thr-OR^{B30} human insulin, Fig. 1, similar to the conversion of porcine insulin to human insulin ester (3,4). Cleavage of the ester bond completes the post fermentation chemical reactions.

Des(B30) single-chain insulin (I) was prepared earlier by tryptic transpeptidation of porcine insulin in the absence of a threonine ester (5). The molecule is recovered in higher yields after a redox cycle than porcine proinsulin, demonstrating that an increased efficacy in folding is obtained by shortening the chain connecting link by 34 amino acids (6). The transpeptidation of precursor I is a very slow reaction, possible due to steric hindrance in the tightly knotted molecule. Two analogues (II & III) having 2 and 3 additional amino acids in the link were expressed in yeast. The kinetics of transpeptidation of precursors I, II & III at conditions optimized with respect to yields, an outline of the manufacture and the properties of the product is reported.

(I)	SCI	:	B(1-29)-A(1-21) insulin
(II)	SCI-SK	:	B(1-29)-Ser-Lys-A(1-21) insulin
(III)	SCI-AAK	:	B(1-29)-Ala-Ala-Lys-A(1-21) insulin

Peptides 1986

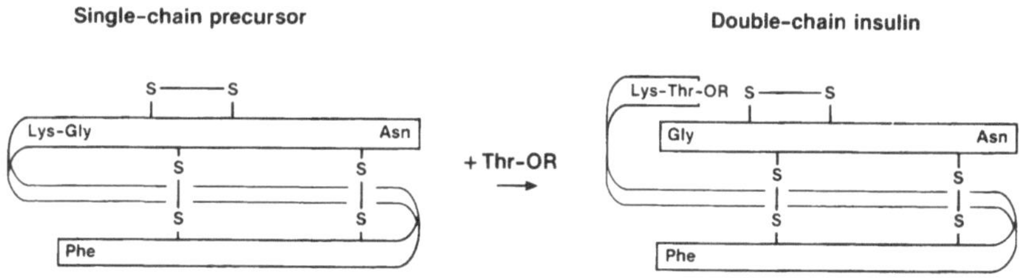

Fig. 1. Transpeptidation of a single-chain precursor to a double-chain human insulin ester by trypsin.

Methods.

The genes used for expression of the single-chain precursors consist of the triose phosphate promotor followed by the mating factor α1 (MFα1) signal-leader sequence joining the single chain precursors via Lys-Arg codons, and ending in the TPI terminator. The α-factor, a yeast mating pheromone, is a tridecapeptide. It is exported by its signal-leader sequence of 89 amino acids ending in Lys-Arg-$(Glu\text{-}Ala)_2$ (7). At least two enzymes are involved in the maturation of the α-factor, a trypsin like protease with specificity for pairs of basic amino acids (7) and a dipeptidyl aminopeptidase with specificity for Glu/Asp-Ala sequences (8). The use of the entire α-signal-leader sequences resulted in secretion of unmaturated precursors having one or two Glu-Ala dipeptides attached to Phe^{B1} (9), indicating that the dipeptidyl aminopeptidase is a limiting factor. Consequently the gene segment encoding $(Glu\text{-}Ala)_2$ were deleted.

The plasmids used for expression of the precursors further comprise the 2μ origin of replication, an ampicillin resistence gene and the Schizosaccharomyces pombe gene for TPI (POT). The latter provides a strong selection for the plasmids because the host is a strain of Saccharomyces cerevisiae carrying deletions in its TPI gene (10). The yeast is grown in continuous fermentation and the precursor is isolated from the broth after removal of cells. The flow sheet is shown in Fig. 2. The final purification is preparative HPLC. Transpeptidation experiments were conducted at 12°C in 21% H_2O in DMF, see Table 1. Analyses of conversion yields were performed by HPLC; the single-chain precursors elute earlier than both human insulin and its esters (3,5).

The purity of the product was determined by HPLC, Fig. 3. The identity of the product with human insulin (ss) was confirmed by amino acid sequence analysis and by peptide mapping after digestion with subtilisin A, Fig. 4.

FLOW	PROCESS	PURITY "Yeast protein" ppm	PURITY HPLC %
↓	Continuous fermentation of yeast strain		
	Centrifugation, cells → waste		
	Adsorption of SCI to ion exchange columns		
	Desorption		
	Crystallisation of insulin precursor	10.000-20.000	~ 90
	Anion exchange chromatography		
	Crystallisation of insulin precursor	< 10	~ 97
	Transpeptidation to human insulin ester		
	Gel filtration at low pH, trypsin → waste		
	Anion exchange chromatography, DAI + SCI → waste		
	Cleavage of ester bond		
	Preparative HPLC, HI-OR → waste		
	Monocomponent human insulin (ge)	< 1	> 99.5

Fig. 2. Flow sheet for production, from continous fermentation to monocomponent human insulin (ge), showing the increasing purity by HPLC and the vanishing of yeast protein contaminants as measured by a yeast protein ELISA.

An ELISA for yeast proteins were developed using total protein from the fermentation broth to immunize rabbits and guinea pigs. The specificities of the sera were evaluated by immunoblotting of total protein separated by SDS-PAGE, and for each species of animal the serum with the broadest specificity was chosen. The rabbit IgG was used as the catching antibody, total yeast protein as the standard and guinea pig immune serum on top. Binding was detected with horseradish peroxidase conjugated anti-guinea pig immunoglobulin serum and o-phenylenediamine as the substrate. The detection limit is 1 ng/ml of yeast protein, Fig. 5.

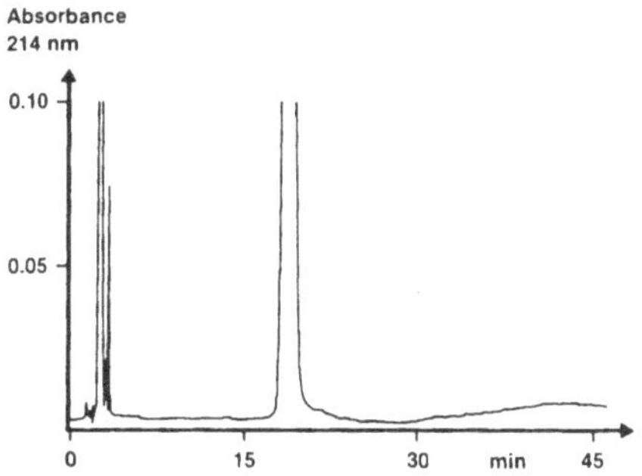

Fig. 3. HPLC chromatogram of human insulin (ge).
Column: Lichrosorb RP-18, 5 µ, 4x250 mm. Oven temp. 50°C. Flow 1 ml/min. Solvent A: 0.04 M H_3PO_4, 0.2 M Na_2SO_4, 10% acetonitrile, pH adjusted to 2.5 with ethanolamine. Solvent B: 50% acetonitrile. Elution: Isocratic using 50% B for 20 min, then a linear gradient to 87.5% B over 20 min, followed by isocratic elution for 6 min. The insulin peak exceeds the scale 13 times.

Results and Discussion.

Expression of human proinsulin in yeast was attempted in the initial studies. C-peptide and insulin-like compounds were found in the broth in a ratio of 10 to 1, indicating that disulfide bond formation is predominantly preceded by cleavage at either of the pairs of basic amino acids (9). The three single-chain precursors were found in the broth in similar concentrations as C-peptide, showing the advantage of deleting the paris of basic amino acids from the connecting link.
The maximum yields and first-order reaction rate constants for transpeptidation of the three precursors to HI-OMe and HI-OBu^t in two different buffers are given in Table 1. The rate increases through the series I<III<II, indicating that steric hindrance may be the cause of slow transpeptidation of I, and that Ser is a better leaving group than Ala in the transpeptidation reaction. Transpeptidation is faster in Thr-OBu^t than in Thr-OMe, in agreement with the earlier observation that Thr-OMe inhibits transpeptidation (11). The water concentration is critical for the specificity, as the side-reaction at Arg^{B22} takes over at water concentrations a few percent higher than 21%.
By-products include desoctapeptide-(B22-B30) insulin, its threonine ester and des-(B30) insulin. Transpeptidation is slower in buffer a (acidic) than in buffer n (neutral), both with regard to the desired transpeptidation at Lys^{B29} and to

Table 1. First order Rate Constants, maximum Yields and By-products formed at indicated times for the Transpeptidation of Single-chain Precursors to Human Insulin esters, viz. HI-OMe^{B30} and HI-OBu^{tB30}.

Precursor	I : SCI				II: SCI-SK				III: SCI-AAK			
Thr-OR	-OMe		-OBu^t		-OMe		-OBu^t		-OMe		-OBu^t	
Buffer	a	n	a	n	a	n	a	n	a	n	a	n
$k_{cat} \times 10^3$, h^{-1}	15	46	43	200	90	280	>370	510	53	200	82	160
Yield, %	61	74	69	75	81	81	84	83	84	86	82	89
By-products, %	3	16	29	25	6	19	16	5	7	9	11	11
Time, h	96	96	96	24	24	24	24	3,5	296	24	96	96

Conditions: solvent DMF; water 21%; temperature 12°C; single-chain precursors 9 mM; porcine trypsin 0.2 mM; Ca^{++} 7 mM. Buffer a: 1.14 M Thr-OR and 2.84 M acetic acid; buffer n: 1.14 M Thr-OR and 1.14 M acetic acid.

cleavage at Arg^{B22} leading to by-products. Yields close to 90% were obtained with precursors II and III, whereas somewhat lower yields were encountered with I due to the slow rate of transpeptidation and the concomitant specificity problem.
The purity as assessed by HPLC was 99.5-99.9%, Fig. 3. The high purity is a result of preparative HPLC in the final stage of purification, combined with the low level of other proteins secreted by yeast during fermentation.
The identity of human insulin (ge) with the WHO standard of human insulin (ss) as analyzed by peptide mapping is shown in Fig. 4. A complete overlap of the many fragments produced by subtilisin A digestion can be seen.

The standard curve of the ELISA for yeast protein contaminants is shown in Fig. 5. The steep curve provides a detection limit of < 1 ng/ml, corresponding to < 1 ppm yeast protein in the product. No contamination could be detected in the product by this ELISA, but the decrease in yeast proteins through the processing, Fig. 2, demonstrates the usefullness of such contaminant assays.

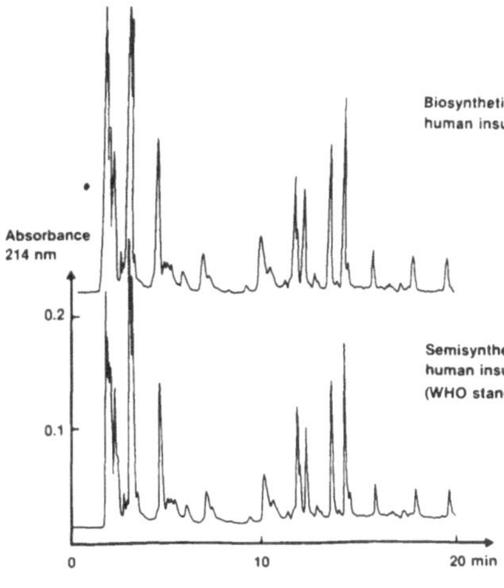

Fig. 4. Peptide mapping of subtilisin A digests of human insulin (ge) compared to human insulin (ss).
Digestion: 0.8 mM human insulin in 50 mM Tris-HCl buffer, pH 7.4, 12 mM EDTA using 0.024 mM subtilisin A at 37°C for 21 h.
HPLC. Column: Nucleosil RP C18, 5 µ, 4x200 mm. Oven temp. 30°C. Flow 1 ml/min. Solvent A: 5% (v/v) acetonitrile, 60 µl H_2SO_4 in 900 ml H_2O. Solvent B: 45% (v/v) acetonitrile, 0.15 M $(NH_4)_2SO_4$, 1.5 mM H_2SO_4. Elution: Linear gradient from 0 to 45% B over 20 min.

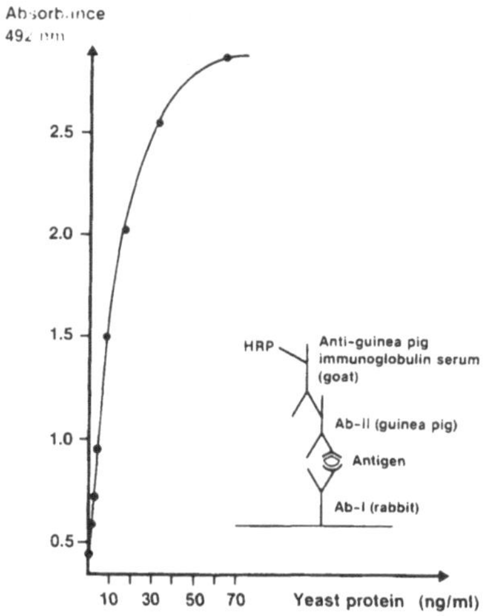

Fig. 5. ELISA standard curve for yeast proteins. The insert shows the principle, the yeast proteins in the standard and in samples being sandwiched between 2 antibodies and thereby giving rise to color formation by the HRP.

References.

1. Chance, R.E., J.A. Hoffmann, E.P. Kroeff, M.G. Johnson, E.W. Schirmer, W.W. Bromer, M.J. Ross, R. Wetzel. 1981. In: Peptides. Synthesis-Structure-Function. Proceedings of the Seventh American Peptide Symposium (D.H. Rich and E. Gross, eds). Pierce Chemical Company, p. 721

2. Frank, B.H., J.M. Pettee, R.E. Zimmerman, P.J. Burck. ibid. p. 729

3. Markussen, J., US patent 4.343.898

4. Markussen, J. 1984. In: Methods in Diabetes Research 1 (J. Larner and S. Pohl eds.). Wiley & Sons, p. 403

5. Markussen, J., K.H. Jørgensen, A.R. Sørensen, L. Thim. 1985. Int.J.Peptide Protein Res. 26, 70

6. Markussen, J. 1985. Int.J.Peptide Protein Res. 25, 431

7. Kurjan, J., Herskowitz, I. 1982. Cell 30, 933

8. Julius, D., L. Blair, A. Brake, G. Sprague, J. Thorner, Cell 32, 839

9. Thim, L., Hansen, M.T., Norris, K., Hoegh, I., Boel, E., Forstrom, J., Ammerer, G., Fiil, N.P. 1986. Proc. Natl. Acad. Sci. USA 83, in press.

10. Kawasaki, G. 1984. 12th Int. Conf. on Yeast Genetics and Molecular Biology, abstract p. 15

11. Markussen, J., Aa. Vølund. 1985. In: Enzymes in Organic Synthesis (R. Porter and S. Clark, eds.). Ciba Foundation Symposium 111, Pitman, p. 188

STUDIES ON SYSTEM PROPERTIES INFLUENCING THE PARTITION CONSTANT IN KINETICALLY CONTROLLED ENZYMATIC PEPTIDE SYNTHESIS

V. Schellenberger, H.-D. Jakubke
Karl Marx University, Section of Biosciences, Department of Biochemistry, DDR-7010 Leipzig, German Democratic Republic

Introduction

Protease-catalyzed peptide synthesis can be performed either in equilibrium or kinetically controlled reactions (1). The latter process is characterized by short reaction times and low enzyme requirements. In addition to the small amount of experimental data which has been obtained on the influence of temperature (2) and pH (3) in kinetically controlled peptide synthesis, systematic studies are necessary on the effects of other system properties which can influence the partition constant (4). We report here the effects of organic solvents and various salts, respectively, on the partition constant of a chymotrypsin-catalyzed model reaction.

Results

Using maleyl(Mal-)-L-phenylalanine methyl ester as acyl donor and the nucleophile H-Leu-NH_2 both the reactants and the product are soluble in 0.1 M borate buffer (pH 8.5). The experiments were carried out using Radiometer pH-assemblies in 2 ml buffer containing 20 mM Mal-Phe-OMe and 50 mM H-Leu-NH_2 in the presence of α-chymotrypsin (EC 3.4.21.1.) in concentrations indicated in the Tables. After 2 and 3 min, respectively, samples were monitored by HPLC (254 nm) using Liquochrom 307 equipment of Labor MIM (Budapest, Hungary) and a 100 x 4 mm HTS column. The partition constant was calculated from the HPLC data.

To obtain basic information on the inhibiting effect of organic solvents on the formation of maleyl-L-phenylalanine-α-chymotryp-

Peptides 1986

sin we have studied the hydrolysis of Mal-Phe-OMe in the presence of various concentrations of dimethyl formamide (DMF). The kinetic constants are listed in Table 1.

Table 1. Kinetic Constants of the Chymotrypsin-Catalyzed Hydrolysis of Mal-Phe-OMe in the Presence of DMF

DMF (%;v/v)	K_M (mM)	k_{cat} (s^{-1})	k_{cat}/K_M ($M^{-1}s^{-1}$)
0	4.85 ± 0.3	99.4 ± 4.1	20.5 ± 1.2
5	7.50 ± 1.4	68.9 ± 5.2	9.2 ± 2.3
10	13.10 ± 3.0	54.8 ± 4.8	4.4 ± 0.6

Due to the strong inhibition in the presence of 20 % (v/v) DMF K_M and k_{cat} could not be determined separately.

Table 2 shows the influence of methanol on the synthesis of Mal-Phe-Leu-NH_2. Due to the decrease in the rate of acyl-enzyme formation with increasing methanol concentration the amount of chymotrypsin must be increased dramatically.

Table 2. Influence of Methanol on the Chymotrypsin-Catalyzed Synthesis of Mal-Phe-Leu-NH_2

Methanol (%;v/v)	Chymotrypsin (mg)	Content of Mal-Phe-OMe after 2 min (%)	p (mM)
0	0.1	55	20
5	0.2	45	18
10	0.4	53	18
20	1.0	33	17
30	2.0	44	18
50	4.0	78	22

It is seen from the data in Table 2 that methanol does not change the ratio between hydrolysis product and peptide product since the p values are nearly constant. Interestingly, the results in Table 3 indicate that beside methanol even other organic solvents with different properties do not influence the partition constant, whereas these solvents inhibit the reaction in a different manner.

Table 3. Influence of Various Organic Solvents on the Chymotrypsin-Catalyzed Synthesis of Mal-Phe-Leu-NH_2

Organic solvent (%;v/v)	Enzyme (mg)	Content of Mal-Phe-OMe after 2 min (%)	p (mM)
None	0.1	55	20
DMF (10)	0.2	51	20
DMF (30)	2.0	34	20
Diglyme (30)	2.0	16	29
Acetonitrile (30)	2.0	3	19
Dioxan (30)	2.0	75	26
Glycerol (30)	2.0	2	27
Isopropanol(30)	2.0	24.5	36

In contrast to dioxan, both glycerol and acetonitrile show only slight inhibiting effect on the reaction. Generally, the reduction of the activity coefficient of water or the water content seems not to be important for the improvement of the partitioning ratio of the acyl-enzyme in favour of peptide product.

Since enzyme-catalyzed reactions are strongly influenced by the ionic strength of the medium, we have studied the influence of various salts on the partitioning ratio of the chymotrypsin-catalyzed model reaction. Using a constant ionic strength (I = 2) the differences observed in these systems should then depend on the effect of the ions as a partition controlling factor. The results in Table 4 indicate that with the exception of NaSCN all added salts cause a significant acceleration of the reaction. Furthermore, due to the chaotropic properties of NaSCN the ability of chymotrypsin to bind the nucleophile and to exclude H_2O, which competes with H-Leu-NH_2 in the deacylation process, is diminished significantly. This effect seems to be the main reason for the increased p value. On the other hand, in the presence of Na_2SO_4 the partitioning of the acyl-enzyme is shifted in favour of peptide product formation.

We have shown in this study that, in contrast to equilibrium controlled peptide synthesis, organic solvents do not increase the yield of peptide product in kinetically controlled synthesis. To carry out kinetically controlled peptide synthesis in aqueous solution, the introduction of solubilizing protecting groups into

Table 4. Influence of Salts on the Chymotrypsin-Catalyzed Synthesis of Mal-Phe-Leu-NH_2+)

Salt concentration	(M)	Content of Mal-Phe-OMe after 3 min (%)	p (mM)
Without		75	20
$CaCl_2$	(0.67)	8.5	29
Na_2SO_4	(0.67)	11.5	9
Li_2SO_4	(0.67)	20	16
CH_3COONa	(2.00)	12	18.5
NaCl	(2.00)	10	19
NaSCN	(2.00)	78	55

+) 0.02 mg chymotrypsin was used in all experiments

the reactants is necessary in order to overcome the poor solubility of the reactants and the product. From the preparative point of view it is important that the addition of suitable salts leads to a significant acceleration of the reaction. Beside the lower enzyme requirement in nearly all cases, the addition of Na_2SO_4 results in a decreased p value, providing higher yields of peptide product.

References

1. Jakubke, H.-D., P. Kuhl, A. Könnecke. 1985. Angew. Chem. Int. Edn. Engl. 24, 85.
2. Riechmann, L., V. Kasche. 1985. Biochim. Biophys. Acta 830,164.
3. Schellenberger, V., A. Könnecke, H.-D. Jakubke. 1984. In: Peptides 1984(U. Ragnarsson,ed.). Alquist & Wiksell, p. 201.
4. Schellenberger, V., H.-D. Jakubke. 1986. Biochim. Biophys. Acta 869, 54.

ENZYMATIC SYNTHESIS OF ASPARTAME. A STUDENT EXPERIMENT.

Gunnar Lindeberg
Institute of Biochemistry, University of Uppsala, Biomedical Center, Box 576, S-751 23 Uppsala, Sweden

Introduction

Experiments in peptide synthesis suitable for inclusion in a general chemistry/biochemistry course are not easily found. The preparation of most biologically active peptides is too laborious and time-demanding. The synthesis of a simple di- or tripeptide, on the other hand, may well demonstrate different procedures and techniques, but the product as such is rarely of particular interest. However, the dipeptide sweetener aspartame, H-Asp-Phe-OMe, can be conveniently prepared in a couple of days using a minimum of equipment. The physiological effects are easily measured and, if the synthesis of analogues is included, some structure - activity relationships can be studied.

Z-Asp-OH + 2 H-Phe-OMe

↓ Thermolysin, pH 7

Z-Asp-Phe-OMe·H-Phe-OMe

↓ H^+

Z-Asp-Phe-OMe

↓ $HCOONH_4$, Pd/C, MeOH

H-Asp-Phe-OMe

Scheme 1. Synthesis of aspartame.

Peptides 1986

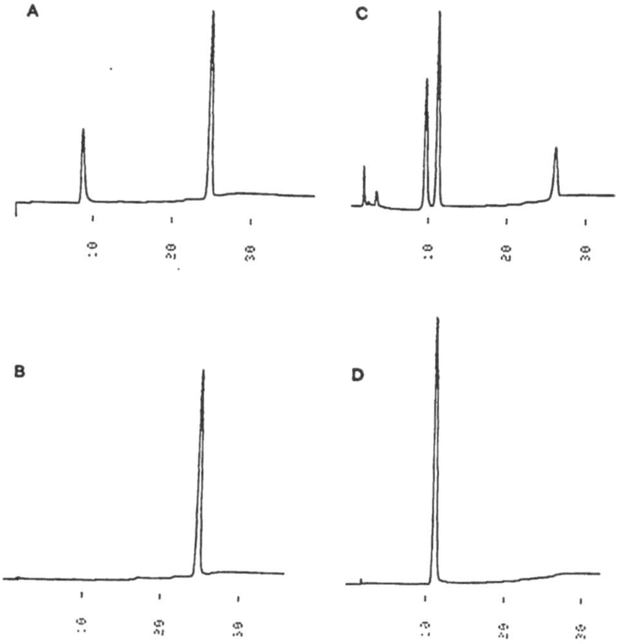

Figure 1. HPLC of A) Z-Asp-Phe-OMe·H-Phe-OMe, B) Z-Asp-Phe-OMe, C) crude, and D) purified H-Asp-Phe-OMe.
Support: Spherisorb S10 ODS, 4.6 x 150 mm. Buffer A: 0.1M sodium phosphate, pH 3.0. Buffer B: 0.1M sodium phosphate, pH 3.0, in 50% EtOH. Gradient: 5 - 90% B in 20 min., then isocratic. Flow rate: 1 ml/min. Detection: UV at 210 nm.

Methods

The synthesis of aspartame is outlined in Scheme 1. Z-Asp-Phe-OMe·H-Phe-OMe is obtained by the thermolysin-mediated condensation of Z-Asp-OH and H-Phe-OMe essentially as described by Isowa et al. (1). The salt is decomposed by treatment with dilute HCl, and the dipeptide isolated. The benzyloxycarbonyl group is then removed by catalytic transfer hydrogenation in MeOH using ammonium formate as hydrogen donor and 5% Pd/C as catalyst (2). The progress of the reactions and the purity of the intermediates can be conveniently monitored by HPLC (Fig. 1), if such equipment is available, or by TLC (Fig. 2).
The product is purified by lyophilization and passage through a

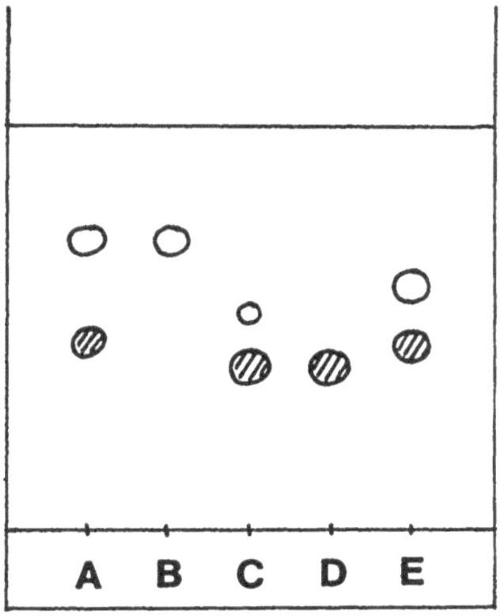

Figure 2. TLC of A) Z-Asp-Phe-OMe·H-Phe-OMe, B) Z-Asp-Phe-OMe, C) crude, D) purified H-Asp-Phe-OMe, and E) a mixture of Z-Asp-OH and H-Phe-OMe.
Support: Silica gel. Solvent: n-BuOH - AcOH - H_2O (3:1:1). Detection: UV and ninhydrin (hatched areas).

small column of anion-exchanger in order to remove diketopiperazine and other negatively charged by-products. The concentration of the eluate is determined, and a dilution series is made over the adequate concentration range. The solutions are tasted and compared to reference solutions of sucrose.

Experimental Procedure

Z-Asp-Phe-OMe·H-Phe-OMe

Z-Asp-OH (267 mg, 1 mmole) and H-Phe-OMe·HCl (431 mg, 2 mmoles) are weighed into a centrifuge tube and suspended in H_2O (3 ml). The pH is adjusted to 7 by addition of 4M NaOH (ca. 0.7 ml) in portions while the mixture is being stirred on a vortex mixer. Thermolysin (2 mg in 0.2 ml H_2O) is added to the resulting clear solution which is then incubated at 40°C overnight. The precipitate is collected by centrifugation and washed with 5 x 4 ml H_2O (Fig. 1A).

Z-Asp-Phe-OMe

The peptide salt is decomposed by trituration in the cold with a mixture of H_2O (3 ml) and 1M HCl (1 ml). After 5 - 10 min. the precipitate is again collected by centrifugation and washed with 5 x 4 ml ice-water (Fig. 1B).

H-Asp-Phe-OMe, Aspartame

Z-Asp-Phe-OMe (still moist product above; less than 1 mmole) and $HCOONH_4$ (126 mg, 2 mmoles) are dissolved in MeOH (2 ml). The catalyst, 5% Pd/C (50 mg), is added and washed down with MeOH (0.5 ml). The mixture is kept at room temperature with occasional swirling until the gas evolution ceases, ca. 20 min. Deprotection is usually complete within 5 min. (TLC). The catalyst is removed by centrifugation and filtration of the supernate through a tightly packed plug of cotton wool in a Pasteur pipet. Catalyst and filter are washed with MeOH (0.5 ml), and the combined filtrates are evaporated in a stream of nitrogen with mild heating. The remaining solid is dissolved in a few ml of warm water and lyophilized (Fig. 1C).

Sweetness evaluation

Part of the crude product (50 mg) is dissolved in H_2O (5 ml) and passed through a column (5 x 40 mm, Pasteur pipet) of QAE-Sephadex A-25 (acetate form) in H_2O. The column is washed with H_2O (2 ml), and the concentration of the pure peptide (Fig. 1D) in the eluate is determined from the UV absorbance (ε_{257} = 200). A dilution series covering the range 0.003 - 0.3% is prepared, and the threshold value of the sweet taste is determined. Sucrose, 0.25 - 4%, serves as a reference. The sweetness is defined as the quotient:

$$S = \frac{\text{threshold value (sucrose)}}{\text{threshold value (peptide)}}$$

References

1. Isowa, Y., M. Ohmori, T. Ichikawa, K. Mori, Y. Nonaka, K. Kihara, K. Oyama, H. Satoh, S. Nishimura. 1979. Tetrahedron Lett. 28, 2611.

2. Anwer, M.K., A.F. Spatola. 1980. Synthesis, 929.

ENZYMATIC SYNTHESIS OF FRAGMENTS OF THE GONADOTROP RELEASING HORMONE MOLECULE WITH LH-RH/FSH-RH ACTIVITY

L.Mladenova-Orlinova, I.Mancheva, J.Calderon Vera
Department of Organic Chemistry, Institute of Chemical Technology, 1156 Sofia, Bulgaria

Introduction

The present paper deals with the enzymatic synthesis of either N-or N-and C-protected fragments having the amino acid sequence 3-4; 5-7; 1-4 and 3-7. The advantage of the enzymatic peptide synthesis is in the suppression of the side-reactions such as modification of the side-chains and racemization. This approach ensures the preparation of both chemically and optically pure peptides (1). The minor protection required as well as the higher purity of the final products makes it possible to plan an improved overall route of the synthesis of LH-RH with respect to yield, purity and biological activity.

Materials and methods

The following enzymes and reagents were used: bovine α-chymotrypsin (Merck), papain (Merck), Thermolysin (Sigma), L-amino acids (Reanal), Kieselgel 60 (Merck) and precoated silicagel plates (DC-Alufolien, Merck).
Z-Trp-Ser-$NHNH_2$ and Z-Tyr-Gly-Leu-OH were synthesized using -chymotrypsin.
Z-Trp-Ser-$NHNH_2$ (1) was made from Z-Trp-OMe and H-Ser-$NHNH_2$ in the medium of carbonate buffer (pH 9-10) containing 10% DMF. The reaction proceeded in the presence of α-chymotrypsin at room temperature for 0,5 - 2 hrs at molar ratio acyl component: amino component 1:2 and acyl component: enzyme (1:0,003) up to the full exhaustion of the acyl component. The obtained precipitate,

Peptides 1986

washed with H_2O, and recrystallized from MeOH.
Z-Tyr-Gly-Leu-OH.H_2O (2) was prepared from Z-Tyr-OMe and HBr.Gly-Leu-OH in the presence of α-chymotrypsin in aqueous carbonate buffer at pH 9-10. The molar ratio acyl component: amino componenet and acyl component: enzyme was in 1. The reaction was run at room temperature for 10-30 min. The mixture was acidified, the precipitate filtered and washed. 2 was purified by column chromatography on Kieselgel, EtOAc:benzene (1:1) being used as eluent.
pGlu-His-Trp-Ser-$NHNH_2$ (3) was obtained from pGlu-His-OH and H-Trp-Ser-$NHNH_2$, dissolved in MeOH and 0,2 Tris/HCl buffer (pH 8), containing 5.10^{-2}M calcium acetate in the presence of thermolysin and ammonium sulfate.
At molar ratio acyl component: amino component 1:1 and temperature 38-40°C the reaction was completed in 15 hrs. The crude product was washed and recrystallized from MeOH.
Z-Trp-Ser-Tyr-Gly-Leu-OH (4) was synthesized from Z-Trp-Ser-OMe and H-Tyr-Gly-Leu-OH (molar ratio 1:1), dissolved in methanol, in the presence of Papain, dissolved in 0,2 M Mellvain-buffer (pH 5) adding small quantities of 2-mercaptoethanol. The reaction was conducted at 38°C in 4-5 hrs. The crude product was purified by column chromatography on Kieselgel using $CHCl_3$:MeOH (9:1) as eluent.

Results and discussion

Two methods for preparation of 1 (yield54% in both cases) have been in literature (2,3). Z-Trp-Ser-$NHNH_2$ was synthesized by us using the described enzymatic method quicker, yield 85%, m.p.177-178°C, the optical purity being higher $[\alpha]_D^{20}$ =40° (c 0.25 in DMF). This is due to the fact that the amino acid hydrazides are more reactive nucleophiles compared with both the free amino acid and their esters in the course of α-chymotrypsin catalyzed amonolysis of the corresponding appropriate acyl derivates (4). On the other hand, the unsolubility of the Z-protected peptide-hydrazides favors the efficiency of the peptide synthesis in aqueous solution as minor excess of the amino component is used, the equilibrium being displaced in the direction of the peptide formation. The

peptide precipitates and can be seperated from the reaction mixture. Peptide 2 was synthesized by us for first time. The peptide BOC-Tyr(Me)-Gly-Leu-OMe has been reported in the literature (5) but its synthesis (70% yields) includes only chemical methods. The preparation of 2 by the described enzymo-chemical approach does not require protection of the tyrosine phenol group which leads to shortening the time for peptide bond formation, and raising both the yield (87%) and the optical purity $[\alpha]_D^{20} = -41^\circ$ (c 0,29 DMF), m.p. 112-114°C.
We used pGlu-His-OH and the appropriate for enzyme synthesis nucleophile H-Trp-Ser-$NHNH_2$ in the preparation of 3. The experiments with papain were unsuccessful. The usage of thermolysin under the conditions described in (6) produced 3 in 71% yield (m.p.185-186°C). The yield is lower when the synthesis was carried out by chemical methods.
The coupling of Z-Trp-Ser-OMe and H-Tyr-Gly-Leu-OMe to 4 was made in the presence of papain under the conditions described in (7) in the medium of 50% MeOH. The yield of 4 was 73% $[\alpha]_D^{20} = -27^\circ$ (c 0,3 in MeOH).The NMR-spectrum of the product confirmed the existence of -COOMe group which is in accordance with earlier drawn conclusion (7) of hydrolysis.
The peptides 1, 2, 3, and 4 were characterized by TLC, elemental analysis, specific rotation and NMR-spectra.

References

1. Jacubke,H.D.,P.Kuhl.1982,Die Pharmazie 2, 89
2. Wittle,E.L.,M.C.Rebstock.1978.US.Patent 4086219
3. Dutta,A.S.,et all.1977.Ger.offen 2720245
4. Petkov,D.,I.Stoineva.1984.Biochem.Biophys.Res.Commun.118.317
5. Künzi,H.,et all.1974.Helv.Chim.Acta 57, 2131
6. Isowa,Y., T.Ichikawa.1979.Bull.Chem.Soc.Japan. 52, 796
7. Isowa,Y., et all. 1977. Bull.Chem.Soc. Japan. 50, 2766

NUCLEOPHILE BINDING IN CHYMOTRYPSIN PEPTIDE SYNTHESIS

D.D. Petkov, E.K. Bratovanova and I.B. Stoineva
Institute of Organic Chemistry with Centre of Phytochemistry,
Bulgarian Academy of Sciences, 1040 Sofia, BULGARIA

Introduction

The asymmetric peptide synthesis, catalysed by proteinases, is determined by the specific binding of the amine component (nucleophile) to the enzyme leaving-group binding site. We report here an observation of a kinetic saturation of the enzyme by the nucleophile. Such saturation kinetics is consistent both with a mechanism involving a direct nucleophile attack on the acylenzyme and a nucleophile binding mechanism. The discrimination between the two mechanisms was achieved by specifically designed experiments demonstrating the competitive inhibition of the chymotrypsin peptide hydrolysis by the nucleophile.

Results and Discussion

The effect of the nucleophile concentration on the initial rate of formation of a peptide during the chymotrypsin aminolysis of acetyl-Tyr-OEt is shown in Fig. 1. The shape of the curves indicates a kinetic saturation of the enzyme by the nucleophile. Ordinal and abscisal intercepts of the linear plot of $1/v_0$ vs. $1/[N_0]$ for different nucleophiles are summarized in Table 1. This saturation phenomenon may be accounted for by two different mechanisms: one that does not, and a second that does, involve a nucleophile binding to the acylenzyme.

<u>A. Mechanism involving direct nucleophile attack on the acylenzyme.</u>

$$E + S \xrightarrow{k_2/K_m} Ac\text{-}E \xrightarrow{k_3^i[N_i]} E + Ac\text{-}N_i \qquad 1.$$

Peptides 1986

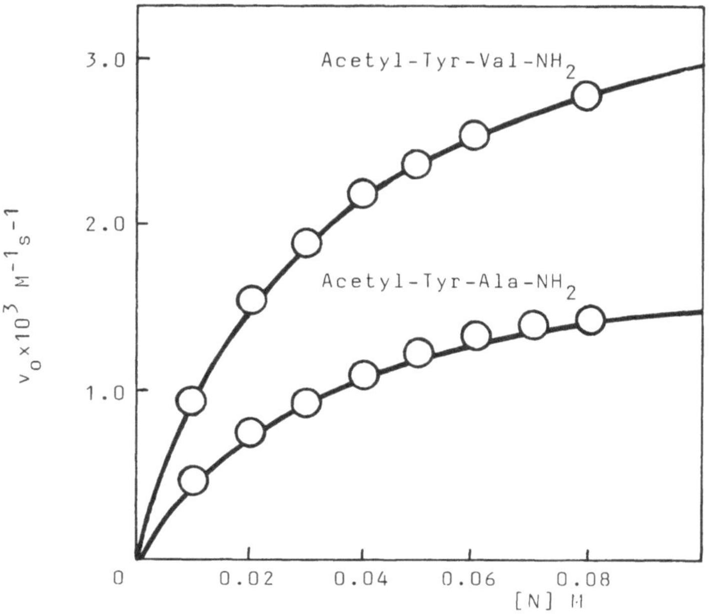

Fig. 1. Initial rate vs. nucleophile concentration studied by HPLC on a RP-18 Bondapack analytical column, Waters solvent delivery system 6000A; flow rate of 1.3 ml/min and detection at 280 nm; eluent mixture MeOH/30 mM KH_2PO_4, pH 4.6 (1:1), v/v).

In mechanism 1. the nucleophile N_i (i=w, water; a, amines) directly attacks the acylenzyme Ac-E during deacylation by parallel reactions of Ac-E with N_i. According to this mechanism, the initial rate expression for the formation of a peptide (i=a, amino acid or peptide derivative) is (1):

$$v_o = \frac{k_2[E_o][N_a]}{\frac{k_2 + k_3^w}{k_3^a} + [N_a]} \qquad 2.$$

This equation describes an initial rate that follows saturation kinetics with respect to the nucleophile concentration as in Fig. 1. At sufficiently high $[N_a]$, v_o reaches the limiting value $k_2[E_o]$, i.e. independent of the nature of the nucleophile, which is not observed (Fig. 1 and Table 1).

TABLE 1.
Kinetic and Binding Parameters of Different Nucleophiles in the Chymotrypsin Aminolysis of Specific Substrates.

Nucleophile	$1/A[E_o]$ [a]	$1/B$ [a]	$B/A[E_o]$	K_i [b]
	s^{-1}	mM	$M^{-1}s^{-1}$	mM
H-Gly-Leu-NH_2	2130	10	213000	19
H-Val-NH_2	3070	29	105860	52
H-Leu-NH_2	1780	38	46840	43
H-Ala-NH_2	1780	42	42380	61
H-Gly-NH_2				369

a) A and B are ordinal and abscisal intercepts of the plot $1/v_o$ vs. $1/[N_a]$.

b) Nucleophile inhibition constant in the chymotrypsin hydrolysis of anthraniloyl-Phe-Leu-NHPh(NO_2).

B. Nucleophile binding mechanism.

Provided $k_2 >> k_3^w[H_2O]$, the concentration of the acylenzyme Ac-E is practically equal to the total enzyme concentration $[E_o]$. This requirement is fulfilled in chymotrypsin hydrolysis of acetyl-Tyr-OEt (2). Hence, it appears that the saturation kinetics we observed (Fig. 1) is consistent with a mechanism involving nucleophile binding to the acylenzyme:

$$\text{Ac-E} + N_a \underset{k_{-4}}{\overset{k_4}{\rightleftharpoons}} \text{Ac-EN}_a \xrightarrow{k_5^a} E + \text{Ac-N}_a \qquad 3.$$

Since $[E] << [\text{Ac-E}] \approx [E_o]$, the initial rate expression for this mechanism is

$$v_o = k_5^a[\text{Ac-EN}_a] = \frac{k_5^a[E_o][N_a]}{K_m^a + [N_a]}, \qquad 4.$$

where $K_m^a = (k_5^a + k_{-4})/k_4$ is the Michaelis constant for the reaction of Ac-E with N_a. If this mechanism holds, the kinetic parameters obtained from a double reciprocal plot of the initial rate data (Table 1) could be assigned to the deacylation rate constant $k_5^a = 1/A[E_o]$, and to the nucleophile binding constant $K_n \approx K_m^a = 1/B$.

C. Discrimination between the two mechanisms.

Provided N_a binds to the nucleophile binding site, a competitive inhibition of the leaving group release by N_a is expected for the enzymatic hydrolysis of $Ac\text{-}N_a$ (3):

$$EN_a \underset{}{\overset{K_n}{\rightleftharpoons}} N_a + E + Ac\text{-}N_a' \overset{K_s}{\rightleftharpoons} EAc\text{-}N_a' \xrightarrow{k_2} Ac\text{-}E + N_a' \qquad 5.$$

Actually, the amine components (N_a) in chymotrypsin peptide synthesis do inhibit the chymotrypsin hydrolysis of anthraniloyl-Phe-Leu-$NHPh(NO_2)$. This has been registered fluorimetrically by monitoring the release of the fluorescence of the anthraniloyl group during the hydrolysis of the intact intramolecularly quenched substrate. The graphically determined K_i-values are summarized in Table 1. They are of the same order of magnitude as those of 1/B. These values we assign to the nucleophile binding constants $K_n \approx K_m^a \approx K_i$ and conclude that the results obtained are compatible with the nucleophile binding mechanism (eq. 3).

References

1. Bender, M.L., G.E. Clement, C.R. Gunter, F.J. Kezdy. 1964. J. Amer.Chem.Soc. 86, 3697-3702.
2. Berezin, I.V., N.F. Kazanskaya, A.A. Klyosov. 1971. FEBS Lett. 15, 121-124.
3. Riechmann, L., V.Kasche. 1984. Biochem.Biophys.Res.Commun. 120, 686-691.

SYNTHESIS AND PROPERTIES OF [Hse^{65},Ala^{81}]-, [Hse^{65},Leu^{81}]- and [Hse^{65},Val^{81}]-CYTOCHROME c

H.A. Boots and G.I. Tesser

Laboratory for Organic Chemistry, Catholic University, Toernooiveld, NL 6525 ED Nijmegen, the Netherlands

Introduction

The strategy for the preparation of unprotected 65-homoserine cytochrome c using the principle of semisynthesis, has been developed by Boon (1). The total syn-

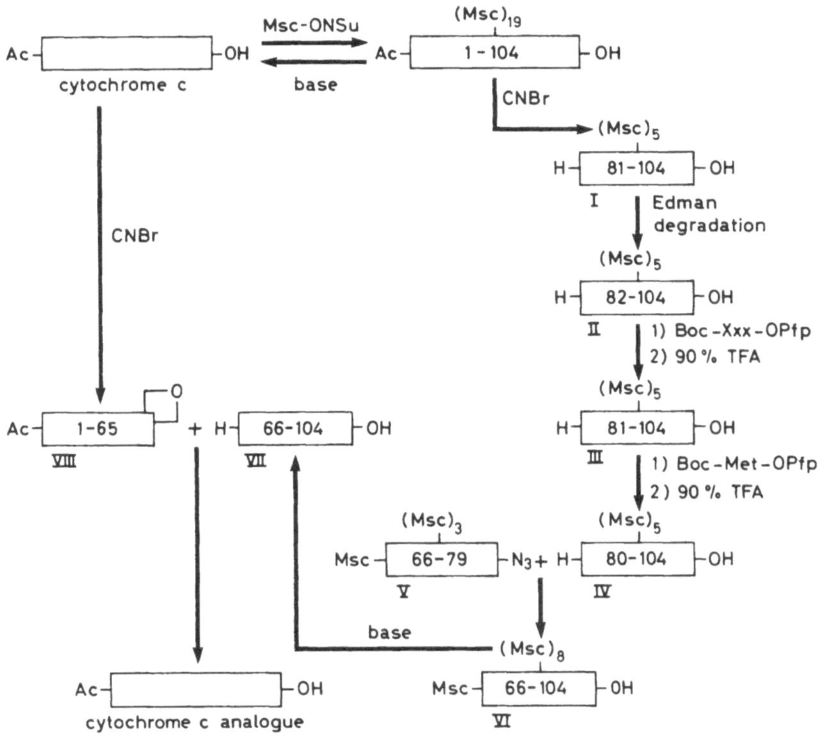

Fig. 1: The semisynthesis of cytochrome c analogues bearing variations in pos.81.

thesis of cytochrome c-(66-104)-nonatriacontapeptide has been worked out by Ten Kortenaar (2,3). Both investigators applied their strategies to the synthesis of analogues of the protein to evaluate some aspects of the role of phylogenetically converved amino acid residues (3,4). The end products of their syntheses contained free, unprotected ammonium and guanidinium groups. This could be achieved by the use of the 2-methylsulphonylethyloxycarbonyl group for protection of the pertinent amino groups. Full details and proof of the applicability of this function as a protective group in work with cytochrome c have recently been given (5). Presently, analogues have been obtained using the strategy of Boon (Fig. 1) with insertion of one Edman cyclus on one of the intermediates (compound I) and subsequent replacement of the eliminated amino acid (isoleucine, pos. 81) by alanine, leucine or valine.

Methods & Results

To prove the completeness of the degradation, aliquots of compound II (purified solely by gelfiltration) were subjected to amino acid analysis prior to and following arylation with 2,4-dinitrofluorobenzene (*cf.*5). The degraded sequence (II) was recovered in 93% yield. Samples of compound II were subsequently acylated with ten equiv. of Boc-Xxx-OPfp (6; Xxx symbolizing one of the three aliphatic amino acids). The condensates (III) were purified by gelfiltration (Fig.2); main peaks were invariably preceded by a small peak (Fig.2, Xxx = Leu,Val) of higher

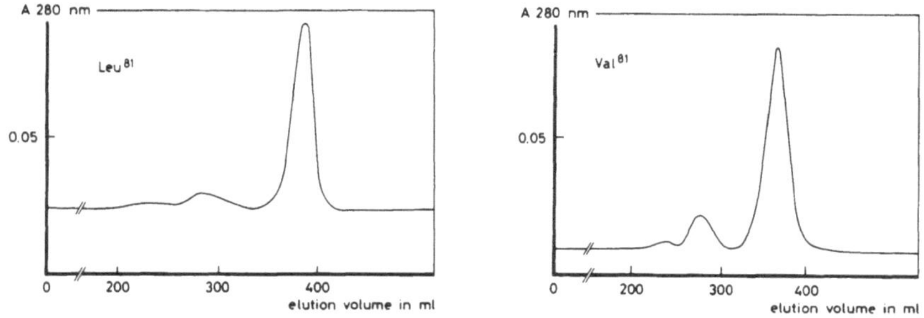

Fig. 2: Gelfiltration (Sephadex G 50, 70% HCOOH) of two compounds of type III.

molecular weight. These compounds originate from the activation (6) by the active ester of a side chain carboxylic group and analysed for a heptatetracontapeptide derivative (composed of one Xxx residue and two indentical tricosapeptide moieties). The aspartyl residue of pos.93 has been suspected for this (4). Analysis of compounds of type III as indicated for compound II, proved that the N-terminus had been acylated exclusively. For the semisynthesis of the analogous cytochrome c fragments IV-VII the original route of Fig.1 was followed. Full details of this part of the synthesis will appear shortly (7), exemplified in the native sequence. The amino acid composition of the final nonatriacontapeptides (VII) is given in Table 1. The analogous cytochromes c corresponding to the compounds of type VII arose under standard conditions (1) from the spontaneous condensation of each nonatriacontapeptide in the unprotected state with the reduced form of VIII.

Table 1: Amino acid composition of the compounds of type VII.

Amino acid	Ala^{81}	Leu^{81}	Val^{81}	Amino acid	Ala^{81}	Leu^{81}	Val^{81}
Asp	2.93	3.07	2.91	Met	0.99	1.03	0.96
Thr	3.06	3.03	2.98	Ile	2.97	3.08	3.09
Glu	5.00	5.07	4.83	Leu	3.03	4.11	3.13
Pro	1.94	1.93	1.98	Tyr	2.97	2.93	2.90
Gly	2.09	2.11	2.04	Phe	1.07	1.06	1.06
Ala	4.08	2.98	3.09	Lys	8.14	7.93	7.91
Val	-	-	1.05	Arg	1.05	1.01	0.97

Although no changes in charge were to be expected, the analogues differed from the native compound in their adsorbtion to carboxymethyl cellulose, which appeared impossible at buffer concentrations 0.05 M (native cytochrome c 0.1-0.2 M). The spectral (UV,VIS) properties of the analogues were very similar to those of the native protein (maxima, ferro-form: 416, 520 and 550 nm; ferri-form: 409

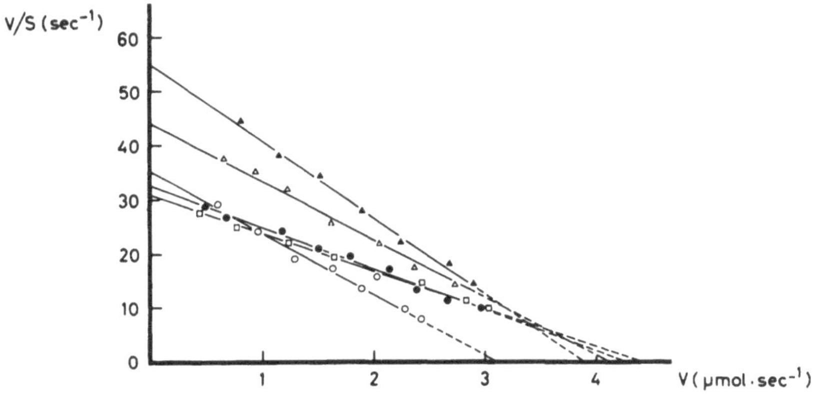

Fig. 3: Eadie-Hofstee plots of the reaction of the cytochromes c with cytochrome c oxydase; □ Native; ● [65-Hse]-; △ [65-Hse, 81-Ala]-; ○ [65-Hse, 81-Leu]-; ▲ [65-Hse, 81-Val]-cytochrome c.

and 528 nm). The analogues exhibited the charge-transfer band at 695 nm. Each analogue transferred electrons from L-ascorbic acid to cytochrome c-oxydase (9). Kinetic data are expressed in Fig.3. The redox potentials were measured spectrophotometrically (8) and did not vary widely. They are listed in Table 2.

Table 2: Redox-potentials (mV) of cytochrome c and the pertinent analogues.

Compound	Em (± 10 mV)
Cytochrome c	265 mV
65-Homoserine cytochrome c	262 mV
65-Homoserine,81-Alanine cytochrome c	256 mV
65-Homoserine,81-Leucine cytochrome c	257 mV
65-Homoserine,81-Valine cytochrome c	253 mV

The present analogues were prepared to investigate the function of the highly conserved isoleucyl residue of position 81. The results given here show that replacement by another structurally related (aliphatic) amino acyl residue is not prohibitive for the transfer of electrons or the tuning of the redox potential at a value of about 260 mV.

Acknowledgement

This investigation was carried out under the auspices of the Netherlands Foundation for Chemical Research (SON) and with financial aid of the Netherlands Organization for the Advancement of Pure Research (ZWO).

References

1. Boon, P.J., G.I. Tesser, R.J.F. Nivard. 1979. Proc.Natl.Acad.Sci. USA 76, 61.
2. Tesser, G.I., P.B.W. ten Kortenaar, H.A. Boots. 1984. In: Peptides 1984 (U. Ragnarsson, ed.). Almquist & Wiksell International, Stockholm Sweden, p. 221.
3. Ten Kortenaar, P.B.W., P.J.H.M. Adams, G.I. Tesser. 1985. Proc.Natl.Acad.Sci. USA 82, 8279.
4. Boon, P.J., A.J.M. van Raay, G.I. Tesser, R.J.F. Nivard. 1979. FEBS Lett. 108, 131.
5. Boon, P.J., G.I. Tesser. 1985. Int. J. Peptide Protein Res. 25, 510.
6. Bodanszky, M., J. Martinez. 1983. In: The Peptides. Analysis, Synthesis, Biology 1 (E. Gross & J. Meienhofer, eds.). Acad. Press, New York, p. 111.
7. Boon, P.J., J.F.M. Mous, P.B.W. ten Kortenaar, G.I. Tesser, Int. J. Peptide Protein Res. (in press).
8. O'Reilly, J.E. 1973. Biochem. Biophys. Acta 292, 509.
9. Ferguson-Miller, S., D.L. Brautigan, E. Margoliash. 1976. J. Biol. Chem. 253, 149.

ENZYMATIC SEMISYNTHESES WITH INSULIN: PREPARATION OF (ARG B31)INSULIN

E.Krause, K.-D. Kaufmann
Institute of Drug Research, Academy of Sciences of GDR
DDR-1136 Berlin, Alfred-Kowalke-Str. 4

Introduction

In commercial insulin preparations minor components can be proved by polyacrylamide gel electrophoresis or reversed phase HPLC,which are necessary as reference substances for insulin standardization by analytical methods. One of the minor components, /Arg B31/insulin, formed by incomplete cleavage of proinsulin in the biosynthesis of insulin can hardly be isolated from natural sources. But semisynthetic methods, developed for the conversion of porcine insulin into human insulin, are suitable to replace the C-terminal alanine B30 in porcine insulin by a dipeptide Ala-Arg (1 - 5).
Based on this type of reactions, we examined the conversion of porcine insulin into /Arg B31/insulin by reaction of insulin with trypsin and an Ala-Arg dipeptide. Two different versions for the preparation of the insulin derivative were used.

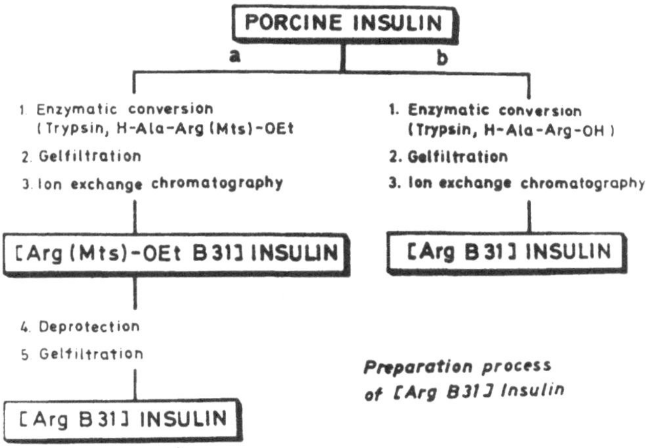

Preparation process of [Arg B31] Insulin

Peptides 1986

In the first case the trypsin catalyzed conversion was done by means of a protected dipeptide as amino component, with the guanidino function of arginine having been protected by the mesitylene-2-sulphonyl group (Mts) introduced by Yajima (6) and with the carboxyl group as ethyl ester. In the second version the enzymatic transpeptidation was performed by means of unprotected Ala-Arg.

Conversion with the protected dipeptide

Transpeptidation with porcine insulin(solution concentration 16 mM), H-Ala-Arg(Mts)-OEt (0,8 M) and trypsin (0,35 mM) was carried out in a mixture of 1,4-butanediol/dimethyl sulphoxide (1:1) and water (15%) at pH 7 and 20°C (5). Under the specified reaction conditions 55% of porcine insulin could be converted into /Arg(Mts)-OEt B31/insulin, controlled by PAAGE. After precipitation of proteins by acetone, trypsin was removed by gel filtration on Sephadex G 50 and /Arg(Mts)-OEt B31/insulin separated from unconverted porcine insulin by anion exchange chromatography on Sephadex DEAE A-25. The arginine insulin derivative was homogeneous in PAAGE and showed full biological activity in mouse convulsion test.

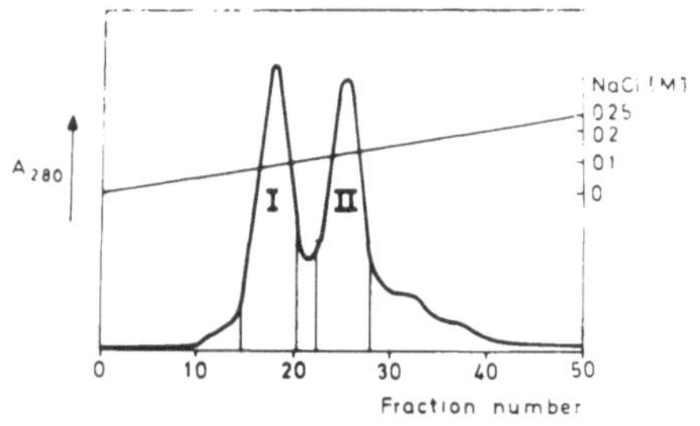

Chromatography on DEAE Sephadex A-25

Column (1.6 x 16 cm)
Buffer. 0.025 M Tris/HCl, 45% 2-propanol, pH 8
Elution NaCl gradient (0-0.25 M, 250 ml/250 ml 20 ml/h)

Fract I [Arg(Mts)-OEt] insulin
Fract II Porcine insulin

The deprotection of arginine side chain was carried out with 1 M methanesulphonic acid and 1 M thioanisole in trifluoroacetic acid containing 5% of m-cresol. The arginine carboxyl group was liberated by alkaline saponification at pH 10,5 for 24 hours. After a final gel chromatography on Sephadex G50 /Arg B31/insulin could be isolated. The product showed a correct amino acid composition. By means of PAAGE, however, 3o% of a minor component formed by incomplete deprotection could be proved.

Conversion with unprotected Ala-Arg

In a further variant we tried to avoid the problems of incomplete deprotection by using the unprotected dipeptide H-Ala-Arg-OH. The trypsin catalyzed transpeptidation was performed under the same conditions as described for protected dipeptide.

The course of reaction was controlled by PAAGE. Beside the main product /Arg B31/insulin having a lower electrophoretic mobility than insulin, an even stronger basic component was formed, the product of trypsin catalyzed coupling of excessive dipeptide to arginine B31 of /Arg B31/insulin. The dependence of conversion rate on reaction time is shown in the following table:

Reaction time	porcine insulin	[Arg B31] insulin	[Arg-Ala-Arg B31-33] insulin
	% yield		
16 h	75.6	20.5	3.9
21 h	66.9	28.1	5.0
40 h	60.6	32.7	6.7
3 d	47.4	40.8	11.8
4 d	47.5	41.8	10.7

Trypsin-catalyzed reaction of porcine insulin with H-Ala-Arg-OH

It becomes obvious that the yield of more basic /Arg-Ala-Arg B31-B33/insulin rises with the increasing reaction time. 40% of insulin could be converted into /Arg B31/insulin.
The applied trypsin was separated by gel filtration on Sephadex G50 and the /Arg B31/insulin further purified by anion exchange chromatography on DEAE Sephadex A-25 and desalting by gel filtration on Sephadex G50.
The product showed the correct amino acid analysis with the ratio of lysine to arginine being 1:2. Due to additional arginine in the molecule, the product has a lower mobility than insulin in PAAGE. The biological activity of /Arg B31/insulin was determined in mouse convulsion test and was 100% compared to insulin standard. In reversed phase HPLC the prepared /Arg B31/insulin

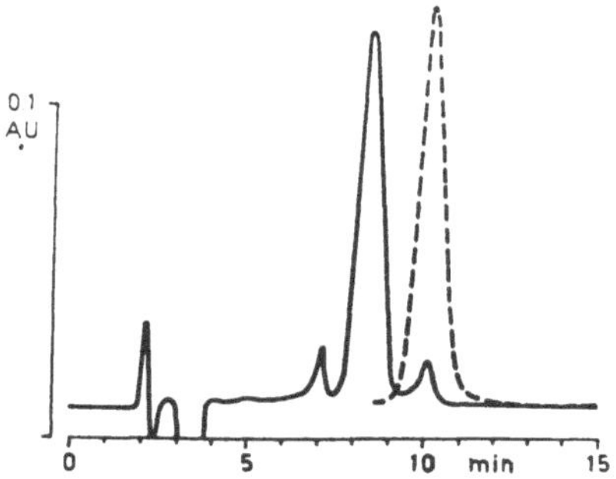

HPLC of [Arg B31] Insulin

Column: 4,0 x 150 mm, RP 18, 5 µm
Mobile Phase: 28% acetonitril, 72% 0.1m NaH_2PO_4 + 0.1m Na_2SO_4, pH 2.2
Temperature: 43°C
Detection: 215 nm
Flow rate: 0.54 ml/min
Sample: 25 µl, 0.5 mg/ml

----- porcine insulin

was eluted as more polar compound with a shorter retention time than insulin.

/Arg-Ala-Arg B31-33/insulin was isolated by anion exchange chromatography in an analogous procedure. This insulin derivative showed the expected ratio of the amino acids Lys 0,95(1):Arg 3,0 (3): Ala 3,2(3) and a shorter retention time in HPLC than /Arg B31/insulin.

References

1. Markussen,J. 1980, DK 80574; 1981, DE 3104949
2. Beyermann,M., M.Beyermann, E.Krause 1980, DD 160282
3. Obermeier,R., J.Ludwig, G.Seipke 1981, DE 3101 382
4. Jonczyk,A., H.-G.Gattner 1981, Hoppe-Seyler's Z. Physiol. Chem. 382,1591
5. Offord,R.E., K.Rose 1982,PCT WO 8302772; GB Appl. 82/3526
6. Yajima,H., M.Takeyama, J.Kanaki, K.Mitani, 1978, J.C.S. Chem. Comm.482

ENZYME-ASSISTED SEMISYNTHESIS OF POLYPEPTIDE ACTIVE ESTERS FOR SUBSEQUENT SPONTANEOUS COUPLING

Keith Rose, Carlos Herrero, Amanda E.I. Proudfoot, Carmichael J.A. Wallace and Robin E. Offord

Département de Biochimie Médicale, C.M.U., 9 avenue de Champel, 1211 Geneva 4, Switzerland

Introduction

Protein semisynthesis [1] has made great progress recently due to the employment of enzymes to form peptide bonds (for reviews see e.g. [2,3]). While the exopeptidase carboxypeptidase Y is appropriate for stepwise semisynthesis, the use of endopeptidases permits, also, fragment condensation to be achieved with great specificity and minimum use of protecting groups. In general, endopeptidases are used under conditions of thermodynamic control, and a rather high concentration of one (usually the amino) component is required in order to obtain good coupling yields with respect to the other component.

The goals of enzyme-assisted semisynthesis are becoming ever more ambitious, and a serious problem is now evident, that of the coupling together of two large polypeptide fragments. This problem arises because (i) larger peptides are more likely to possess potential sites of undesirable cleavage by the endopeptidase used for the coupling; (ii) it is generally more difficult to obtain solutions of high molarity with larger peptides than with small peptides or derivatives of amino acids, so it is difficult to obtain a favourable equilibrium, and (iii) larger peptides may present steric constraints (two large polypeptides and the enzyme have to come together) under the non-denaturing conditions of enzymic resynthesis (see e.g. [4]). Indeed, points (ii) and (iii) lead to long reaction times and this, in turn, amplifies point (i). The effects of all three of these points may be attenuated by first employing a high concentration of a nucleophilic agent in a rapid, preliminary, enzyme-assisted coupling to a carboxyl component, and then using the resulting modified polypeptide in a subsequent spontaneous chemical coupling to an amino component. Such an approach, that of enzyme-assisted specific carboxyl-terminal activation, has the advantage over chemical activation of complete specificity while avoiding the need for extensive protection and deprotection of side-chain carboxyl groups, difficult to accomplish satisfactorily when natural peptides are used as acyl components. It has the advantage over direct enzymic coupling of large fragments that the activation is rapid and proceeds in good to high yield: steric constraints are reduced in magnitude since a small nucleophile is used in the activation step, and the enzyme is not present when the two large fragments are brought together in the subsequent coupling step.

Peptides 1986

Enzyme-assisted specific carboxyl-terminal activation may be obtained in several ways:

(i) Coupling of hydrazine, either directly [5] or via an initial enzymic esterification (see (iv) below) to yield a peptidyl hydrazide for subsequent azide coupling.

(ii) Coupling of protected (tert-butyloxycarbonyl-) hydrazine [5-7] prior to or after amino protection, selective deprotection of the hydrazide, and an azide coupling.

(iii) Coupling of phenylhydrazine [5,8], for subsequent oxidation to a diimide.

(iv) Direct esterification, using the esterase activity of certain proteases. This is restricted by thermodynamic considerations to the formation of "unactivated" esters such as methyl and ethyl esters, and is usually a prelude to stepwise elongation with carboxypeptidase Y or to hydrazinolysis (see (i) above; such an approach to the formation of hydrazides is much less attractive than (ii) [6]). Direct aminolysis of the methyl ester by the alpha-amino group of another large peptide is unlikely to be very efficient except in cases where anchimeric assistance is considerable (as it is for some cases of aminolysis of polypeptidyl homoserine lactones, e.g. [9]). The production of a polypeptide methyl ester may of course also be achieved by the coupling of an amino acid methyl ester (see (v) below).

(v) Coupling of a preformed amino acid active ester, acting as nucleophile, to the carboxyl component. The resulting polypeptide active ester, freed from excess amino acid active ester, is then allowed to react with a suitable amino component. While enzymic formation of polypeptide esters which are not "active" (such as methyl esters) is commonplace, we believe we are here the first to describe the enzymic formation of polypeptide "active" esters (4-nitrophenyl or 2,6-dichlorophenyl esters). Full details will be published elsewhere.

Results

Preliminary experiments on the stability at 22°C of amino acid active esters in solutions of aqueous butane-1,4-diol at various values of pH showed that N-hydroxysuccinimide esters and 8-hydroxyquinoline esters are unstable at a pH above 4. The 2,6-dichlorophenyl ester of alpha-L-alanine is sufficiently stable at pH 5.5 to be useful as a nucleophile. For example, alanine 2,6-dichlorophenyl ester (0.44M) was coupled to fully acetimidylated tryptic fragment 1-38 of horse heart cytochrome-c̲ (18.4 mg) in 3.1 ml 88% butane-1,4-diol using TPCK-treated porcine trypsin at pH 5.5 (uncorrected glass electrode; acetate buffer). After 2h, the reaction mixture was quenched with an equal volume of formic acid, diluted with deionized 8M urea to starting conductivity, and applied to an ion exchange column (Trisacryl SP, 8cm x 1cm diam.) equilibrated with a buffer consisting of 40mM sodium acetate, pH 4.9, and containing NaCl, 50mM, and urea, 7M. After elution of unretained solvent and reagents, the column was washed with 135mM NaCl in the same buffer to elute trypsin and then eluted with a linear NaCl gradient from 135 to 500mM, also in the same buffer. In this way, a good separation was achieved between solvent and reagents, trypsin, uncoupled product (acetimidyl (1-38)) and coupled product (acetimidyl (1-38)-Ala-dichlorophenylester). Any unesterified acetimidyl (1-38)-Ala coelutes with uncoupled product.

The coupling yield estimated from u.v. monitoring (280 nm) of the effluent was 61%, and the overall recovery after desalting and lyophilisation was 7.3 mg. The product was characterised by reversed phase HPLC (it elutes significantly later than starting material) and by digestion with trypsin at pH 6 (alanine dichlorophenyl ester was liberated and identified by t.l.c., and the digested peptide then ran on HPLC with the acetimidyl (1-38) standard, as expected). The polypeptide active ester was stable at 22°C for many hours at pH below 6 in aqueous solution containing non-nucleophilic buffers. Aminolysis of the peptide active ester (ca. 15mM) was achieved in DMSO solution with L-Thr(Bu^t)-OBu^t as nucleophile (1M) at 22°C within 2h to give ca. 64% yield acetimidyl (1-38)-Ala-Thr(Bu^t)-OBu^t, which was characterised by reversed phase HPLC, and by identification by mass spectrometry of the dipeptide Ala-Thr(Bu^t)-OBu^t released by digestion with trypsin. Since trypsin was able to hydrolyse the ^{38}Arg-^{39}Ala bond rapidly and quantitatively, it is probable that racemisation was insignificant, but we have not yet established this directly. The coupling yield was apparently limited by hydrolysis of the ester due to residual water. Aminolysis using cytochrome-c fragment ε-acetimidyl (40-104) (ca. 11mM) was much less successful: although SDS-PAGE showed some apparent product formation, HPLC analysis indicated that aminolysis was not able to compete effectively with hydrolysis of the ester. More effective removal and exclusion of water would be expected to improve matters.

We have preferred to work with porcine trypsin, which retains its activity [6] in semi-aqueous solution even at low pH, but it should be possible to obtain similar results using other enzymes which retain their proteolytic activity at relatively low pH. The 2,6-dichlorophenyl ester proved to have excellent characteristics for our purposes. It is sufficiently stable to be used at a pH where trypsin is active and where thermodynamics favours good yields of polypeptide active ester, yet it is sufficiently active to give good yields upon aminolysis under appropriate conditions. The 4-nitrophenyl ester was also employed in our experiments, but was less convenient since it led to moderate yields of polypeptide active ester which was rather unstable.

Conclusion

A simple route to the preparation of polypeptide active esters has been described. Since no protection and deprotection of side-chain carboxyl groups is required, this route should already lend itself to (particularly enzymic) peptide synthesis involving a minimal protection strategy. Now that conditions of activation and isolation have been established, it remains to study the aminolysis reaction in order to obtain good yields of coupled product from two large fragments.

Acknowledgements

We thank the Fonds National Suisse de la Recherche Scientifique, the Stanley Thomas Johnson Foundation and the Schmidheiny Foundation for support.

References

[1] Offord, R.E. (1969) Nature 221 37-47

[2] Jakubke, H.-D., Kuhl, P. and Könnecke, A. (1985) Angew. Chem. Int. Ed. Engl. 24 85-93

[3] Breddam, K., Widmer, F. and Johansen, J.T. (1981) Carlsberg Res. Commun. 46 121-128

[4] Proudfoot, A.E.I., Offord, R.E., Rose, K., Schmidt, M. and Wallace, C.J.A. (1984) Biochem. J. 221 325-331

[5] Jones, R.M.L. and Offord, R.E. (1982) Biochem. J. 203 125-129

[6] Yagisawa, S. (1981) J. Biochem. (Tokyo) 89 491-501

[7] Zhang, Y.S., Cao, Q.P, Li, Z.G. and Cui, D.F. (1983) Biochem. J. 215 697-699

[8] Riemen, M.W., Pon, L.A. and Carpenter, F.H. (1983) Biochemistry 22 1507-1515

[9] Boon, P.J., Tesser, G.I. and Nivard, R.J.F. (1979) Proc. Natl. Acad. Sci. USA 76 61-65

CHARACTERIZATION OF SYNTHETIC PEPTIDE HORMONE ANALOGS AND OF BYPRODUCTS

L. Caporale, R. Nutt, J. Levy, M. Rosenblatt, J. Smith, B. Arison, W. Randall, C. Bennett, G. Albers-Schonberg. S. Pitzenberger, and R. Hirschmann

Merck Sharp & Dohme Research Laboratories
West Point, PA 19486 and Rahway, NJ 07065 (USA)

Introduction

The need for a battery of independent methods for structure confirmation of synthetic peptides is well recognized. There exists an equally important need to recognize byproducts and to determine their structure in order to facilitate yield improvement. In addition, in a program involving the synthesis of closely related analogs of peptides of medium and large size, methodology is desirable which simplifies the characterization of such analogs.

Results

We describe herein the use of spectroscopy and of peptide mapping to help achieve the above goals. The application of NMR spectroscopy to peptide chemistry in this laboratory has been previously reported (1). In this report we describe its application to synthetic peptides of higher molecular weight. We also report the use of Fast Atom Bombardment (FAB) mass spectroscopy pioneered by Sedgwick and his associates (2), as a practical, important tool for the synthetic peptide chemist. The peptide [$Nle^{8,18}$-Tyr^{34}]b PTH(3-34) amide (I), a biologically active analog of parathyroid hormone (3), was synthesized on solid support. The sidechain protecting groups were Tos(Arg), Chx(Asp), BOM(His), ClZ(Lys), BrZ(Tyr), For(Trp), and Bzl(Ser,Glu). The peptide was removed from the p-methylbenzhydrylamine resin by the "low-high" HF procedure (4). After purification by gel filtration and HPLC, the product gave the correct amino acid analysis (AAA) within 5%. FAB mass spectroscopy gave a molecular ion $(M+H)^+$ of 3918.00 (Calc: 3918.65) daltons. FAB measurements were carried out on a modified MAT-731 mass spectrometer at 3KV accelerating voltage (mass

Peptides 1986

range 5300 daltons) and - to maximize sensitivity - a resolution of 1:1000. Ion abundances recorded at unit mass intervals were used to calculate the center of gravity of the $(M+H)^+$ peak envelope consisting of the unresolved isotope peaks. Cesium iodide cluster ions were used for calibration. Molecular weight estimates by this method, in our experience, are accurate within ± 0.5 to ± 0.7 daltons. Depending on the direction and degree of deviation of the observed from the calculated value, even our instrument, with only modest resolving power, may permit one to conclude that unwanted conversion of Asn and Gln to Asp and Glu, respectively, did not occur during the synthesis of the peptide in question (5). The longer peptides [$Nle^{8,18}$-Tyr^{34}]b PTH(1-34)amide (II) and the corresponding 3,5-diiodinated tyrosine derivative (III) gave $(M+H)^+$, Calc: 4088.84; Fd: 4088.71, and $(M+H)^+$, Calc: 4349.66; Fd: 4350.23, respectively.

Peptide mapping of tryptic digests of I and II on HPLC gave, as expected, four fragments, three identical, for each peptide. All fragments had the expected amino acid compositions.

The synthesis of I afforded also an equal amount of a byproduct IV, eluting prior to the desired material on reverse phase HPLC in acetonitrile. AAA (Trp not determined) did not distinguish between I and IV. NMR spectroscopy of IV revealed formyl protons at δ 9.10 and 9.45. These could be shown to represent formyl rotomers as in the case of Boc·Trp(For), since heating to 70° caused the two peaks to coalesce at δ 9.23 in a temperature related, reversible manner. The C-7 proton of Trp was seen at δ 7.84 and 8.27, coalescing at 8.0. The FAB mass spectrum was consistent with the proposed structure (Calc: 3946.62; Fd: 3946.28). Treatment of IV with hydrazine afforded I. Formation of byproduct IV is now avoided by increasing p-thiocresol in the "low" <u>and</u> in the "high" HF procedures.

AAA suggested truncation at Glu^{19} for another important byproduct, V, from the synthesis of I. Trifluoroacetylation and pyroglu formation were considered possible causes for this chain termination. Fluorine NMR spectroscopy, however, excluded the former. FAB mass spectroscopy excluded also pyroglu formation as the explanation of chain termination, but strongly suggested acetylation of Glu^{19} ($(M+H)^+$ Fd: 2155.94; Calc: pyroglu 2092.0; acetylglu 2155.46). The NMR spectrum of V confirmed the presence of an acetyl, δ 2.04. Indeed, acetic acid was found as a contaminant in the purchased specimen of the Boc-Nle oil, which had been employed to form the symmetric anhydride for the addition of Boc-Nle^{18}. V also gave the expected amino acid analysis. Having been able to characterize V unequivocally to be acetyl [Tyr^{34}]b PTH(19-34) amide avoided our engaging in a pointless endeavor to prevent pyroglu formation, and led in a straightforward way to yield improvement. In a similar manner, we were able to show by FAB mass spectroscopy, NMR spectroscopy and AAA that a minor truncated component VI is acetyl [Tyr^{34}]b PTH(24-34) amide (Acetyl δ 2.02, $(M+H)^+$ Calc: 1455.70; Fd: 1455.91).

NMR spectroscopy also proved useful to distinguish Met from the corresponding sulfoxides. The Met methyl signal appears as a singlet at δ 2.14, and at 2.74 in the oxides. In PTH analogs such as [Tyr^{34}]b PTH(7-34) amide, $Met^{8,18}$ show singlet S-methyl peaks at δ 2.08 and at 2.10, whereas the sulfoxides exhibit peaks at δ 2.70. Unlike HPLC, this method does not require the synthesis of authentic specimen to determine the presence or absence of unwanted oxidation of Met. HPLC, however, was more effective in separating the 8-Met sulfoxides from the 18-Met sulfoxides than NMR.

Tyr, and especially Trp, are likely to be involved in side reactions. A facile method to assess Tyr/Trp ratios (e.g. when evaluating chromatographic fractions) would, therefore, be of practical value. It is well known that Trp has a λ_{max} at about 280 nm. The spectrum is only slightly pH dependent. Tyr has a pronounced maximum near 244 nm as the phenolate anion, but a minimum at this λ at neutral or acidic pH. For a variety of peptides, the ratio of absorbance at 280 nm (a measure of Trp content) to the change in absorbance with pH at 244 nm (a measure of Tyr content) is a fairly constant value for a given ratio of Trp to Tyr (0.623 ± 0.062 for a 1:1 ratio, and ca. 1.2 for a 2:1 ratio, etc.).

NMR spectroscopy also provides a sensitive technique for the detection of retained blocking groups or of unwanted, carbonium-ion-generated side products. In addition, we have found that the sensitivity of NMR at very high fields is adequate to permit quantitative analysis of the aromatic amino acids. Examination of the NMR spectrum of a sample of glucagon showed the ratios of the aromatic amino acids to be in excellent agreement with theory. Setting the area of the four tyrosine proton signals found at δ 6.76 and at 6.96 as 4.0, the areas assigned to the His, Trp and Phe protons were found to be as follows: Trp-4,7 plus His-5 protons: 3.0; Trp-2,4,5 protons plus the 10 phenylalanine protons: 13.4.

Unexpected signals in an NMR spectrum may arise from trivial, low molecular weight contaminants such as glycerol or acetic acid as well as from synthetic byproducts. We have extensively applied the Water-Eliminated-Fourier-Transform (WEFT) technique (6) to distinguish trivial from peptidal impurities.

Conclusion

Fourier Transform NMR spectroscopy, FAB mass spectroscopy, and U.V. spectroscopy are practical and powerful tools in the characterization of synthetic peptides in the 4500 dalton range. These techniques proved particularly useful in permitting structure assignment to byproducts which could not have been readily characterized by more conventional methods.

Acknowledgement

We wish to thank Ms. Joan Murphy for NMR spectra and Ms. Sue Fitzpatrick for amino acid analyses.

References

1. Daniel F. Veber. 1976. In: Peptides: Chemistry, Structure and Biology (R. Walter and J. Meienhofer, eds.). Ann Arbor Science, Michigan, p. 307-16.

2. Barber, M., R.S. Bordoli, R.D. Sedgwick, A.N. Tyler. 1982. Biomedical Mass Spectrometry 9, 208.

3. Habener, J.F., M. Rosenblatt, J.T. Potts, Jr. 1984. Physiol. Reviews 64, 985.

4. Tam, J.P., W.F. Heath, R.B. Merrifield. 1983. J. Amer. Chem. Soc. 105, 6442.

5. Jenkins, S.R., R.F. Nutt et al. 1969. J. Amer. Chem. Soc. 91, 505.

6. Patt, S.L., B.D. Sykes. 1972. J. Chem. Phys. 56, 3182.

DDZ-, A HIGHLY EFFICIENT PROTECTING GROUP IN LARGE SCALE SYNTHESES OF THYMOSIN-alpha 1 RELATED PEPTIDES.

Nguyen-Trong Hao, Wolfgang Heinzel and Christian Birr
ORPEGEN, Med.-Molekularbiologische Forschungsgesellschaft mbH, Heidelberg, F.R.G.

Introduction

The alpha,alpha-dimethyl-3,5-dimethoxybenzyloxycarbonyl (Ddz) residue was introduced as a N-protecting group for amino acids in 1972 (1). Nowadays all Ddz-amino acids are commercially available (2). Here we describe for the first time the large scale (kg) utilization of Ddz-amino acids.
Since the N-Ddz residue can be easily and selectively cleaved from the Ddz-amino acids under weakly acidic conditions (1-5% trifluoroacetic acid in dichloromethane, 15-45 minutes), it is possible and very convenient to synthesize peptides with Ddz-amino acids in combination with tert.butyl side chain protecting groups. Ddz-amino acids and the Ddz- residue are easily monitored on the tlc plate under u.v.light. Ddz-amino acids and -peptides on tlc plates become ninhydrin positive after brief warming of the tlc. The tert. butyl side chain protecting groups are stable under the Ddz- removal condition. A highly efficient synthesis in large scale of thymosin alpha 1 - and twin alpha 1 - fragments was therefore designed:

Results

Starting C-terminally, we prepared five peptide fragments of Thymosin alpha 1 and Twin alpha 1 in large scale (400 g amount and more!) in solution using Ddz- in combination with tert.butyl side chain protecting groups and L-lysine(Z), Figures 1-5. We activated the carboxyl group 30 minutes with DCCI/HOBT. Peptide synthesis was performed overnight with the amino component at room temperature

Peptides 1986

and pH 7.5. After basic, acidic, neutral washing and evaporation under reduced pressure the combined peptide is usually pure for

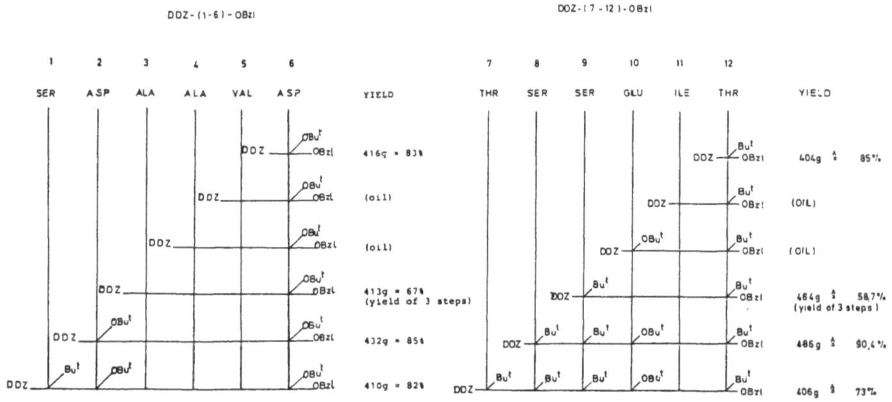

Fig. 1-2, Large Scale Synthesis of Fragment I and II of Thymosin alpha 1

the next coupling. The Ddz-residue of the peptide intermediates is now removed by stirring at room temperature for 20 minutes with 5% TFA in dichloromethane (v/v). After neutralization with N-methylmorpholine the solution is directly used as amino component for the next step. Up to the tripeptide stage, in most cases we obtained the peptide intermediates as oily products. On the tlc plate the removed Ddz- residue can be monitored under u.v.light. The Ddz- cleavage product is easily washed-out when the peptide is obtained as a solid. Washing with diethyl ether is the best method even in large scale. Assemblage of the synthesized intermediate fragments led to the preparation of the fully protected key product desacethyl-thymosin alpha 1 (3,4), Fig.6. To produce H-(20-28)-OBzl fragment IV and V were coupled by the azide method ensuing protected nonapeptide VI (86%) followed by removal of Ddz. Coupling of framgent III also by the azide procedure afforded the protected 16-peptide VII (56,8%) and by cleavage of the Ddz- residue, H-(13-28)-OBzl. Catalytic hydrogenolysis of fragment II and CDI/HOBT coupling with H-(13-28)OBzl resulted in the 22-peptide VIII (73,4%).

Hydrogenolysis of fragment I in trifluoroethanol gave Ddz-(1-6)-OH (98%), which was coupled with the amino component VIII, H-(7-28)-OBzl in N-methylpyrrolidone/dimethylformamide by the DCCI/HOBT procedure. We obtained the fully protected Ddz-thymosin alpha 1 with 84,4%. The product was purified by gel chromatography (Sephadex LH 20- column) with trifluoro ethanol and shown to be homogeneous on tlc. The amino acid analysis was in full agreement with the theoretical values.

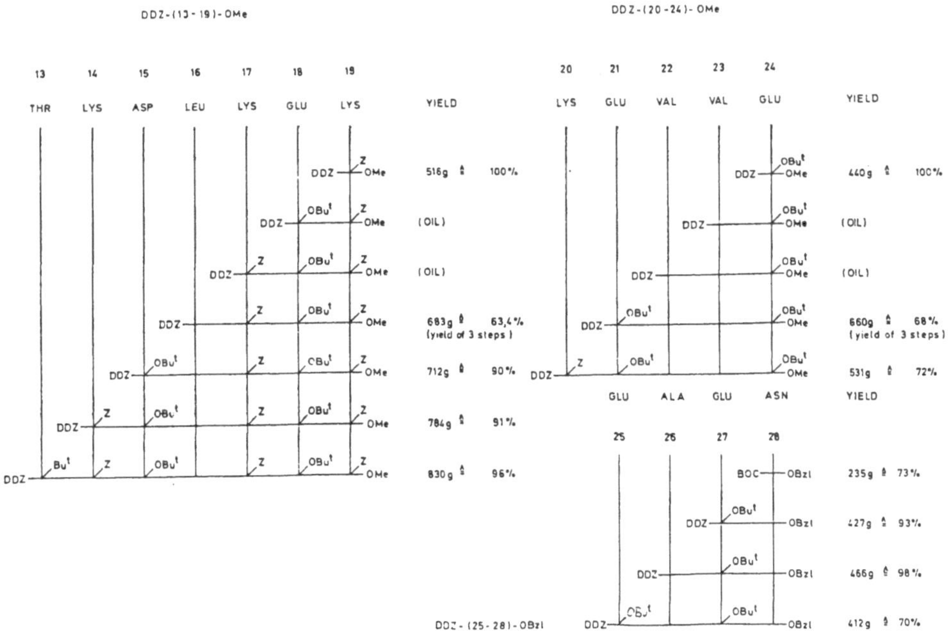

Fig. 3-5, Large Scale Synthesis of Fragment III-V of Thymosin alpha 1.

The key product H-(1-28)-OBzl was obtained by cleavage of the Ddz-residue and used for further syntheses of Thymosin alpha 1 and Twin alpha 1, the head-to-head dimer of Thymosin alpha 1, for clinical evaluations (5).

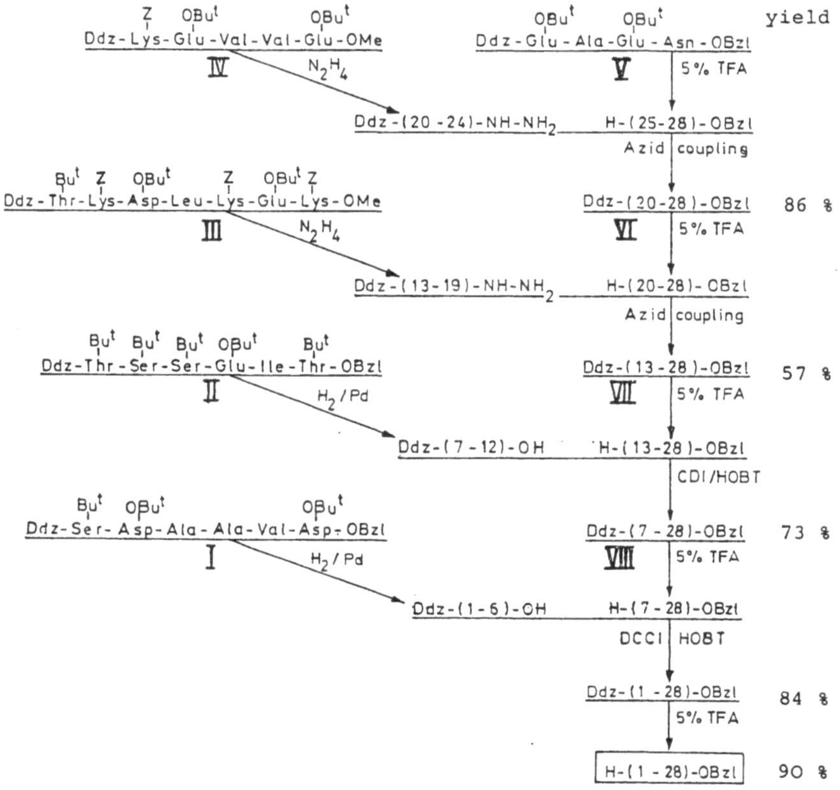

Fig. 6, Large Scale Fragment Condensations for Thymosin alpha 1

References

1. Chr.Birr, W.Lochinger, G.Stahnke and P.Lang, Liebigs Ann.Chem. 763, 162 (1972)

2. SHELL Fine Chemicals, Frankfurt/Eschborn, F.R. Germany and ORPEGEN, Heidelberg, F.R. Germany

3. Chr.Birr and U.Stollenwerk, Angew.Chem. 91, 422 (1979); Angew.Chem., Int.Ed.Engl. 18, 94 (1979).

4. Chr.Birr and I.Krueck, In PEPTIDE CHEMISTRY 1984 (N.Isumiya, ed) Protein Res.Foundation, Osaka (1985), 171-176.

5. Chr.Birr, W.Heinzel, C.Th.Nebe, A.Ho and B.Stehle, In PEPTIDE CHEMISTRY 1985 (N.Isumiya, ed) Protein Res.Foundation, Osaka (1986), 39-44.

CHEMICAL SYNTHESIS OF TRICHOTOXIN, A MEMBRANE ACTIVE POLYPEPTIDE MYCOTOXIN

H. Brückner
Institute of Food Technology, University of Hohenheim,
D-7000 Stuttgart 70, F.R.G.

Introduction

The mold *Trichoderma viride* NRRL 5242 produces a mixture of closely related, membrane active octadecapeptides named trichotoxins which have manifold biological activities (1). We have demonstrated that trichotoxin consists of neutral and acidic components (distinguished by an exchange of Glu/Gln and named A-40 and A-50 according to their r_f-values in TLC, both of which revealed, in turn, a great microheterogeneity in HPLC (2). We also elucidated the complete sequences of highly purified trichotoxins using FABMS and GC/MS (3,4). In the following we report the synthesis of one component which has been named trichotoxin A-50/E according to its retention time in the HPLC elution pattern (2,4).

General strategy of the Synthesis

We used, with one exception, the Z/OBu^t strategy throughout the syntheses of all intermediates as in contrast to the Z/OMe, Boc/OMe or Boc/OBzl strategy. It has the following advantages (for a discussion of the various synthetic approaches involved in the alamethicin syntheses, see reference (5) and the literature cited therein): The synthesis of the starting material Z-Aib-OH, which is required in large amounts, proceeds rapidly and in high yields, in contrast to the formation of Boc-Aib-OH. Stable H-Aib-OBu^t is easily available through esterification with isobutylene and subsequent hydrogenolytic deprotection and distillation. The tert. butyl esters of the fragments used in the syntheses are completely and rapidly cleaved with TFA or TFA/CH_2Cl_2 with 1 h, as compared to saponifications of methyl esters which, in the case of Aib

Peptides 1986

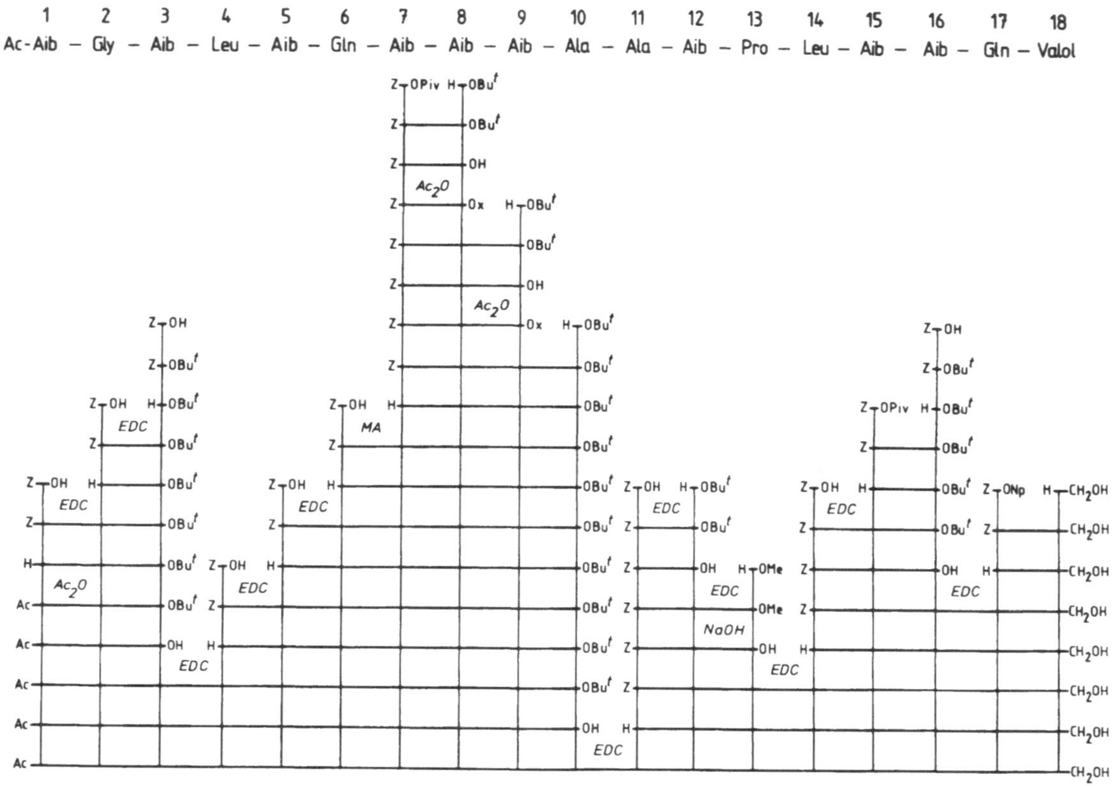

SCHEME OF THE SYNTHESIS OF TRICHOTOXIN A-50/E. Abbreviations: Ac, acetyl; Z, benzyloxycarbonyl; OBu^t, tert. butylester; OMe, methyl ester; OPiv, pivaloyl mixed anhydride; MA, mixed anhydride with isobutyl chloroformate; Ox, oxazolone; ONp, p-nitrophenylester; EDC, N-ethyl-N'-(3-dimethylaminopropyl)carbodiimide x HCl (1-hydroxybenzotriazole as additive); Ac_2O, acetic acid anhydride; Aib, α-aminoisobutyric acid; Valol, valinol. Deprotection was carried out with hydrogen and 10% Pd on charcoal (Z) or trifluoroacetic acid, TFA (OBu^t). Chiral constituents: L-config.

requires prolonged reaction time and/or elevated temperature. Furthermore, the hydrogenolytic cleavage of benzyl esters of Aib are known to proceed slowly and do not go to completion. This is just the reverse of the behaviour of Z-protected Aib-peptides. As a further advantage, the tert. butyl esters of dipeptides, for example H-Gly-Aib-OBu^t and H-Aib-Aib-OBu^t, are remarkably stable and show no tendency to form diketopiperazines. Water soluble carbodiimide (EDC) was found to give purer products compared to dicyclohexylcarbodiimide and was used exclusively.

Synthesis of the N-terminal Decapeptide Ac-1-10-OH

The key segment H-Aib$_3$-Ala-OBu^t was obtained in good yield from Z-Aib$_3$-Ox (6) and H-Ala-OBu^t and subsequent hydrogenolytic deprotection. As a result of the low reactivity of peptides with N-terminal Aib, the synthesis of the segment Z-4-10-OBu^t was the most difficult and required a stepwise elongation of H-7-10-OBu^t with Z-protected amino acids using the mixed anhydride or EDC/HOBt procedure. However, coupling of Ac-1-3-OH and H-4-10-OBu^t proceeded with a yield of 89%. Furthermore, kinetic studies using HPLC revealed that the deprotection of Ac-1-10-OBu^t by TFA was completed within 15 min without any peptide bond cleavage.

Synthesis of the C-terminal Octapeptide H-11-18-ol

To prevent acidolytic cleavage of the Aib-Pro bond (1,4) in the course of the synthesis of the segment Z-11-13-OH, coupling of Z-Ala-Aib-OH and H-Pro-OMe was chosen, followed by alkaline saponification. The utilization of C-terminal Pro in this segment has the further advantage that no racemization during saponification and subsequent coupling procedures are to be expected. Furthermore, condensation of Z-11-13-OH and H-14-18-ol proceeds in very high yields (93%). For the synthesis of Z-14-18-ol, Z-14-16-OH and an excess of H-Gln-Valol and the EDC/HOBt procedure were used.

Synthesis of the Octadecapeptide Ac-1-18-ol (Trichotoxin A-50/E)

Because no sterically hindered amino acids (7) were encountered in the final condensation step, coupling of Ac-1-10-OH and H-11-18-ol proceeded rapidly and in very high yield (88%) with EDC/HOBt and gave almost pure octadecapeptide as shown after the usual extraction procedure (ethyl acetate/1-butanol) by TLC and HPLC. Beautiful crystals were formed immediately after addition of acetonitrile to a methanol solution of the synthetic peptide.
A bioassay of this material revealed antibacterial activity toward *Bacillus subtilis* and hemolytic activity in agreement with the behaviour of the natural product (1). Moreover, the synthetic compound also exhibited ichthyotoxic properties against *Poecilia ciliata*.

Acknowledgements

This work has been supported in part by the Ministerium für Wissenschaft und Kunst Baden-Württemberg and the Deutsche Forschungsgemeinschaft.

References

1. Jung, G., Brückner, H., Schmitt, H. 1982. In: Structure and Activity of Natural Peptides (W. Voelter and G. Weitzel, eds.). Walter de Gruyter, Berlin · New York. pp. 75-114.

2. Brückner, H., Przybylski, M. 1984. J. Chromatogr. 296, 263-275.

3. Brückner, H. König, W.A., Aydin, M., Jung, G. 1985. Biochim. Biophys. Acta 827, 51-62.

4. Przybylski, M., Dietrich, I. Manz, I., Brückner, H. 1984. Biomed. Mass Spectrom. 11, 569-582.

5. Schmitt, H., Jung, G. 1985. Liebigs Ann. Chem. 321-344.

6. Jones, D.S., Kenner, G.W., Preston, J., Sheppard, R.C. 1965. J. Chem. Soc. 6227-6239.

7. Brückner, H., Bosch, I., Graser, Th., Fürst, P. 1986. J. Chromatogr. (in press).

CHEMICAL PHOSPHORYLATION OF SERINE PEPTIDES RELATED TO SUBSTRATES OF CYCLIC AMP-DEPENDENT PROTEIN KINASE

L. Grehn, B. Fransson, U. Ragnarsson
Institute of Biochemistry, University of Uppsala, Biomedical Center, Box 576, S-751 23 Uppsala, Sweden

Ö. Zetterqvist
Department of Medical and Physiological Chemistry, University of Uppsala, Biomedical Center, Box 575, S-751 23 Uppsala, Sweden

Introduction

At the last European Peptide Symposium we preliminarily reported on some novel work related to the synthesis, analysis and purification of substrates of cyclic AMP-dependent protein kinase (1). Further details on this project are now available (2-4). This report describes our most recent experiments related to the chemical synthesis of the corresponding phosphopeptides. Some preliminary results of HPLC analysis of such peptides are also mentioned.

Peptide Synthesis

The syntheses of the starting peptides 1a, 1b and 1d have been described elsewhere (3). Compound 1c was readily prepared in excellent yield from Boc-Ala-ONSu and Ser-Val-Ala-OBu^t under standard conditions.
The protected oligopeptides 1a-d were conveniently phosphorylated on serine with diphenylphosphorochloridate to give the fully protected phosphoserine peptides 2a-d in nearly quantitative yields. In all cases the phosphorylation step afforded a very pure product and traces of impurities were conveniently removed by routine procedures. The main advantage using diphenylphosphorochloridate as phosphorylating agent is that it is very reactive, and therefore gives a complete and selective reaction, especially under the influence of catalytic amounts of 4-dimethylaminopyridine (DMAP) (5).

Peptides 1986

Scheme 1. Synthesis of Phosphopeptides

Protected Ser(OH)-peptide $\xrightarrow{i}$ Protected Ser(OPO_3Ph_2)-peptide
<u>1</u> <u>2</u>

$\xrightarrow{ii,\ iii}$ Free Ser(OPO_3H_2)-peptide
<u>3</u>

Reagents: i, diphenylphosphorochloridate, DMAP, dry pyridine, 15 h, 0 °C; ii, TFA-CH_2Cl_2 (1:2, v/v), 1 h, RT; iii, H_2 (PtO_2), 80 % aqueous HOAc, 400 kPa, 50 h, RT.

Furthermore, it is commercially available in sufficient purity and can be handled conveniently without special precautions. For details on reagents and reaction conditions, see Scheme 1. Physical data for all peptides under discussion are given in Table 1.
Other methods of phosphorylation tried were less successful in these cases. Thus, treatment of <u>1d</u> with dibenzyl phosphoric acid in dry pyridine using excess dicyclohexylcarbodiimide or 1-(mesitylene-2-sulfonyl)-3-nitro-1,2,4-triazole as condensing agent gave complex, inseparable mixtures containing the starting material (6). Similarly, preliminary attempts to phosphorylate <u>1c</u> by a suitably designed

Table 1. Synthetic Phosphopeptides and Their Precursors

No	Peptide formulas	mp/°C	$[\alpha]_D^{25}/°$	(c)[a]
1a	Z-[Arg(Adoc)$_2$]$_2$-Ala-Ser-Val-Ala-OBu^t	225-30	-4.5	1
1b	Z-Arg(Adoc)$_2$-Ala-Ser-Val-Ala-OBu^t	230-35	-6.8	1
1c	Boc-Ala-Ser-Val-Ala-OBu^t	199-200	-18.8	1
1d	Z-Ser-Val-Ala-OBu^t	165-66	-7.3	1
2a	Z-[Arg(Adoc)$_2$]$_2$-Ala-Ser(OPO_3Ph_2)-Val-Ala-OBu^t	138-39	-12.4	0.25[b]
2b	Z-Arg(Adoc)$_2$-Ala-Ser(OPO_3Ph_2)-Val-Ala-OBu^t	~140	-7.8	0.5
2c	Boc-Ala-Ser(OPO_3Ph_2)-Val-Ala-OBu^t	146-47	-15.1	1
2d	Z-Ser(OPO_3Ph_2)-Val-Ala-OBu^t	132-33	-7.9	1
3a	Arg-Arg-Ala-Ser(OPO_3H_2)-Val-Ala	0.22	(0.38)[c]	
3b	Arg-Ala-Ser(OPO_3H_2)-Val-Ala	0.31	(0.47)[c]	
3c	Ala-Ser(OPO_3H_2)-Val-Ala	0.35	(0.54)[c]	
3d	Ser(OPO_3H_2)-Val-Ala	0.35	(0.53)[c]	

[a] In DMF, when not otherwise stated. [b] In $CHCl_3$. [c] R_f-values in n-BuOH-pyridine-HOAc-H_2O (15:10:3:12) on cellulose plates. Values in brackets refer to the corresponding non-phosphorylated peptides.

Figure 1. Separation of phosphoserine peptides 3b (1), 3c (2) and 3d (3) by reversed-phase ion-pair chromatography. Support: Spherisorb C_{18} (ODS-2), 10 μm, 4.6x150 mm. Mobile phase: Sodium bicarbonate buffer (pH 7.44)/ethanol (86:14), 1 mL/min. Counter ion: DMOA, 0.0086 M. Detection: UV at 220 nm.

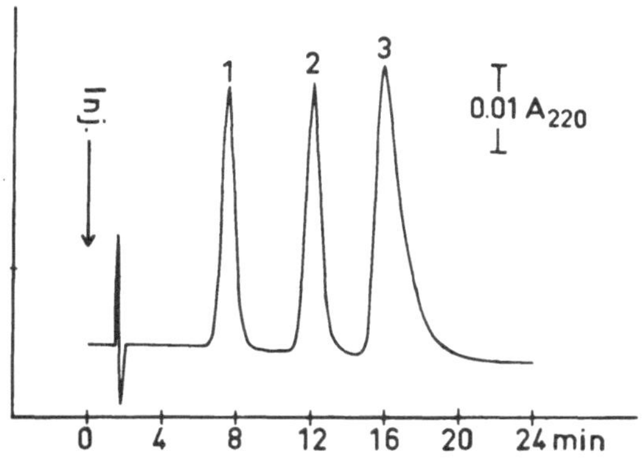

cyclic enediol phosphoryl approach gave an incomplete reaction together with considerable transesterification (7).

The sequential deprotection of the phosphorylated products 2a-d offered no special problems. First, the acid-labile protecting groups were smoothly removed by trifluoroacetic acid (TFA) in dichloromethane as usual. The resulting intermediates were subjected to catalytic hydrogenation in 80 % aqueous acetic acid using PtO_2 as catalyst. Thus, crude, essentially pure phosphopeptides 3a-d were obtained in satisfactory yields. Earlier investigators have carried out similar deprotections in one step by catalytic hydrogenation in TFA-acetic acid mixtures (8). We have found that our stepwise procedure offers a favourable alternative in this respect since the acid-labile groups are sometimes not removed completely in TFA-acetic acid even after several hours at ambient temperature.

Peptide Analysis

Previously we applied reversed-phase ion-pair HPLC to analysis (1-3, 9) and purification (1,2) of protein kinase substrates. The systems originally used exploited ion-pair formation between Arg-residues and hydrophobic sulfonates, used as additives, and could be applied also to analysis and purification of Arg-containing phosphopeptides (3). The phosphopeptides 3c and 3d, being acidic, hydrophilic peptides, required different conditions for their analysis. We have developed an HPLC system in which these peptides are chromatographed as anions with an amine, typically N,N-dimethyloctylamine (DMOA), as counter ion under isocratic conditions with ethanol as the organic modifier. This system appears to behave optimally around pH 7.5. Details on it are given in the legend to Figure 1. This system could be applied to assess the purity of 3b-d. The basic peptide 3a elutes in front of 3b under these conditions and can with advantage be chromatographed with sulfonate as counter ion (3).

Further work on the analysis and purification of acidic peptides, particularly phosphopeptides, by ion-pair technique is in progress.

Acknowledgement

This research was supported by grants from the National Swedish Board for Technical Development and the Swedish Natural Science Research Council. The cyclic enediol phosphorylation experiments were carried out in cooperation with Prof. I. Ugi and Dr. P. Lemmen.

References

1. Grehn, L., Ragnarsson, U., Fransson, B., Zetterqvist, Ö. 1984. In: Peptides 1984 (U. Ragnarsson, ed.). Almqvist & Wiksell International, p. 497.

2. Fransson, B., Grehn, L., Ragnarsson, U. 1985. J. Chromatogr. 328, 396.

3. Grehn, L., Fransson, B., Ragnarsson, U. 1986. J. Chem. Soc. Perkin Trans. 1 (in press)

4. Bålöw, R.-M., Tomkinson, B., Ragnarsson, U., Zetterqvist, Ö. 1986. J. Biol. Chem. 261, 2409.

5. Kellner, H.A., Schneiderwind, R.G.K., Eckert, H., Ugi, I.K. 1981. Ang. Chem. 93, 581; Ang. Chem. Int. Ed. Engl. 20, 577.

6. Jones, S.S., Rayner, B., Reese, C.B., Ubasawa, A., Ubasawa, M. 1980. Tetrahedron 36, 3075.

7. Ramirez, F., Marecek, J.F. 1985. Synthesis 449.

8. Perich, J.W., Alewood, P.F., Johns, R.B. 1986. Tetrahedron Lett. 27, 1373.

9. Fransson, B., Ragnarsson, U. 1983. In: Peptides 1982 (K. Bláha and P. Maloň, eds.). de Gruyter, p. 415.

COMPARISON OF FOUR APPROACHES TO THE SYNTHESIS OF RAT ATRIAL NATRIURETIC POLYPEPTIDE (ANP)

Nils Langeland Johansen, Peter Faarup, and B.F. Lundt,
Lab. of Medicinal Chemistry III, NOVO Pharmaceuticals R&D, Novo Alle, DK-2880 Bagsvaerd, Denmark.

Introduction

Our work with the synthesis of ANP analogues has involved four different approaches. These four syntheses, as applied to ANP(5-28), will be summarised briefly below and a comparison of the methods will be presented with the emphasis put on efficiency (i.e. no. of moles of protected amino acids used to prepare 1 mole crude ANP analogue), yield and purity of the crude product.

Synthesis in solution

In this synthesis Boc, But, OBut, Acm and protonation (Arg) were used for permanent protection: Z and Fmoc (in presence of Cys) were used for temporaty protection. The following four fragments were prepared mainly by the mixed anhydride procedure: Boc-(5-10)-OH, H-(11-16)-OH, Fmoc-(17-22)-OH, and H-(23-28)-OBut. Boc-(5-16)-OH and Fmoc-(17-28)-OBut were synthesized by fragment condensation utilising the mixed anhydride method and then coupled to Boc-(5-28)-OBut by EDAC/HOBt.
The protecting groups were cleaved by TFA and the resulting di-Acm-ANP(5-28) was converted to ANP(5-28) by iodine in 80% acetic acid.

SPPS using the Fmoc polyamide method

The synthesis was performed essentially as described in (1.). But, OBut, Acm and Mtr were used for the side chain protection. In the resin valine was employed as an internal standard and dialkoxy benzyl ester was utilised as the linker to the resin (2.). Complings were performed in DMF using symmetrical anhydrides (5 equiv.) for all amino acids except for Arg, Asn, and Gln for which HOBt esters (5 equiv.) were used. Couplings were checked by the qualitative ninhydrin test and the coupling was repeated when required. The synthesis was monitored by HPLC

Peptides 1986

analysis of protected fragments which were cleaved from resin samples by TFA/CH_2Cl_2 (1:99). After the assembly was compleated di-Acm-ANP(5-28) was liberated with TFA/phenol 95:5. Di-Acm-ANP(5-28) was converted to ANP(5-28) as described in the solution synthesis.

Combined solution and solid phase synthesis

It is appearent that a series of ANP analogues modified in the (17-28) fragment can be synthesized most efficiently by a combination of SPPS and solution synthesis. Boc-(5-16)-OH (1,5 equiv.) was synthesized by the solution method just described and coupled using EDAC (3.4 equiv.) and HOBt (3 equiv.) in DMF to the resin bound (17-28) fragment which was prepared according to the Fmoc polyamide method. Deprotection, cleavage from the resin and ring closure was performed as described previously for the Fmoc polyamide method.

SPPS using the Boc polysterene method

The synthesis was performed on an Applied Biosystems 430 A peptide synthesizer according to the methodology described in (3.). Bzl, OBzl, Br-Z, Mob, and Tos were used for side chain protection and the resin was of the Pam polystyrene type. Couplings were performed in DMF. Amino acids were activated as symmetrical anhydrides except for Asn, Arg and Gln where HOBt esters (4 equiv., two times) were used. The coupling yields were determined by the quantitative ninhydrin procedure. Deprotection and cleavage from the resin was effected by HF either by the low high procedure (4.) or by HF/anisole/CH_3-S-CH_2CH_3, 90:5:5 (5.). Formation of the disulfide bridge was performed by a ferricyanide oxidation.

Results and discussion

As expected the synthesis in solution exhibits the best efficiency and the highest crude product purity (HPLC) but it is of cource a rather time consuming method.
The crude products from the two SPPS methods clearly reflects coupling problems in the (10-15) sequence as observed from the ninhydrin test during synthesis. HPLC shows many impurities

Fig. 1. HPLC analysis of crude products

Synthesis in solution
Yield 1) 6.6%
Efficiency 2) 219 mol/mol ANP

3)

SPPS, Fmoc polyamide
Yield 8.2%
Efficiency 2900 mol/mol ANP

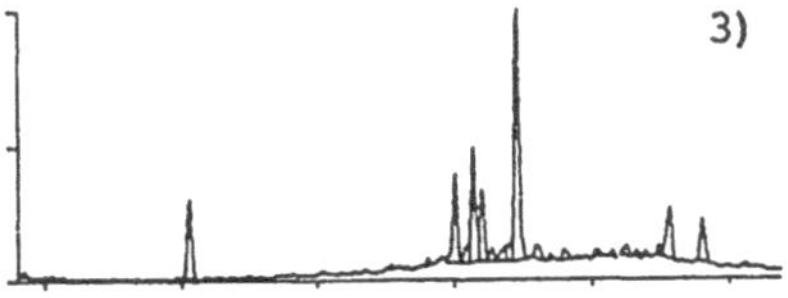
3)

Combined solution synthesis and SPPS
Yield 20%
Efficiency 825 mol/mol ANP

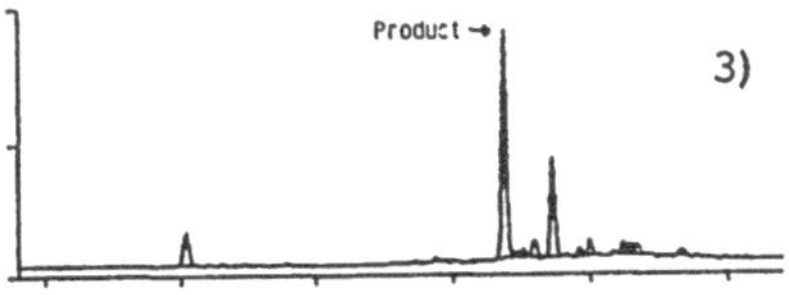

3)

SPPS Boc polystyrene
Yield 15%
Efficiency 880 mol/mol ANP

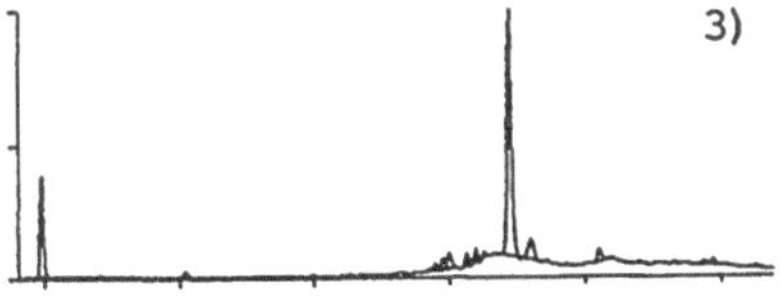
3)

1) Yields are calculated from the ANP(5-28) content in the crude products (HPLC). For the solution method the yield was based on Gly^{22}.
2) Efficiency is the total amount of protected amino acids used to prepare 1 mole crude ANP.
3) Conditions: Collumn:Nucleosil 5 μ C 18, flow: 1 ml/min. aquous buffer: 0.02 M, heptanesulfonate 0.05 M $(NH_4)H_2PO_4$, pH 6.0, gradient: 15% - 35% CH_3CN during 25 min.

eluting close to the ANP(5-28) peak. Purification of the products from these methods,by HPLC seems to be unavoidable. The combined solution and solid phase synthesis circumvents the coupling problems as the (10-15) sequence is not assembled by SPPS. On the other hand a major impurity of a still unknown structure is formed during cleavage of the peptide from the resin. Yet we have found this method well suited for synthesis of a series of ANP analogues with modifications in the (17-28) sequence. We have tried to improve the coupling yields in the (10-15) sequence for the Boc polystyrene method by changing the coupling solvent from DMF to 1) DMF/TFE, 4:1 with no significant improvement and to 2) DMF/NMP, 1:1 with even worse results.

A comparison of the crude product obtained from the low high HF procedure with that from the HF/anisole/CH_3-S-CH_2CH_3 procedure exhibits exactly the same yield and purity.
Despite the low coupling yields in the (10-15) sequence the Boc polystyrene procedure has been an extremely efficient method for the preparation of ANP analogues.

In Fig. 1 the purity of each of the crude products from the four synthetic methods is illustrated by the HPLC chromatograms. The yields given are calculated from the ANP(5-28) content in the crude products as determined by HPLC analysis. For the solution method the yield is based in the amount of Gly in position 22 used in the synthesis.

References

1. Brown, E., Sheppard, R.D., Williams, B.J.: J. Chem.Soc.Perkin I,p. 75-82 (1983).
2. Sheppard, R.C., Williams, B.J.: J.Chem. Soc.Chem.Comm. p. 587-588(1982).
3. Kent, S.B.H., et al.: Proc. 18th European Peptide Symposium, Stockholm, Ragnarson, J. ED., p.185-188 (1984).
4. Tam, J.,Heath, W.F., Merrifield, R.B.: J. Amer. Chem. Soc. 105, p. 6442-6455 (1983).
5. Schiller, P.W., et al., Biochem. Biophys. Res. Commun. 131. p. 1056-1062 (1985).

SEQUENCE ANALYSIS OF THE ANTIBIOTIC EPIDERMIN, A HETERODET TETRACYCLIC 21-PEPTIDE AMIDE

G.Jung, H.Allgaier
Institut für Organische Chemie, Universität Tübingen, D-7400 Tübingen

R.G.Werner
Abteilung Biotechnische Verfahren, Dr.Karl Thomae GmbH, D-7950 Biberach

U.Schneider, H.Zähner
Institut für Biologie, Mikrobiologie I, Universität Tübingen, D-7400 Tübingen

Epidermin is a new, large and complex polypeptide antibiotic exhibiting high antimicrobial activities against gram-positive bacteria including Propionibacterium acnes, Staphylococci and Streptococci involved in acne disease. Insertion of the basic amphiphilic polypeptide into the lipid membrane and interaction with various membrane constituents induces lysis via blocking of the bacterial cell wall synthesis and liberation of autolysines. Epidermin stops the murein synthesis via interaction with undecaprenyl-PP-GlcNac-MurNac-pentapeptide as observed with nisin.

A ribosomal biosynthesis via precursor proteins is indicated from experiments with translation and transcription blockers, and isolation of precursor proteins from cell extracts of the producer Staphylococcus epidermidis Tü 3298. Producing and nonproducing strains obtained by UV mutagenesis and curing agents show different plasmid patterns.

Postulated precursor sequences for subtilin (a), nisin (b) and epidermin (c):

a W K S E S L C T P G C V T G A L Q T C F L Q T L T C N C K I S K
b I T S I S L C T P G C K T G A L M G C N M K T A T C H C S I H V S K
c I A S K F I C T P G C A K T G S F N S Y C C

The enzymatic dehydration of Ser and Thr to Dha and Dhb residues and addition of the thiol groups of Cys would result in four sulfide bridges of epidermin, e.g. the conservative region TPGC would lead to the (2S,3S,6R)-3-methyllanthionine containing sulfide ring 8-11 present in all three antibiotics.

As the C-terminus of epidermin we elucidated the so far unique amino acid S-(2-aminovinyl)-D-cysteine, Cys(Avi), which may arise from oxidative decarboxylation of an intermediate meso-Lan ring in position 22.

Peptides 1986

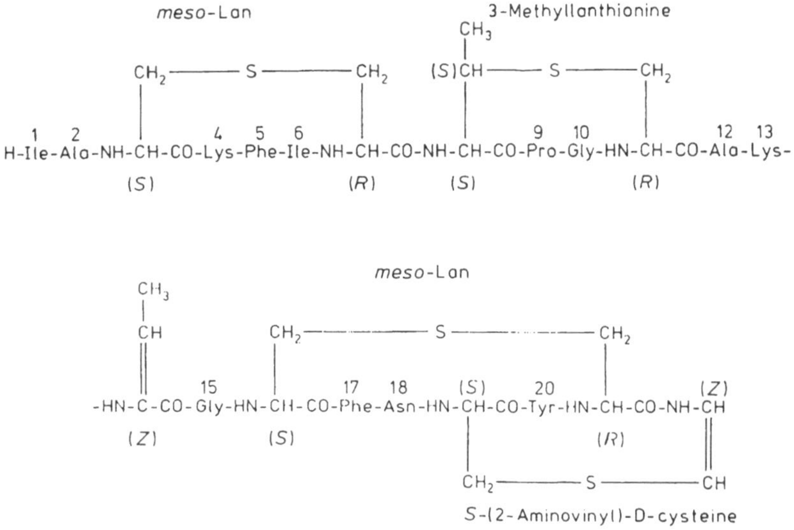

Tryptic cleavage of the Lys^{13}-Dhb^{14} bond yielded the water-soluble fragment P1 and the insoluble 2-oxobutyryl-fragment P2, which were elucidated as outlined in Scheme 1. Experimental details are described elsewhere (1,2). All fragments obtained by enzymatic and acidolytic degradations and by desulfurization with Raney-nickel W2 were purified by HPLC on Nucleosil 7C_{18}. The sequences were determined via a puzzle of end group determinations, Edman degradations, gas-phase sequencing, configurational assignments by GC on Chirasil-Val (3), FAB- and FD-MS, ^{1}H- and ^{13}C-NMR, and UV spectroscopy.

Hydrogenation of epidermin on Pd/C gave small amounts of hydrogenation products with antibiotic activity against Micrococcus luteus. GC-MS revealed, that D-Cys(Avi) was converted to the more stabile S-(2-aminoethyl)-L-cysteine (4-thialysine, D-Cys(Aet)), whereas treatment with Raney-Ni led to C-terminal L-alanine N-ethylamide and meso-lanthionine N-ethylamide on partially desulfurized and inactive fragments respectively.

The CD spectra of nisin are strongly solvent dependent in methanol, methanol/HCl, trifluoroethanol and hexafluoroacetone sesquihydrate, whereas those of epidermin show no significant variations. Compared to nisin the antibiotic epidermin exhibits a more rigid conformation. The restricted mobility of the tetracyclic 21-peptide amide also demonstrated by comparison of the ^{13}C-NMR spectra of the native antibiotic with those of its tryptic fragments P1 and P2. The three peptides show almost exactly coinciding signals of corresponding

Scheme 1.Sequencing of N-terminal epidermin fragment P1.

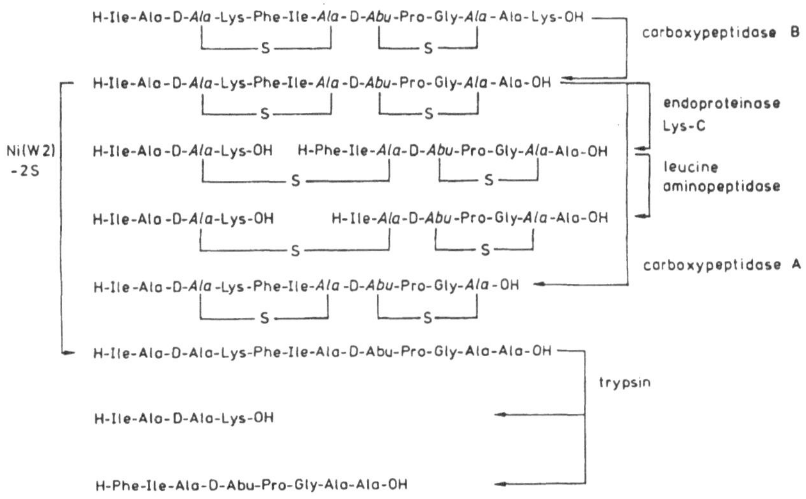

Sequencing of C-terminal epidermin fragment P2.

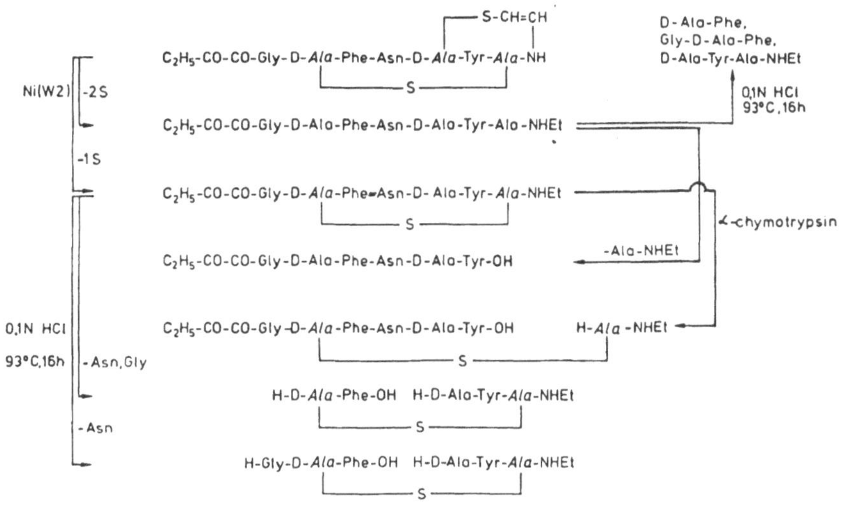

comparable carbon atoms, except for a small solvent shift (P1 in DMSO, epidermin and P2 in DMF).

Cys(Avi) is destroyed under the conditions of acidic total hydrolysis, and it can not be cleaved enzymatically. However its presence is indicated in the UV spectra of the native antibiotic and the tryptic fragment P2, since the bicyclic C-terminal part with the thioenol group of D-Cys(Avi) contributes to a UV maximum at 267 nm and the high extinction coefficient of 11 000 in water. This group gives also rise to a positive Cotton effect at 269 nm, θ = 22 300 deg cm^2 $dMol^{-1}$ (methanol). This long-wavelength transition is disappeared in all desulfurized or hydrogenated fragments and derivatives not containing D-Cys(Avi). The configuration of the S-aminovinyl group is Z($J_{H,H}$ = 7.25 Hz, $\delta(^{13}C)$ 102.3 (C_δ) and 131.7 (C_ε) ppm), and that of the Dhb residue is also Z($J_{H,H}$ = 6.9 Hz; $\delta(^1H)$ 1.7 and 6.44. The pyruvyl group in the tryptic fragment P2 is indicated by characteristic ^{13}C-NMR absorptions at 199.5(C-2), 162.2(C-1), 30.7(C-3), and 7.2(C-4) ppm.

References

1. Allgaier,H., Jung,G., Werner,R.G., Schneider,U., Zähner,H. 1985. Angew. Chem. 97, 1052-1054; Angew. Chem. Int. Ed. Engl. 24, 1051-1053.

2. Allgaier,H., Jung,G., Werner,R.G., Schneider,U., Zähner,H. Eur.J.Biochem. (in press).

3. Küsters,E., Allgaier,H., Jung,G., Bayer,E. 1984. Chromatographia 18, 287-293.

ISOLATION AND STRUCTURE IDENTIFICATION OF A NEW MAST CELL DEGRANULATING PENTADECAPEPTIDE FROM *PARAVESPULA VULGARIS*

W.A. König, U. Schulze, K.D. Schrader
Institut für Organische Chemie, Universität Hamburg,
D-2000 Hamburg 13 (FRG)

R. Chen
Max-Planck-Institut für Biologie, D-7400 Tübingen (FRG)

L. Grotjahn
Gesellschaft für Biotechnologische Forschung mbH,
D-3300 Braunschweig (FRG)

Introduction

Oligopeptides have been identified as toxins in bee and snake venom. About 60 % of the dry matter of bee venom can be attributed to peptides. In the case of wasps two types of physiologically active peptides have been described: kinins (1,2) with smooth muscle stimulating activity, which can be detected by a delayed contraction of the Guinea pig ileum and mastoparans (3), which have mast cell degranulating properties. In this process toxic and allergenic mediators, contributing considerably to the toxicity of wasp venom, are released from mast cells.
In our investigation of the venom of *Paravespula vulgaris*, the most common European wasp species, the presence of kinins could indeed be proved in the biological assay, however, their isolation was impossible due to the very low concentration in the venom. Though we succeeded in isolating a new mast cell degranulating peptide, which we named mastoparan V. It was investigated by quantitative amino acid analysis, Edman degradation and fast-atom-bombardment mass spectrometry (FAB-MS).

Peptides 1986

Results

Isolation of mastoparan V

Wasp venom was collected after freezing complete nests in liquid nitrogen and by preparation of individual venom glands under a stereomicroscope. Enzymes were precipitated by trichloroacetic acid and removed by centrifugation. The venom of 3991 wasps was fractionated on a Sephadex G 25 column and the fractions submitted to the biological tests (Guinea pig ileum test, mast cell degranulation test).

A substantial part of biological activity of wasp venom is certainly caused by catecholamines (histamine, serotonine, putrescine, dopamine, β-phenylethylamine, tyramine), which were identified by gas chromatography and mass spectrometry. These compounds are considered to be, at least in part, responsible for the pain caused by wasp stings.

Peptides were located by differential amino acid analysis of hydrolyzed and non-hydrolyzed samples and by screening biological activity (mast cell degranulation). The peptide containing fractions were pooled and investigated by HPLC on RP-18 columns with water, acetonitrile and 0.1 % trifluoroacetic acid as eluent and UV-detection at 215 nm. Surprisingly only 4.5 % of the total venom of *Paravespula vulgaris* can be attributed to peptides. A major fraction in the HPLC chromatogram proved to have a strong mast cell degranulating activity. This peptide was isolated by preparative HPLC on an analytical column (250 x 4 mm) and further investigated.

Structure investigations

Quantitative amino acid analysis of a hydrochloric acid total hydrolyzate revealed the following amino acids: met-sulfoxide (1), asx (2), ser (1), ala (2), ile (4), lys (4). A strong UV absorption at 254 nm pointed to the presence of tryptophane. Amino acid analysis after enzymatic hydrolysis with pronase E not only confirmed the presence of 1 equivalent of trp and met but also indicated the absence of asp. The asx moieties eluted with the retention time of asparagine.

Tryptic digestion of mastoparan V resulted in 5 fragments, which were isolated by electrophoresis and thin-layer chromatography (polygram cel 300). The fragments were submitted to the micro-

4. König, W.A., M. Aydin, U. Schulze, U. Rapp, M. Höhn, R. Pesch, V.N. Kalikhevitch. 1983. Int. J. Mass Spectrom. Ion Phys.46,403.

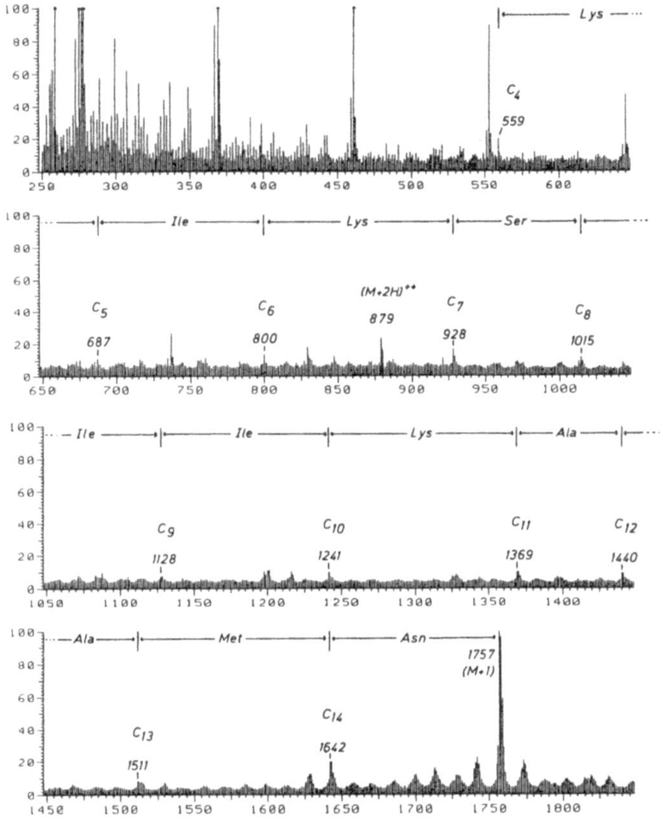

Figure 1. Positive ion FAB mass spectrum of mastoparan V. Kratos MS 50 instrument; collision gas: xenon; matrix: glycerol

Figure 2. Mode of fragmentation of a free peptide under formation of sequence specific positive ions of the "C-series" under fast atom bombardment

dansyl-Edman technique and identified as:

T 1 = ala-ala-met-asn
T 2 = ile-asn-trp-lys
T 3 = ser-ile-ile-lys
T 4 = ile-lys
T 5 = lys

For unambiguous identification of T 3, which was contaminated by trp, further investigations by quantitative amino acid analysis and FAB-MS were necessary to prove the presence of two ile residues.

In a separate experiment the complete peptide chain was degraded by the Edman technique and, in agreement with the tryptic fragments, the following sequence of a pentadecapeptide was derived:

ile-asn-trp-lys-lys-ile-lys-ser-ile-ile-lys-ala-ala-met-asn

The only ambiguity remains the C-terminal residue, which could either be asn or asp-α-amide.

Fast-atom-bombardment mass spectrometry of mastoparan V showed a protonated molecular ion at m/z 1757 (Figure 1), which confirms the structure derived by amino acid analysis and Edman degradation. In addition to molecular weight information the positive ion FAB-spectrum showed an almost complete set of N-terminal sequence specific ions, originating from cleavage of the peptide bonds (4) according to Figure 2.

So far 5 different mastoparans were isolated from different wasp species and described in the literature. They are all basic tetradecapeptide amides with a lys-lys partial sequence and a rather long hydrophobic part interrupted by lysine residues. Except for the chain elongation the new mastoparan V has a closely related overall structure.

References

1. Holdstock, D.J., A.P. Mathias, M. Schachter. 1957. Brit. J. Pharmacol. 12, 149.
2. Mathias, A.P., M. Schachter. 1958. Brit. J. Pharmacol. 13, 326.
3. Hirai, Y., T. Yasuhara, H. Yoshida, T. Nakajima, M. Fujino, C. Kitada. 1979. Chem. Pharm. Bull. 27, 1942.

PEPTIDES CONTAINING α,α-DISUBSTITUTED AMINO ACIDS. EXPERIMENTS RELATED TO α-HYDROXYMETHYLSERINE

K. Kociołek, U. Słomczyńska and M.T. Leplawy

Institute of Organic Chemistry
Technical University, 90-924 Łódź, Poland

Introduction

Efficient incorporation of α,α-disubstituted amino acids into peptide chain requires synthetic approaches serviceable for overcoming difficulties arising from steric hindrance. Reports from our own and other laboratories show the need for systematic investigation of each particular α,α-disubstituted amino acid and method of coupling. We describe here for the first time an examination of routes to peptides containing α-hydroxymethylserine (Hms), sterically hindered and highly hydrophilic α,α-disubstituted amino acid which may be expected to gain wider interest in biological and stereochemical studies on peptide analogues.

Methods and Results

Accessibility of α-hydroxymethylserine (Hms). Syntheses of Hms have been reviewed by O´Connor et al. (1) who obtained this amino acid by two step procedure involving copper(II) ion assisted hydroxymethylation of serine to the copper(II) complex of dihydro-1H,3H,5H-oxazolo(3,4-c)oxazole-7a-carboxylic acid (1) (2) followed by removal of copper, and subsequent ring cleavage. We found it equal efficient to use glycine instead the serine in the copper(II) assisted hydroxymethylation step by means of aqueous formaldehyde.

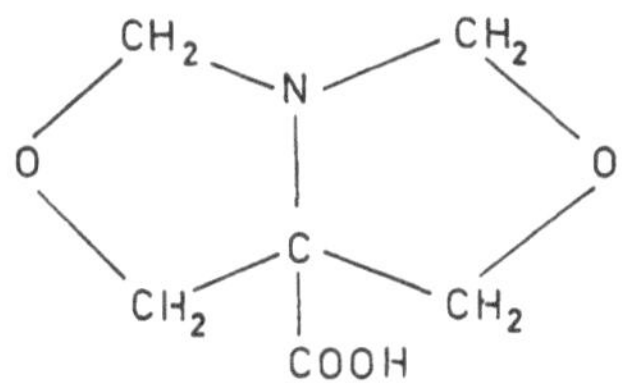

1

Peptides 1986

Thioacetamide proved to be more convenient and more efficient for removal of copper(3) than hydrogen sulphide used by australian authors. Overall yield 54-72%. Synthesis of Hms is easy to carry out, and isolation of pure product is simple, however a drawback of the method is high dilution of reactants (0,1 mol of glycine in 1l of water) and long time (5 days) needed for completion of hydroxymethylation.

N- and C-protection. As could be expected, urethane-type N-protection proceeds less satisfactory with Hms than with protein amino acids. N-Benzyloxycarbonyl (Z) derivative could be prepared in 41% yield after prolonged (3 days) treatment of Hms water/acetone solution with an excess (3 equiv.) of benzyloxycarbonyl chloride added in portions. The most efficient reagent for introducing the Boc-group is di-t-butyl dicarbonate. Under forcing conditions (50°, 48h) a 25-38% yield was obtained. Deprotection of Z-Hms by means of 5% Pd/C in the presence of ammonium formate (4), and acidic deprotection of Boc-Hms (4N HCl in ethyl acetate, 30% CF_3COOH) proceeded normally.

We encountered difficulties in attempted C-protection of Hms. Reactions are time consuming and yields are rather low. Crystalline Hms-OMe.HCl is accessible by the treatment (4 days) with thionyl chloride-methanol mixture albeit only in low yield (35%). Attempts to prepare carboxamidomethyl ester of Z-Hms by the method of Martinez et al. (5) failed. Treatment of Z-Hms with diazomethane led to a complex mixture of products. Even at the temperature -30°, more than one methyl group was introduced. The method of choice for converting the Z-Hms into methyl ester involves refluxing of methanolic solution of Z-Hms in the presence of p-toluenesulfonic acid (6). Yield 70%, crystalline product. The methyl group of Z-Hms-OMe could not be removed selectively by alkaline hydrolysis. Under such conditions decomposition of Z-derivative occured. Above results show the need for further search for efficient methods of C-protection and selective deprotection of Hms.

O-Protection. While the presence of two OH groups in Hms in 1,3 positions diminishes the value of the customary O-protection methods used in peptide work, it favours the isopropylidene approach from carbohydrate research, because formation of the 1,3-dioxane ring would enable simultaneous protection of both OH groups. Dipeptide derivative Z-Hms-Ser(Bzl)-OMe was treated with 2,2-dimethoxypropane (6 equiv., r.t., 16h) in the presence of p-toluenesulfonic acid (0.1 equiv.), and thus converted into O-isopropylidene derivative 2 in 77% yield. Deprotection could be smoothly achieved in 92% yield by heating of 2 in methanol containing 1 equivalent of p-toluenesulfonic acid. Stability of 2 under different peptide synthesis conditions is at present being examined.

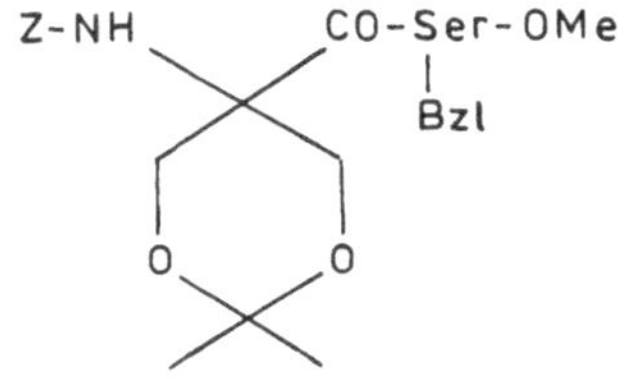

2

Dipeptides with N-terminal α-hydroxymethylserine. There are no restrictions as to the coupling methods which may be used to link the carboxyl group of N-protected Hms to the amino group of esters of protein amino acids. Examples: Z-Hms-Ser(Bzl)-OMe (78%.HOBt/DCC--method), Z-Hms-Phe-OMe (79%, HOSuc/DCC), Boc-Hms-Gly-OEt (67%, mixed anhydride method using isobutyl chloroformate).

Dipeptides with C-terminal α-hydroxymethylserine. We developed an indirect synthetic approach to dipeptides with C-terminal Hms avoiding the use hardly accessible C-protected Hms as coupling component. This approach involves transformation of dipeptide C-terminal serine into α-hydroxymethylserine by selective α -hydroxymethylation. The general synthetic sequence illustrating the transformation of C-terminal amino acid into α-hydroxymethylated analogue has been published (7).

Examples:

Boc-Gly-Ser(Bzl)-OH ⟶ ⟶ Boc-Gly-Hms(Bzl)-OH (56%)

Boc-Tyr(Bzl)-Ser(Bzl)-OH ⟶ ⟶ Boc-Tyr(Bzl)-Hms(Bzl)-OH (47%)

Conclusions

1. α-hydroxymethylserine is readily accessible by two step procedure involving copper(II) ion assisted hydroxymethylation of glycine to the copper(II) complex of 1 followed by removal of copper and subsequent ring cleavage.

2. Urethane-type N-protection proceeds less satisfactory with α -hydroxymethylserine than with protein amino acids.

3. Preparative value of hardly accessible methyl ester of α -hydroxymethylserine as amino component is doubtful because methyl group could not be removed selectively.

4. Isopropylidene-type O-protection enables simultaneous protection of both OH groups in α-hydroxymethylserine.

5. There are no restrictions as to the coupling methods which may be used to link the carboxyl group of N-protected α -hydroxymethylserine to the amino group of esters of protein amino acids.

6. Useful route to dipeptides with C-terminal α-hydroxymethylserine involves transformation of dipeptide C-terminal serine into α -hydroxymethylserine by selective α-hydroxymethylation.

Acknowledgement

This research was supported by grant MR I.12.1.6.10 awarded by the Polish Academy of Sciences.

References

1. O´Connor, M.J., J.R. Brush, S. B. Teo. 1977. Aust. J. Chem. 30, 683.
2. Brush, J.R., R.J. Magee, M.J. O´Connor, S. B. Teo. R.J. Geue, M.R. Snow. 1973. J. Amer. Chem. Soc. 95, 2034.
3. Taylor, U.F., D.F. Dyckes, J.R.Cox. 1982. Int. J. Peptide Protein Res. 19, 158.
4. Anwar, M.K., A.F. Spatola. 1980. Synthesis 929.
5. Martinez, J., J. Laur, B. Castro. 1985. Tetrahedron 41, 739.
6. Bodanszky, M. 1984. Int. J. Peptide Protein Res. 23, 111.
7. Kamiński, Z.J., M.T. Leplawy, Urszula Słomczyńska, Peptides 1982. Proc. 17th Eur. Peptide Symposium, p. 323. Walter de Gruyter, Berlin - New York.

PROTAMINES - VI. TOTAL SYNTHESIS OF [PERORNITHINE]-THYNNINE Z1 AND AMIDINATION STUDIES

F. MARCHIORI, G. BORIN, G. CHESSA, A. CALDERAN
Biopolymer Research Centre, C.N.R., Department of Organic Chemistry, Via Marzolo 1. 35131 Padova, Italy

Introduction

The substitution of somatic histones by protamines or specific basic histones in sperm cells is related with the condensation and repression of chromatine occurring in the male gametes during spermatogenesis. It seems quite obvious to correlate the capacity in compacting DNA with the high arginine content of protamines; however very little is known in regard to the features of the DNA-protamines complexes. Synthetic protamines or better suitably modified analogs will be very useful in exploring the structural features responsible for their biological function.

The presence of clusters of arginine, a peculiar character of the primary structure of protamines, represents the major difficulty in their synthesis. For this reason we proposed a methodology based on the amidination of synthetic perornithine analogs (1). In this communication we report the last results of our approach in the preparation of thynnine Z1 (H-Pro-Arg_5-Ser_2-Arg-Pro-Val-Arg_5-Tyr-Arg_2-Ser--Thr-Val-Ala-Arg_5-Val_2-Arg_4-OH). The choice of this protamine to test the methodology was determined by the presence of a chromophoric tyrosine, which can be very useful as a spectroscopic probe for conformational and DNA interaction studies.

Methods and results

To synthesize the [perornithine]-thynnine Z1 analog the classical

Peptides 1986

fragment condensation method in solution was employed. The strategy for the synthesis and selective protection of different functions was previously explained (2,3) and is inferable from scheme 1 where the final assembling of three fragments covering the entire sequence is reported. The hydroxyl functions of the two serines in fragment 1-12 were blocked as O-benzyl derivatives in order to prevent the distruction of these residues during the amidination reaction.

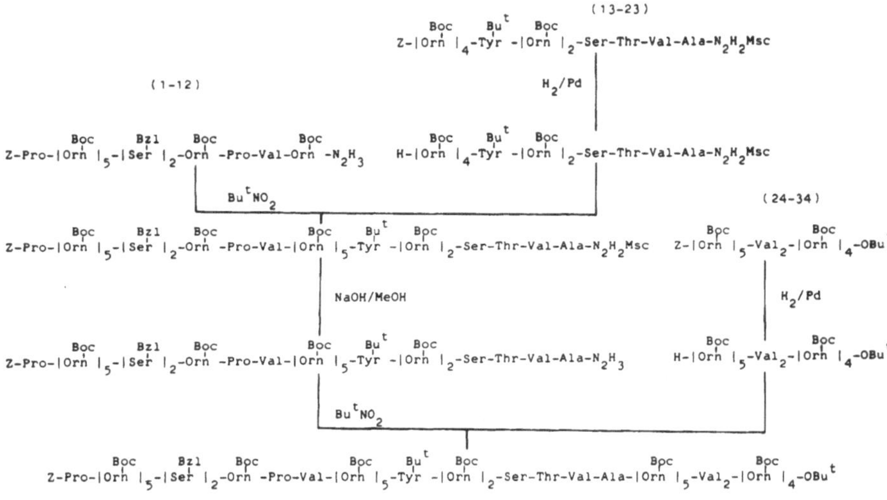

The protection of the C-terminal carboxyl function in the 24-34 fragment was necessary for solubility reasons; Z-Orn(Boc)-OBu^t was satisfactory prepared by esterification of Z-Orn(Boc)-OH with tert--butanol via mixed anhydride. The introduction "ab initio" of the Msc-protected hydrazide function on the pentaornithine peptide in this fragment (Scheme 2) presents some advantages in the purification of the intermediate hydrazide.

The last condensation step of Z-(1-23)-N_2H_3 with H-(24-34)-OBu^t occurs satisfactory and the crude product after treatment with TFA was purified on Sephadex G 50 SF (170x3 cm) column eluted with 10^{-2}M HCl. 350 mg (condensation and purification yield of 27%) of pure Z-Pro-Orn_5-Ser(Bzl)-Ser(Bzl)-Orn-Pro-Val-Orn_5-Tyr-Orn_2-Ser-Thr-Val-

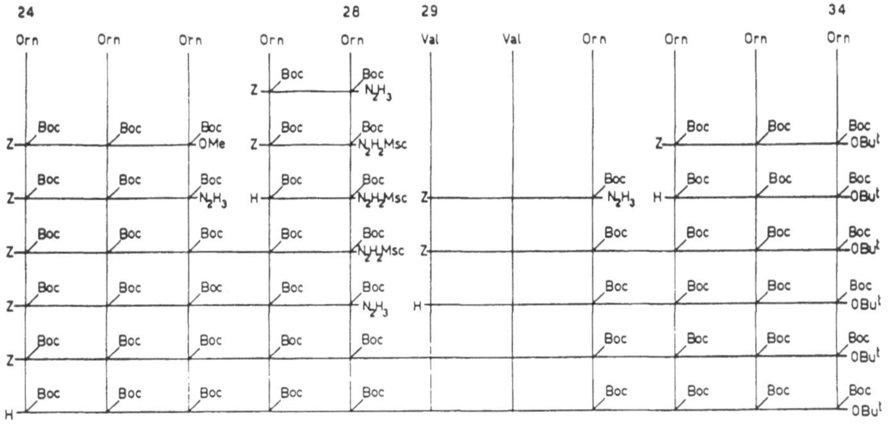

Scheme 2

-Orn$_5$-Val$_2$-Orn$_4$-OH were obtained. Amino acid ratios of the acid hydrolysate (48 h, 6N HCl in presence of 0.2% phenol): Thr 0.95 (1), Ser 2.80 (3), Pro 2.10 (2), Ala 1.10 (1), Val 4.10 (4), Tyr 1.15 (1), Orn 22.10 (22). In order to obtain the completely free perornithine thynnine Z1 analog, a fraction of this partially protected derivative was treated with liquid HF in the presence of 10% anisole; the acid hydrolysate contained the constituent amino acids in the theoretically expected ratios, but the peptide was not digested by AP-M. The amidination reaction represents a critical step of the proposed strategy. A relatively good compromise for the conditions of amidination of Z-(1-34)-OH, with the Ser^7 and Ser^8 still benzylated, was found: pH 9.4, 37°C for 3-4 days under nitrogen in the presence of 30-40 equivalents of 1-guanyl-3,5-dimethyl-pyrazole nitrate. The reaction mixture was dialyzed (mw cutoff 2000), lyophilized and the product was isolated passing through a Sephadex G 25 column (200x2 cm) eluted with 10% aqueous acetic acid. Amino acid ratios of the acid hydrolysate: Thr 1.09 (1), Ser 2.88 (3), Pro 1.99 (2), Ala 1.16 (1), Val 3.97 (4), Tyr 0.82 (1), Orn 2.33 (0), Arg 19.78 (22). The complete removal of protecting groups (N^α-benzyloxycarbonyl and the two O-benzyl of serines) was performed by treatment with HF and the product was further purified by chomatography in the same conditions reported above. The lyophilized product shows correct amino acid ratios in the acid hydrolysate [Thr 1.13 (1),Ser 2.81 (3),Pro 1.98 (2),Ala 1.19 (1),Val 4.06 (4),Tyr 1.08 (1),Orn 2.27 (0),Arg 19.45 (22)], but in the enzymic digest (AP-M) proline is absent and about one serine is lost. The presence of two significant extra picks suggest an incomplete hydrolysis,probably due to the resistence of particular peptide bonds to the enzyme. On the other hand a loss of one proline residue was found also in the enzymic hydrolysis of native clupeines YI and YII.

In the acid hydrolysate of synthetic peptides 5-10% of ornithine is always present and any attempt of increasing the conversion of ornithine into arginine modifying the reaction conditions did not give better results, even in the presence of KJ (30-40 equivalents) as reported by Zakirov (4). However it is very difficult to determine the real amount of unreacted ornithine since we are within the uncertainty of the analytical methods; the possibility of a little reconversion of arginine into ornithine during the acid hydrolysis must be also considered.

References

1. Marchiori, F., G. Borin, B. Filippi, V. Moretto, G.M. Bonora and C. Toniolo. 1979. Int. J. Peptide Protein Res. 14, 143-152.

2. Marchiori, F., G. Borin, D. Stivanello, V. Moretto and G. Chessa. 1982. Hoppe-Seyler's Z. Physiol. Chem. 363, 1483-1491.

3. Marchiori, F., G. Borin and G. Chessa. 1984. In : Peptides 1984, (V. Ragnarsson, ed.). Almqvist and Wiksell Int., Stockholm, Sweden. pp 275-278.

4. Zakirov, A.V.. 1973. Dokl. Akad. Nauk Tadzh. SSR. 18, 38-41.

MULTIPLE POSITIONS ^3H-LABELLING OF GROWTH HORMONE RELEASING FACTORS

R. Genet[1], J. Diaz[2], P. Perreaut[2], A. Sasaki[3], P. Fromageot[1] and J.L. Morgat[1].

1. Service de Biochimie, Département de Biologie, Centre d'Etudes Nucléaires de Saclay, 91191 Gif-sur-Yvette Cédex (France).
2. Centre de Recherches CLIN MIDY, rue du Pr. J. Blayac, 38082 Montpellier Cédex (France).
3. Institut de Chimie des Substances Naturelles, CNRS, 91190 Gif-sur-Yvette (France).

1. Introduction

Growth hormone releasing factor, GRF (1-44) NH_2 isolated from human pancreatic tumors is a highly potent and specific secretagogue of pituitary growth hormone (1). All structural requirements for GH release are present in the first 29 aminoacids. The recent availability of synthetic replicates has prompted us to label several members of this family with tritium. Binding, metabolism studies and autoradiographic localization of GRF required compounds selectively labelled of high specific radioactivities. The most frequent procedure for tritiation is iodination of tyrosinyl residue(s) followed by catalytic hydrogenolysis with tritium. GRF (1-44) NH_2 and GRF (1-29) NH_2 were tritiated with specific radioactivities of 1850 - 2.035 TB_q/mmole (50-60 Ci/mmole), respectively. Recently, it was reported that replacement of methionyl residue by its isostere norleucyl enhances its biological properties (2). This finding encouraged us to synthetize GRF (1-29) NH_2 in which the lysine 12 and methionine 27 were replaced by two acetylenic precursors. By reductive tritiation, Lys^{12}, Nle^{27}-^3H GRF (1-29) NH_2 was labelled in two positions with a specific radioactivity close to 1.665 TB_q/mmole (45 Ci/mmole).

2. Methods and results

2.1. Iodinations

GRF (1-44) NH_2 was iodinated with ICl according to a method described elsewhere (3) and GRF (1-29) NH_2 tetra-iodo derivative (a generous gift of Dr. Brazeau) was obtained by peptide synthesis.

2.2. Acetylenic precursor synthesis.

As a precursor of the tritiated Nle_{27} GRF (1-29) NH_2, GRF (1-29) NH_2 acetylenic derivative has been synthesized by a solid-phase technique (4). The two actylenic analogs of N_α-Boc-N_ε-Z-L-lysine and N-Boc-L-norleucine were prepared by Sasaki et al (5)(6).

Peptides 1986

The synthesis of this peptide is conducted in a stepwise manner using a Beckman 990 peptides synthesizer with 3 g. of paramethyl benzhydrylamin resin(M BHA) prepared as described by Rivier (7) from a Bio-Beads S-X_1 styrene divinylbenzene resin (200-400 mesh). This M BHA resin has a substituant range of about 0,4 to 0,5 mmoles/g. resin. The N-amino function is protected exclusively with the Boc group (t-butyloxycarbonyl). Side-chain functional groups are protected as follows : 2,6-dichlorobenzyl for tyrosine ; benzyl ester for glumatic acid, aspartic acid, serine and threonine ; tosyl for arginine ; 2 chlorobenzyloxycarbonyl for lysine ; benzyloxycarbonyl for amino L-2,6-diamino-4-hexynoic acid. The coupling reagent used is N,N'-dicyclohexylcarbodiimide (DCC) in presence of two molar equivalents of 1-hydroxybenzotriazol in the case of asparagine and glutamine. In general, one mole of Boc-amino acid is used per gram. of resine. The Boc-group is removed by 45 % TFA and 5 % $HS-(CH_2)_2SH$ in CH_2Cl_2. The built peptide is finally cleaved from the resin and the protecting groups removed with 0,5 ml anisole 1,5 ml thioanisole and 10 ml HF per gram of peptide-resin at -5° for one hour. The crude cleaved and deprotected peptide is then dissolved in 30 % acetic acid and subjected to Sephadex G 50 fine gel filtration. The peptide is then further purified by CM trisacryl M cation-exchange chromatography. Final purification is carried out using partition chromatography on Sephadex G 25 fine support with a n-BuOH : Acetic acid : water (4:1:5) solvent system. The chromatographic fractions are carefully monitored by HPLC, and only the fractions showing a purity > 99 % are pooled. HPLC conditions are: Column: RP 300 (35 x 0,39 cm, 10 µm particle size); eluent : CH_3CN (26 %) - TEAP (pH : 3) ; Flowrate : 2 ml/mn ; detection : 195 nm. Amino acid analysis was carried out after chemical hydrolysis, the amino acid ratios corresponded to the theoretical values.

2.3. Tritiations

Previous studies in our laboratory have permitted to define many parameters required for labelling (catalysts, pressure, low-absorption reacting vial, total deshalogenation and reduction). (diiodo 3,5-$Tyr^{1,10}$) GRF (1-44) NH_2 (0,46 µmole)(I), (diiodo 3,5-$Tyr^{1,10}$) GRF (1-29) NH_2 (0,51 µ mole)(II) and (Dah^{12}, Aha^{27}) GRF (1-29) NH_2 (Dah : L-2,6 diamino-4 hexynoic acid, Aha : L-2-amino-4-hexinoic acid) (0,58 µmole)(III) were respectively dissolved in 1 ml of a mixture (ACOH-MeOH-DMF-ACN : 1:1:1:1), pure methanol and 0,5ml of water; the solution was then frozen. Catalyst: Pdo : 20 mg(I), 15 mg(II) and Pd/Al_2O_3 : 8,5 mg(III), was disposed on the surface and the reacting vial (Minisorb Nunc) was connected to the automatic gas transfer device (8).

When the vacuum of 10^{-3}Torr was reached, 2,590 TBq (70 Curies) of pure tritium gas was introduced and compress to obtain a working pressure (I : 1 Bar, II : 1.1 Bar, III : 1.42 Bar) and the dispersed catalyst spread in the solution still frozen was flushed 15 minutes. The solution was then thawed and brought at room temperature under magnetic stirring (I : 2 hours, II : 2.5 hours, III : 3 hours). The adsorption of tritium gas gave a pressure reduction of about : 0.6(I) 0.2(II) 0.07 Bar(III). Catalysts were removed by centrifugation (3000 RPM) and labiles tritium atoms eliminated by rotovapor with 100 ml of dilute acetic acid or by two successive lyophilisations. The three ^{3}H-peptides were analyzed and purified by HPLC monitored with U.V. detector (Spectral LKB 2140) and radioactivity control (Berthold LB 506 A). HPLC conditions:

(I) RP.P 300 column, 30 cm x 3.5 mm, flow rate 1 ml/min., solvent TEAF 0.25 M, pH : 3, 29 % to 50 % ACN (60 minutes), RT : 860 sec.(I) and 1390 sec. (tetraiodinated GRF (1-44) NH_2) ;

(II) and (III): 18_C μBondapak, 30 cm x 7.8 mm, 2 ml/min. solvent system: TEAF 0.25 M pH3, 20 % to 40 % ACN (25 min.), RT : 1800 sec (II), 2040 sec (tetraiodinated GRF (129) NH_2), 1881 sec. (III) and 1801 sec. (acetylenic derivative (Nle^{27}) GRF (1-29) NH_2). ^{3}H-peptides retained both their chemical and biological properties. The ability of GRF to stimulate GH relase was measured by RIA (Dr. Haour Institut Pasteur, Paris). After hydrolysis of an aliquot of the labelled peptides (HCl 6N, 110°C, 17-18 hours) quantitative and qualitative estimations were consistent for the calculations of specific radioactivities.

Conclusion

Two tritiation methods applied to three different GRF species are reported. ^{3}H-labelling of peptides can generally be performed by catalytic hydrogenolysis from iodinated tyrosine-containing peptide obtained either directly or by synthesis. To label GRF' family on various residues, we have endeavoured to develop several methods to increase the specific radioactivity or to make a selective labelling for metabolism studies. Nevertheless, the specific radioactivities attained, indicate a significant isotopic dilution and the difficulty to control the parameters which favor the exchange between $^3H_2$ and the proton sources : catalyst, solvent, helium amount, labelling positions accessibility, etc... We are trying to develop new techniques allowing tritiations of methionine 27 as described elsewhere (9) and C-terminal residue by enzymic routes.

Acknowledgements

We thank Mrs. Odette Lebourguais for excellent technical assistance

References

1. Guillemin, R., Brazeau, P., Böhlen, P., Esch, F., Ling, N. and Wehrenberg, W.B.
Science, Vol. 218, 5 November 1982.

2. Rivier, J., Spiess, J. and Vale, W.
Proceedings of the 8th American Peptide Symposium. 1983.

3. Morgat, J.L., Hung, L.T. and Fromageot, P.
Biochim. Biophys. Acta. 207. 1970.

4. Merrifield, R.B. 1962 : Fed. Proc. Amer. Soc. Exp. Biol. 21, 412 and Merrifield, R.B. 1963 : J. Am. Chem. Soc. 85, 2149.

5. Sasaki, A. and Bricas, E.
Tetrahedron Lett. 1980, 21, 4263

6. Sasaki, N.A., Morgat, J.L. and Potier, P.
Int. J. Peptide Protein Res. 1986, 27, 360

7. Rivier, J.
J. Med. Chem. 16, 545. 1973.

8. Morgat, J.L., Desmares, J. and Cornu, M.
J. Label Compounds Vol. XI, 2. 1975.

9. Chassaing, G., Lavielle, S., Marquet, A., Genet, R., Fromageot, P. and Morgat, J.L.
in "Peptides Proceedings" of 9th Am. Peptide Symposium, Toronto. 1985.

THE SYNTHESIS OF CYSTEINE-CONTAINING FRAGMENTS (RESIDUES 6-12 AND 13-20) OF *CLOSTRIDIUM BUTYRICUM* APOFERREDOXIN

Brian Ridge and George Stavropoulos*

Department of Chemistry, University of Exeter, Exeter, EX4 4QD, England.
[*Present address: Department of Chemistry, University of Patras, Patras, Greece]

Introduction

The ferredoxin from *Clostridium butyricum* is an important electron transfer protein involved in bacterial metabolism (1). It contains two [4Fe-4S] cubic clusters which are ligated by cysteine thiolate groups of the enfolding protein chain (2). The synthesis of a series of cysteine-containing sequences of the apoprotein has been undertaken in order to study the influence of the peptide sequence on cluster properties.

This paper reports the synthesis, using conventional solution phase techniques, of two cysteine containing peptides, Z·Asp[OBu^t]·Ser[Bu^t]·Cys[Bzl(OMe)]·Val·Ser[Bu^t]·Cys[Bzl(OMe)]·Gly·NHNH·Boc and Z·Ala·Cys[Bzl(OMe)]·Ala·Gly·Glu[OBu^t]·Cys[Bzl(OMe)]·Pro·Val·NHNH·Boc, corresponding to residues 6-12 and 13-20 in the amino-acid sequence of the protein.

Results and Discussion

The syntheses are outlined in Schemes I and II.

A previously reported synthesis of related peptides used the benzyloxycarbonyl group for N-protection throughout (3). In the case of cysteine-containing peptides the group was removed by hydrogenolysis over freshly prepared palladium black in liquid ammonia (4). However, when the cysteine residue occupied an N-terminal position the hydrogenolysis was incomplete, the reaction

Peptides 1986

Scheme I Synthesis of Fragment (6-12)

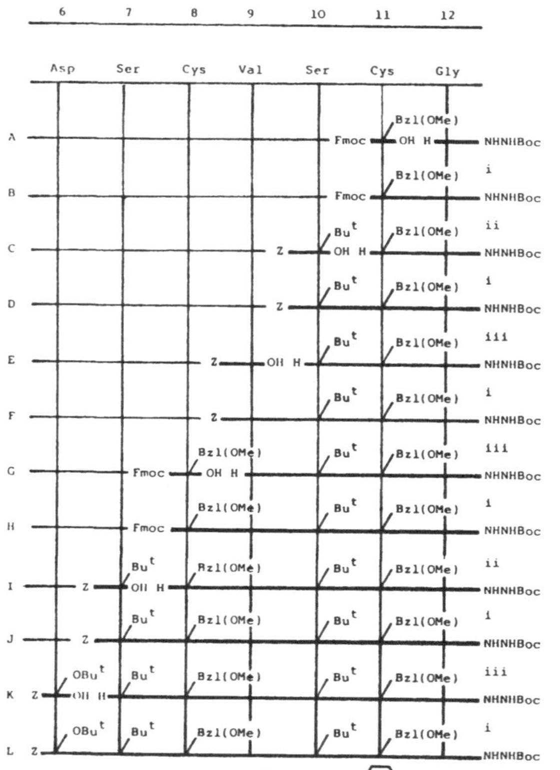

Reagents and Conditions: i, $Bu^iO_2C{\cdot}Cl$/ Me-N O , 2 min activation at -15°C, DMF; ii, liquid NH_3; iii, H_2-Pd black-liquid NH_3.

Scheme II Synthesis of Fragment (13-20)

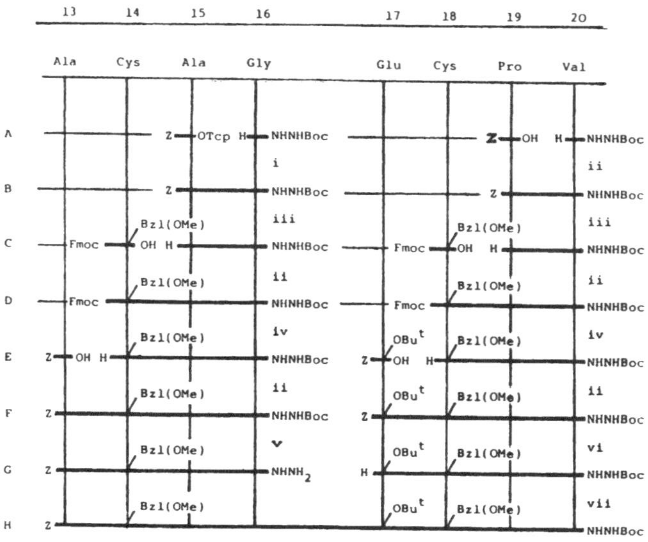

Reagents and Conditions:

i, active ester coupling, DMF;

ii, $Bu^iO_2C{\cdot}Cl$/MeN O, 3 min activation at -15°C, DMF;

iii, H_2-Pd/$BaSO_4$;

iv, liquid NH_3

v, $CF_3{\cdot}CO_2H/CH_2Cl_2$;

vi, H_2-Pd black-liquid NH_3;

vii, Bu^tONO/HCl, DMF.

coming to a standstill after about 70% conversion, and this led to purification difficulties. Accordingly, in the present synthesis, the 9-fluorenylmethoxycarbonyl group (5) was used to protect the amino-group of cysteine, whereas *N*-benzyloxycarbonyl was retained for all other residues. The fluorenylmethoxycarbonyl group was subsequently removed smoothly by treatment with liquid ammonia, and the latter group by the aforementioned hydrogenolysis.

The side chains of aspartic acid, glutamic acid, and serine, were protected as t-butyl esters or ethers, respectively. The side chain of cysteine was protected with the *S*-*p*-methoxybenzyl group (6), removable with mercury (II) acetate in trifluoroacetic acid (7). The t-butoxycarbonylhydrazide was used for C-terminal protection.

The fragment peptide (I L 6-12; Scheme I) was assembled by a stepwise synthesis by elongation from the C-terminus using the mixed anhydride method of coupling using isobutyl chloroformate and *N*-methylmorpholine with short activation time.

The fragment peptide (II H 13-20; Scheme II) was prepared from two subfragments (II G 13-16) and (II G 17-20) which were prepared by a stepwise mixed anhydride strategy, and coupled together using the Rudinger modification of the azide method.

Compound (*1*), a key intermediate in the syntheses, was prepared according to Scheme III.

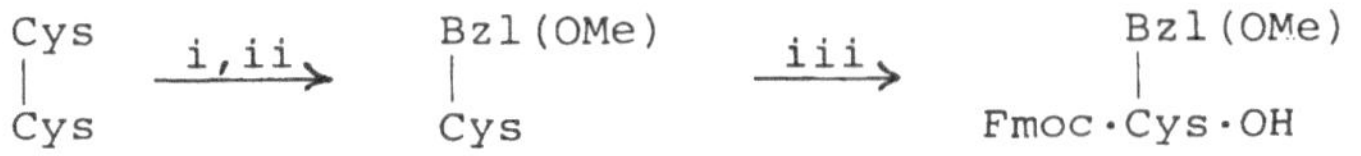

(*1*)

Scheme III

Reagents and Properties: i, Na/liq.NH_3, ii, (*p*)$MeO{\cdot}C_6H_4{\cdot}CH_2Cl$, iii, $Fmoc{\cdot}Cl/Na_2CO_3$.

(*1*), mp 141.5-143^oC, $[\alpha]_D^{25}$ -38.9 (C=1 in DMF).

Conclusion

The synthetic strategy outlined in this paper successfully overcomes the problem of the lack of reproducibility encountered during the deprotection of *N*-benzyloxycarbonyl cysteine peptides by palladium black catalyzed hydrogenation.

Acknowledgements

This research was funded by an SERC Grant, no GR/B 14142. The excellent technical assistance of Robin Holman is gratefully acknowledged.

References

1. Benson, A.M., H.F.Mower, K.T.Yasunobu. 1966. Proc.Natl.Acad.Sci. USA, 55, 1532.

2. Adman, E.T., L.C.Sieker, L.H.Jensen. 1973. J.Biol.Chem., 248, 3987; Adman, E.T., K.D.Watenpaugh, L.H.Jensen. 1975. Proc.Natl. Acad.Sci. USA, 72, 4854.

3. Gunter, M.J., B.Ridge, H.N.Rydon, R.Sharpe. 1979. Bioorg.Chem., 8, 371.

4. Meienhofer, J., K.Kuromizu. 1974. Tetrahedron Lett., 3259; J.Am. Chem.Soc., 96, 4978; Felix, A.M., M.H.Jimenez, J.Meienhofer. 1979. Org.Synth., 59, 159.

5. Carpino, L.A., G.Y.Han. 1972. J.Org.Chem., 37, 3404.

6. Akabori, S., S.Sakakibara, Y.Shimonishi, Y.Nobuhara. 1964. Bull. Chem.Soc.Jpn., 37, 433; Sakakibara, S., Y.Nobuhara, Y.Shimonishi, R.Kiyoi. 1965. Bull.Chem.Soc.Jpn., 38, 120.

7. Nishimura, O., C.Kitada, M.Fujino. 1978. Chem.Pharm.Bull., 26, 1576.

LABELLED [pGlu^{4}, Cyt^{6}]AVP-(4-8): FAILURE TO LABEL PROLINE WITH TRITIUM BUT SUCCESSFUL INTRODUCTION OF [^{35}S]CYSTEINE

J.W. van Nispen, J.A.J. Hannink, F.M. Kaspersen and E.M.G. Sperling
Organon Scientific Development Group, P.O. Box 20, 5340 BH Oss, The Netherlands

Introduction

Fragments of arginine vasopressin (AVP) containing a pyroglutamic acid residue (pGlu) in position 4 and cystine (Cyt) in position 6 have been isolated upon incubation of AVP with brain enzymes (1). In order to support and extend studies with these behaviourally very active fragments, labelled peptides were needed. The most obvious candidate amino acid to function as precursor for tritium incorporation in this case is proline, as the double bond in L-3,4--dehydroproline (Dhp) may readily be converted into the corresponding [3,4-$^{3}H_{2}$]Pro analogue (2) (specific activity several tens of Ci/mmol). In addition, the Dhp^{7}-containing analogue is an interesting analogue for pharmacological evaluation.
A second approach is the incorporation of ^{35}S with a very high specific activity (order of magnitude of 1000 Ci/mmol); therefore we decided to incorporate [^{35}S]cysteine into the AVP-(4-8) fragment.

Synthesis of [pGlu^{4}, Cyt^{6}, Dhp^{7}]AVP-(4-8) and attempts to reduce the double bond.

The fragment condensation approach as outlined previously (3,4) was used for this synthesis, the only difference being the use of Fmoc-Dhp-OH instead of Z-Pro-OH. Conversion of the Cys(Trt)-containing peptide into the Scm derivative and the subsequent reaction with cysteine gave the protected asymmetrical disulfide. Removal of the protecting groups with MSA/TFA/thioanisole and purification by chromatography on silica gel (4) gave pGlu-Asn-Cyt-Dhp-Arg-OH

Peptides 1986

with an HPLC purity of 95.0%, a correct NMR spectrum and a correct amino acid composition (except for Dhp which is not stable under the applied hydrolysis conditions with 5.7 N HCl). In the experiments to reduce the double bond with hydrogen or deuterium gas we used a mixture of DMF and water (9:1, v/v) as the solvent, and Pd/C, Rh/$CaCO_3$ or PdO (also mixtures were used) as the catalyst in a 1:1, m/m, ratio with the peptide. All experiments showed that the desulfurization reaction to give pGlu-Asn-Ala-Dhp-Arg-OH proceeded much faster than the Dhp ⟶ Pro conversion. Even the second, subsequent reaction pGlu-Asn-Ala-Dhp-Arg-OH ⟶ Pro analogue was faster than the appearance of the desired compound. Attempts to reduce the double bond in the protected pentapeptide pGlu-Asn-Cys(Trt)-Dhp-Arg(Pms)-OtBu (Pms stands for pentamethylbenzenesulfonyl) in DMF with Pd/C as the catalyst also failed.
The addition of BF_3-etherate, reported by Losse et al. (5) to improve yields of hydrogenolytical removal of the benzyloxycarbonyl group in S-protected cysteine-containing peptides was not beneficial in our hands. According to Losse's hypothesis (5) the double bond in Dhp in our case may be too close to the sulfur atom(s).

[^{35}S]cysteine-containing [pGlu^4, Cyt^6]AVP-(4-8)

In order to limit the number of steps to be carried out with the labelled peptide we wanted to perform the final reaction with cysteine on pGlu-Asn-Cys(Scm)-Pro-Arg-OH. Therefore, pGlu-Asn-Cys(Trt)-Pro-Arg(Pms)-OtBu was treated with TFA/thioanisole (4 h at 50°C) and the solution was evaporated in the presence of triphenylmethanol for retritylation (6). The reaction product was then directly treated with methoxycarbonylsulfenyl chloride (Scm-Cl) and cysteine; TLC and HPLC confirmed that the main product formed coincided with the desired asymmetrical disulfide. Characterization of a large batch of purified intermediate showed, however, the absence of the Trt function; we had obtained the symmetrical dimer of [pGlu^4]AVP-(4-8) which could be converted into the Cys(Scm)-peptide using Scm-Cl.
The conversion of the dimer of [pGlu^4]AVP-(4-8) into the Scm-derivative and the subsequent reaction with cysteine was optimized on a small scale. Two mg of dimer in 0.2 ml of trifluoroethanol (TFE) were treated with 2 µl Scm-Cl at 0°C for 20 min. and the

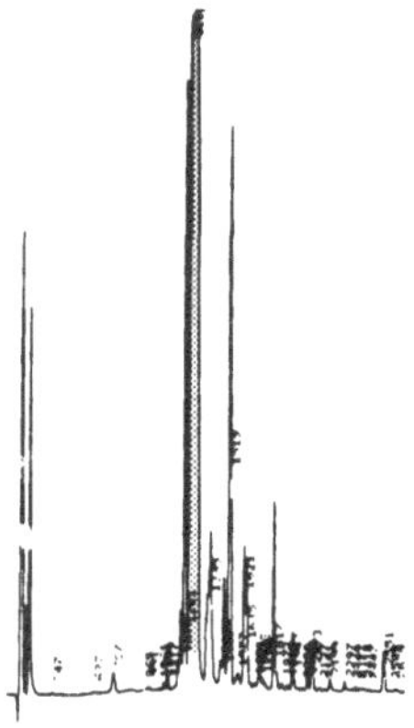

Supelcosil:LC-18-DB
Gradient :0.5% TFA/CH_3CN
Detection :210 nm

Fig.1 HPLC purification of [pGlu^4, Cyt^6]AVP-(4-8), obtained from the corresponding pentapeptide dimer.

reaction product isolated by the addition of ether. After several washings with ether the dried material was treated with 0.6 mg of H-Cys-OH.HCl.H_2O in TFE-MeOH (2:1, v/v) at room temperature for 3-4 h and the formed asymmetrical disulfide was precipitated with ether. Purification by HPLC (see Fig.1) gave 1.23 mg of [pGlu^4, Cyt^6]AVP-(4-8) (purity 98.4%) as was confirmed by co-injection with authentic material (3,4) and by amino acid analysis. In view of the high specific activity of [^{35}S]cysteine (approx. 1200 Ci (44 x 10^3 GBq) $mmol^{-1}$) and its delivery as an aqueous solution, we performed the final step with 15 µg of H-Cys-OH.HCl quantities in both TFE/MeOH and TFE/H_2O mixtures in polypropylene Eppendorf microtubes, and checked the reactions with HPLC. Similar patterns were obtained in all cases with the desired peptide present for 74-80%. Since Amersham Int. p.l.c. could only provide a solution of [^{35}S]cysteine containing a large amount of the reducing agent dithiothreitol (DTE), we first investigated the removal of DTE with unlabelled material (chromatography on SP-Sephadex) and then carried out this separation with labelled material. Reaction with the Cys(Scm)-containing peptide gave a disappointingly low yield of the correct peptide (20% in the presence of 100 µg of carrier and only 1% in the absence of carrier), which could only partly be ascribed to the quality of the labelled amino acid (3 radioactive peaks upon HPLC). Small-scale experiments in which DTE was added to the reaction mixture were then carried out; they showed that

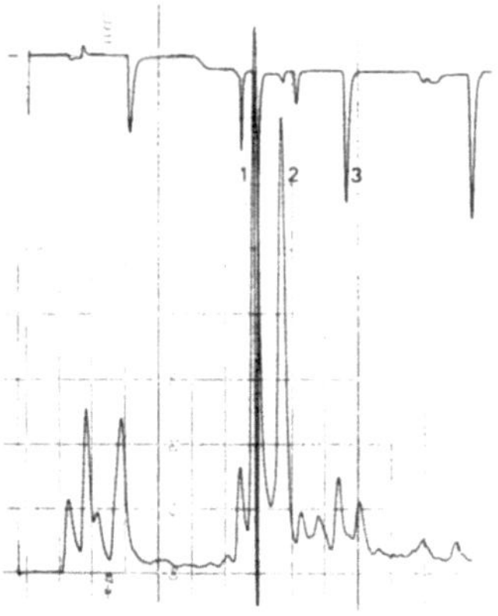

Supelcosil:LC-18-DB
Gradient :0.1% TFA/CH_3CN
Detection :UV 210 nm (upper trace)
radioactivity (bottom)

1. [pGlu^4, C$\overset{\star}{y}$t^6]AVP-(4-8)
2. unknown by-product
3. pentapeptide dimer

Fig.2 HPLC profile of crude [^{35}S]cysteine-containing [pGlu^4, Cyt^6]AVP-(4-8)

the presence of even a large excess (160 equiv.) of DTE did not interfere too much with formation of the desired asymmetrical disulfide. With another batch of [^{35}S]cysteine (4 mCi; still containing DTE) this could be confirmed; after HPLC purification (using a Supelcosil LC 18-DB column and a TFA/CH_3CN/H_2O gradient), the labelled peptide (540 µCi) could be isolated (see Fig.2). Its calculated specific activity was 1070 Ci/mmol.

References

1. Burbach, J.P.H., G.L. Kovács, D. de Wied, J.W. van Nispen, H.M. Greven. 1983. Science 221, 1310.

2. Felix, A.M., C.-T. Wang, A.A. Liebman, C.M. Delaney, T. Mowles, B.A. Burghardt, A.M. Charnecki, J. Meienhofer. 1977. Int. J. Pept. Protein Res. 10, 299.

3. Van Nispen, J.W., J.A.J. Hannink, H.M. Greven. 1983. In: Peptides: Structure and Function (V.J. Hruby and D.H. Rich, eds.). Pierce Chem. Co., p. 421.

4. Van Nispen, J.W., J.A.J. Hannink, M.S. Schoffelmeer, W.P.A. Janssen, J.P. Polderdijk, H.M. Greven. 1984. Recl. Trav. Chim. Pays-Bas 103, 68.

5. Losse, G., H.-U. Stiehl, B. Schwenzer. 1982. Int. J. Pept. Protein Res. 19, 114.

6. Barber, M., J.H. Jones, A.V. Stachulski, G.W. Bisset, H.S. Chowdrey, A.L. Hudson. 1979. Int. J. Pept. Protein Res. 14, 247.

CONTRIBUTION TO LANTHIONINE AND CYCLOLANTHIONINE CHEMISTRY

A. Yiotakis, A. Hatgiyannacou and [+]I. Photaki
University of Athens, Laboratory of Organic Chemistry
13A Navarinou Street, Athnes 106 80 Greece

Introduction

The interest on amino acid lanthionine has increased since it has been found in natural peptides isolated from bacterial cultures. Gross et al. (1) have shown that the antibiotics nisin and subtilin are built up from cyclolanthionine and β-methylcyclolathionine. Zervas and Ferderigos (2) showed that when the N-carbobenzoxy-S-benzoyl-L-cysteinyl-O-benzoyl-L-serine methyl ester was subjected to methanolysis, in the absence of air, two compounds I and II were isolated out of a mixture of other cyclo-lanthionyl diastereoisomers (fig. 1). As they demonstrated, compound I was a derivative of L-cyclolanthionyl and compound II a dimer of cyclo-(L→D)-lanthionyl, containing a 14-member ring. Photaki et al.(3) synthesized directly the monomeric cyclo-L-lanthionyl and cyclo-(D→L)-lanthionyl and studied the properties of the direct and indirect compounds.

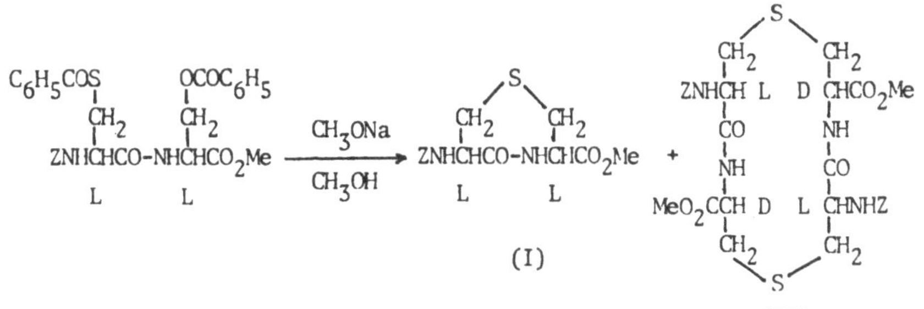

(I)

(II)

Fig. 1: Indirect synthesis of cyclolanthionine derivatives.

In continuing these investigations we describe the direct synthesis of the dimer cyclo-(L→D)-lanthionyl derivative.

Synthesis and methods

The dimer cyclo-(L→D)-lanthionyl derivative (II) was synthesized via known methods. Compounds III, IV, V, VI, and VII were prepared according to the methods described by Photaki et al. (3) with minor modifications. Compound VI was obtained by partial saponification of compound V. Because of the steric

Peptides 1986

hindrance exerted by the a-N-triryl group the saponification of methyl ester farther from the trityl group proceeds more rapidly than the saponification of the methyl ester adjacent to the trityl group (4). During the esterification of compound III (fig. 2a) it was observed that the formation of monomethylester was more rapid than the formation of dimethylester. By interrupting the reaction at a certain time, the monomethyl ester derivative could be isolated in almost pure state. As expected, this monomethyl ester was not the desired monomethyl ester VI, but the reverse.

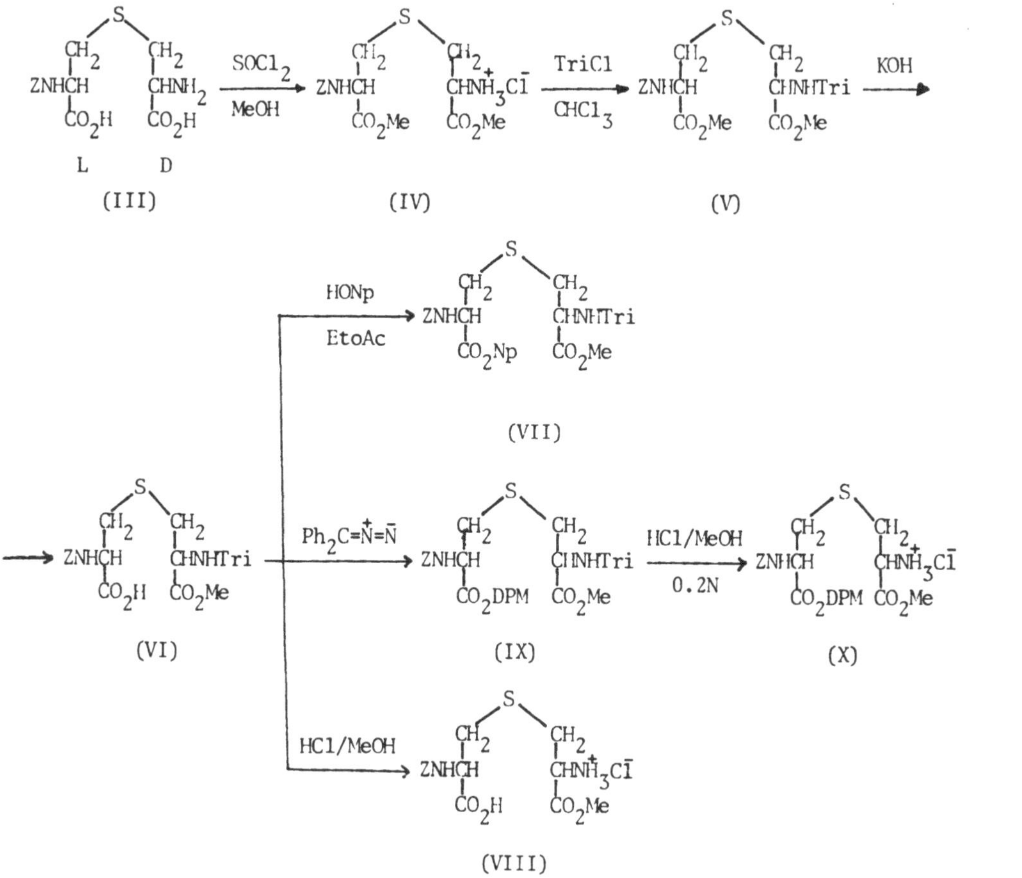

(III) (IV) (V) (VI) (VII) (VIII) (IX) (X)

Fig. 2a: Direct synthesis of the dimer cyclo-(L→D)-lanthionyl derivative.

The dimer cyclo-(L→D)-lanthionyl derivative was synthesized by three different processes (fig. 2b). In the first process the differenciation in the two carboxyl groups was achieved by converting one to methyl ester and the other to diphenylmethyl ester, compound X. After detritylation and dediphenylmethyla-

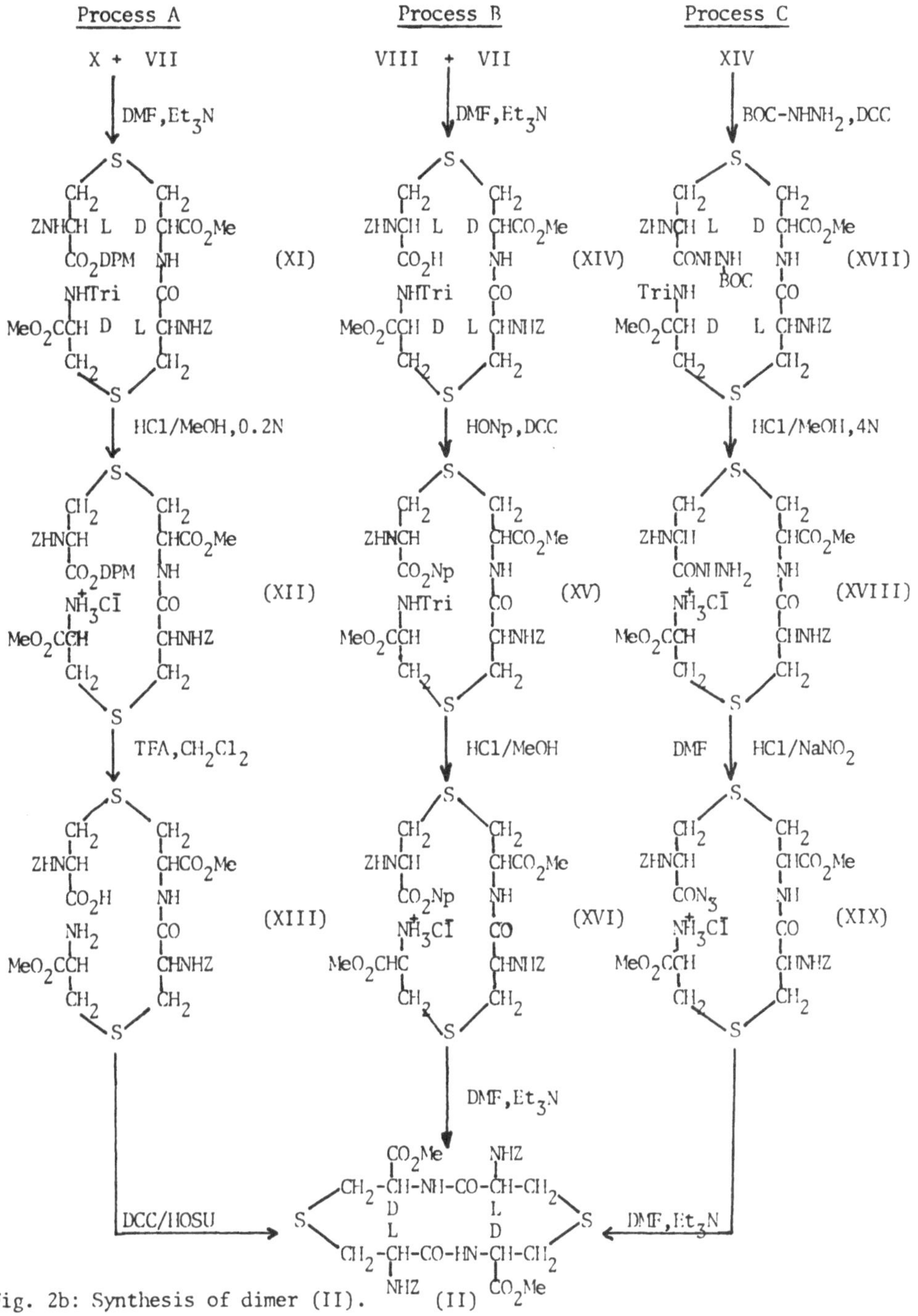

Fig. 2b: Synthesis of dimer (II).

tion, compound XI was converted to cyclic compound II, by means of dicyclohexyl-carbodi-imide (5) in the presence of N-hydroxysuccinimide.

In the second process, the cyclization was succeeded by p-nitrophenyl ester method (6). The yields of these two processes, were very low.

Much better yields have been obtained by the third process. In this case, compound XIV was transformed to the protected hydrazide XVII by reacting with t-Butoxycarbonyl hydrazine (DCC method). After removing the protecting groups (BOC and Tri) compound XVIII was converted to the reactive azide XIX and upon addition of base (Et_3N) and dilution (DMF), cyclization takes place to yield the final cyclic compound II. This product, prepared directly, proved to be identical with that prepared indirectly (2). This was in good agreement with the suggestion of Zervas and Ferderigos concerning the optical homogeneity of indirect cyclic compound.

All intermediates were purified either by recrystallization or by column chromatography and well characterized by tlc, elemental analysis and NMR spectroscopy.

In conclusion, we can note that these results not only conform and supplement previous investigations in our laboratory, but contribute farther to the lanthionine and cyclolanthionine chemistry.

Acknowledgement

The authors would like to thank Dr. F. Toma and Dr. A. Aumelas for performing and interpreting the NMR spectra.

Reference

1. Gross, E. and L. J. Morell. 1971. J.Amer.Chem.Soc. 93, 4634. Gross, E., H.H. Kiltz and E. Nebelin. 1973. Z.Physiol.Chem. 354, 810.
2. Zervas, L. and N. Frederigos. 1973. Experientia. 29, 262. Zervas, L. and N. Ferderigos. 1974. Isr.J.Chem. 12,139.
3. Photaki, I., I. Samouilidis, St. Caranikas and L. Zervas. 1979. J.C.S. Perkin I. 2599.
4. Zervas, L. and D.M. Theodoropoulos. 1956. J.Amer.Chem.Soc. 78, 1359.
5. Vogler, K., R.O. Studer, P. Lanz, W. Lergier, E. Böhni and B. Fust. 1963. Helv.Chim.Acta. 46, 2823.
6. Bodanszky, M. and V. du Vigneaud. 1959. J.Amer.Chem.Soc. 81, 2504.

LARGE-SCALE SYNTHESIS OF GRF(1-29) AND (1-44) AMIDES

Nagana A. Goud, Victor Garsky, Elizabeth Huelar, Firouz Shakoori, Mustaqim Syed and Michael S. Verlander

Bachem, Inc., 3132 Kashiwa Street, Torrance, California 90505

Introduction

It is now well-established that regulation of the secretion of growth hormone by the pituitary is controlled by two hypothalamic hormones: a growth hormone release-inhibiting factor characterized in 1973 as somatostatin (1); and a growth hormone-releasing factor (GRF) which was first isolated in 1982 (2) from a pancreatic tumor and later (3) shown to be identical to the molecule isolated from hypothalamic tissue. The native sequence of GRF is a 44-amino acid peptide with the following structure:

H-Tyr-Ala-Asp-Ala-Ile-Phe-Thr-Asn-Ser-Tyr-Arg-Lys-Val-Leu-Gly-Gln-Leu-Ser-Ala-Arg-Lys-Leu-Leu-Gln-Asp-Ile-Met-Ser-Arg-Gln-Gln-Gly-Glu-Ser-Asn-Gln-Glu-Arg-Gly-Ala-Arg-Ala-Arg-Leu-NH_2

The fragment, GRF(1-29)NH_2, has also been shown to possess full biological activity (4).

Because of the potential clinical significance of these molecules in the diagnosis and treatment of growth-related disorders, it is necessary to develop procedures for the production of substantial quantities of pharmaceutical-grade material for clinical studies. Although a large-scale synthesis of GRF(1-44)NH_2 has been reported (5) the approach is based on solution-phase methodology which is both tedious and unlikely to prove economical. We wish to report the synthesis of both GRF(1-44)NH_2 and GRF(1-29)NH_2 by conventional solid-phase methodology -- an approach which is both rapid and economical and has provided excellent yields (20-37 grams) of high-purity, pharmaceutical-grade materials.

Peptides 1986

Synthesis

The synthesis of both molecules was accomplished by standard, solid-phase methodology on benzhydrylamine resins. Conventional sidechain protecting groups were utilized, including tosyl on arginine, xanthyl on asparagine and glutamine, 2-chlorobenzyloxycarbonyl on lysine and 2,6-dichlorobenzyl on tyrosine. Couplings were mediated by dicyclohexylcarbodiimide/1-hydroxybenzotriazole, with repeat couplings where necessary. Ethanedithiol (0.5% v/v) was used as a scavenger during all acidolysis steps. Details of the synthesis of the two peptides are summarized in Table 1. After batchwise cleavage of the peptide-resins the crude products were purified by a rapid, two-step,

Table 1. Synthetic Details for GRF(1-29)NH_2 and GRF(1-44)NH_2

	GRF(1-29)NH_2	GRF(1-44)NH_2
Starting resin/subst.	150 g/0.4 meq/g	200 g/0.4 meq/g
Yield of peptide-resin	450 g	846 g
Yield of crude product	114 g	207 g
Yield of final product	20.25 g	36.90 g
HPLC purity	>99%	>99%

preparative HPLC procedure on Vydac C_{18} reverse-phase silica. The products were first purified utilizing a triethylammonium phosphate/acetonitrile mobile phase, followed by a second step which utilized 0.1% aqueous trifluoroacetic acid/acetonitrile. This procedure gave excellent yields of very pure products (Figure 1). The following analytical data were obtained for each product:

GRF(1-29)NH_2. Amino acid analysis: Ala, 3.18; Arg, 2.83; Asx, 3.07; Glx, 1.93; Gly, 0.95; Ile, 1.89; Leu, 4.00; Lys, 2.03; Met, 0.84; Phe, 1.00; Ser, 2.86; Thr, 0.83; Tyr, 2.19, Val, 0.87. Peptide content = 84%. $[\alpha]_D^{\sim 20} = -80.63^\circ$ (c = 0.1, 1% HOAc; corrected for peptide content).

GRF(1-44)NH_2. Amino acid analysis: Ala, 5.18; Arg, 5.84; Asx, 4.05; Glx, 7.26; Gly, 2.99; Ile, 1.74; Leu, 5.00; Lys, 2.02; Met, 0.93; Phe, 0.94; Ser, 3.32; Thr, 0.83; Tyr, 1.95; Val, 0.95.

Peptide content = 85%. $[\alpha]_D^{\sim 20} = -78.62^{\circ}$ (c = 0.1, 1% HOAc; corrected for peptide content).

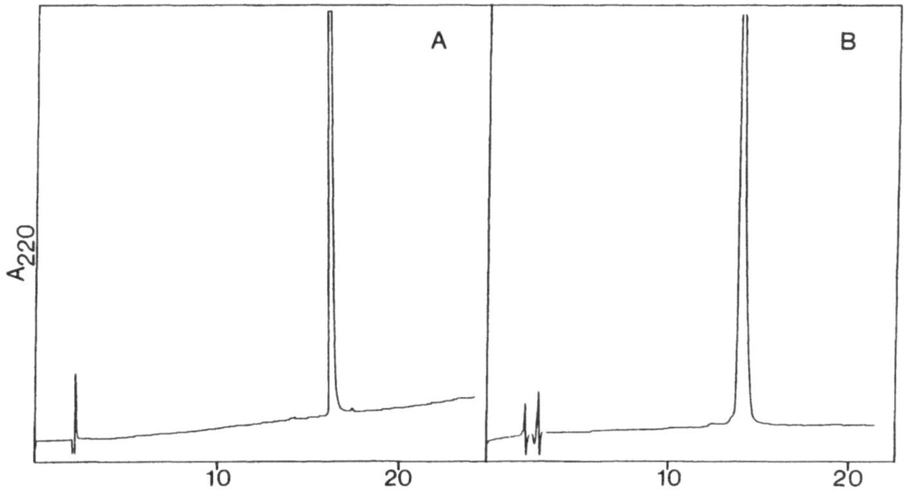

Figure 1. Analytical HPLC of purified GRF(1-29)NH_2 (A) and GRF(1-44)NH_2 (B). Column: Vydac C_{18}; gradient: 15 to 30% acetonitrile in 0.1% aqueous trifluoroacetic acid over 25 minutes.

Characterization

Both products were characterized by analytical HPLC (Figure 1), when the products were shown to co-elute with authentic reference samples obtained from the Salk Institute. Amino acid analysis (see above) and sequencing confirmed the identity of the products. Both products were shown to be fully biologically active.

Conclusion

Our studies have demonstrated that large peptides such as the growth hormone-releasing factors can be synthesized conveniently and economically on a large scale by standard solid-phase methods. The syntheses described were each completed in 8 to 10 weeks and scale-up of the procedures should be possible without significant modifications. Thus, most of the peptides which we now know to be pharmaceutically important may be prepared by this method.

Acknowledgement

The authors wish to thank Dr. Jean Rivier of the Salk Institute for providing samples of the authentic materials, Dr. Henry Keutmann of Harvard Medical School for sequencing the products and Rao Makineni for his constant encouragement.

References

1. Brazeau, P., W. Vale, R. Burgus, N. Ling, M. Butcher, J. Rivier and R. Guillemin. 1973. Science 179, 77-79.

2. Guillemin, R., P. Brazeau, P. Bohlen, F. Esch, N. Ling and W.B. Wehrenberg. 1982. Science 218, 585-587.

3. Ling, N., F. Esch, P. Bohlen, P. Brazeau, W. Wehrenberg and R. Guillemin. 1984. Proc. Natl. Acad. Sci. USA 81, 4302-4306.

4. Spiess, J., J. Rivier, M. Thorner and W. Vale. 1982. Biochemistry 21, 6037-6040.

5. Diaz, J., R. Guegan, P. Cabrera, M. Mellet, Y. Muneaux, P Perreaut, M. Vedel and R. Roncucci. 1985. In: Peptides Structure and Function. Proc. Ninth American Peptide Symp. (C.M. Deber, V.J. Hruby and K.D. Kopple, eds.). Pierce Chemical Company, p. 297-300.

PRESS-STUD PROTEIN CONJUGATES

R.E. Offord, S. Pochon & K. Rose

Département de Biochimie Médicale, Centre Médical Universitaire,
9, avenue de Champel, 1211 Geneva 4, Switzerland.

Protein conjugates are of great value in modern biology and medicine, and are likely to become even more so in the future. Protein labelling techniques must reconcile the wish to have as extensive a substitution of the protein as possible with the need to preserve the authenticity of its behaviour. We describe briefly in the present paper some outline approaches toward reconciling these two factors. Experimental details will be published elsewhere.

Specific activation

If the target protein does not possess a single side chain of unique chemical reactivity, we believe that it should be labelled, not by random attack on groups of side chains (eg. iodination of tyrosines, acylation of lysines) but by using some other means to direct the coupling to one, or a very few, chosen sites. We have therefore used the specificity of proteases working in reverse to fix activating groups exclusively at the carboxyl terminus of proteins and of protein fragments. The reversal of proteolysis is now frequently used in the semisynthesis and even total synthesis of polypeptides (for reviews see eg. Refs. 1 & 2). In the work under discussion, the activating groups are potential partners in Schiff-base couplings. We describe below some of the results that we have obtained in our chosen model system: conjugation to the C-terminus of the B-chain of the insulin molecule.

For satisfactory yields, we replace the majority of the water in the reaction solution by an organic cosolvent. We use butane-1,4-diol under the conditions described in Ref. 3. To couple, for example, the acetal of *m*-aminobenzaldehyde in place of Ala^{B30}, we use a concentration of the aminoaldehyde acetal of approx. 1.2M. Yields of 60-70 % of the wanted product can be obtained after isolation from a digest carried out at 37° C for 6-8 hours. Since the acetal is aromatic, its acid deprotection is over in a matter of a few minutes in dilute aqueous solution, pH 3.5-4. The dimethyl acetal is slightly more stable than the diethyl acetal (A. Martin, private communication).

Alternatively, we introduce an aromatic amine in the chosen position, using the same conditions. With, for example, the methyl ester of *p*-amino phenylalanine, the enzyme-catalyzed coupling proceeds with a strong preference for the aliphatic amino group. There is no need to protect the aromatic group.

Peptides 1986

We have also, using the same conditions, coupled to the C-terminus the synthetic amino-acid amide $CH_2OH-CHOH-CH_2-S-CH_2-CH(NH_2)-CONH_2$ (this compound is readily obtained by alkylation of L-cysteine methyl ester with 1-bromo-2,3-dihydroxypropane in aqueous solution at pH 8 and subsequent treatment of the dried product with strong ammonia solution). The protein-diol derivative obtained by this technique can be converted to the protein aldehyde (this time an aliphatic one) by periodate oxidation under mild conditions.

While endopeptidases seem simple to use with this technique, we can only apply them to the activation of proteins and protein fragments that have favoured sites for such coupling at their C-termini. Whilst there are several such molecules of clinical importance, it is desirable to seek methods using exopeptidases, since these could in principle be applied to most polypeptides that do not have residues essential for activity in the immediate terminal region. We find carboxypeptidase Y, even without co-solvents such as butane-1,4,-diol, couples p-aminophenylalanine amide to the C-terminus of peptide chains. However, in return for the more general applicability of this enzyme, it is necessary to optimize the conditions for its use with each new protein chosen for study.

Proteins activated in the above way can be stored and coupled whenever required by the formation of a Schiff base, followed by reduction to a stable form with cyanoborohydride. Cyanoborohydride can be used at mildly acid pH, which is best for Schiff-base formation.

Typically, coupling between complementary partners when both are aromatic goes to 90 % completion or better in few minutes at mild acid pH (acetate buffers 3.5-5.5) with concentration of reactants of 0.5-5mM.

Substituents

Appropriate complementary partners for such Schiff-base couplings are simple to make. For instance, we have prepared the following compounds: m-aminobenzoyl-Defeoxamine (a powerful, cyclic chelon for Fe and Ga), an aromatic aldehyde derivative of DTPA (diethylenetriaminepentaacetic acid) and $N^{(alpha)}$-m-aminobenzoyl-penta-[$N^{(epsilon)}$DTPA-alanyl-]pentalysine.

These compounds couple easily to the activated proteins described above and the protein derivatives can be purified, characterized, and stored indefinitely. They can be labelled (radioactively or, in principle, with paramagnetic ions for n.m.r) at any time under conditions that are easily repeatable from batch to batch, simply by mixing the protein-chelon conjugate with a solution of the appropriate metal.

We have prepared in this way insulin labelled with ^{111}In at a specific radioactivity of 5,000 Ci/mmole, and with the positron emitter ^{68}Ga at a theoretical specific activity 2.6 x 10^7 ci/mmole. (The latter figure is so large because the half-life of the isotope is only 68 min). We stress that these are simply model experiments for more clinically relevant labelling operations.

As has previously been suggested by others (Refs. 4 & 5), multiple labelling might not damage the protein if all the labelling groups are fixed to a single site in the form of a polymer, and the penta-lysine derivative mentioned is an example of a short polymer of this type. Our methods should work well even with much longer polymers, since we find the reactivity of the activating groups is high enough to permit even ligands that are macromolecular in size to couple rapidly (unpublished work by the authors and G. Sundaram, private communication).

The press-stud

The above principles of activation for coupling are those of the press-stud used on clothing. Neither side can fasten to another of its own kind; each will fasten only to its complementary partner. To confine the site of conjugation to the C-terminue will in many proteins keep it safely away from sites important to biological activity. Activation for Schiff-base coupling is not the only means of putting this principle into practice, although since aldehydes are not found in proteins, this particular strategy has a high degree of specificity.

References

1. Breddam, K., Widmer, F. and Johansen, J.T. (1981) Carlsberg. Res. Commun. 46 121-128.

2. Jakubke, H.-D., Kuhl, P. and Könnecke, A. (1985) Angew. Chem. Int. Ed. Engl. 24 85-93.

3. Rose, K., De Pury, H. and Offord, R.E. (1983) Biochem. J. 211 671-676.

4. Alam, F., Soloway, A.H., Barth, R.F., Johnson, C.W., Carey, W.E. and Knoth, W.H. (1984) in "Proceedings of the First International Symposium on Neutron Capture Therapy" (Fairchild, R.G. and Brownell, G.L. (eds). pp 229-236.Brookhaven National Laboratory New York 11973 USA.

5. Kunio, S., Akira, Y., Takahashi, K., Ueda, N. and Hazne, M. (1983) European Patent Application 0 131 361.

Acknowledgments

We thank the Ligue suisse contre le Cancer, the Luzerner Krebsliga, Hoffmann-La Roche S.A., and the Schmidheiny Foundation. In addition, the Fonds National suisse pour la Recherche Scientifique provided much of the general instrumentation on which this work was carried out.

INTERNAL MOBILITY OF A CYCLIC PEPTIDE, A ROTATING FRAME RELAXATION STUDY

K.D. Kopple* and Y.-S. Wang

Department of Chemistry, Illinois Institute of Technology
Chicago, Illinois 60616, U.S.A.
*Present address: L-940, Smith Kline & French Laboratories,
P.O. Box 7929, Philadelphia, Pennsylvania 19101, U.S.A.

When conformation interchanges occur near 10^5 s.$^{-1}$, the nuclear magnetic resonance spectrum of a peptide will still be a narrow line average over the exchanging species. In this rate range, however, conformational fluctuations can contribute significantly to the rate of spin-lattice relaxation in the rotating frame, $1/T_{1\rho}$.

We have been exploring the potential of proton $T_{1\rho}$ measurements to report the internal mobility of peptides, using a series of cyclic octapeptides, cyclo(D or L-Ala-Gly-Pro-D or L-Phe)$_2$. In these experiments we noted differences in the exchange contribution to $1/T_{1\rho}$ that parallel differences in internal mobility suggested by other observations (1). Here we report results of a more detailed study of a bridged analog, cyclo(Cys-Gly-Pro-Phe)$_2$, including temperature and spin-locking field dependence of the exchange contribution.

The chemical exchange contribution to the $T_{1\rho}$ relaxation of a given nuclear spin is determined by the exchange rates and the chemical shift differences between sites (2). For a given exchange rate, larger chemical shift

Peptides 1986

fluctuations have larger effects. Because fluctuations in chemical shift have maximal effect at the frequency of precession about the spin-locking field, the exchange contribution can be dependent on the strength of the spin locking field. For exploratory purposes, it is easy to apply a two-site exchange model to the spin locking field dependence, thereby to extract a characteristic time for exchange and an estimate of the chemical shift difference between sites. According to the two site model,

$$1/T_{1\rho} = \rho_A \rho_B (\delta\omega)^2 \frac{\tau_{exch}}{1 + (\omega_1 \tau_{exch})^2} \qquad (1)$$

where ρ_A and ρ_B are the probabilities at the sites, ω is the chemical shift difference between the sites, ω_1 is the precession frequency about the spin locking field, and τ_{exch} is $1/(k_{AB} + k_{BA})$. Although it is probably not realistic, this model can provide useful indications.

A crystal structure for cyclo(Cys-Gly-Pro-Phe)$_2$ (3) is shown below.

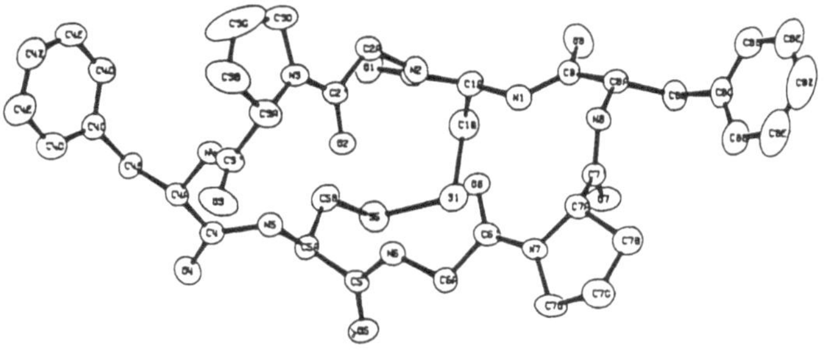

The peptide backbone in the crystal is unsymmetrical, containing both type I and II ß turns at Pro-Phe, and two different dispositions of the Cys side chains in the disulfide bridge. In DMSO solution the peptide shows C_2 symmetry in the NMR average. Analysis of the HNCH coupling constants suggests that, whether or not unsymmetrical structures corresponding to the crystal conformation are present and exchanging in solution, a component with two type I ß turns in the backbone is also present. Two modes of internal mobility are thus suggested for the peptide in solution: rotation of the plane of the Pro-Phe peptide bond relative to the octapeptide ring, and rotations about bonds of the Cys side chains.

$T_{1\rho}$ measurements of cyclo(Cys-Gly-Pro-Phe)$_2$ in 1:1 DMSO-chloroform were made between 15° and 55°, at spin locking fields from 3 to 9 kHz. Attention was focused on the N-H proton resonances, which are expected to have more conformationally dependent chemical shifts and are well separated in the spectrum. For each measurement, the chemical exchange contribution to $1/T_{1\rho}$ was isolated from the total rotating frame relaxation rate as already described (1), and plots of $T_{1\rho}$ vs. $(\omega_1)^2$ gave the straight lines predicted by Eq. 1. The effective exchange lifetimes and chemical shift differences obtained from Eq. 1 for the observed average N-H signals are below:

	τ_{exch}, s. x 10^5 ($\Delta\nu$, ppm)		
T (°)	Cys	Gly	Phe
55	1.1 (.59)	1.1 (.81)	1.0 (.66)
45	1.5 (.54)	1.6 (.79)	1.3 (.64)
30	1.9 (.56)	2.9 (.81)	2.4 (.62)
15	2.0 (.53)	2.8 (.71)	2.3 (.56)

The effective chemical shift differences are of the magnitude to be expected for conformation changes in cyclic peptides. The interesting observation is that although the effective lifetimes increase as the temperature is lowered from 55 to 30°, at a rate of change that corresponds to about a 7 kcal/mole activation barrier, the effective lifetimes do not increase further between 30° and 15°. This carefully checked result indicates that the two-conformation exchange model is incorrect, that there are at least two largely independent exchange processes affecting the $T_{1\rho}$ relaxation, one with the 7 kcal barrier, and the other with a much smaller barrier. Consistent with this indication, the effective chemical shift differences tend to decrease at lower temperatures.

Line broadening becomes apparent at temperatures below 30° in spectra of cyclo(Cys-Gly-Pro-Phe)$_2$ (but not in cyclo(Ala-Gly-Pro-Phe)$_2$), indicating that some conformational exchange process is becoming slow enough to effect the transverse relaxation rate, $1/T_2$. Presumably this is the process with the larger barrier. Greatest broadening at the lower temperatures occurs in the resonances of the Gly NH and the Cys β carbon and protons. One Gly α proton is broadened significantly more than the other, and the Pro α proton and carbon broaden relatively little. Of the two modes of mobility suggested by the crystal structure, these effects are more consistent with the slower process being reorientation of the Cys side chains, although other backbone motions are undoubtedly coupled to this.

This work was supported by a grant from the U.S. Public Health Service, National Institutes of Health, GM 26071.

References

1. Kopple, K.D., Bhandary, K.K., Kartha, G., Wang, Y.-S. and Parameswaran, K.N. (1986), J. Am. Chem. Soc. 108, 4637-4642.

2. Deverell, C., Morgan, R.E. and Strange, J.H. (1970), Mol. Phys. 18, 553-559.

3. Bhandary, K.K., Kopple, K.D., and Go, A. (1986) Abstracts American Crystallographic Assn. Annual Meeting, June 1986, Ser. 2, Vol. 14. abstract PA47.

SYNTHETIC HELICAL PEPTIDES

Eunice J. York and John M. Stewart
Department of Biochemistry,
University of Colorado School of Medicine,
Denver, Colorado 80262

Robert L. Baldwin and Kevin R. Shoemaker
Department of Biochemistry,
Stanford University Medical Center,
Stanford, California 94305

Introduction

Peptides shorter than about twenty amino acids have not generally been found to possess any helical structure in aqueous solution, and energy calculations have suggested that such ordered structure in short peptides is not likely. In ribonuclease A (RNase) the N-terminal 13 residues are helical, as shown by X-ray crystallography; this structure is thought to be stabilized by a Glu-2 to Arg-10 salt bridge and by Phe-8 to His-12 aromatic stacking interactions (1). The "C-peptide" of RNase (structure [1] in Table 1) forms about 30% α-helix in water near 0°C, pH 5.2, contrary to the Zimm-Bragg and "host-guest" predictions. This unexpected stability was thought to be due to a Glu-9 to His-12 salt bridge between side chains in addition to the interactions seen in the RNase crystal structure. Hydrolysis of the lactone [2] destroyed helical structure, and a competing salt bridge between this new carboxyl and His-12 was thought to disrupt the helix (2). Analogs of RNase(1-13)-amide have been synthesized to test these proposals. We report that none of these proposed interactions is necessary for helix stabilization, and that facilitation of formation of the induced dipole of the helix appears to be most important in stabilization of short helices in aqueous solution. Parallel H-bonds in the α-helix induce a +0.75 charge at the amino-terminus and a -0.75 charge at the C-terminus (3); free terminal ammonium and carboxylate groups in short peptides oppose

Peptides 1986

establishment of these induced charges, and hence oppose formation of the helix. Charged side chains have a similar effect when they are located within three residues of the helix termini. In proteins, acidic residues cluster near the N-terminal (and basic residues near the C-terminal) of helical segments, neutralizing (and thus stabilizing) the helix dipole.

Gln-9 and Met-13 of the RNase sequence were replaced by Ala to avoid chemical problems. Peptide amides were synthesized by solid phase methods on methylbenzhydrylamine resins and were purified by countercurrent distribution and hplc. Helix content of peptide solutions was estimated by the magnitude of the CD negative band at 222nm; the baseline was assumed to be +2,000 deg.cm^2/dmole (4). Proton NMR confirmed the extent of bonding.

Results and Discussion

Replacement of Gln-11 and Met-13 by Ala residues and masking of the C-terminal carboxyl as the amide gave analog [3], which showed only 17% helix. However, replacement of Glu-9 by Leu enhanced helix formation [4], demonstrating that a Glu-9 to His-12 salt bridge was not necessary for helix stabilization. The series of analogs 4, 5, 6 and 7 demonstrates the helix-enhancing effect of changing the N-terminal charge from +1 through neutral to -2; helix content increases from 27 to 60%.

In analog 8 the Phe-8 of analog 6 is replaced by Ala, thus destroying any possible aromatic-aromatic interaction between this position and His-12; helix content was not significantly reduced by this change. The importance of a Glu-2 to Arg-10 salt bridge was examined with analogs 9 and 10. In analog 10, inversion of the residues at positions 2 and 10 reduced helix content almost by half, as compared to analog 6; this is evidently mainly due to the deleterious effect of the basic Arg residue at position 2, since a Glu residue would be expected to be acceptable in the helix at position 10 near the C-terminus of the helix (5). In analog 9, the Arg-10 residue of analog 7 is replaced by Ala, with a striking concomitant increase of helix content from 60 to 68%.

Table 1. Helix Content of Ribonuclease Peptides.

PEPTIDE	STRUCTURE	$[\Theta]_{222}$[a]	HELIX
[1]	K-E-T-A-A-A-K-F-E-R-Q-H-Hse-lactone	-7,000	32%
[2]	K-E-T-A-A-A-K-F-E-R-Q-H-Hse-COOH	-1,500	5
[3]	K-E-T-A-A-A-K-F-E-R-A-H-A-NH_2	-2,800	17
[4]	K-E-T-A-A-A-K-F-L-R-A-H-A-NH_2	-5,700	27
[5]	A-E-T-A-A-A-K-F-L-R-A-H-A-NH_2	-8,900	38
[6]	Ac-A-E-T-A-A-A-K-F-L-R-A-H-A-NH_2	-13,150	53
[7]	Suc-A-E-T-A-A-A-K-F-L-R-A-H-A-NH_2	-15,150	60
[8]	Ac-A-E-T-A-A-A-K-A-L-R-A-H-A-NH_2	-12,300	50
[9]	Suc-A-E-T-A-A-A-K-F-L-A-A-H-A-NH_2	-17,400	68
[10]	Ac-A-R-T-A-A-A-K-F-L-E-A-H-A-NH_2	- 6,200	29
[11]	Suc-A-A-T-A-A-A-K-F-L-A-A-H-A-NH_2	-18,900	73

[a]Measured in 0.1M NaCl solution at 3°C, pH 5.2. 100% Helix is assumed to be -26,500 deg.cm^2/dmol and the baseline +2,000. Ac=acetyl; Suc=succinyl.

Elimination of the Arg-10 residue suggests that a single positively charged residue near the C-terminus is adequate to maintain high helicity. In analog 11, the same question is asked with respect to multiple charges at the N-terminus. Again, helix content is increased to 73%. This analog has the highest helix content yet seen in these 13-residue peptides related to the N-terminal sequence of RNase and demonstrates that the single negative charge of the succinyl carboxylate is adequate to stabilize the helix. The shape of the helix versus pH plot shows that both these ionizable groups must be ionized for maximum helix stabilization; helix is maximum at pH 5.2. The plot also suggests that the charged group on the side chain of Lys-7 is not important for maintenance of the helix; this group increases solubility and perhaps inhibits aggregation of the peptides in solution. All solutions obeyed Beer's law in the concentrations used for CD studies.

These results suggest that the types of helix-stabilizing interactions between residues seen (or presumed) in the N-terminal sequence of RNase are not necessary for maintenance of high helical content of small peptides in isolated solution. The high helical content of the native sequence in position in the RNase molecule is apparently due in large measure to interactions with the S-protein moiety. Application of the principles deduced in this work to analogs of S-peptide (1-15) has yielded peptides which bind to S-protein [RNase (21-124)] with very high affinity and effectively regenerate RNase enzymic activity (6).

Conclusions

Charged group effects play a major role in stabilization of helical structure in short peptides in water. Modification of the structure of the RNase C-peptide has led to analogs of high helix content that support the helix dipole model. (Supported by NIH grant GM-31475 and NSF grant GP-23633 to RLB.)

References

1. Rico, M., J. Santoro, F.J. Bermejo, J. Herranz, J.L. Nieto, E. Gallego, M.A. Jimenez. 1986. Biopolymers 25, 1031.
2. Kim, P.S., A. Bierzynski, R.L. Baldwin. 1982. J.Mol.Biol. 162, 187.
3. Perutz, M.F., A.M. Gronenborn, G.M. Clore, J.H. Fogg, D.T-b. Shih. J.Mol.Biol. 183, 491.
4. Shoemaker, K.R., P.S. Kim, D.N. Brems, S. Marqusee, E.J. York, I.M. Chaiken, J.M. Stewart, R.L. Baldwin. 1985. Proc.Nat.Acad.Sci.USA 82, 2349.
5. Blagdon, D.E., M. Goodman. 1975. Biopolymers 14, 241.
6. Mitchinson, C., R.L. Baldwin. 1986. Proteins 1 (in press).

CONFORMATION-ACTIVITY RELATIONSHIPS IN THE SERIES OF MURAMYL-PEPTIDES

S. FERMANDJIAN*, B. PERLY**, P. SIZUN**, M. LEVEL° and P. LEFRANCIER°

* Département de Biologie, C.E.N. Saclay, F 91191 GIF-sur-YVETTE ; **Département de Chimie-Physique, C.E.N. Saclay, F 91919 GIF-sur-YVETTE ; ° Institut Choay, 10, rue Morel, F 92120 MONTROUGE

INTRODUCTION

N-acetyl-muramyl-L-alanyl-D-isoglutamine (MDP, for muramyl-dipeptide), the minimum part of bacterial peptidoglycan endowed with immunomodulating properties (1), is the leader of over 500 analogs which have been designed and synthesized in order to discover new drugs aiming at modifying the efficiency of the disturbed or normal immune system. N-acetyl-muramyl-L-alanyl-D-glutamine n-butylester (Murabutide) is one of the most promising molecule (2,3). Accumulation of analogs has resulted in structure-activities relationship (1) : many analogs have close chemical structures, but display dissociated pharmacological activities. The present work rests on the ^{1}H-NMR studies of a number of MDP analogs in order to search a conformational explanation to these results.

MATERIAL and METHODS

MDP, Murabutide and their analogs have been synthesized according to previously described methodology (4-6). ^{1}H-NMR spectra were obtained from a Bruker WM 500 spectrometer. Complete spectral assignment was achieved by two dimensional techniques. Dimethyl sulfoxide solutions of the compounds were prepared at the final concentration of 10^{-2} M.

RESULTS and DISCUSSION

Various pharmacological properties of MDP are separately affected by chemical modifications. The unwanted pyrogenic effect is for instance suppressed by replacing D-isoglutamine residue with D-glutamine alkylester moieties, as for Murabutide, without loss of immunopotentiating activities (table 1).

Peptides 1986

TABLE 1. PHARMACOLOGICAL PROPERTIES OF MDP AND MURABUTIDE

	M D P	Murabutide
Adjuvant activity (sc administration in PBS to mice with BSA)(1)		
Primary response	3.64	4.21
Secondary response	8.34 ± 1.34	8.43 ± 1.52
Protective activity (sc administration of 100 µg to mice - 24 hr before an intramuscular infection with 10^4 K.pneumoniae)(2)	58 %	63 %
Pyrogenic activity(3)		
iv administration	25	) 10^4
sc administration	25	) 10^4
o administration	3.6 x 10^3	) 10^4
icv administration	0.0013	) 100

(1) Antibody titers are given as log of the highest dilution of serum able to agglutinate BSA-coated SRBC
[controls : primary response : (1.64. Secondary response : 3.57 ± 1.78]
(2) Protection % : difference between percentages of survivors in treated group and its respective control group at day + 15.
(3) Minimal pyrogenic dose (µg/kg).

Is there a conformational explanation to these effects ? ^{1}H-NMR analysis indicates that MDP and Murabutide have different conformations in solution (7). For MDP, the α-anomeric form adopts a "S" shaped conformation (figure 1).

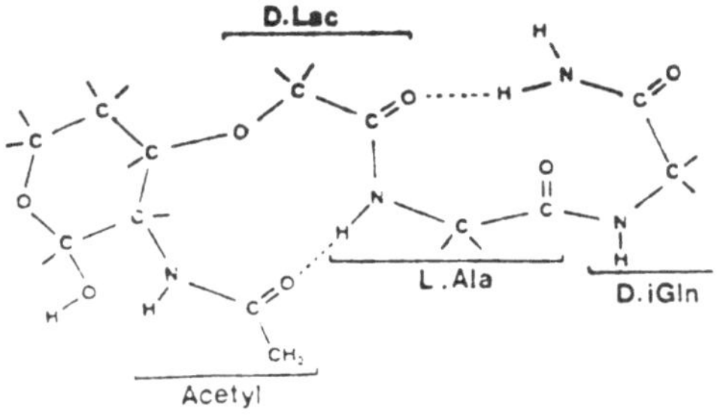

Figure 1. "S" shaped conformation of MDP

This is characterized by two adjacent C_{10} hydrogen bonded turns. The first one is similar to β-turns of type II found in (L)-(D) depsipeptide. Only the i + 1 amino acid is replaced by the $-N-C_2-C_3$ motif belonging to N-acetyl group and pyranosyl ring [the N and C_3 atoms replace the Nα (amino) and the C' (carbonyl) atoms, such as the C_2 atom behaves like the Cα-atoms in L-amino acids].

The proposed angular values in the first β-turn are as follows : ϕ_1, ψ_1 (NAG) : - 86°, + 101° ; ϕ_2, ψ_2 (lactoyl) : 90°, - 10°. This allows a C_{10} hydrogen bond involving the NH group of L-alanyl residue and the CO group of N-acetyl moiety.

The second turn is created around L-alanyl and D-isoglutaminyl residues and appears as a more classical β-turn of type II found in (L)-(D) amino acid sequences. Possible values ϕ_1, ψ_1 (L-Ala) are -80°, + 120°, and ϕ_2, ψ_2 (D-iGln) are + 90°, -15°. The amide proton E -anti to the oxygen carbonyl- in the α-$CONH_2$ group interacts with the CO group of D-lactoyl. The hydrogen bond lacks in Murabutide, thus destabilizing the second β-turn. Alanyl residue has furthermore rotated around its N-Cα bonds, such as the ϕ-Ala angle is larger in Murabutide than in MDP. In the former, the "S" structure is thus distorted, even if the first β-turn is exactly the same as in MDP.

The defection of the second turn with subsequent increase of the ϕ-Ala angle may explain the dissociation of activities in Murabutide (table 1). This is ascertained by the observation of a similar effect in other MDP analogs, such as MDPA and MDPG. In these compounds, where D-isoglutamine residue is replaced with D-glutamic and D-glutamine residue respectively, the absence of an α-$CONH_2$ function at the C-terminal position, excludes the formation of the second C_{10} hydrogen bonded structure and provocks the aperture of the ϕ-Ala angle.

In contrast, monomethylation of the α-$CONH_2$ function in MDP, as for MDPA-$NHCH_3$, permits the C_{10} hydrogen bond : a correlated conservation of the physico-chemical and biological MDP-type properties is observed with this compound.

Modulation of the protective activity against bacterial infection by modifications performed on the L-alanyl site (X) is another striking point of our systematic analysis. When (X) is a L-valyl, D-alanyl, glycyl or N-Me-L-alanyl residue, a correlation is observed between the stability of the first β-turn (tested from the angular values and the ability of hydrogen bond formation) and the anti-infectious activity : the MDP or Murabutide analogs with (X) being the above mentioned residues, show a strong destabilization of the first β-turn and are deprived of anti-infectious activity.

In conclusion, in MDP analogs, maintenance of the totalities of pharmacological activities requires the "S" shaped structure. Deletion of the second turn abolishes the unwanted pyrogenic activity, but preserves the useful protective activity against bacterial infection. This latter needs the elimination of the first turn to be abolished. It must be pointed out that adjuvanticity of the specific immune response is only weakly affected by the varied substitutions performed.

REFERENCES

1. ADAM A. PETIT J.F., LEFRANCIER P. and LEDERER E., Mol. Cell. Biochem., **41** (1981) 27.

2. CHEDID L., PARANT M., AUDIBERT F., RIVEAU G., PARANT F., LEDERER E., CHOAY J. and LEFRANCIER P., Infect. Immun., **35** (1982) 417.

3. TELZAK E., WOLFF S., DINARELLO C.A., CONLON T., CHOAY J., MORIN A., BAHR G. and CHEDID L., J. Infect. Dis., **153** (1986) 628.

4. LEFRANCIER P. and LEDERER E., Prog. Chem. Org. Nat. Prod., **40** (1981) 1.

5. LEFRANCIER P., CHOAY J., DERRIEN M. and LEDERMAN I., Int. J. Peptide Protein Res., **9** (1977) 249.

6. LEFRANCIER P., DERRIEN M., JAMET X., CHOAY J., LEDERER E., AUDIBERT F., PARANT M., PARANT F. and CHEDID L., J. Med. Chem., **25** (1982) 87.

7. FERMANDJIAN S., PERLY B., LEVEL M. and LEFRANCIER P., Carbohydr. Res., to be published.

NMR STUDY OF THYMULIN, A THYMIC NONAPEPTIDE HORMONE. INFLUENCE OF Zn(II) COMPLEXATION ON THE CONFORMATIONAL PROPERTIES

M.T. Cung, M. Marraud
CNRS-UA-494, ENSIC - INPL, 54042 Nancy, France

J.P. Laussac, M. Pasdeloup, R. Haran
CNRS-UP-8241, 205 Route de Narbonne , 31400 Toulouse, France

P. Lefrancier
Institut Choay, 46, Av. Théophile Gauthier, 75016 Paris, France

M. Dardenne, J.F. Bach
INSERM-U-25, CNRS-UA-122, 161, rue de Sèvres, 75730 Paris, France

Thymulin-induced T-cell maturation requires the presence of Zn(II) ions (1). The existence of different monoclonal antibodies against the free or Zn(II)-complexed hormone is in favor of a Zinc(II)-specific conformation of the thymulin molecule (<Glu^1-Ala^2-Lys^3-Ser^4-Gln^5-Gly^6-Gly^7-Ser^8-Asn^9) (2). We briefly report here the preliminary results of a 1D and 2D-NMR study of thymulin in DMSO solution for different peptide/metal ratios.

Free Thymulin

The ^1H and ^{13}C resonances of thymulin in DMSO were assigned by $^1H,^1H$ and $^1H,^{13}C$-COSY. Similar spin systems were attributed by $^1H,^1H$-NOESY, on the basis of the connectivity between consecutive C^{α}_iH and $N_{i+1}H$ protons. The absence of long-range NOE effects and the average value ($\sim$ 7Hz) of the NH-$C^{\alpha}H$ coupling constants are in favor of a rather fluctuating conformation. However, the low temperature coefficients for three N^1H proton resonances, and the low sensitivity of four carbonyl ^{13}C resonances to the solvent (DMSO - water mixtures) indicate some structuration of the C-terminal heptapeptide sequence with three N-H...O=C hydrogen bonds (Fig. 1).

Peptides 1986

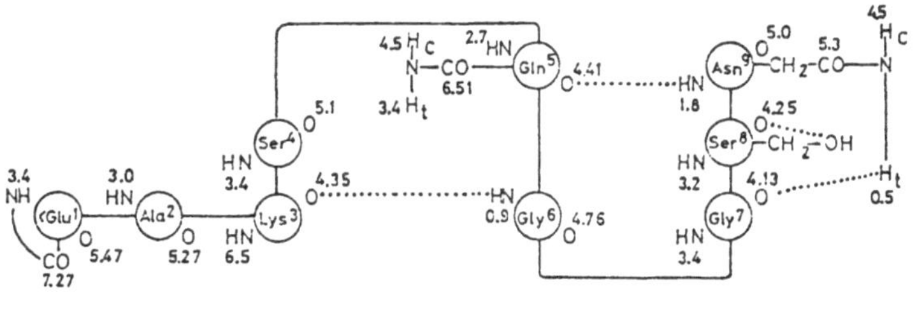

Fig. 1 : Schematic representation of the structure of thymulin in DMSO. The N ^{1}H proton temperature coefficients (10^{-3} ppm/deg) and the carbonyl ^{13}C shift (ppm) from DMSO to water solutions are also indicated.

Thymulin/Zn(II) complex

Most ^{1}H and ^{13}C resonances are progressively shifted by addition of $Zn(ClO_4)_2$ (Fig. 2), indicating that thymulin is in fast exchange on the NMR time scale between its free and complexed states. These variations cannot be accounted for by the existence of a simple 1:1 or 1:2 Zn(II)/thymulin complex, but quite good agreement with the experimental data is obtained assuming a mixture of both complexes (Fig. 2). An optimization by Rosenbrock's method affords us the association constants, together with the ^{1}H and ^{13}C NMR contributions of each complex. On the basis of the two following equilibria, the association constants have been adjusted to :

$$Zn(II) + Thymulin \rightleftarrows 1:1 \text{ complex}, \quad K_1 = 3540 \pm 80 \ M^{-1}$$
$$Zn(II) + 2(Thymulin) \rightleftarrows 1:2 \text{ complex}, \quad K_2 = (5.0 \pm 0.1)10^6 \ M^{-2}$$

It results that the 1:2 complex is predominant up to the equimolar composition, and that 70 % of the hormone is 1:1 complexed for a twofold excess of the Zn(II) ion.

The persistance of all the N^1$\underline{H}$ resonances upon addition of Zn(II) indicates that this ion is not liganded to a nitrogen atom after N-deprotonation. The analysis of the metal-induced changes in the NMR spectra gives direct information on the metal binding site and the sequences with altered conformation. The large shift and the broadening of the Asn^9-$^{13}\underline{C}O_2^-$ signal agree fairly well with the implication of this carboxylic group as a ligand to Zn(II). Two other candidates are the Ser^{4,8}-O^{γ}(H) atoms, as

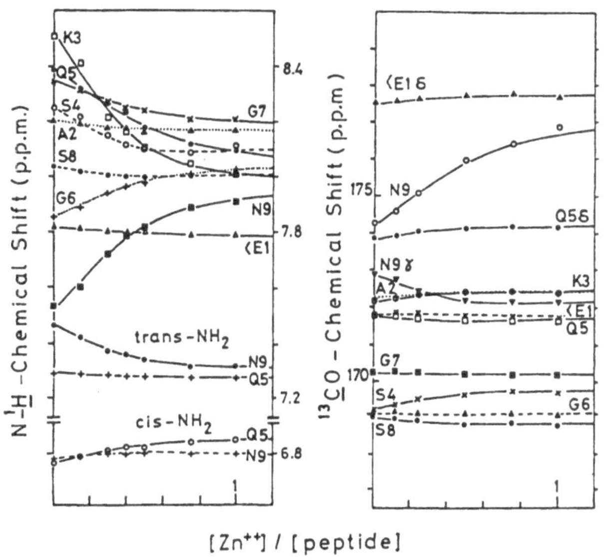

Fig. 2 : Influence of the Zn(II) ion on the N$^{1}\underline{H}$ and $^{13}\underline{C}O$ resonances of thymulin in DMSO solution. The experimental data are compared to the variations calculated on the basis of a mixture of 1:1 and 1:2 Zn(II)/thymulin complexes.

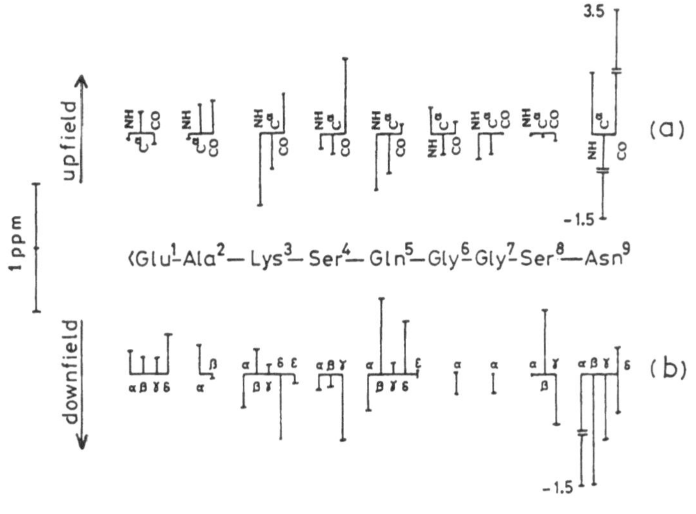

Fig. 3 : Upfield and downfield perturbations of the ^{1}H and ^{13}C resonances of 1:1 complexed thymulin with reference to the free hormone (peptide backbone, a) and side-substituents, b).

demonstrated by the strong upfield shift of both $O^{\gamma}H$ protons, and the slowing down of their proton exchange rate with residual water molecules.

Therefore it appears that the Ser^4-$O^{\gamma}(H)$, Ser^8-$O^{\gamma}(H)$ and Asn^9-CO_2^- sites are liganded in the 1:1 complex, and this is confirmed by the perturbations of the ^1H and ^{13}C resonances of the Lys^3-Ser^4-Gln^5 and Ser^8-Asn^9 sequences (Fig. 3). The fourth ligand could be a residual water molecule.

The main difference between the NMR contributions of the 1:1 and 1:2 complexes concerns those of the Ser^8 residue, suggesting that the Ser^8-$O^{\gamma}(H)$ site is not liganded in the latter complex.

Discussion

In agreement with the suggestion of Bach et al.(1), and Ikeyama et al. (3), it appears that the C-terminal heptapeptide sequence is necessary for the expression of the biological activity of thymulin through Zn(II) complexation, whereas the N-terminal dipeptide sequence is exposed to the solvent. Moreover, the existence of two thymulin/Zn(II) complexes with nearly equal stability suggests the possibility of a partial Zn(II) ligand exchange at the receptor site.

The existence of multiple and intense through-space NOE effects upon addition of Zn(II) argues for a structuration of the thymulin molecule around the metal ion. Work is in progress to obtain more precise knowledge of the conformations assumed by the thymulin molecule in the complexes.

References

1. Bach, J.F. 1983. In: Clinics in Immunology and Allergy, Vol. 3, W.B. Saunder Co., Philadelphia, U.S.A.

2. Dardenne, M., W. Savino, S. Berrih, J.F. Bach. 1985. Proc. Natl. Acad. Sci. U.S.A. 82, 7035-7038.

3. Ikeyama, S., K. Kato, M. Takaoli, M. Takeuchi, A. Kalinuma. 1984. J. Immunoassay 5, 87-97.

OXYGEN-17 NMR STUDIES OF ENKEPHALINS IN AQUEOUS SOLUTION

I.P. Gerothanassis, T. Karayannis, M. Sakarellos-Daitsiotis, C. Sakarellos

Department of Chemistry, University of Ioannina, 451 10 Ioannina, Greece

M. Marraud

Laboratoire de Chimie-Physique Macromoléculaire, ENSIC-INPL, 1, rue Grandville, 54042 Nancy Cédex, France

Introduction

Leu- and Met- enkephalins are brain pentapeptides (Tyr-Gly-Gly-Phe-$^{Leu}_{Met}$) whose conformational state in solution is still controversial despite numerous investigations with a variety of physico-chemical techniques. Thus evidence exists in the literature of 5 → 2, 4 → 1, 5 → 1 and 3 → 1 intramolecular hydrogen bonds. An intramolecular interaction between the terminal $^{+}NH_3$ and COO^- groups (bridge ionic bond) has also been suggested (1). ^{17}O-NMR spectroscopy can be considered as an obvious method for elucidating inter-molecular, intra-molecular and ionic bridge hydrogen bonding interactions and through space field effects (2,3). However, most of the ^{17}O-NMR studies have been limited to small peptides with molecular weights well below 500 (2,3). In this work we have been investigating the ^{17}O chemical shifts and relaxation times of the ^{17}O selectively enriched in carboxyl oxygens of Leu- enkephalin and, for comparison, of [^{17}O]Leu-OH, Phe-[^{17}O]Leu-OH, Gly-Phe-[^{17}O]Leu-OH and Gly-Gly-Phe-[^{17}O]Leu-OH in an effort to elucidate the existence of bridge ionic hydrogen bonding interactions and through space field effects in the dipolar form in aqueous solution.

Experimental Part

The ^{17}O selectively enriched compounds were synthesized by saponi-

Peptides 1986

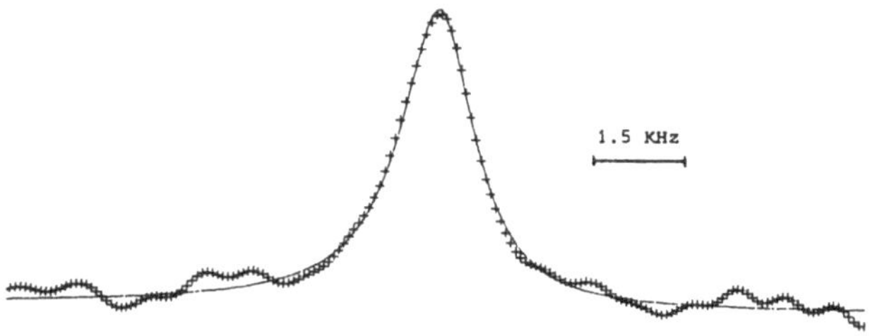

Fig. 1. Simulation of the ^{17}O NMR spectrum of Leu-enkephalin ($\Delta\nu_{1/2} \simeq 1.2$ KHz). Experimental and data acquisition parameters : pH 5.1, temperature 30° C, concentration 10^{-2} M in the presence of EDTA, 90° pulse 29 μs, spectral width 50 KHz, preacquisition delay 50 μs, number of scans 260,000, exponential multiplication of the FID (LB = 300 Hz), total experimental time c.a. 22 min.

fication of the corresponding methyl esters with $CH_3CH_2ONa/H_2{}^{17}O$ (20 or 40 atom % ^{17}O). The ^{17}O-NMR measurements were obtained by using a Bruker AM-400 MHz multinuclear FT spectrometer with resonance frequency for ^{17}O 54.48 MHz (Fig. 1).

Results and Discussion

Chemical shifts. Deprotonation of the carboxyl group results in a chemical shift to a higher frequency by 16.5 ppm for the Leu and 18.0-20.0 ppm for the peptides studied. Deprotonation of the amino group results in a low frequency shift by ∿ 3 ppm for Leu-OH, which can be attributed to counterbalancing inductive and through space field effects (2). This inductive effect is closely related to the number of intervening bonds (2) and it is expected to vanish in the peptides of the present work. Through space field effects have been shown to play an important role in the ^{17}O nuclear screening constants of the carboxyl oxygens and chemical shift differences of up to 5 ppm were observed, depending on the average distance of the carboxyl oxygens from the neighboring polar or ionic group (2,3). Additionally, varying degrees of hydrogen bonding interactions have been shown to result in significant variations of the ^{17}O screening constants of the carboxyl oxygens (3). However, a bridge ionic bond or a through space proximity (field effects) of the terminal $^+NH_3$

and COO^- groups of Leu-enkephalin should be excluded. Firstly, no titration shift of the carboxyl group at high pH (deprotonation of the terminal amino group) is observed. Secondly, the titration shift on deprotonation of the carboxyl group is similar to that of the tripeptide, which is assumed to exist in random conformations.

Relaxation time measurements. In diamagnetic compounds the ^{17}O nucleus relaxes through the quadrupolar mechanism. Under the extreme narrowing condition $r_c^2 \omega_o^2 << 1$ (where r_c is the correlation time for molecular tumbling and ω_o is the resonance frequency) which is expected to be valid for small molecules like enkephalins, the transverse relaxation rate (T_2^{-1}) and therefore the linewidth of the resonance absorption at half height ($\Delta\nu_{1/2}$) is given by :

$$\Delta\nu_{1/2} = \frac{1}{\pi T_2} = \frac{12\pi^2}{125} \left(1 + \frac{\varepsilon^2}{3}\right) \left(\frac{\chi}{h}\right)^2 r_c \qquad (\underline{1})$$

where ε is the asymmetry parameter and χ/h is the nuclear quadrupole coupling constant (NQCC) in Hz. In most cases $\varepsilon < 0.6$, therefore variations of the $\varepsilon^2/3$ term can be ignored. On the other hand, it has recently been shown that the oxygen NQCCs of the carboxyl groups change insignificantly with the pH. Therefore linewidth measurements of the carboxyl oxygens can provide an excellent means for an estimation of r_c at the site of the carboxyl group.

i) Effect of pH on the linewidths. The linewidths and, therefore, rotational correlation times for molecular tumbling of the carboxyl oxygens of all the peptides studied were found to be independent of the pH, thus excluding the participation of the carboxyl-group in pH-dependent conformational changes.

ii) Effect of molecular weight on the linewidths. For spherical molecules the correlation time can be expressed as a function of molecular volume and therefore molecular weight, M_w, as follows :

$$r_c = 0.74 \frac{\eta f_r}{kTN_o\rho} M_w \qquad (\underline{2})$$

where η is the viscosity of the solution, f_r is the microviscosity factor (<1 depending on the relative sizes of the solute and solvent), k is the Boltzmann constant, T is the absolute temperature, N_o is the Avogadro number and ρ is the density of the solute. This linear relationship between the linewidth at half height and molecular weight (eqns. (1) and (2)) was verified in simple aminoacids (η, f_r and ρ in eqns. (1) and (2) were assumed to be constant) but not in the molecules of the present work, where it was found that the linewidth varies relatively little among the tri-, tetra- and penta- peptides (the solution viscosity was found to be practically independent of the peptide). This result indicates the importance of the segmental motions of the end residues and the internal rotation of the carboxyl group around the C-COOH bond which would have been hindered by the presence of a bridge ionic hydrogen bond or through space field effects. This confirms the conclusions from the chemical shift data : the existence of a bridge ionic hydrogen bond or long range through space field effects should be excluded in the dipolar form in aqueous solution of Leu-enkephalin.

Acknowledgement

This work was supported by the C.E.C. (Grant ST2J-0184-1-GR), the Greek-French Committee of Research and Technology, the Research Committee of the University of Ioannina and EMBO (short-term fellowship to I.P. Gerothanassis).

References

1. Schiller, P.W. 1984. In: The Peptides, Vol. 6 (S. Udenfriend and J. Meienhofer, eds.). Academic Press, pp. 219-268.
2. Gerothanassis, I.P., R.N. Hunston and J. Lauterwein. 1985. ^{17}O-NMR chemical shifts of the twenty protein aminoacids in aqueous solution. Magn. Reson. Chem. 23, 659-665.
3. Hunston, R.N., I.P. Gerothanassis and J. Lauterwein. 1985. A study of L-proline, sarcosine, and the cis/trans isomers of N-acetyl-L-proline and N-acetylsarcosine in aqueous and organic solution by ^{17}O-NMR. J. Am. Chem. Soc. 107, 2654-2661.

CONFORMATIONAL STUDY OF NEUROKININ A : COMPARISON WITH SUBSTANCE P AND PHYSALAEMIN

G. Chassaing*, O. Convert° and S. Lavielle*
*Laboratoire de Chimie Organique Biologique, CNRS UA 493
°Laboratoire de Chimie Organique Structurale, CNRS UA 455
Université P. et M. Curie, 4 Place Jussieu 75230 Paris, France

Three major mammalian tachykinins have been identified : Substance P, Neurokinin A (or Substance K, Neurokinin α) and Neurokinin B (Neurokininβ, Neuromedin K).

Substance P (SP)	Arg-Pro-Lys-Pro-Gln-Gln-Phe-Phe-Gly-Leu-Met-NH_2
Neurokinin A (NKA)	His-Lys-Thr-Asp-Ser-Phe-Val-Gly-Leu-Met-NH_2
Neurokinin B (NKB)	Asp-Met-His-Asp-Phe-Phe-Val-Gly-Leu-Met-NH_2
Physalaemin (PHY)	pGlu-Ala-Asp-Pro-Asn-Lys-Phe-Tyr-Gly-Leu-Met-NH_2

Up to now, three tachykinin binding sites have been characterized : two in the central nervous system and one in the peripheral nervous system of the rat. In rat brain, a SP-type binding site and a NKB-type binding site have been found. NKA has the same low affinity (Ki = 1.2 10^{-7} M) for these both binding sites (1). Recently, using ^{125}I-BHNKA (2) and ^{3}H-NKA, (3), a NKA-type binding site has been evidenced on crude membrane preparations of rat duodenum. SP and NKB have weak relative affinities for the NKA-type binding site (2), 0.4% and 2.5% that of NKA, respectively. This selectivity could be the result of different backbone foldings of SP, NKB and NKA. SP and PHY, two substrates of SP-type binding site, adopt very close spatial structure in methanol (4,5). In this paper, we report the 1D- and 2D-NMR studies of the conformational behaviour of NKA in three solvents : $(C^2H_3)_2SO$, C^2H_3OH and H_2O.

Results and Discussion

The assignments (shown in Tables 1-3) were done at 500 MHz by using 2D-NMR methods : COSY,

Peptides 1986

COSY long range and RELAYED COSY. 1D-NOE difference spectra were used for the NOE experiments

Table 1 : NH Chemical Shifts, Temperature Coefficients, Coupling Constants, $d_{\alpha N}$ and d_{NN} NOE in $(C^2H_3)_2SO$

	His	Lys	Thr	Asp	Ser	Phe	Val	Gly	Leu	Met	NH_2
δ NH (ppm)	-	8.72	8.23	8.18	7.80	8.09	7.91	8.17	7.96	7.96	
δ NH/ΔT (ppb per K)		1.7	3.7	1.5	2.2	2.3	2.9	3.1	2.9	2.9	
$^3J_{CH_\alpha\text{-}NH}$ (Hz)		6.0	7.5	7.5	7.0	7.0	7.0	Σ11	7.0	8.0	
$^a d_{NN}$	-		W	-	W	W	W	-	-	-	
$^a d_{\alpha N}$	S		S	W	S	S	S	M	M	-	

a) W : weak; S : strong; M : medium

Table 2 : NH Chemical Shifts, Temperature Coeeficients, Exchange Rate, Coupling Constants, d_{NN} NOE in C^2H_3OH

	His	Lys	Thr	Asp	Ser	Phe	Val	Gly	Leu	Met	NH_2
δ NH (ppm)		8.83	8.39	8.43	8.14	8.23	7.83	8.19	8.14	8.05	
δ NH/ΔT (ppb)		3.5	7.4	5.1	4.9	4.9	5.8	5.8	5.1	5.4	
^{b}Exchange rate		VF	VF	VF	F	F	S	S	S	S	
$^3J_{CH_\alpha\text{-}NH}$ (Hz)		6.0	7.5	7.5	7.0	7.0	7.0	Σ11	7.0	8.0	
$^a d_{NN}$							W	W			

a) W :weak ; b) VF : very fast; F : fast; S : slow

Table 3 : NH Chemical Shifts, Temperature Coefficients, and Coupling Constants in H_2O

		His	Lys	Thr	Asp	Ser	Phe	Val	Gly	Leu	Met	NH_2
δ NH	(ppm)		8.76	8.40	8.47	8.13	8.20	7.91	7.95	8.06	8.26	
δ NH/ΔT	(ppb)		4.7	5.3	5.0	5.0	4.3	4.0	3.3	5	4.7	
$^3J_{CH_\alpha\text{-}NH}$	(Hz)		6.5	7.5	7.0	7.2	7.2	7.7	Σ11	6.7	7.5	

Conformation in $(C^2H_3)_2SO$

The lack of variation of the amino proton chemical shifts, in function of temperature, indicates the presence of an internal salt bridge. The C2H imidazole proton chemical shift changes weakly with the temperature. Moreover, the $\delta NH/\Delta T$ coefficients of NH Lys and Asp protons are small, showing their involvement in hydrogen bonding. The d_{NN} NOE observed are weaker than the $d_{\alpha N}$ NOE. All these data support the existence of a γ turn in the N-terminal part of NKA, stabilized by an internal salt between the carboxylate of Asp^4 and either the imidazole proton of His^1 or the ammonium group of Lys^2.
The NMR parameters of the C-terminal amino acids from Ser^5 to Met^{10} ($^3J_{CH_\alpha-NH}$, $d_{\alpha N}$, $\delta NH/\Delta T$) can be interpreted as an equilibrium between β-pleated sheet aggregates and C_7 and C_5 conformers.

Conformation in C^2H_3OH

The conformational behaviour of NKA in this solvent is rather similar to that observed in dimethylsulfoxide. The equilibrium between two internal salt links (His^1 -- Asp^4 and Lys^2 -- Asp^4) is confirmed by the titration of the C2H imidazole proton. At pH 6, the specific broadening of this proton can be explained by an interconversion of the imidazole between two states :the protonated species which is involved in an internal salt bridge and the deprotonated form. At higher pH, the complete deprotonation of the imidazole ring induces an aggregation of NKA. Furthermore, the $\delta NH/\Delta T$ coefficients of the NH Lys and Asp protons are weak, according to the presence of a γ turn in this N-terminal sequence. However, since the exchange rates of the NH Lys and Asp protons with deuterons of $C^2H_3O^2H$ are fast, the participation of another turn in this N-terminal part has to be envisaged. On the other hand, all the NH Val, Gly, Leu, Met, protons exchange very slowly with the deuterons in $C^2H_3O^2H$ solution in the presence of 0.5% $C^2H_3COO^2H$ at -10°C. However, no strong d_{NN} NOE are observed for these amino acids. Thus, the weight of an helicoidal structure in the conformational equilibrium of the C-terminal part appears to be very low. As in DMSO, the simultaneous presence of C_5, C_7 and β-pleated sheet conformers could explain the NMR data.

Conformation in H_2O

The NH proton chemical shifts for Lys, Thr, Asp are very similar to those observed in DMSO and methanol. The titration of the imidazole proton is influenced by the deprotonation of the carboxylic group of Asp^4. Thus, this internal salt bridge in the N-terminal part is maintained in water. As in the presence of methanol, the deprotonation of the imidazole proton (pH 7.6) induces the aggregation of NKA. Therefore, a salt bridge between Lys^2 and Asp^4 cannot be detected. The large $\delta NH/\Delta T$ coefficients of all the backbone NH protons indicate the presence of a complex conformational equilibrium.

Comparison with Substance P and Physalaemin

As for the majority of short peptides, the NKA is flexible in water except in the N-terminal part where an internal salt bridge is detected. In DMSO, two forms of this salt bridge are present and the C-terminal part seems to adopt a relatively extended structure. The aggregative tendency of NKA in these two solvents is similar to that of SP and related to the amphiphilic nature of the peptides of the tachykinin family. However, in a more hydrophobic medium like methanol, the α helical conformation of the 4-8 sequence observed for SP and PHY (4,5) does not take part in the conformational equilibrium of NKA. Thus, the difference of behaviour between the cores of NKA and SP/PHY could play an important role in the recognition process of the different tachykinins for their specific binding sites.

References

(1) Torrens, Y., Lavielle, S., Chassaing, G., Marquet, A., Glowinski, J. and Beaujouan, J.C. 1984. Eur. J. Pharmacol. 102, 381.

(2) Buck, H.S., Burcher, E., Shults, W.C., Lovenberg, W., and O'Donohue, L.T. 1984. Science 226, 987.

(3) Bergström, L., Torrens, Y., Saffroy, M., Beaujouan J.C., Lavielle, S., Chassaing, G., Glowinski, J. and Marquet, A. to be published

(4) Chassaing, G., Convert, O. and Lavielle, S. 1986. Eur. J. Biochem. 154, 77.

(5) Chassaing, G., Convert, O. and Lavielle, S. Biochim. Biophys. Acta in press

THE CONSTRUCTION OF NEW PROTEINS - DESIGN, SYNTHESIS AND CONFORMATIONAL PROPERTIES OF POLYPEPTIDES WITH 'βαβ'-TOPOLOGY

M. Mutter, K.-H. Altmann, T. Vorherr, S. Vuilleumier
Institute of Organic Chemistry, University of Basel,
CH-4056 Basel, Switzerland

Introduction

Recent progress in the elucidation of the structural features of proteins point to the exceptional role of secondary structure in the development of specific folding topologies (1). On the other hand, the knowledge of the relationship between primary sequence and secondary structure formation has been markedly increased by conformational studies on synthetic homo- and cooligopeptides (2). Based on these experimental foundations, a general procedure for the construction of polypeptides having the propensity to fold in a tertiary structure has been delineated (3). According to this strategy, amphiphilic oligopeptide-blocks with stable secondary structures are assembled to structures of higher order, the amphiphilic nature of the peptide-blocks acting as major driving force for the folding in topologically preferred conformations, e.g. βαβ-type folding units. In the present contribution we report on the design, conformational properties and potential applications of synthetic βαβ-folding units as depicted schematically below.

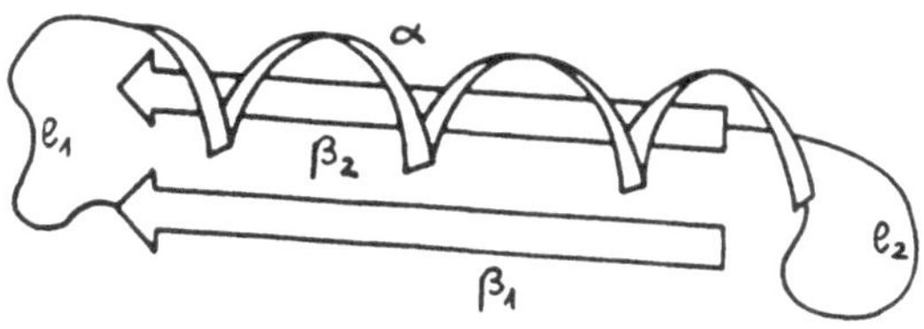

Peptides 1986

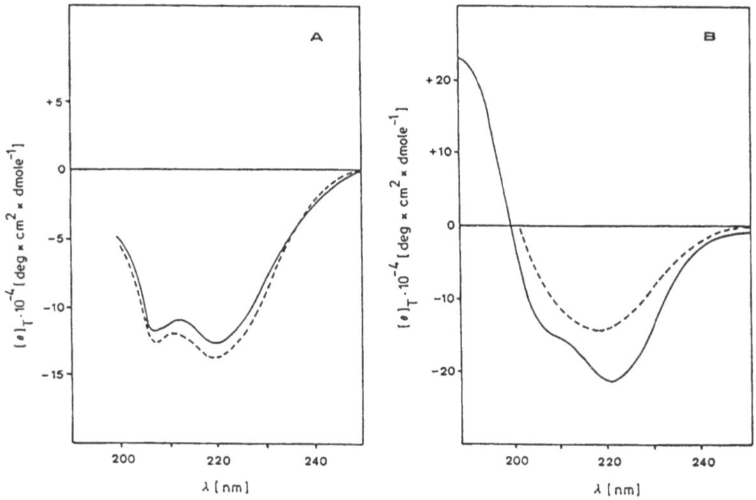

Fig. 1: CD-spectra of βαβ-models. A: I in H_2O pH7, c = $7\cdot10^{-5}$M (--) and $7\cdot10^{-6}$M (—). B: II in H_2O pH8, c = $2\cdot10^{-4}$M (--), and in H_2O/CF_3CH_2OH 1/1, c = $1\cdot10^{-4}$M (—).

Results and Discussion

The β-structure- and helix-forming potential of amphiphilic oligopeptides were investigated using polyethylene glycol as solubilizing C-terminal protecting group for stepwise synthesis (4) and CD-and IR-methods for the elucidation of the preferred conformation in solution. It has been shown that amphiphilic oligopeptides with an alternating sequence of hydrophilic and hydrophobic residues exhibit high β-forming potential in aqueous solution, probably due to the formation of bilayer sheets. The critical chain-lengths for β-structure formation (5), e.g. n = 4 for $(SL)_n$, n = 4 for $(TV)_n$ and n = 3 for $(IQ)_n$, are consistent with the lengths of β-blocks in natural βαβ-folding units. Similarly, the construction of tailor-made helical blocks with chain-lengths of n ~ 14 residues was achieved by incorporation of the strongly helix-inducing Aib residue into the hydrophobic part of the amphiphilic helix (6). For the construction of polypeptides with a

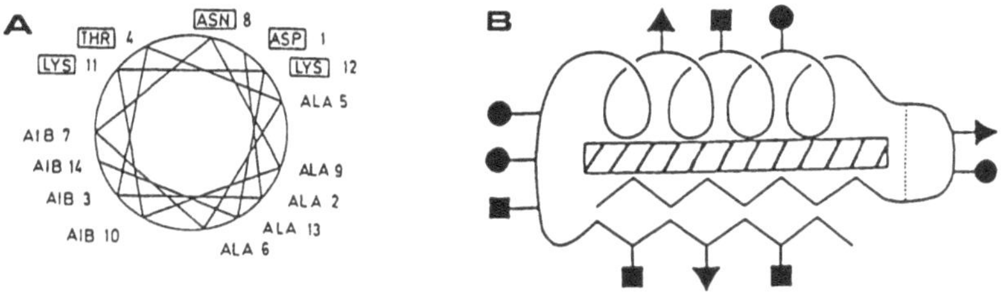

Fig. 2: Epitope stabilization. A: Lysozyme epitope 2 on the hydrophilic side of a stabilized helix (antigenic residues are framed). B: βαβ-folding unit as multivalent antigen. ●,■,▲: antigenic residues. ▨: hydrophobic core.

stable tertiary structure of minimum chain-length, molecules with βαβ-like topologies were designed by assembling these amphiphilic secondary structure blocks via hydrophilic loops in a single linear polypeptide chain of about 40 residues. Syntheses were performed by solid- and liquid-phase techniques according to standard procedures. Fig. 1 shows the CD-spectra of two representative examples of βαβ-models in aqueous solution (7). For model I (Fig. 1A) with the sequence $[LS]_3(\equiv \beta_1)$-AGEAG(l_1)- PAAibEAAAib-EAAAibEAAib (α) -EGPG$(l_2)[SL]_4G(\beta_2)$ the spectrum is characterized by two minima at 207 nm and 220 nm, indicating a highly ordered conformation even in very dilute solution. The same conclusions can be drawn for model II KVQIQVK(β_1)-GSGE(l_1)-AAibQQLAibEAib-LQQLQQ(α)-SGSG(l_2)-EIQVQIG(β_2) (Fig. 1B). The IR-spectra of the models exhibit two distinct bands in the amide I region at about 1630 cm^{-1} and 1660 cm^{-1}, pointing to the coexistence of β-structure and α-helix in the solid state as well as in aqueous solution. These findings indicate that significant amounts of a folded conformation are present in the conformational equilibrium.

Due to their well defined spatial geometry amphiphilic secondary structures as well as folding units can serve as hosts for the incorporation of antigenic determinants matching the antigenic structure in the native protein (8). This is achieved by fitting the antigenic residues of the protein to the hydrophilic surface of the model peptide (Fig. 2). Preliminary immunological studies indicate that amphiphilic helices and βαβ-folding units carrying various lysozyme-epitopes (8) elicit high affinity antibodies.

Acknowledgement

This work was supported by the Swiss National Science Foundation, Hoffmann-La Roche Inc. and the Ciba Research Fund, Basel.

References

1. Richardson, J.S., 1981. Adv. Protein Chem. 34, 167.
2. Maser, F., K. Bode, V.N.R. Pillai, M. Mutter. 1984. Adv. Polym. Sc. 65, 177.
3. Mutter, M., 1985. Angew. Chem. Int. Ed. Engl. 24, 639.
4. Mutter, M. and E. Bayer. (1979) "THE PEPTIDES", Vol. 2, Meienhofer, J. and Gross, E., Eds., Wiley, New York, pp. 285-333.
5. Altmann, K.-H., A. Flörsheimer, M. Mutter. 1986. Int. J. Pept. Protein Res. 27, 314.
6. Mutter, M., K.-H. Altmann, A. Flörsheimer, J. Herbert. 1986. Helv. Chim. Acta 69, 786.
7. Mutter, M., K.-H. Altmann, T. Vorherr: Z. Naturforsch. (in press).
8. Mutter, M., K.-H. Altmann, K. Müller, S. Vuilleumier, T. Vorherr. 1986. Helv. Chim. Acta 69, 985.

CONFORMATIONAL STUDIES ON THREE CYCLIC HEXAPEPTIDES RELATED TO THE SEQUENCE 7-11 OF SOMATOSTATIN

G. Bovermann, L. Moroder, E. Wünsch
Max-Planck-Institut für Biochemie, Abt. Peptidchemie,
D 8033 Martinsried, Germany

T. Tancredi, A. Motta, P.A. Temussi
Dipartimento di Chimica, Università di Napoli, Italy

D. Mierke, P. Lucietto, M. Goodman
Department of Chemistry, University of California,
San Diego La Jolla, USA

As part of our program to study the structure-function relationships of peptides we are developing an integrated approach based upon synthesis, nmr spectroscopy and computer simulations (1) using as target molecules constrained structures which are closely related topochemical analogs. Here we report a prelimininary study on the two cyclic hexapeptides **2** and **3** related to the central sequence 7-11 of somatostatin in comparison to the known compound **1** (2). Peptide **2** utilizes thiazolidine-4-carboxylic acid (Thz) as the bridging unit while compound **3** contains a retro-inverso peptidomimetic structure (see fig. 3). The cyclic peptides were synthesized by stepwise methods in solution and purified by well established techniques. The details will be reported elsewhere.

400 MHz 2D-HOHAHA spectroscopy (3) allowed all the assignments throughout the coupled spin network for each amino acid. The aromatic residues could be assigned from the sequential NOE's and independently from a COSY experiment optimized for small couplings. The peptidomimetic bridged structure was analyzed with the aid of 2D correlated measurements including COSY-Relayed (4) and Double Quantum experiments (5) at 500 MHz.

		7	8	9	10	11
1	cyclo(-	Phe-	D-Trp-	Lys-	Thr-	Phe-Pro-)
2	cyclo(-	Phe-	D-Trp-	Lys-	Thr-	Phe-Thz-)
3	cyclo(-	Phe-	D-Trp-	Lys-	Thr-	gPhe-mAla-)

Peptides 1986

Table 1. Chemical shift values (δ) and temperature coefficients (Δδ/T) of backbone NH's for the cyclic hexapeptides **1** - **3**

	1		**2**		**3**		
	δ [ppm]	Δδ/T [ppb/K]	δ	Δδ/T	δ	Δδ/T	
Phe^7	7.25	(-1.6)$^+$	7.74	-3.5	7.87	-4.0	
Trp^8	8.42	-5.8	8.41	-4.5	8.69	-5.3	
Lys^9	8.70	-6.2	8.72	-6.8	8.51	-4.8	
Thr^{10}	7.02	(-0.3)$^+$	7.09	-0.4^{++}	7.64	-1.0	
$X_{xx}{}^{11}$	8.33	-3.6	8.54	-3.4	7.47	-1.3	("normal"NH)
					8.38	-4.0	("gem"-NH)

$^+$cf. ref. (6); $^{++}$determined by difference

The temperature dependence of the backbone NH's (table 1) showed a low temperature shift for Thr^{10} in compound **2** in agreement with the findings on the Veber peptide (6). Conversely only a moderate effect is seen for Phe^7 as well as for Phe^{11} in the Thz peptide. Similarly for the peptidomimetic bridged analog the shift for Thr-NH is low and that for Phe^7 medium, but the 'normal' NH of $gPhe^{11}$ is now solvent shielded.

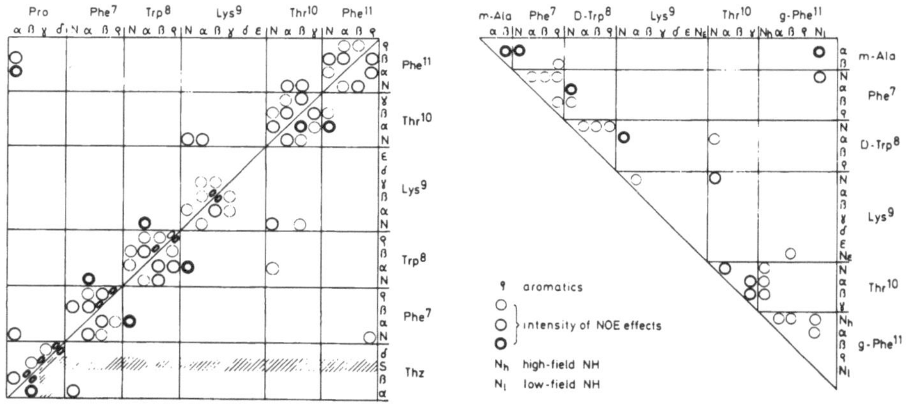

a b

Figure 1. NOE diagrams for the cyclic hexapeptides: **a**, compound **1** (upper triangle) and **2** (lower triangle); **b**, compound **3**

Useful NOESY spectra for the cyclic hexapeptides **1** and **2** at 400 MHz were obtained at 275 K in d_6-DMSO/d_8-sulfolane (77:23; v/v; fig. 1a), whereby misinterpretations due to temperature changes were avoided using rotating frame NOE measurements (CAMELSPIN (7)) at 295 K. Comparison of the interresidue NOE's indicates, that the intense crosspeak between Pro-H_α and Phe^{11}-H_α and (weaker) Phe^{11}-H_β is completely missing in the cyclic hexapeptide **2**. While these crosspeaks are indicative of the presence of a cis Phe-Pro bond (6), their absence as well as a weak NOE observed between the upfield H of Thz and Phe^7-NH strongly suggest that the trans peptide bond is favored in the Thz-peptide (fig. 2a). This is in agreement with the findings from the temperature measurements and is further supported by earlier results showing the preference for the trans peptide bond in Thz-polypeptides (8).

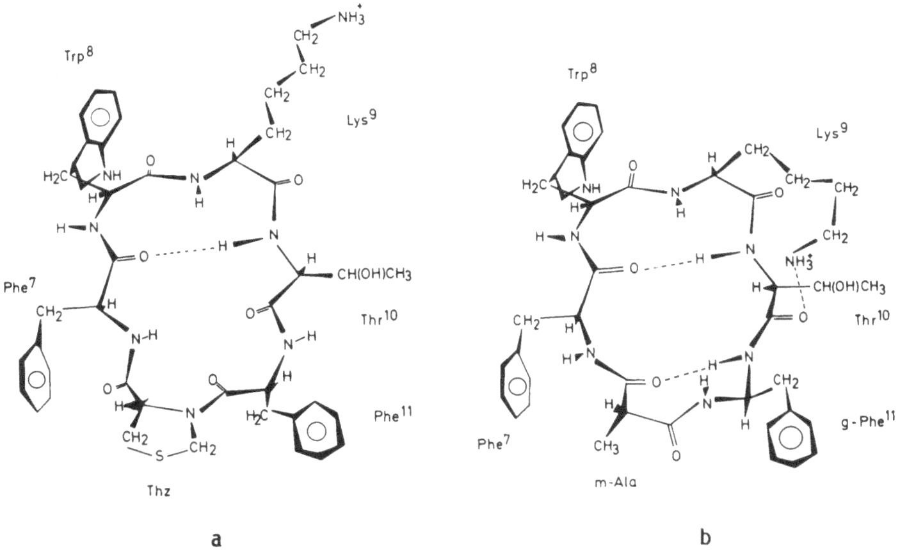

Figure 2. Schematic models for the cyclic hexapeptides consistent with data from nmr study: **a**, compound **2**; **b**, compound **3**

NOESY results for peptide **3** containing the peptidomimetic structure (d_6-DMSO; 295 K; 500 MHz; fig. 1b) indicate a similar folding as shown for the Thz containing peptide (fig. 2b). Conformational probing of this peptide by a 10 ps molecular dynamics simulation utilizing the programs of Hagler et al. (9) show that the Thr^{10}-NH is intramolecularly H-bonded in agreement with above NMR data. The computer simulation reveals that the H-bond receptor can involve the Trp-CO (7 membered ring) or other carbonyls (10 membered rings). In addition the peptidomimetic gPhe can form 8 (and 6) membered intramolecular H-bonded rings.

From these results the conclusion can be drawn that the structural changes afforded in the bridging part of the cyclic molecules lead to substantial effects in this region. But they do not necessarily alter the folding of the 'business' end of the molecule. From the experimental results conformational changes in this moiety can only be minor.

ACKNOWLEDGEMENT

M. Goodman thanks the Alexander-von-Humboldt-Foundation for the receipt of a Senior Scientist Award 1986.

REFERENCES

1. Mammi, N.J.; Hassan, M.; Goodman, M.. 1985. J. Am. Chem. Soc. **107**, 4008-4013

2. Veber,D.F.; Holly,F.W.; Nutt,R.F.; Bergstrand,S.J.; Brady,S.F.; Hirschman,R.; Glitzer,M.S.; Saperstein,R.. 1979. Nature **280**, 512-514

3. Bax, A.; Davis, D.G.. 1985. J. Magn. Res. **65**, 355-360

4. Wagner, G.. 1983. J. Magn. Res. **55**, 151-155

5. Mareci, T.H.; Freeman, R.. 1983. J. Magn. Res. **51**, 531-536

6. Kessler, H.; Bernd,M.; Kogler, H.; Zarbock, J.; Sorensen, O.W.; Bodenhausen, G.; Ernst, R.R.. 1983. J. Am. Chem. Soc. **105**, 6944-6952

7. Bothner-By,A.A.; Stephens,R.L.; Lee,J.; Warren,C.D.; Jeanloz, R.W.. 1984. J. Am. Chem. Soc. **106**, 811-813

8. Goodman, M.; Su, K.C.. 1972. Biopolymers **11**, 1773-1778

9. Dauber,P.; Osguthorpe,D.; Sharon,R.; Stern,P.; Goodman,M.; Hagler,A.T.. 1981. In: Supercomputers in Chemistry (Lycas,T.; Shavitt,I., eds.). p. 161-191

STRUCTURAL VERSATILITY OF PEPTIDES FROM $C^{\alpha,\alpha}$-DIALKYLATED GLYCINES: Acc^5- AND Acc^6-CONTAINING PEPTIDES

E. Benedetti, V. Barone, A. Bavoso, B. Di Blasio, F. Lelj, V. Pavone and C. Pedone

Department of Chemistry, University of Naples, 80134 Naples, Italy

C. Toniolo, M. Crisma, and G.M. Bonora

Biopolymer Research Center, C.N.R., Department of Organic Chemistry University of Padova, 35131 Padova, Italy

Introduction

$C^{\alpha,\alpha}$-dialkylated glycyl residues have been shown to represent a versatile new type of conformational constraint in peptides (1,2). In particular, our current interest in the structural preferences of peptides containing the $C^{\alpha,\alpha}$-dialkylated glycines with cyclic side chain Acc^5 (α-aminocyclopentane carboxylic acid), previously termed Cle (cycloleucine), and Acc^6 (α-aminocyclohexanecarboxylic acid) was stimulated by the recently published incorporation of these residues in analogues of aspartame, the formyl-methionyl chemotactic tripeptide, angiotensin II, enkephalin, oxytocin, glutathione, and ACE and peptidyltranferase inhibitors.

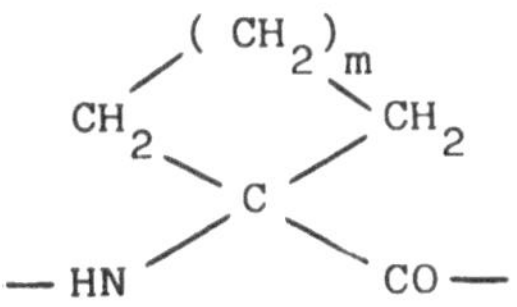

m=2 Acc^5

m=3 Acc^6

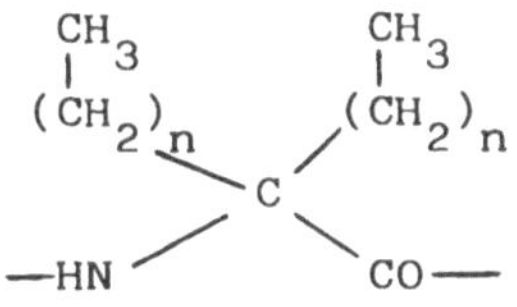

n=0 Aib

n=1 Deg

n=2 Dpg

We report here our theoretical and experimental results on : (i) conformational energy computations of $Ac(X)_nNHMe$ (n=1,2; X= Acc^5, Acc^6) as a function of the relevant $N-C^{\alpha}-C'$ (τ) bond angle; (ii) X-ray diffraction analyses of $Y-(X)_n$-OtBu (Y= pBrBz, Z; n=4-6; X=Acc^5, Acc^6); and (iii) IR absorption and ^1H NMR investigations of $Z-(X)_n$-OtBu (n=1-6; X=Acc^5, Acc^6) in $CDCl_3$ solution.

Peptides 1986

Results and Discussion

Conformational energy computations have shown that the minimum energy conformation for $Ac(Acc^5)_{1,2}NHMe$ and $Ac\text{-}Acc^6\text{-}NHMe$ corresponds to the region where both α- and 3_{10}-helices are found. A comparison with Ac-X-NHMe [X=Aib (α-aminoisobutyric acid), Deg ($C^{\alpha,\alpha}$-diethylglycine), and Dpg ($C^{\alpha,\alpha}$-di-n-propylglycine)] indicate that the τ-independent preference for helical structures of the cyclic residues closely resembles that shown by Ac-Aib-NHMe (1-5). On the other hand, the helical structures of Ac-Deg-NHMe and Ac-Dpg-NHMe are preferred for the standard value of the τ angle, while the fully-extended C_5 conformation is favoured when this angle is smaller than 107° , as experimentally observed (2,4,6,7). The results of the theoretical conformational analysis of Acc^5 and Acc^6 peptides compare well with their conformational preferences in the solid state, as assessed by X-ray diffraction. As a representative example, Fig. 1 shows the molecu-

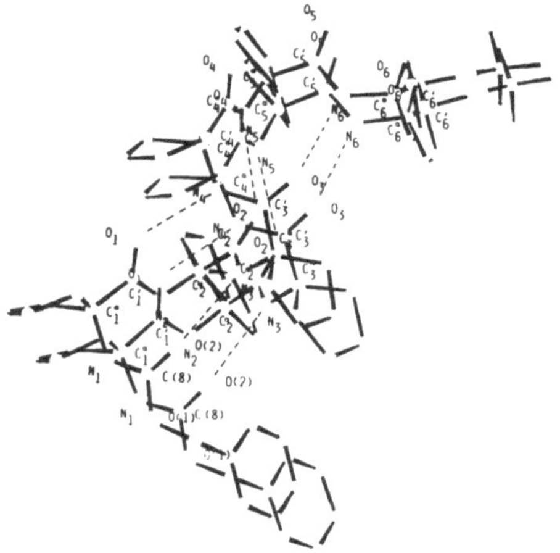

Fig. 1. Molecular structure of $Z\text{-}(Acc^5)_6\text{-}OtBu$. Dashed lines indicate intramolecular H-bonds.

lar structure of $Z\text{-}(Acc^5)_6\text{-}OtBu$. In the six structures, so far solved, the 3_{10}-helix is always observed independently of the nature of the N -blocking group. The refined structures are: $Z\text{-}(Acc^5)_n\text{-}OtBu$ with n=4,5,6 , $pBrBz\text{-}(Acc^6)_4\text{-}OtBu$, and $Z\text{-}(Acc^6)_4\text{-}OtBu$ in both the triclinic and monoclinic forms.

The solution conformational preferences of the $Z\text{-}(X)_n\text{-}OtBu$ (n=1-6; X=Acc^5, Acc^6) homo-peptide series have been determined in $CDCl_3$ by IR absorption and ^1H NMR as a function of concentration, temperature, and addition of perturbing agents. In both series, below $2x10^{-3}$M concentration, intramolecular H-bonding, the extent of which increases with increasing chain length, is the predominant factor. Fig. 2 illustrates the IR absorption spectra (NH stretching region) of Z-$(Acc^5)_n$

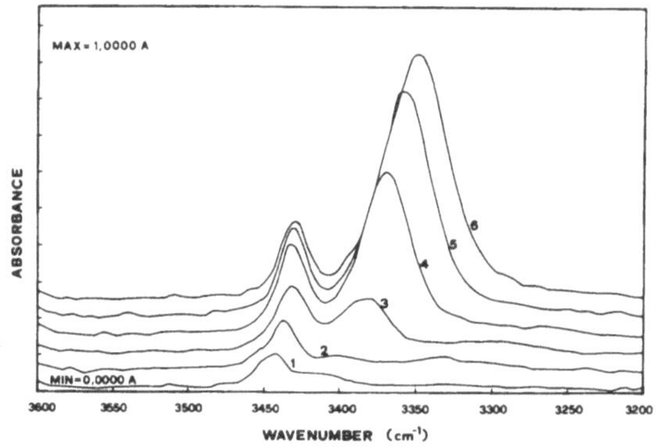

Fig. 2. IR absorption spectra in the N-H stretching region of Z-$(Acc^5)_n$-OtBu (n=1-6) homo-peptides in $CDCl_3$ solution (10^{-3} M conc.).

peptides (the band associated with the intramolecularly H-bonded N-H....O=C groups is that observed near 3350 cm^{-1}).
In addition, the ^{1}H NMR results (not shown) indicate that in the higher homologues of both series only the N(1)H and N(2)H protons are solvent exposed. This latter finding strongly supports the view that the 3_{10}-helix, not the α-helix, is the structure preferred by Acc^5 and Acc^6 homo-peptides in solution, as found in the solid state.
It may be concluded that the results of our theoretical and solid-state and solution experimental analyses of Acc^5 and Acc^6 homo-peptides unequivocally indicate their propensity to adopt a 3_{10}-helical structure, as already shown by the homo-peptides from Aib, the prototype of $C^{\alpha,\alpha}$-dialkylated glycines (1-5, 8). By contrast, unfavourable steric interactions prevent the Deg and Dpg homo-peptides from adopting helical structures, forcing them in the unusual, fully-extended, multiple C_5 conformation (2,4,6,7,9).

Acknowledgement

This research was supported by grants from the C.N.R. and the Ministry of Education of Italy. We also wish to thank Dr. Antonello Santini for his technical assistance.

References

1. Toniolo, C., Bonora, G.M., Bavoso, A., Benedetti, E., Di Blasio, B., Pavone, V., and Pedone, C., Biopolymers 22 (1983) 205.
2. Benedetti, E., Di Blasio, B., Pavone, V., Pedone, C., Bavoso, A., Toniolo, C. Bonora, G.M., Leplawy, M.T., and Hardy, P. , J. Biosc. suppl., 8 (1985) 253.
3. Paterson, Y., Rumsey, S.M., Benedetti, E., Nemethy, G., and Scheraga, H.A., J. Am. Chem. Soc. 103 (1981) 2947.
4. Barone, V., Lelj, F., Bavoso, A., Di Blasio, B., Grimaldi, P., Pavone, V., and Pedone, C., Biopolymers 24 (1985) 1759.
5. Toniolo, C., Bonora, G.M., Barone, V., Bavoso, A., Benedetti, E., Di Blasio, B., Grimaldi, P., Lelj, F., Pavone, V., and Pedone, C., Macromolecules 18 (1985) 895.
6. Benedetti, E., Toniolo, C., Hardy, P., Barone, V., Bavoso, A., Di Blasio, B., Grimaldi, P., Lelj, F., Pavone, V., Pedone, C., Bonora, G.M., and Lingham, I., J. Am. Chem. Soc. 106 (1984) 8146.
7. Bavoso, A., Benedetti, E., Di Blasio, B., Pavone, V., Pedone, C., Toniolo, C., Bonora, G.M., Leplawy, M.T., Redlinski, A., and Kaczmarek, K., Peptides: Structure and Function (eds. C.M. Deber, V.J. Hruby, and K.D. Kopple), p. 193 Pierce Chem. Co., Rockford, Illinois (1985).
8. Toniolo, C., Bonora, G.M., Bavoso, A., Benedetti, E., Di Blasio, B., Pavone, V., and Pedone, C., Macromolecules 19 (1986) 472.
9. Bonora, G.M., Toniolo, C., Di Blasio, B., Pavone, V., Pedone, C., Benedetti, E., Lingham, I., and Hardy, P., J. Am. Chem. Soc. 106 (1984) 8152.

SYNTHESIS AND X-RAY ANALYSIS OF OLIGOPEPTIDES CONTAINING α-AMINOISOBUTYRIC ACID (AIB)

M. Kokkinidis, D. Tsernoglou

EMBL, Meyerhofstrasse 1, Postfach 10.2209, D-6900 Heidelberg, FRG

H. Brückner

Institute of Food Technology, University of Hohenheim, D-7000 Stuttgart 70, FRG

Introduction

The sterically hindered amino acid α-aminoisobutyric acid (Aib), a constituent of the peptide antibiotic alamethicin and related microbial peptides has provoked a great deal of interest in the conformational parameters of these molecules. From the biological point of view an insight into these conformations is important for understanding the interactions of these microbial antibiotics with biological membranes and explaining their ability to form voltage-gated channels in black bilayer membranes. Energy calculations have shown that the Aib residue favours α- or 3_{10}-helical conformations, depending on the geometry of the substituents on the Aib C^{α} atom. So far, the crystallographic investigations of Aib containing peptides have established a preference for these molecules to form β-bends, 3_{10}- or α-helices or to adopt conformations consisting of α- and 3_{10}-helical domains. However, despite the growing number of known structures of Aib containing peptides, the influence of parameters such as chain length, amino acid constitution, or protecting end-groups on the molecular conformation is not yet clear.

Syntheses and Crystallographic Investigations

We report the syntheses and X-ray structure analyses of four Aib containing oligopeptides: H-Aib-Aib-OH (**I**), H-Aib-Aib-Aib-OH (**II**), Z-Aib-Aib-Aib-Ala-Ala-Aib-OtBu (**III**) and H-Pro-Leu-Aib-Aib-Glu-Valol (**IV**) (Z=benzyloxycarbonyl, OtBu=tertiary butylester,

Peptides 1986

Valol=valinol). Peptides (**III**) (1) and (**IV**) correspond to residues 7-12 and 13-18 (C-terminus) respectively, of several sequence analogues of the antibiotic trichotoxin from *Trichoderma viride*. The title compounds were synthesized via the oxazolone and carbodiimide/1-hydroxybenzotriazole route and deprotection of (**I**) and (**II**) by trifluoroacetic acid followed by hydrogenation with palladium on charcoal. Elemental analysis, mass spectrometry and ^{13}C nmr spectroscopy were used for the characterization of the compounds. Circular dichroism data suggest a helical conformation for **III** and a non-helical one for **IV**. Single crystals of the title compounds were obtained by controlled evaporation from organic solvents. X-ray intensities were collected on a CAD-4 diffractometer (Cu K_α radiation). The structures were solved by direct methods using the programs MULTAN80 and SHELX76. Least squares refinement on F was performed using SHELX76.

Molecular Conformations

The conformations of the peptides are shown in Figs. 1-3. **I** and **II** are clearly non-helical; their conformations deviate significantly from those of their protected counterparts (2) Z-Aib_2-OH and Z-Aib_3-OtBu (incipient 3_{10} helix), both in the conformational angles ϕ, ψ and in the formation of intramolecular hydrogen bonds. The folding and the intramolecular hydrogen bonding pattern of the trichotoxin fragments **III** and **IV** correspond to a right-handed 3_{10} helix (with the exception of Aib 6 in **III** which lies in the left-handed helical region). In **III** the helix results from four consecutive 10-atom hydrogen bonded β-turns of type III. **IV** consists of two β-turns of type III and one of type I. Despite the overall 3_{10} character of **III** and **IV**, the conformational angles ϕ, ψ of most of the residues deviate significantly from the ideal values for a 3_{10} helix and in many cases are close to the α-helical case. The hydrogen bonding pattern in **III** and **IV** (Figs. 2 and 3) is typical for a 3_{10} helix . Two of the hydrogen bonds in **IV** (Aib 4→Pro 1, Valol 6→Aib 3) are remarkably weak. The non-helical conformation of **IV** observed by circular dichroism, may be explained by the flexibility of the weakly hydrogen bonded regions of the molecule. Most of the Aib residues in **I-IV** show extensive non-planar distortions of the peptide units. The importance of these distortions for the conformational stability of poly(Aib) peptides is discussed in (3). The valency geometry around the C^α atoms is asymmetric for most of the Aib residues. If one designates as C^β_L and C^β_R the C^β atoms in Aib that correspond to the C^β and H substituents respectively in L-amino acids, then for a right-handed helical conformation the bond angles C^β-C^α-N and C^β-C^α- C are systematically smaller for C^β_L as compared with C^β_R. An opposite asymmetry is found for Aib 6 in **III** which lies in the left-handed helical region. This asymmetry is consistent with conformational energy calculations (3).

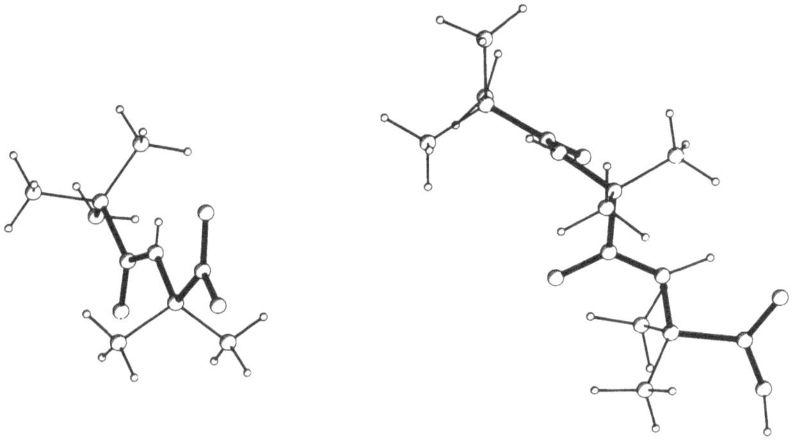

Fig. 1.: The molecular structure of **I** (left) and **II** (right). The backbone is indicated by thick lines.

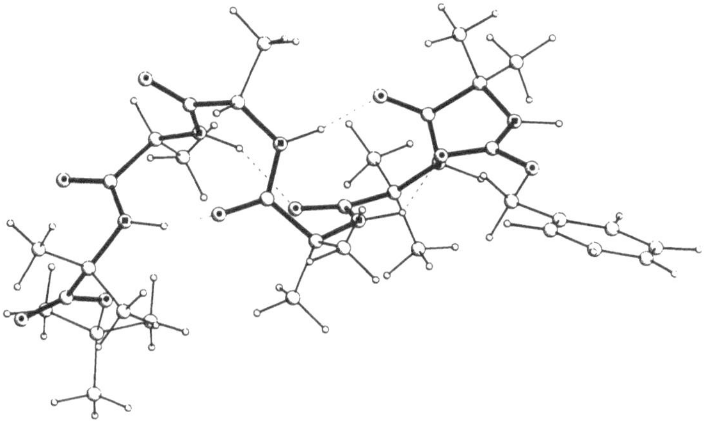

Fig. 2.: The molecular structure of **III**. Intramolecular hydrogen bonds are indicated by broken lines.

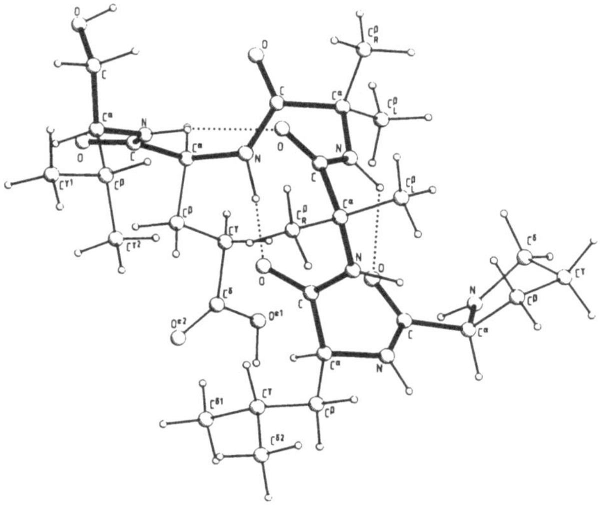

Fig. 3.: The molecular structure of **IV**. Intramolecular hydrogen bonds are indicated by dotted lines.

Conclusions

a) Main chain length is an important parameter for the establishment of helical conformations in Aib containing peptides.

b) Backbone dihedral angles alone are insufficient for the assignment of a particular helix type in Aib containing peptides.

c) Aib residues show non-planar distortions of the peptide units and asymmetric valency geometries around the C^{α} atoms which serve to relieve steric strain; the crystallographic results are consistent with theoretical energy calculations.

References

1. Kokkinidis, M., Tsernoglou, D. and Brückner, H. (1986) Biochem. Biophys. Res. Comm. 136, 870-875

2. Benedetti, E., Bavoso, A., Di Blasio, B., Pavone, V., Pedone, C., Crisma, M., Bonora, G.M. and Toniolo, C. (1982) J. Am. Chem. Soc. 104, 2437-2444

3. Paterson, Y., Rumsey, S.M., Benedetti, E., Némethy, G. and Sheraga, H.A. (1981) J. Am. Chem. Soc. 103, 2947-2955

RENIN INHIBITORS AND SUBSTRATES: CONFORMATIONAL EFFECT OF PEPTIDE BACKBONE MODIFICATIONS

G. Précigoux, Y. Barrans, S. Geoffre, P. Picard and M. Hospital
Laboratoire de Cristallographie, UA 144, Université de Bordeaux I,
351 cours de la Libération, 33405 Talence Cedex, France.

Introduction

Research has been accelerating in recent years on the discovery, synthesis, mechanism of action of renin effectors (1,2). Every conformational information on peptide substrates or inhibitors can provide new ways for potential therapeutic developments. Among several methods, one of the most efficient is X-ray analysis, and the greatest part of our knowledge about details of conformations comes from peptide crystal structures.
Our approach involves first, conformational studies and computer graphic modeling of natural or synthetic substrates and, then, synthesis of new compounds which might mimic the postulated conformation of angiotensinogen (endogenic substrate) in the enzymatic active site.
Amino acid exchanges and main chain chemical modifications are of interest to elaborate new active drugs, more resistant to the enzymatic damage. A well chosen peptide bond modification (trans C=C double bond) does not modify the main geometrical characteristics of the polypeptide, but an elongation of the backbone by adding one or several atoms can lead to dramatic changes in the conformation.
We will discuss some X-ray structures, already solved in our laboratory, and concerning such modified peptides (3,4,5). We present here one of the last result, the crystal structure of the pentapeptide Boc - leu - ψ (C=C,trans) - D leu - sta - ala -sta - OMe.

Results

The molecule crystallizes in the monoclinic space group C2 with a=24.489(6), b=10.075(3), c=18.728(5)Å, β= 100.62(2)° and Z=4. The structure was solved by means of the direct methods program MITHRIL (6) and refined to a residual of

Peptides 1986

0.10. Due to the limited number of observed data (1349), hydrogen atoms were not refined. The conformation of the peptide molecule is shown in Figure 1.

Fig. 1: Observed conformation for Boc-leu=D leu-sta-ala-sta-OMe.

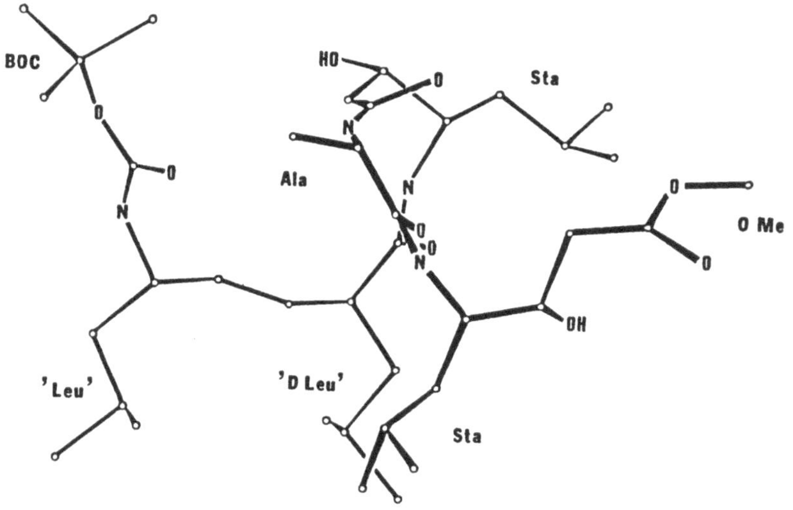

Two intramolecular hydrogen bonds produce a folded conformation in the sequence leu-D leu-sta-ala-sta. These bonds occur between the amino nitrogen of ala and the carbonyl oxygen of Boc (2.96Å) as well as between the amino nitrogen of sta.2 and the carbonyl oxygen of D leu (2.89Å). The observed conformation orients all side chains in exposed positions on the surface of the molecule. If the backbone C-terminal part of the peptide is thought of as approximating a plane, the leu and D leu side chains are located on the same half plane and the peptide main chain is on the other half of this hypothetical plane.

The main chain makes an abrupt turn after the hydroxyl atom of the central statine. Consequently, the peptide chain is folded back at the sta and ala residues to form a twelve membered ring with an intramolecular hydrogen bond. Such a turn is very similar to the one observed in the crystal structure of an analogue peptide:Boc - leu - Ψ(C=C, trans) leu - sta - ala - sta - OMe (3). By analogy with the classical O_i-N_{i+3} (ten membered ring), such a sta turn is of type 1.

The title compound belongs to the pepstatin family. The main difference is the leu and D leu residues as a substitute for val and val. Insertion of a D amino acid leads to a conformation with the leu, D leu and sta side chains situated on the same side of the peptide main chain. This geometry is then very different to the zig-zag conformations observed for regular amino acid sequences in references (3) and (8).

By comparison with a classical oligopeptide backbone, the molecule involves two major modifications. The first one, in the sequence, is the substitution of a C=C, trans double bond, in the place of the leu-D leu peptide bond. The second one is the presence of statine residues including two additional main chain carbon atoms. The main features involved in the molecular conformation are shown in Figure 2.

Fig. 2: Conformational effects of peptide backbone modifications. (a):Comparison between the observed C=C, trans double bond geometry and the classical trans peptide bond (7). (b): Dihedral angles describing the sta-turn.

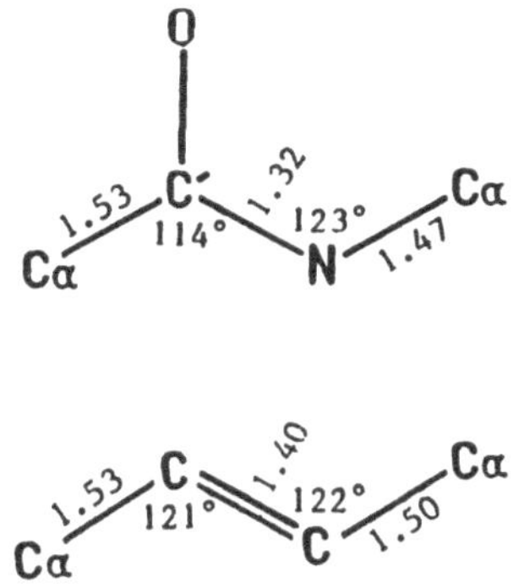

(a)

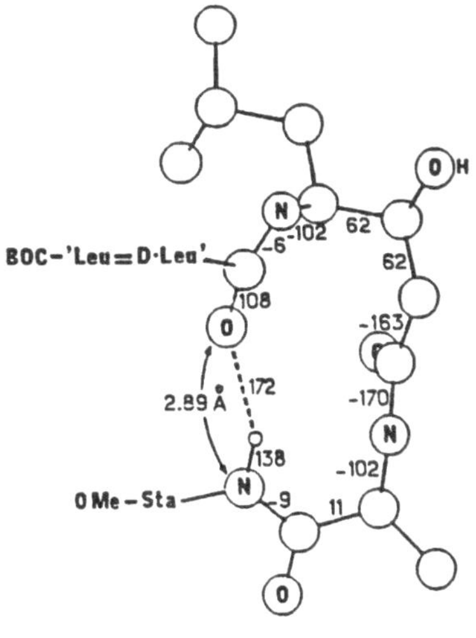

(b)

The bond lengths and angles in the C=C double bond are in general agreement with the corresponding standard values for a peptide bond, and, the observed Cα-C=C-Cα dihedral angle of 179° is not significantly different of the expected 180°

value. At the leu-D leu bond level, the main geometrical characteristics of the peptide are therefore preserved.

The remaining significant conformational feature is due to the addition of carbon atoms in the statine chain. The sta residue seems to be able to induce a very unusual type of turn. This turn can be compared with the classical β turn of type I with a difference in the Ψ_{i+1} value (62°) far from the theoritical value of -30°.

The present study gives precise information about such an uncommon conformation. Work is in progress in this serie, including molecular graphic modeling of unusual peptide bonds and synthesis of corresponding molecules.

Acknowledgement

This work was supported in part by a grant from Roussel-Uclaf. We are grateful to Dr Gasc for the gift of the sample.

References

1. Bock, M.G., R.M. Dipardo, B.E. Evans, R.M. Freidinger, W.L. Whitter, L.S. Payne, J. Boger, E.H. Ulm, E.H. Blaine, D.F. Veber, M. Poe. 1985. in Peptides Structure and Function (C.M. Deber, V.J. Hruby, K.D. Kopple, eds.). Pierce Chemical Co., Rockford, Ill., 751-754.

2. Jones, M., J. Sueiras-Diaz, M. Szelke, B. Leckie, S. Beattie. 1985. in Peptides: Structure and Function (C.M. Deber, V.J. Hruby, K.D. Kopple, eds.). Pierce Chemical Co., Rockford, Ill., 759-762.

3. Precigoux, G., E. Ouvrard, S. Geoffre. 1985. in Peptides: Structure and Function (C.M. Deber, V.J. Hruby, K.D. Kopple, eds.). Pierce Chemical Co., Rockford,Ill., 763-766.

4. Geoffre, S., F. Leroy, G. Precigoux. 1986. Int. J. Peptide Protein Res. 27, 454-460.

5. Precigoux, G., S. Geoffre, E. Ouvrard. 1986. Acta Cryst. C42, 721-724.

6. Gilmore, C.J. 1984. J. Appl. Cryst., 17, 42-46.

7. Pauling,L., R.B. Corey, H.R. Branson. 1951. Nat. Acad. Sci. USA., 37, 205-211

8. Bott, R.R., D.R. Davies. 1983. in Peptides: Structure and Function (V.J. Hruby, D.H. Rich, eds.). Pierce Chemical Co., Rockford, Ill., 531-540.

COMMON STRUCTURAL FEATURES FOR CYTOPROTECTION ACTIVITY OF CYCLIC RETRO ANALOGUES OF SOMATOSTATIN

H. Kessler, A. Haupt, M. Klein, K. Wagner
Institut für Organische Chemie, Johann Wolfgang Goethe-Universität Frankfurt, D-6000 Frankfurt am Main, Germany

Previous studies have shown that cyclic peptides containing the retro sequence of somatostatin 1 exhibit high biological activity in the inhibition of cholate uptake in rat hepatocytes (1). This effect parallels protection of liver cells against intoxication by cell toxins like phalloidin (2). Typical hormonal effects of somatostatin (3) are completely suppressed in these retro analogues. Here we concentrate on four cyclic hexapeptide analogues containing the retro sequence 7-10 of somatostatin with D- or L-Trp in position 8. Amino acid residues 7 and 10 were bridged by D-Pro^6 and Phe^{11} (5 and 6) or Phe^6 and D-Pro^{11} (7 and 8). The cyclic hexapeptides were synthesized according to methods described previously (4). The cytoprotection activity of these compounds is shown in Table 1 in comparison with some known cyclic peptides.

Table 1. *Concentration needed for a 50 % Inhibition of Cholate Uptake in Isolated Hepatocytes (testing described in ref. 2). References for peptides* ***1-4*** *are given in ref. 5.*

		CD_{50}
1	H-Ala-Gly-Cys-Lys-Asn-Phe-Phe-Trp a) HO-Cys-Ser-Thr-Phe-Thr-Lys	220 µmol
2	cyclo(-Pro-Phe-D-Trp-Lys-Thr-Phe-) b)	100 µmol
3	cyclo(-Val-Pro-Pro-Ala-Phe-Phe-Pro-Pro-Phe-Phe-) c)	8.0 µmol
4	cyclo(-Pro-Pro-Phe-Phe-Leu-Ile-Ile-Leu-Val-) d)	3.0 µmol
5	cyclo(-Phe-Thr-Lys(Z)-Trp-Phe-D-Pro-)	1.5 µmol
6	cyclo(-Phe-Thr-Lys(Z)-D-Trp-Phe-D-Pro-)	2.0 µmol
7	cyclo(-D-Pro-Thr-Lys(Z)-Trp-Phe-Phe-)	3.2 µmol
8	cyclo(-D-Pro-Thr-Lys(Z)-D-Trp-Phe-Phe-)	2.1 µmol

a) Somatostatin; b) MSD-peptide; c) Antamanide;d) Cyclolinopeptide

Peptides 1986

To investigate the structural features responsible for the inhibition of cholate uptake we determined the conformations of these peptides in solution by NMR-spectroscopy.As an example we discuss here the conformational analysis of <u>6</u>, cyclo(-Phe-Thr-Lys(Z)-D-Trp-Phe-D-Pro-), in d_6-DMSO solution.

The identification of the ^1H, ^{13}C and ^{15}N NMR signals was achieved by correlated spectroscopy (H,H-COSY; H,X-COSY) as usual (6). The discrimination of the Phe residues was done by sequencing the peptide using heteronuclear long range couplings to the carbonyl carbons obtained from a COLOC experiment (7).

Conformational analysis of the backbone is based on the temperature dependency of NH chemical shifts, the $^3J(NH\text{-}C_\alpha H)$ coupling constants and NOE-effects. Due to the small NOE's of a molecule of this size, NOE's were determined in the rotating frame by using the CAMELSPIN experiment (8) (Fig.1). To suppress contributions from TOCSY transfer (J-coupling) small flip angles (18°) were used for the mixing pulse train $(\beta\text{-}\tau)_n$ (9). Integration of the NOE cross-peaks has been used to get semiquantitative proton distances that are translated into molecular conformation via model building.

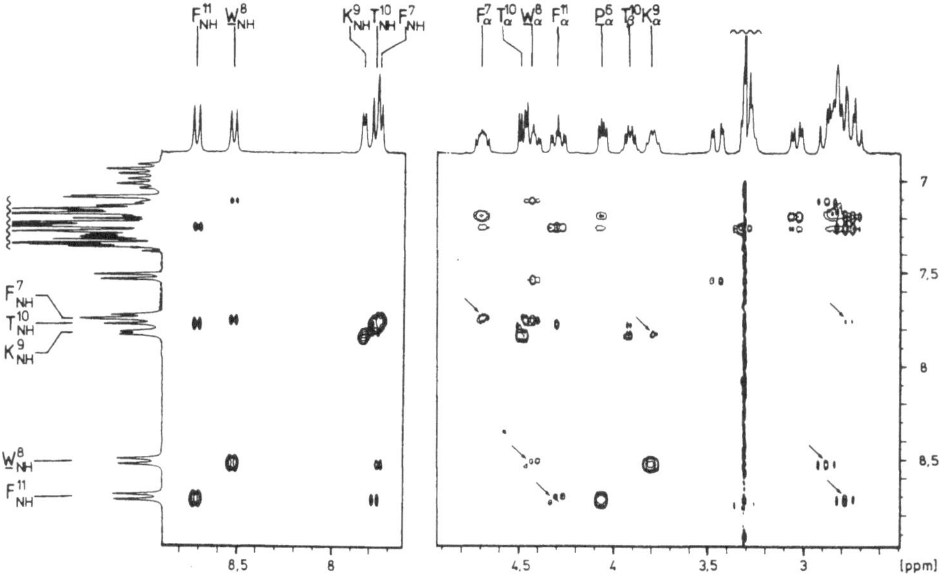

Fig.1 *Section of the CAMELSPIN spectrum of <u>6</u> showing cross-peaks from NH protons. For experimental conditions see ref.9; but $\beta=18^o$. Cross-peaks containing J-contributions are indicated by arrows.*

The population of the rotamers for the side chains of the aromatic amino acid residues were determined according to Pachler's procedure (10). The discrimination between the two rotamers P_I and P_{II} was done in all peptides by using the relatively strong heteronuclear long range coupling of the anti-periplanar oriented *pro-R* β proton and the carbonyl carbon in the conformation P_{II} of Phe^7 in 5 and 6 and Phe^6 in 7 and 8. This information is easily obtained from the cross peaks between those nuclei in routinely performed COLOC spectra (11). In summary we propose the following conformations for the four cyclic hexapeptides (Fig. 2):

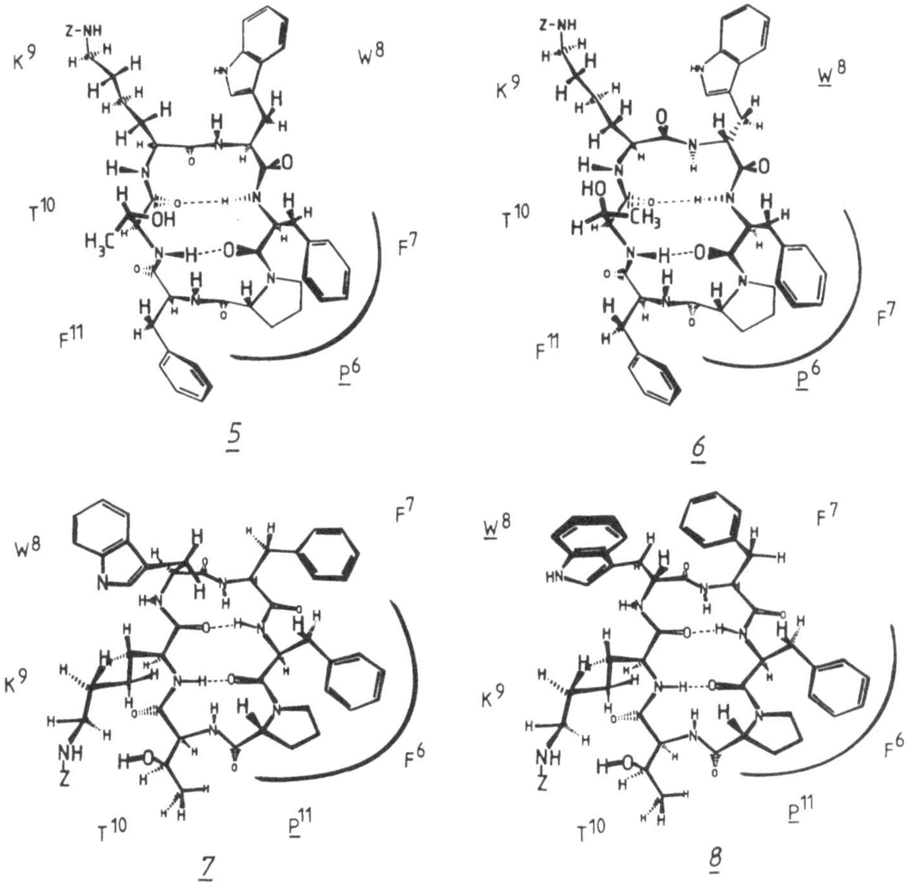

Fig. 2 *Conformations of 5, 6, 7 and 8 in d_6-DMSO solution. The region with a common structure is indicated.*

The common structural feature in all compounds consists of two ß-turns where proline is always in position i+1 of a ßII'-loop. The side chains of the amino acid residues preceeding D-proline (Phe^7 in 5 and 6; Phe^6 in 7 and 8) show the same spatial arrangement in all compounds discussed. We assume this region to be essential for cytoprotection activity.

Acknowledgement

We wish to thank Prof. M. Frimmer and Dr. K. Ziegler (University of Giessen) for performing the biological tests. We greatfully acknowledge financial support by the Deutsche Forschungsgemeinschaft and the Fonds der Chemischen Industrie.

References

1. Kessler, H.; Klein, M.; Müller, A.; Wagner, K.; Bats, J.W.; Frimmer, M.; Ziegler, K.. Angew. Chem. in press.

2. Ziegler, K.; Frimmer, M.; Kessler, H.; Damm, I.; Eiermann, V.; Koll, S.; Zarbock, J.. 1985. Biochim. Biophys. Acta 845, 86-93.

3. Szabo, S.; Usadel, K.H.. 1982. Experientia. 38, 254-256.

4. Kessler, H.; Bernd, M.. 1985. Liebigs Ann. Chem. , 1145-1167.

5. Kessler, H.; Gehrke, M.; Haupt, A.; Klein, M.; Müller, A.; Wagner, K.. Klinische Wochenschrift. Supplement. In press.

6. Kessler,H.; Bermel,W.; Müller,A.; Pook,K.H..1985.In: The Peptides, Vol. 7 (V.J. Hruby, ed.). Academic Press.pp. 437-473.

7. Kessler, H.; Griesinger, C.; Zarbock, J.; Loosli, H.R..1985. J. Magn. Reson.57, 331.

8. Bothner-By, A.A., Stephens, R.L., Lee, J., Warren, C.D., Jeanloz, R.W..1984.J. Am. Chem. Soc. 106, 811.

9. Kessler, H.; Griesinger, C.; Kerssebaum, R.; Wagner, K., Ernst, R.R.; submitted for publication.

10. Pachler, K.G.R..1964. Spectrochimica Acta 20, 581-587. Kobayashi, J.; Nagai, U.; Higashijima, T.; Miyazawa, T.; 1979. Biochim. Biophys. Acta. 577, 195-206.

11. Kessler, H.; Griesinger, C.; Wagner, K.; submitted for publication; Griesinger, C.. doctoral thesis, Frankfurt, may 1986.

SOLUTION CONFORMATION OF CONOTOXIN GI

Y. Kobayashi, T. Ohkubo, Y. Kyogoku
Institute for Protein Research, Osaka University
Suita, Osaka 565, Japan

Y. Nishiuchi, S. Sakakibara
Peptide Institute, Protein Research Foundation
Minoh, Osaka 562, Japan

N. Gō
Department of Physics, Faculty of Science, Kyushu University
Fukuoka, 812, Japan

The determination of the structure of polypeptide in solution is essential to study the relationship between structure and function of polypeptide. In order to realize the way of this determination we have developed a new technique using two dimensional NMR measurement and distance geometry algorithm.[1,2] The results of the conformational analysis of conotoxin GI in a DMSO-d_6 solution have been preliminarily reported.[3,4] Conotoxin GI is a neurotoxic peptide contained in the venom of the marine snail, <u>Conus geographus</u>, which blocks the acethlcholine receptor at the neuromuscular junction. It consists of thirteen amino acid residues with two disulfide bonds and has the primary structure as follows.[5]

The small size and the high toxicity of this molecule should provide a better understanding of the relation between the spatial structure of meurotoxins and their modes of action.
Recently, we have made some improvements in the algorithm to derive a molecular conformation and applied it on conotoxin GI to reinvestigate its conformation.

Methods

We did not consider the fact that we could no differentiate stero-

Peptides 1986

specifically two individual C_β proton resonances. Here, we solve the problem as follows. When an NOE is observed between any proton and C_β proton, the distance is replaced by that measured from the C_β carbon atom. The upper bound of the distance from the proton to the replaced C_β carbon is given by the sum of the upper distance constraint deriven from the NOE and the maximum distance between the C_β carbon and C_β hydrogen atoms. Such a treatment has been also applied in the cases of methyl and methylene groups, and aromatic ring protons which cannot be distinguished stereospecifically. After these considerations, we reexamined the conformation of conotoxin GI.

Results and Discussions

Pauling and Petcher investigated nueromuscular blocking agents with low molecular weights which are antagonists to the nicotinic cholinergic receptor by X-ray analysis and model building. They elucidated the correlation between their three dimensional structure and physiological activity and pointed out the structural requirements as follows, (1) two quatery nitrogen atoms separated by 10.8 ± 0.3 Å (2) rigidity of molecular conformation (3) a sufficient degree of lipophilicity in the area between the cationic center (4) oxygen atoms suitably disposed about the molecule to aid orientation.[6] Dufton and Hider showed that the active tip of erabutoxin has such an atomic arrangement to fit these requirements, i.e. the side chains of Arg-33 and Lys-47, and Phe-32 and Tyr-29 might be so proximate that these residues could mimic the essential features of such a curare-like alkaloids[7]. Furthermore Gray et al. have studied extensively structure-activity relationships of conotoxins using chemical modification and model building. They assumed the existence of β-turns initiating Asn-4 and Gly-8. Comparing CPK models of conotoxin GI and active tip of erabutoxin, they showed a strong resemblance between them and a correlation of their amino acid sequence in the sence of "reverse homology", i.e. Gly-Arg-His-Tyr-Ser in conotoxin GI and Ser-Asp-His-Arg-Gly in snake toxins. Thus they have made an assumption that the requirements listed by Pauling and Petcher should be common in these neurotoxins.[8]

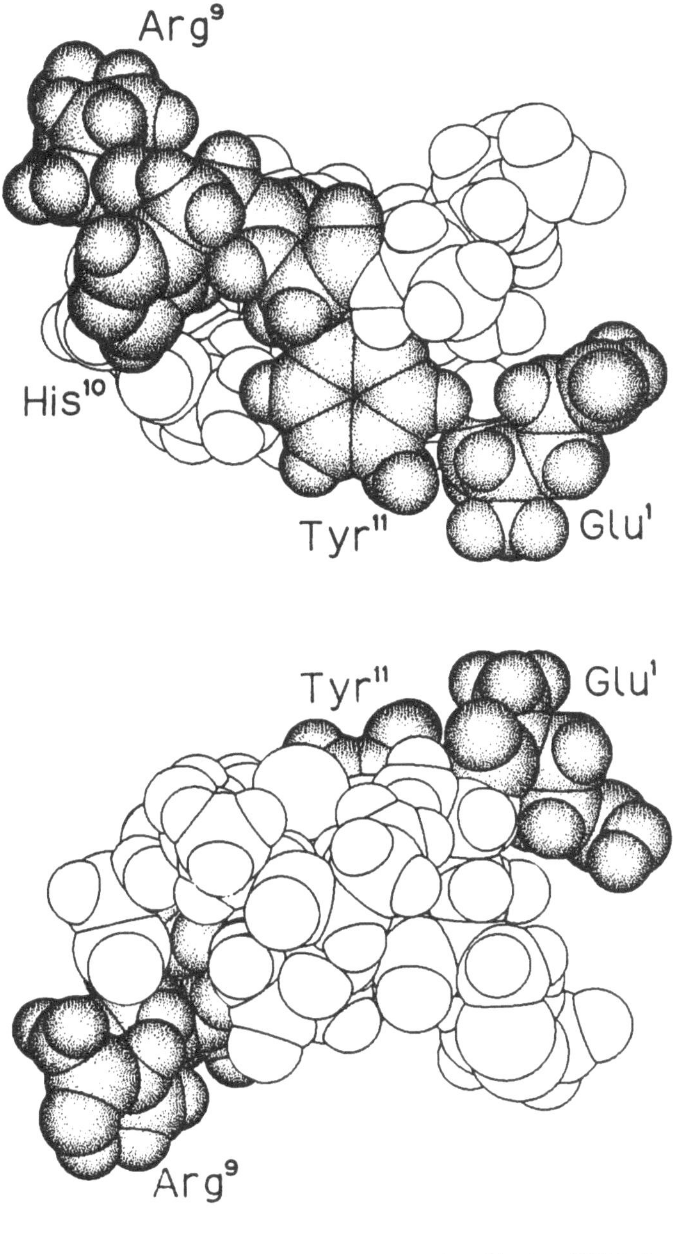

Fig. 1 Solution Conformation of Conotoxin GI in DMSO-d_6
Upper is the fron side and lower is the reverse side.
Cationic site (Glu-1, Arg-11) and lipophilic site (Tyr-10, His-11) are shadowed.

The solution conformation obtained here is the good system to testify experimentally these assumptions. The result shown in Fig. 1 reveals obviously the resemblance between atomic arrangements of conotoxin GI and the active tip of erabutoxin. Two cationic groups which consist of the quatenary nitrogen atoms of the amino group of Glu-1 and the side chain of Arg-9 stand separately. Two aromatic side chains are located between these two cationic groups in one side of the molecule. Almost all remaining residues including two disufide groups localized in the reverse side. While we have not detected the existence of the salt bridge between side chains of Glu-1 and Arg-9 to stabilize the conformation which was postulated by Gray et al., such an interaction might be possible in an aqueous solution. These results confirm that the molecular model of conotoxin GI proposed by Gray et al. explains the essential feature of the molecule.

Now we are in the course of determination of the conformation in solution of the other α-neurotoxins, another conotoxin and α-bubungarotoxin. These results should provide the key information to elucidate the structure-activity relationships of neurotoxins.

References

1. Braun, W., N. Gō. 1985. J. Mol. Biol. 186, 611.

2. Ohkubo, T., Y. Kobayashi, Y. Shimonishi, Y. Kyogoku, W. Braun, N. Gō: Biopolymers (in press).

3. Kobayashi, Y., T. Ohkubo, Y. Kyogoku, Y. Nishiuchi, S. Sakakibara, W. Braun, N. Gō. 1985. In: Peptides (C.M.Deber, V.J. Hurby and K.D. Kopple, eds.) p.101.

4. Kobayashi, Y. T. Ohkubo, Y. Kyogoku, Y. Nishiuchi, S. Sakakibara, W. Braun, N. Gō. In: Peptide Chemistry, 1985 (Y. Kiso, ed.) p.259.

5. Nishiuchi, Y., S. Sakakibara, 1982. FEBS Lett. 148, 260.

6. Pauling P., T.J. Pecher, 1973. Chem.-Biol. Interactions, 6, 351.

7. Dufton, M.J., R.C. Hider. 1980. Trends Biochem. Sci., 53.

8. Gray, W.R., D.M. Middlemas, R. Zeikus, B.M. Olivera, J. Crug. 1985. In: Peptides (C.M. Dever, V.J. Huby and K.D. Kopple, eds.) p.823.

PROTON MAGNETIC RESONANCE STUDIES OF ANGIOTENSIN II CONFORMATION: CHEMICAL SHIFTS FOR THE AROMATIC PROTONS OF SELECTED ANALOGUES IN DMSO-d_6 AT 400Mz.

John M. Matsoukas, Mahesh H. Goghari and Graham J. Moore

Dept. of Medical Biochemistry, University of Calgary
Calgary, Alberta Canada T2N 4N1

Introduction

Previous conformational studies on angiotensin II (ANG II, Asp-Arg -Val-Tyr-Ile-His-Pro-Phe), using electron spin resonance spectroscopy techniques in particular (1), have suggested the existence of an ion-dipole interaction between the α-carboxylate and the imidazole side-chain in ANG II. Such an interaction would result in partial abstraction of the single ionizable proton from the imidazole by the α-carboxylate, which in turn could lead to the imidazole group seeking proton donation from an alternative source. Based on the inappropriately elevated nucleophilicity of the Tyr hydroxyl group in ANG II and, perhaps more relevantly, the unusual pH dependence of the reactivity of this group, an interaction of the phenoxyl proton with the imidazole side-chain in ANG II has been proposed. (2). The result is a tripartite network interaction in which the negative charge originating from the α-carboxylate can be delocalized to the imidazole ring and perhaps the Tyr ring (3). This interaction is analogous to the charge relay system believed to be present at the active site of serine proteases (4) although, since the mechanism of action of the hormone is presently unknown, we prefer to adopt the term "charge delocalization network" (CDN) to describe the proposed interaction in ANG II

Results

We rationalized that an interaction of the phenoxyl proton with the His side-chain should manifest itself in the NMR spectrum of ANG II. However, the effect would probably only be detectable if

Peptides 1986

the chemical shifts for the His ring protons of ANG II were compared directly with the same protons in analogues lacking the phenoxyl proton. On the assumption that DMSO represents a reasonably facsimile of the microenvironment at membrane receptors, we have examined the ^{1}H-NMR spectra for the natural agonist ANG II, the superagonist [Sar^{1}]ANG II and the antagonists [Sar^{1}Tyr(Me)^{4}] ANG II and [Sar^{1}Phe^{4}]ANG II (5).

Table 1. Chemical shifts for the aromatic protons of angiotensin analogues and reference peptides in DMSO-d_6 at 400 Mz.[1]

Peptide	His		Tyr[2]		Phe
	C_2	C_4	meta	ortho	
ANG II	7.47	6.86	6.99(d)	6.60(d)	7.18 - 7.12
[Sar^{1}]ANG II	7.48	6.87	6.98(d)	6.59(d)	7.18 - 7.11
[Sar^{1}Tyr(Me)^{4}]ANG II	7.50	6.90	7.15(d)[3]	6.70(d)[3]	7.12 - 7.06
[Sar^{1}Phe^{4}]ANG II	7.52	6.92	--	--	7.21 - 7.13
[Sar^{1}D-Pro^{7}]ANG II	7.60	6.84	7.06(d)	6.64(d)	7.24
Amino Acid[4]	7.68[5]	7.04[5]	7.07(d) 7.20(d)[3]	6.70(d) 6.86(d)[3]	7.30
CF_3-Gly-Gly-X-Ala-OCH_3[6]	8.35	7.35	7.02(d)	6.64(d)	7.26
Oxytocin[7]	--	--	7.12(d)	6.68(d)	--
[Arg^{8}]Vasopressin[7]	--	--	6.91(d)	6.62(d)	7.32 - 7.22

[1]Chemical shifts are in ppm relative to TMS at ambient temperature.

[2]Meta and ortho refer to the Tyr hydroxyl group; d= doublet.

[3]Tyr(Me). [4]At 200 Mz. [5]For imidazole, His is insoluble in DMSO. [6]Linear reference peptides (6) in which X = His (protonated), Tyr or Phe.[7] **After D_2O exchange.**

The chemical shifts (Table 1) demonstrate a small shielding influence on the His C_2 and C_4 protons in ANG II and [Sar^{1}] ANG II which is absent in both [Sar^{1}Tyr(Me)^{4}]ANG II and [Sar^{1}Phe^{4}]ANG II.

Apparently a shielding influence on the His ring by the phenoxyl group in ANG II is dependent on the presence of the single ionizable phenoxyl proton. In a previous report (7), we suggested that a shielding influence on the His ring of ANG II might originate from a stacking interaction with the Phe ring. In the light of the present findings, it appears that part of this shielding influence on the His ring originates from the phenolic hydroxyl group. This is consistent with the original contention (2), based largely on chemical reactivity studies, that the Tyr hydroxyl in ANG II is involved in an intramolecular interaction, probably by hydrogen bonding to the His ring.

In order to investigate the proposed occurrence of an ion dipole bond (1) between the α-carboxylate and the His ring, we examined the ^{1}H-NMR spectrum of the analogue [Sar^{1}D-Pro^{7}]ANG II which, according to molecular modelling experiments, would be unable to form this bond (and thereby the CDN) because of steric constraints. The chemical shifts of the aromatic protons of this biologically inactive analogue (Table 1) demonstrate that the ordered compact tertiary structure of the ANG II molecule becomes completely relaxed when the Pro^{7} residue is switched to the D configuration. With the exception of the His$\cdot C_4$ proton, the aromatic protons of [Sar^{1}D-Pro^{7}]ANG II show no evidence of shielding influences when the chemical shifts are compared to those for a simple reference peptide and the free amino acids (Table 1). In addition the Phe side-chain of [Sar^{1}D-Pro^{7}]ANG II, unlike ANG II and [Sar^{1}]ANG II, demonstrates free rotation. These findings suggest that a proline residue in the L configuration is mandatory for maintaining a restricted biologically active conformation for ANG II, i.e. the Pro residue appears to dictate a specific alignment of the three aromatic side-chains, resulting in shielding of these side-chains. This is in accord with chemical reactivity studies (3), which have suggested that the nucleophilic groups in [Sar^{1}D-Pro^{7}]ANG II, unlike ANG II, freely interact with the solvent.

Finally we thought it of interest to compare the chemical shifts of the aromatic protons of ANG II with those of vasopressin and oxytocin - two hormones whose conformations have been studied in greater depth. Previous studies have indicated that the aromatic rings of the neighbouring Tyr and Phe residues in vasopressin are involved in a stacking interaction in D_2O (8). In DMSO (Table 1) the Phe side-chain in vasopressin shows evidence of restricted rotation without notable shielding, whereas the chemical shifts for the Tyr residue illustrate that the meta protons, but not the ortho protons, are subjected to a strong shielding influence (compare chemical shifts for Tyr in vasopressin with free amino acid, reference peptide and ANG II). Possibly there is only partial overlap of the Tyr and Phe rings in vasopressin in DMSO. It is interesting to note that the Phe ring protons in ANG II generally experience a stronger shielding influence than the Phe ring protons in vasopressin. We have suggested previously that shielding and restricted rotation for the Phe side-chain in ANG II might be due to a stacking interaction with the His side-chain (7). It is somewhat surprising that His/Phe stacking (as in ANG II) should have a greater shielding influence on the Phe ring than Tyr/Phe stacking (as in vasopressin). However, in addition to reduced shielding because of partial stacking in the latter, it is possible that the stronger shielding influence on the Phe ring in ANG II is due to the presence of a delocalized negative charge on the His ring, as predicted by the CDN conformation for ANG II (3). The increased density of the electron cloud generated by a negative charge on the imidazole ring would perhaps be expected to increase the shielding influence on the stacked aromatic ring of Phe in ANG II.

Discussion

In conclusion, we suggest that the present findings are not inconsistent with the proposed CDN/ring stacking conformation for ANG II (3). The data in Table 1 indicate that a shielding influence on the His ring in ANG II could originate not only from a stacking interaction with the Phe ring (7) but also from an interaction with the imidazole phenoxyl proton. The chemical shifts in Table 1 also demonstrate that conformational disruption of the C-terminus of ANG II by the introduction of D-Pro in place of Pro^7, markedly de-shields all three aromatic rings and randomizes the overall conformation. Finally, comparison of the chemical shifts for the aromatic protons of ANG II analogues with those for vasopressin and oxytocin and the free amino acids demonstrates that the Tyr, His and Phe rings in ANG II experience strong shielding influences, suggesting a compact molecular structure for ANG II in neutral solution.

Acknowledgements

This work was supported by the Alberta Heart Foundation, the Alberta Heritage Foundation for Medical Research and the Medical Research Council of Canada. We thank Joanne Ward for typing the manuscript.

References

(1). Weinkam, W.J. and Jorgensen, C. 1971. J. Am. Chem. Soc. 93, 7033.

(2). Moore, G.J. 1985. Int. J. Pept. Prot. Res. 26, 469.

(3). Moore, G.J. and Matsoukas, J.M. 1985. Bioscience Reports 5, 407.

(4). Blow, D.M., Birktoft, J.J. and Hartley, B.S. 1969. Nature 221, 337.

(5). Goghari, M.H., Franklin, K.J. and Moore, G.J. 1986. J. Med. Chem. 29, 1121.

(6). Wuthrich, K. NMR in biological research: peptides and proteins, North Holland Publishing Co., pp 51-53.

(7). Matsoukas, J.M. and Moore, G.J. 1984. Biochem. Biophys.Res. Commun. 122, 434.

(8). Deslauriers, R. and Smith, J.C.P. 1963. Biochem. Biophys. Res. Commun. 98, 976.

CONFORMATIONAL STUDIES OF CARBA-ANALOGS OF OXYTOCIN

M. Lebl, I. Frič

Institute of Organic Chemistry and Biochemistry, Czechoslovak Academy of Sciences, 166 10 Prague 6, Czechoslovakia,

E.E. Sugg, W.L. Cody, V.J. Hruby

The University of Arizona, Tucson, AZ 85721 U.S.A.

Introduction

Substitution of CH_2 for a sulfur atom of the oxytocin disulfide bridge has a significant effect on biological properties of the resulting analogs, depending on the location of the sulfur exchanged (for review see (1)). We have investigated the conformational characteristics of a number of carba analogs (the structures of some of them are shown below) by ^{1}H NMR and CD spectroscopy and by UV fluorescence.

```
CH2---------X---------CH2
|                     |
CO-Tyr-Ile-Gln-Asn-NH-CH-CO-Pro-Leu-Gly-NH2
```

I, X=CH_2-S-S II, X=CH_2-CH_2-S III, X=CH_2-CH_2-CH_2 IV, X=CH_2-S-CH_2

Results

^{1}H NMR spectroscopy: We found that the analogs could be divided into two groups depending on the presence or absence of sulfur in position 6. The first group contains analogs having sulfur in position 6 (I and II) while the other group has a CH_2 group instead (III and IV). Comparison of chemical shift values (determined in DMSO solution) with the values of a similar amino acid residue in model tetrapeptide TFA-Gly-Gly-Xaa-Ala-OMe revealed the following similarities: The first group (sulfur in position 6) displays more shielded amide protons of asparagine (0.4 ppm) and glycine (0.1 to 0.2 ppm), the α-proton of isoleucine (0.3 ppm), and also significantly more deshielded is one of the β-protons of tyrosine. The second group displays a deshielded NH proton of tyrosine (0.2 to

Peptides 1986

0.3 ppm), a shielded NH proton of isoleucine (0.6 to 0.8 ppm), and a more shielded α-proton of asparagine (0.1 to 0.2 ppm). The difference in the chemical shifts of β-protons of tyrosine is greater in the first group (0.52 and 0.65 ppm) than in the second group (0.38 and 0.43 ppm). These two groups also differ in the sensitivity to the change of configuration of the amino acid in position 2. Introduction of a D-amino acid into a compound of the first type leads to a large upfield shift of the glutamine NH (0.9 to 1.0 ppm), a downfield shift of the asparagine NH (0.35 to 0.6 ppm), isoleucine α-proton (0.1 ppm) and glutamine α-proton (0.1 to 0.2 ppm), and to a small upfield shift of the α-proton of asparagine (0.05 to 0.1 ppm). A larger upfield shift is observable in one β-proton of the aromatic amino acid in position 2 (0.5 ppm). Analogs of the second type display a deshielded α-proton of the aromatic amino acid (~0.1 ppm) and an amino acid in position 6 (~0.25 ppm). The change of configuration in position 2 leads in both groups of analogs to an upfield shift of all but α-CH protons of isoleucine and to a downfield shift of NH of the amino acid in position 6 (with the exception of deamino-oxytocin). The difference in chemical shifts of β-protons of the aromatic amino acid is strongly reduced by configurational change (from 0.38-0.7 to 0.12-0.25 ppm). This is accompanied by the change of α-β coupling constants (3-4 and 10-12 Hz for all-L-analog to 5-8 and 6-9 Hz for D-diastereoisomers) of the aromatic residue.

Two groups of analogs may also be recognized according to the temperature dependences of the backbone NH chemical shifts. Analogs of the first group have a quite small temperature coefficient for the proton in position 2 (-1 to -2 ppb/deg) and a **positive** coefficient for the asparagine NH proton (+4 to +6 ppb/deg). Analogs of the second type have a low value of the temperature coefficient for amides in position 3 and 6. The change of the aromatic amino acid configuration does not change the temperature coefficients in the analogs of the second group, while changing it dramatically in the first group: the glutamine NH proton is now temperature independent (slightly positive). However, there are differences inside the first group between II and I, the first analog having temperature-independent amides of residues in position 6 and 9.

A very important feature of the spatial structure of carba-analogs may be the interaction of cystine sulfur in position 6 with the residue in position 2, which was suggested on the basis of their CD spectra (2) and, recently, on the basis of the downfield shift of tyrosine aromatic proton resonances (3) in analogs containing sulfur in position 6. That the interaction is not limited only to the tyrosine residue and that phenylalanine can participate in it as well is evidenced by the downfield shift (~0.06 ppm) of aromatic protons in /Phe^2/dCOT-1 in comparison to the shift of phenylalanine in the model tetrapeptide, or in

/D-Phe^2/dCOT-1.

In the analogs of the first type the intramolecular interaction of the aromatic residue with the sulfur in position 6 may be responsible for the more compact structure in which the aromatic side chain influences the amino acids in positions 4 and 5. The positive temperature coefficient of the asparagine amide proton together with its upfield shift in the analogs of the first type may be explained by hydrogen bonding interaction of this proton with π-electron cloud of the aromatic residue; this interaction is made possible by the above mentioned sulfur-aromatic residue interaction. The conformation of II is significantly more similar to that of I than to those of III or IV. The increased flexibility of the carba-joint allows the formation of more intramolecular hydrogen bonds. Our interpretation of the presented data still needs further refinement and should be taken as tentative only.

CD spectroscopy: The results obtained with analogs provide further support for the existence of a tyrosine - S^6 interaction. CD spectra do not contain features which would directly reflect the interaction in question but they indicate the conformational situation in which the interaction is possible. The middle part of the CD spectrum ranging from 210 to 240 nm appears to be the most sensitive to structural and conformational changes in the region of the heterodetic joint and the residue in position 2. This part of the CD spectrum is formed by superposition of a pair of oppositely signed amide n-π^* bands and an aromatic B_{1u} band of tyrosine chromophore. CD spectra (in neutral buffer solution) of compounds in which the interaction tyrosine - S^6 takes place differ from the spectra of the other compounds by a distinctly higher intensity of the positive B_{1u} aromatic band and lower intensity of the positive n-π^*band below 220 nm. The latter group of compounds also includes all D-Tyr^2 analogs. The typical spectral difference between carba-1 and carba-6 oxytocin is smoothed out in corresponding D-Tyr^2 analogs. On the other hand, the change of the tyrosine configuration differentiates greatly the formerly similar spectra of deaminooxytocin (I) and its carba-1 analog (II). This effect might be explained by the higher flexibility of the heterodetic joint in the carba analog. A comparison of CD spectra of carba-analogs including those with a nonaromatic amino acid in position 2 reveals that the conformational difference between the two types of compounds may include partial changes connected with the interaction itself, with the replacement of a sulfur atom with a methylene group and with the change of configuration in position 2. It appears that the change of configuration of the amino acid in position 2 leads to some enhancement of the molecular rigidity. This effect seems to be especially large in the /D-Phe(4-Et)^2/deamino-carba-6-oxytocin.

Fluorescence: We have also measured the quantum yield of tyrosine aromatic fluorescence in neutral water solution. The tyrosine fluorescence is effectively quenched by the disulfide group (as shown for oxytocin in previous studies (4)) even with analogs where tyrosine is in the D-configuration or where the Cys[1] residue has been replaced by the Pen[1] residue. The thioether group in carba-analogs also quenches the tyrosine fluorescence but far less than does the disulfide group. We have found only a small and insignificant difference in the quantum yield of fluorescence for carba-1 and carba-6 analogs of deamino-oxytocin. The fluorescence study proves the existence of contacts between the side chain of tyrosine and the heterodetic joint of the ring moiety. The flexibility of the molecules and the effectiveness of the quenching mechanism are too high in aqueous solution, so that the method is not able to distinguish between cases of different accessibility of the quencher for the aromatic group.

On the other hand, preliminary results of fluorescence measurement in dimethylsulfoxide solution have shown a significant difference in quenching between carba-1 and carba-6 oxytocin. Both analogs IV and III almost do not influence the fluorescence. These results are in agreement with the ^{1}H NMR data that show a fixed conformation with sulfur-aromatic interaction in DMSO solution.

References

1. Jošt, K. 1986. In: Handbook of Neurohypophyseal Hormone Analogs (K.Jošt, M. Lebl, and F. Brtník, eds.). CRC Press (in press).
2. Frič, I., M. Kodíček, K. Jošt, K. Bláha. 1974. Collect.Czech.Chem. Commun. 39, 1271.
3. Lebl, M., E.E. Sugg, V.J. Hruby. 1986. Int.J.Pept.Prot.Res. (in press).
4. Ross, J.B.A., W.R. Laws, A. Buku, J.C. Sutherland, H.R. Wyssbrod. 1986. Biochemistry 25, 607.

CONFORMATIONAL ANALYSIS OF CHOLECYSTOKININ CCK_{27-33} AND CCK_{26-33} BY ^1H NMR SPECTROSCOPY FLUORESCENCE TRANSFER MEASUREMENTS AND CALCULATIONS.

M.C. Fournié-Zaluski, C. Durieux, B.P. Roques

Département de Chimie Organique, U 266 INSERM - UA 498 CNRS, UER des Sciences Pharmaceutiques et Biologiques, 4 avenue de l'Observatoire, 75270 Paris Cedex06, France

B. Lux

Laboratoire de Physique, UA 91 CNRS, Faculté de Pharmacie, 67000 Strasbourg, France

B. Maigret

Laboratoire de Modèles Informatiques Appliqués à la Synthèse, UA 490 CNRS, Université Louis Pasteur, 67000 Strasbourg, France

Introduction

The C-terminal octapeptide of cholecystokinin, CCK_8, is one of the most abundant peptides in the central nervous system, where it has recently been implicated as a putative neurotransmitter (1). Specific binding sites have been found for CCK in the brain and in the gut, and several features suggested the existence of multiple CCK receptors (2). The unambigous characterization of a receptor heterogeneity requires the development of selective ligands for each binding site. This was achieved by deducing the biologically active structures from an extensive conformational analysis of the endogenous peptide. For this purpose, the conformational behaviour of CCK_7 and CCK_8 in their sulfated and unsulfated forms was studied by three complementary techniques : ^1H NMR spectroscopy, fluorescence transfer measurements and conformational calculations (3).

Results and discussion

1/ NMR studies.

In first series of experiments, the ^1H NMR spectra of CCK_7 and CCK_8 were

Peptides 1986

studied in D_2O solution as a function of pH. The titration of the functional groups induced chemical shift variations on the protons of almost all aminoacids. These results reflect through space effects that are indicative of backbone foldings. This is corroborated by the ionization constants of β-Asp^{32} carboxylate in these two peptides (CCK_7, pka ~ 3.85 ; CCK_8, pka ~ 4.50) that indicate their involvement in a hydrophobic surrounding. The complete analysis of CCK_7 and CCK_8 spectra in D_2O/H_2O solution at pH 6 was performed (fig. 1). The NH temperature dependency shows that Asp^{32}NH and PheNH are buried protons in both peptides ; furthermore, low temperature coefficients were also obtained for Trp NH in CCK_7 and Gly NH in CCK_8.

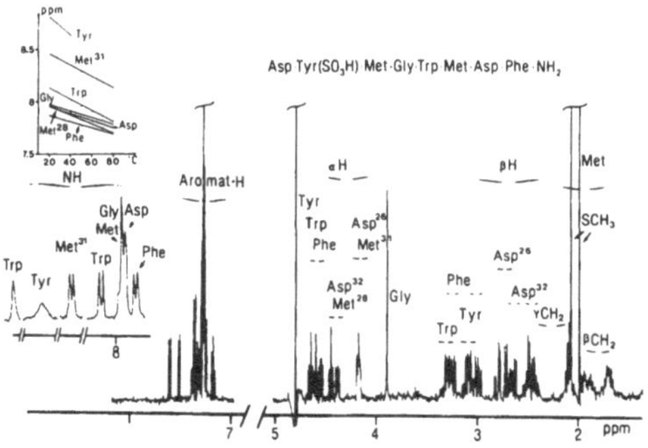

Figure 1 - ^{1}H NMR spectrum of CCK_8 in D_2O/H_2O solution at pH 6

The same spectral characteristics were obtained in $DMSOd_6$, showing the same type of folded conformations in lypophilic medium than in water i) a γ turn around Asp^{32} residue and a β turn around Trp Met, as a common feature for CCK_7 and CCK_8. ii) different folded structures in the N-terminal part of these peptides : a γ turn around Gly residue in CCK_7 (4) and an equilibrium between β or γ turns for CCK_8.

The study of unsulfated peptides in $DMSOd_6$ indicates that in neutral conditions NS CCK_7 and NS CCK_8 present less constrained structures than their sulfated analogues and that in acidic conditions, extended conformations are favoured.

2) Fluorescence studies.

For NS CCK_7 and NS CCK_8, the total emission quantum yield was determined for excitation at 280 nm where both Tyr and Trp absorb and 295 nm where Trp is selectively excited. The spectral parameters indicate that i) Trp is exposed to the solvent and ii) Tyr fluorescence is strongly quenched at neutral pH. This quenching could be due to the interaction of Tyr with the

Asp^{32} COO^- or other quenching group which supposes a folding of the molecule in accordance with the large average distance R = 15 A between the chromophores (table). At pH 4 where the carboxylate is protonated, Tyr fluorescence is enhanced and the average distance between Tyr and Trp is decreased. For CCK_7 and CCK_8 the study is complicated by a very low molar extinction coefficient of $Tyr(SO_3^-)$. However, the transfer efficiency could be measured at pH 7, leading to a R distance identical with that obtained for the unsulfated peptides.

Table - Fluorescence quantum yields of tyrosine and average intramolecular distances between Tyr and Trp in CCK_7 and CCK_8

	pH	Ø Tyr	E Tyr-Trp	R(A)
NS CCK_7 or NS CCK_8	7	0.015	0.10	15.0 ± 0.5
	4	0.025	0.50	10.0 ± 0.5
CCK_7 or CCK_8	7	0.035	0.60	15.0 ± 0.5

3) Conformational calculations

The most stable conformations found from Metropolis calculations on CCK_8 and NS CCK_8 are summarized in figure 2. In the case of CCK_8, an important degree of spatial organization in the N-terminal moiety of the peptide was shown, with β and γ turns including Trp and Gly NH. The folding tendency of the C-terminal part of the peptide observed by NMR is not so clearly shown by calculations : only conformer 2 contains a β turn at the level of Gly^{29}-Asp^{32} sequence.

For the unsulfated peptide, less organized structures are observed.

Conclusion.

In neutral conditions, a folding of peptide backbones was observed in lypophilic solvent and in aqueous medium. (fig. 3). The preferential conformation adopted by the C-terminal part of CCK_7 and CCK_8 was identical with that encountered in smaller fragments (5). Furthermore, the β or γ turns occurring in the N-terminal part of CCK_7 and CCK_8 suggest that the flexible glycine residue may play a role of a hinge between the folded N and C-terminal parts of these peptides. The importance of this aminoacid was recently demonstrated by the complete loss of potency following its replacement by the sterically constrained Aib residue in CCK_8 (6). In acidic medium protonated CCK_7 and CCK_8 exist preferentially as extended structures. The conformational change observed by protonation indicates that extended and folded structures are energetically closely related. Therefore, according to the results of calculations that emphasize the stability of folded forms devoid of internal ionic interactions, one can assume that in the lipidic surrounding of the receptor or during the receptor binding process CCK_8 might adopt folded conformations similar to those described in solution.

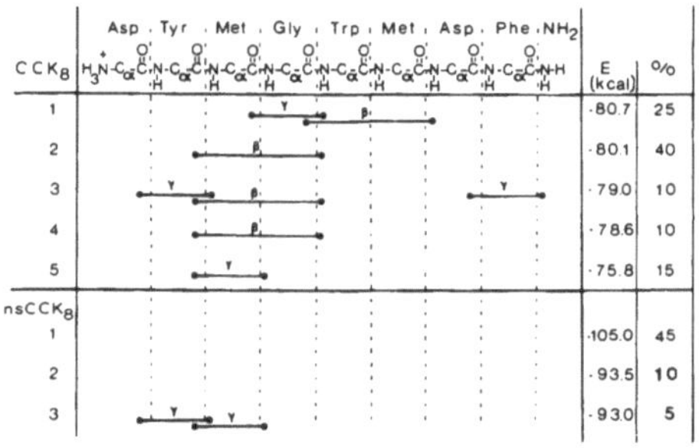

Figure 2 - Schematic representation of the various turns observed in the most stable conformations of CCK_8 and $NSCCK_8$

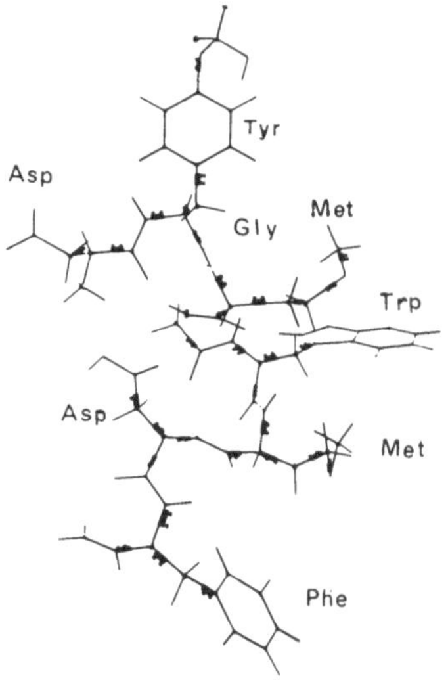

Figure 3 - Dreiding model of the preferential conformation of CCK_8 in solution

References

1. Pinget, M., E. Strauss and R.S. Yalow. 1979. Life Sci. 25, 339-342.

2. Innis, R.B. and S.H. Snyder. 1980. Proc. Natl. Acad. Sci. USA 77, 6917-6921.

3. Fournié-Zaluski, M.C., J. Belleney, B. Lux, C. Durieux, D. Gérard, G. Gacel, B. Maigret and B.P. Roques. 1986. Biochemistry, 25, 3778-3787.

4. Durieux, C., J. Belleney, J.Y. Lallemand, B.P. Roques and M.C. Fournié-Zaluski. 1983. Biochem. Biophys. Res. Commun. 114, 705-712.

5. Fournié-Zaluski, M.C., C. Durieux, B. Lux, J. Belleney, P. Pham, D. Gérard and B.P. Roques. 1985. Biopolymers 24, 1663-1681.

6. Fournié-Zaluski, M.C., J. Belleney, C. Durieux, G. Gacel, B.P. Roques, D. Begue, I. Menant, B. Lux and D. Gérard. 1985. Ann. N.Y. Acad. Sci. 448, 598-600.

STUDY OF THE β-TURN CONFORMATION IN PEPTIDES: AN INTRAMOLECULAR CHARGE-TRANSFER COMPLEXES APPROACH.

M. Sakarellos-Daitsiotis, G. Chassiotis, C. Sakarellos
Department of Chemistry, University of Ioannina, 451 10 Ioannina, Greece.

B. Perly, DPC/SCM, C.E.N. Saclay, 91191 Gif-sur-Yvette, France.

Introduction

The Ultraviolet-Visible (UV-VIS) absorption spectroscopy has been successfully applied in the LRF molecule (p-Glu-His-Trp-Ser-Tyr-Gly-Leu-Arg-Pro-Gly-NH_2), via charge transfer markers, in order to visualize specific side chain-side chain interactions between Trp in position 3 and Nva(Nic^+) substituted for Arg in position 8 and to investigate the extent of folding in various analogues (1,2).

In an attempt to place these studies on a quantitative base and to gain further insight into the characteristics of the type of the backbone folding with the aim of formulating a rather simple way to evaluate the probability of a β-turn conformation in peptides we prepared three model tetrapeptides:

$$\text{Ac-Trp-Tyr-X-Nva(Nic}^+\text{)-NH}_2\cdot\text{Cl}^-$$

where X stands for D-Ala, Gly,and L-Ala, and reflects a gradual decrease of the folding. These molecules were analysed by UV-VIS and ^{1}H-NMR spectroscopies.

Experimental Part

The synthesis of the peptides was carried out by the solid phase methodology and their purification was achieved by partition chromatography. The desired labelled analogues were obtained by quaternization of the ornithyl side chain with dinitrophenylnicotinamidium chloride. The UV-VIS spectra were recorded on a Hitachi

Peptides 1986

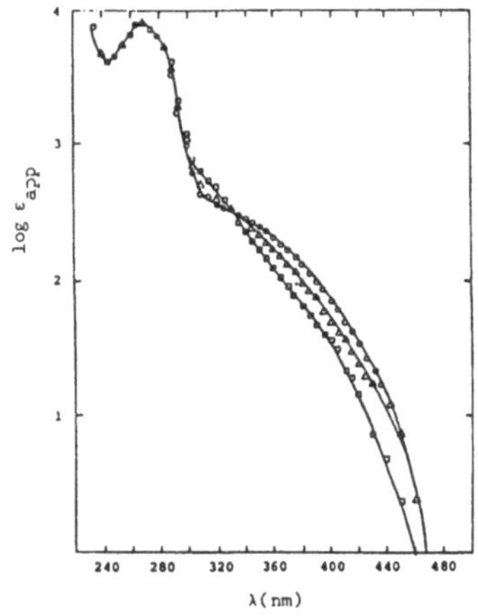

Fig. 1. Apparent molar extinction coefficients as a function of wavelength (H_2O, T = 25°C)
(o) Ac-Trp-Tyr-D-Ala-Nva(Nic$^+$)-NH_2, Cl^-
(△) Ac-Trp-Tyr-Gly-Nva(Nic$^+$)-NH_2, Cl^-
(□) Ac-Trp-Tyr-L-Ala-Nva(Nic$^+$)-NH_2, Cl^-

spectrophotometer fitted with a thermostated cuvette cell holder. The NMR spectra were obtained in a 500 MHz Bruker apparatus and selective decouplings, COSY techniques, as well as the separation method by inversion recovery (3) were applied in order to assign each absorption and to collect all the necessary NMR parameters.

Results and Discussion

UV-VIS studies. All the three analogues showed a characteristic charge-transfer band as a shoulder in the long wavelength region of the UV-VIS spectra (Fig. 1). The differences in the intensity at 380 nm reveal an increased side chain-side chain interaction in the order of L-Ala < Gly < D-Ala for X, which would correspond to an apparent minimum folding of 17 %, 25% and 29 %, respectively, using the following definition of an apparent folding constant for the peptide backbone (1):

$$K_{intra} = \frac{[AD]}{[A]} = \frac{\overline{OD}}{\overline{OD}_{max} - \overline{OD}}$$

where [A] is the molar concentration of the uncomplexed acceptor molecules, which is equal to the total population of all the peptide conformers which do not allow complex formation, and $\overline{OD}_{max}$ is the absorbance of the charge-transfer band which would be observed if all peptide molecules were in the complexed state ($\overline{OD}_{max} = \varepsilon_{intra} \times [A_0] \times 1$, $\varepsilon_{380} = 472$ L mol^{-1} cm^{-1}).

Table 1

Ac-Trp-Tyr-X-Nva-(Nic^{+})-$NH_2 \cdot Cl^-$

Vicinal Coupling Constants ^3J NH-$C^{\alpha}H$, in DMSO-d_6, pH 3 (Hz)

	Trp	Tyr	X	Nva
X = D-Ala	7.60	6.98	7.13	8.39
X = Gly	8.10	7.70	5.70 5.30	7.85
X = L-Ala	7.92	8.08	6.60	8.40

Vicinal Coupling Constants $^3J\ H^{\alpha}$-$H^{\beta}2$ and $^3J\ H^{\alpha}$-$H^{\beta}1$ in DMSO-d_6 and D_2O, pH 3 (Hz)

	DMSO-d_6, pH 3			D_2O, pH 3		
	Trp	Tyr	Nva	Trp	Tyr	Nva
X = D-Ala	4.76 9.62	5.89 8.46	4.59 8.54	6.77 7.48	7.80 7.80	4.60 9.48
X = Gly	4.48 9.50	4.38 8.75	4.81 9.07	6.71 7.38	6.18 7.80	4.65 9.30
X = L-Ala	4.73 9.35	4.35 8.76	4.68 8.80	6.33 8.95	5.64 5.90	3.20 10.40

<u>^{1}H-NMR studies.</u> A detailed ^{1}H-NMR analysis has been performed in aqueous solution as well as in DMSO-d_6 and the vicinal coupling constants ^3J NH-$C^{\alpha}H$, $^3J\ H^{\alpha}$-$H^{\beta}1$ and $^3J\ H^{\alpha}$-$H^{\beta}2$ (Table 1) were analysed in terms of changes on the angle φ of the Tyr and X amino acids (determining the β-turn formation) and the rotamers distribution for the Trp and the Nva. The temperature dependence of the chemical shifts of the NH and the nicotinamidium ring protons was also evaluated for their participation in some interaction, while selected NOE experiments have also been performed.

All the NMR data show that substitution in position 3 with D-Ala, Gly, or L-Ala causes important changes in the structure of the peptide not only locally in position 3 but also further in the molecule. An open reverse β-turn occurs, of the type II when X = D-Ala and of the type I when X = L-Ala (The term of "open" according to Smith and Pease (4) is used to indicate that no hydrogen bonding is observed as it is confirmed by the large negative slopes from the temperature dependence of the chemical shifts all the NH). The type of the turn is further supported by selective NOE measurements between the NH of the X amino acid and the H^{α} of the Tyr, beeing maximum when X = D-Ala (and characterizing a β-turn as type II) and minimum when X = L-Ala (in agreement with a type I β-turn). On the other hand, the positive slopes from the temperature dependence of the chemical shifts of the nicotinamidium ring protons suggest their participation in some interaction.

Conclusion

From the above UV-VIS spectra backed by the ^{1}H-NMR data, which are in agreement with a backbone folding and an increased side chain-side chain interaction from L-Ala to D-Ala passing through Gly for X, it can be concluded that such complexes can be used as selective conformational probes, since all the conformers, which do not allow the formation of the complex, will not be detected. Thus valuable information about the statistical occurence of the folding conformer may be obtained in a simple and precise way. In addition such complexes can be applied to recognize specific and selective interactions.

References

1. Donzel, B., J. Rivier, and M. Goodman. 1977. Study of the Secondary Structure of the Luteinizing Hormone Releasing Factor Using Intramolecular Charge Transfer Complexes. Biochemistry 16, 2611-2618.
2. Donzel, B., C. Sakarellos, and M. Goodman. 1977. In: Peptides: Proc. 5th Amer. Pept. Symp. (M. Goodman and J. Meienhofer, eds.). John Wiley and Sons, Inc., New York. pp. 171-175.
3. Perly, B., Y. Chevalier, and C. Chachaty. 1981. NMR and ESR study of the Conformations and Dynamical Properties of Poly (L-lysine) in Aqueous Solutions. Macromolecules 14, 969-975.
4. Smith, J.A. and L.G. Pease. 1980. Reverse turns in peptides and Proteins. CRC Critical Reviews in Biochemistry, 315-399.

SPECTROSCOPIC STUDIES OF SYNTHETIC PEPTIDE CHEMOATTRACTANTS

G. Vertuani, M. Boggian and A. Scatturin
Dipartimento di Scienze Farmaceutiche, Università di Ferrara, Italy

S. Spisani and S. Traniello
Istituto di Chimica Biologica, Università di Ferrara, Italy.

Introduction

Directed cells migration in response to a chemical gradient, termed chemotaxis, is one of the major processes bringing polymorphonuclear leukocytes (PMNs) from the blood to an inflammatory site whether the inflammation is caused by an infectious agent or an allergic stimulus (1). A large number of substances are chemotactic: formyl amino acids, native and denaturated proteins and various kinds of peptides such as synthetic N-formylmethionyl peptides which are potent chemoattractants for neutrophils and macrophages. One of the most active chemotactic peptides is the tripeptide CHO-Met-Leu-Phe (FMLP). On the basis of extensive structure-activity studies (2) a working model of the rabbit PMNs receptor has been proposed. This model suggests that FMLP exists on the receptor in a ß-pleated sheet. In order to further explore the relationship between chemotactic peptide conformation in solution and biological activity, we have synthesized the methyl ester analogs of FMLP, where the L-Leu residue was replaced by D-Leu and L-OLeu (α-amino-oxyleucine) and L-Phe residue was replaced by cyclohexylalanine (Cha). These analogs have been compared with FMLP and FMLP(OMe) for their ability to induce chemotaxis. Their conformational preferences have been studied in solution by CD and IR absorption.

Peptides 1986

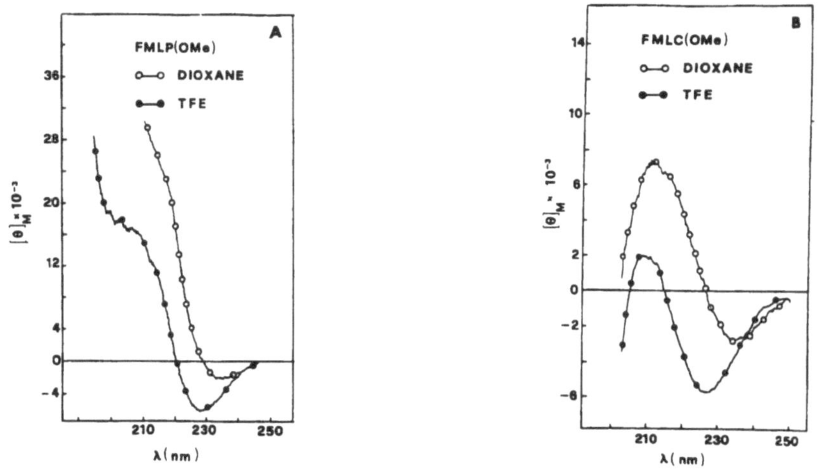

Fig. 1

Results and Discussion

The peptides CHO-Met-Leu-Phe-OMe, FMLP(OMe); CHO-Met-Leu-Cha-OMe, FMLC(OMe); CHO-Met-D-Leu-Phe-OMe, FMLP(OMe)(LDL) were prepared step by step VIA the mixed-anhydride method with a 60-80% yield. The peptide CHO-Met-OLeu-Phe-OMe, FM(O)LP(OMe) was obtained using N-hydroxysuccinimide as an active ester and by using the DCC method. The formyl group was introduced as described by Lajoie et al. (3). The chemoattractant potency of each compound against human PMNs was evaluated in a modified Boyden chamber (4). As previously found (5) FMLP(OMe) has the same activity as FMLP. The most interesting result is that the analog FMLC(OMe) acts as a chemotactic factor with the effective concentration very similar to that of FMLP. The CD spectra of FMLP(OMe) in TFE and dioxane are reported in Fig.1A. In the lesser polar solvent dioxane the intensity and the position of the negative band are respectively decreased and red shifted if compared with the spectrum in TFE characterized by a minimum at 230 nm and a shoulder at 200-215 nm. In order to avoid any spectroscopic complication due to the presence of phenylalanine side chain we have replaced this residue with the cyclohexylalanine. In Fig.1B

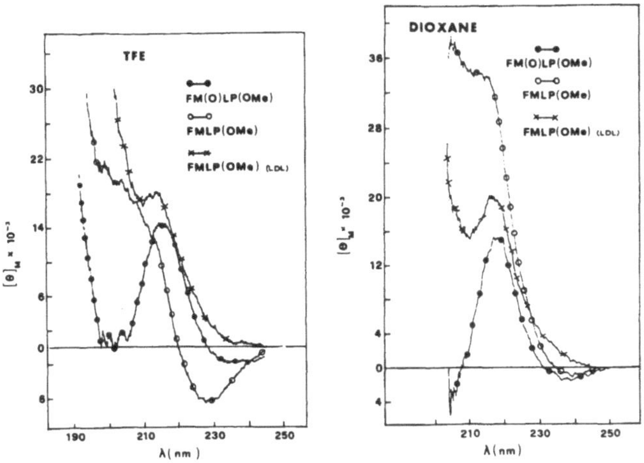

Fig. 2

the CD spectra of the "active" analog FMLC(OMe) are reported. They exibit a positive band centered at 209-210 nm. The position and the intensity of the minimum and maximum in both solvents used and the sharp increase in dioxane of the ratio between ellipticity values of the positive and negative bands indicate that FMLC(OMe) and most probably FMLP(OMe) and FMLP, seem to prefer a folded conformation (β- or γ-turn) in diluted solution (<3mM where any aggregation can be excluded) over extended or unordered structures. The stereospecific preference of the formylpeptide receptor on human PMN cells was pointed out considering two analogs of FMLP(OMe) in which the L-Leu residue was substituted by D-Leu and L-OLeu. The CD spectra of FMLP(OMe), FMLP(OMe)(LDL) and FM(O)LP(OMe) are reported in Fig.2. In both solvents used the shoulder present in the spectra carried out on FMLP(OMe) is nearly resolved in a positive band for FMLP(OMe)(LDL) and completely resolved for FM(O)LP(OMe); the negative band is absent in the spectrum of FMLP(OMe)(LDL) and present in the spectrum of FM(O)LP(OMe). The two analogs do not possess any significant chemotactic activity. As found for FMLC(OMe), the position of the maximum and minimum (215-217 nm and 235-237 nm respectively), the high ratio value between the ellipticities of the positive and negative bands suggest that a flexible folded conformation is favoured over other structures for FM(O)LP(OMe) in solution. The folded conformation of FM(O)LP(OMe), unlike the active peptide FMLP(OMe), prevalently contributes to the shape of CD spectrum, even if the aromatic chromophore is still present. The i.r. spectra in the NH stretching region of all tripeptides synthesized carried out at a very low concentration (≃ 0.6 mM) show that the analog containing the OLeu residue exhibits two intense bands at 3420 cm^{-1} and 3310 cm^{-1} respectively (these bands are ascribed to free and hydrogen-bonded NH vibrators) and that FMLP(OMe), FMLP(OMe)(LDL) and FMLC(OMe) exhibit an intense band at 3420 cm^{-1} and a broad weak band centered at 3340-3360 cm^{-1}. On the basis of our spectroscpic studies we suggest that the peptides FMLP(OMe), FMLC(OMe) and most probably FMLP prefere to assume a folded "active" conformation in very dilute solution probably a γ-turn. The non active analog FM(O)LP(OMe) also prefers a folded conformation which seems to be closer (very intense band at 3310 cm^{-1}) if compared with a γ-turn structure.

References

1. Becker, E.L. and H.Showell, 1980. In: Annaul Reports in Medicinal Chemistry (Hans-Jürgen Hess, ed.) Academic Press, New York. p. 224.
2. Freer, R.J., A.R.Day, N.Muthukumaraswamy, D.Pinon, A.Wu, H.J. Showell and E.L.Becker. 1982. Biochemistry 21, 257.
3. Lajoie, G. and J.L.Kraus. 1984. Peptides 5, 653.
4. Wilkinson, P.C. 1974. Chemotaxis and Inflammation. Churcill-Livingston, Edinburgh. London. p.33.
5. Sukumar, M., P.A.Raj, P.Balaram and E.L.Becker. 1985. Biochem. Biophys.Res.Commun. 128, 339.

A 2D PROTON CONFORMATIONAL ANALYSIS OF TWO SYNTHETIC AMATOXINS ANALOGS

Giancarlo Zanotti

Centro per lo Studio della Chimica del Farmaco del CNR, Istituto di Chimica Farmaceutica, Università "La Sapienza", P.le Aldo Moro, 00185, Roma, Italy.

Gabriella D'Auria, Livio Paolillo

Dipartimento di Chimica dell'Università, via Mezzocannone 4, 80134, Napoli, Italy

Enrico Trivellone

Istituto per la Chimica delle Molecole di Interesse Biologico del CNR, Arco Felice, Napoli, Italy

Introduction

Amatoxins, the most poisonous components of "Amanita" mushrooms, are potent inhibitors of RNA-polymerase B (1). Structure-activity relashionship studies in Amatoxins (2), indicate that glycine-5 should be directly involved in the binding with the enzyme, since its replacement with an L- or D-alanine residue strongly decreases the biological activity of the analogs. Besides a steric hindrance in the binding region of the toxin, the replacement of glycine-5 with the chiral alanines can cause conformational changes in the bicyclic structure of Amatoxins. We report the synthesis and the n.m.r. conformational analysis of S-deoxo-(Ile^3)-(L-Ala^5)-Amaninamide (Ia) and of S-deoxo-(Ile^3)-(D-Ala^5)-Amaninamide (Ib).

Results

The synthesis of the Amatoxins analogs (Ia) and (Ib), follows the already reported scheme, based on the use of the thioether cycli-

Peptides 1986

zation, by the Savige-Fontana reaction (3) (Scheme 1)

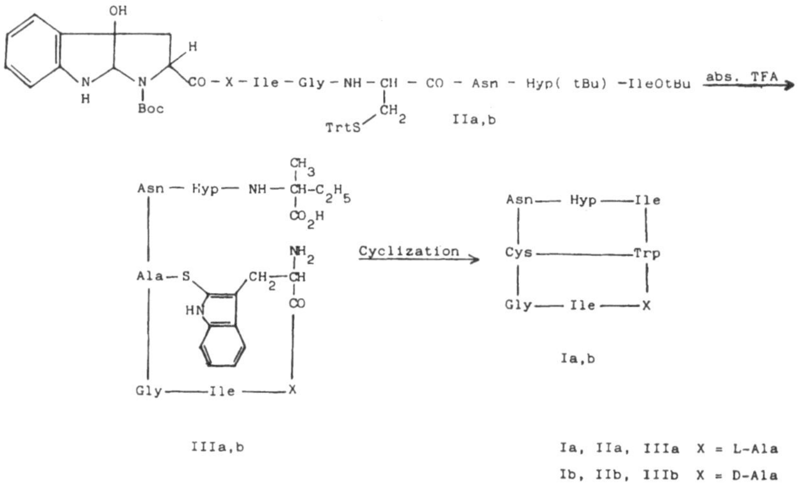

Scheme 1. Synthesis of Amatoxins analogs (Ia) and (Ib)

The inhibitory effect of the analogs has been measured at RNA-polymerase B from embryos of Drosophila melanogaster. The inhibition constants Ki (50 percent inhibition at Ki molarity) of the analogs and of some significative natural Amatoxins are shown in Fig. 1.

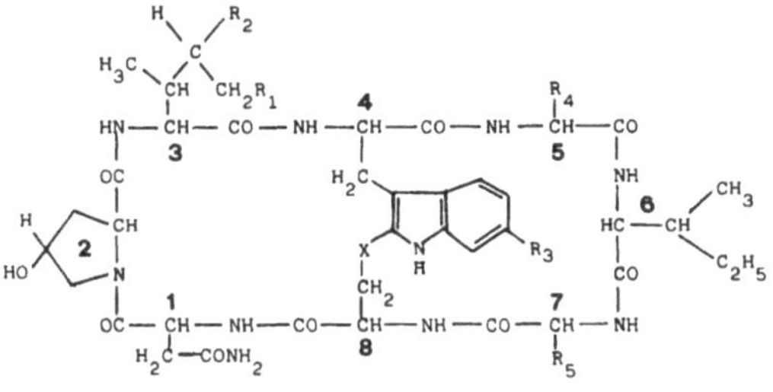

	R_1	R_2	R_3	R_4	R_5	X	$Ki(10^{-7}M)$
α-Amanitin	OH	OH	OH	H	H	SO(R)	0.023
Amaninamide	OH	OH	H	H	H	SO(R)	0.05
Amanullin	H	H	OH	H	H	SO(R)	0.1
S-deoxo-(Ile^3)-Amaninamide	H	H	H	H	H	S	0.8
S-deoxo-(Ile^3)-(L-Ala^5)-Amaninamide (Ia)	H	H	H	CH_3(L)	H	S	200
S-deoxo-(Ile^3)-(D-Ala^5)-Amaninamide (Ib)	H	H	H	CH_3(D)	H	S	20-40
S-deoxo-(Ile^3)-(D-Ala^7)-Amaninamide (Ic)	H	H	H	H	CH_3(D)	S	0.3-1.2

Fig. 1. Structural formula of some natural and synthetic Amatoxins and their inhibitory effect on RNA-polymerase B.

500 Mhz proton spectra of compounds (Ia) and (Ib) show significant differences both in terms of chemical shifts and J NH-αCH coupling constants values. The most relevant changes are reported in Table 1 and are compared to the corresponding data obtained for S-deoxo--(Ile^{3})-(D-Ala^{7})-Amaninamide (Ic) (4).

Table 1. NH Chemical Shifts and NH-αCH Coupling Constants for (Ia), (Ib) and (Ic)

		Asn^{1}	Ile^{3}	Trp^{4}	Gly^{5}	Ala^{5}	Ile^{6}	Gly^{7}	Ala^{7}	Cys^{8}
Ia	NH	8.4	8.45	7.83		7.63	8.86	9.13		8.01
	αCH-NH	3.7	8.4	9.6		3.7	5.2	6.0		10.3
Ib	NH	8.7	8.18	8.35		8.25	8.55	8.94		9.37
	αCH-NH	3.1	10.1	6.0		9.6	3.6	6.5		9.4
Ic	NH	8.56	8.11	8.12	8.14		8.54		8.72	8.16
	αCH-NH	3.1	10.2	6.5	8.7		4.1		7.5	10.0

2D spectra such as COSY, NOESY, J resolved and COSY relayed were obtained for both compounds thus allowing the assignment of all resonances. The coupling constants were evaluated and used to determine the ϑ angles for the NH α-CH fragment. In Table 2 the torsion angles for compounds (Ia), (Ib) and (Ic) are compared with the crystallographically determined angles for βAmanitin.

Table 2. Torsion Angles for βAmanitin and Analogs (Ia), (Ib) and (Ic).

	Asn^{1}	Ile^{3}	Trp^{4}	Gly^{5}	Ala^{5}	Ile^{6}	Gly^{7}	Ala^{7}	Cys^{8}
βAmanitin	-172	-79	-109	140		-59	87		-121
Ia	-177	-90	-100		183	-70	95		-105
Ib	-183	-105	-76		150	-65	80		-120
Ic	-183	-105	-76	151		-65		84	-125

Comparison of the data reported in Tables 1 and 2 shows that the replacement of the Gly-5 with an L- and D-Ala respectively, affects the general features of the n.m.r. spectra only in the case of L-Ala. In particular the resonances of Ile-3, Trp-4 and Ala-5 as well as their coupling constants are involved. These results demonstrate local conformational changes in the Ile^{3}-Trp^{4}-Ala^{5} loop. More information is gained through the NOESY spectrum. While the NOESY spectrum of (Ib) closely resembles that of compound (Ic), the corresponding NOESY spectrum of the L-Ala^{5} analog (Ia) shows different through-space contacts for the Trp-4 and Ile-5 NHs and α-CHs (Table 3).

		Ia	Ib
Ile-3 α-CH	Trp-4 α-CH	present	absent
"	Ile-3 γ-CH_3	absent	present
Trp-4 NH	Ile-3 α-CH	absent	present
"	Ile-3 γ-CH_3	absent	present

Table 3. NOE Contacts between Ile-3 and Trp-4 in Compounds (Ia) and (Ib).

This confirms that a conformational change has occurred in the L-Ala^{5} analog relatively to the Ile^{3}-Trp^{4}-Ala^{5} loop.

Conclusions

In the case of the analog (Ib) (D-Ala^{5}), where no substantial structural modification is observed in solution, the reduction of its affinity to the RNA-polymerase B can be explained only by the steric hindrance in the binding region caused by the introduction of the Ala-5 CH_3 group. In the case of (Ia) (L-Ala^{5}), in addition to the steric effect of the CH_3 group, the observed local conformational change in the Ile^{3}-Trp^{4}-Ala^{5} region is responsible for the further pronounced decrease of the biological activity of the analog.

References

1. Wieland T., H. Faulstich, (1978), CRC Crit. Rev. Biochem. 5, 185.
2. Wieland T., C. Götzendörfer, G. Zanotti, A. Vaisius, (1981). Eur. J. Biochem. 177, 161.
3. Zanotti G., C. Birr, T. Wieland, (1981). Int. J. Peptide Protein Res. 12, 204.
4. Zanotti G., G. D'Auria, L. Paolillo, E. Trivellone (1986). Biochim. Biophys. Acta, 807, 454.

SURFACTANT-INDUCED ORDERED STRUCTURES OF GASTRIN RELATED PEPTIDES

N. Mammi, S. Mammi, E. Peggion

Biopolymer Research Center, Department of Organic Chemistry, University of Padova, Italy

L. Moroder, G. Bovermann, E. Wünsch

Max-Planck Institut für Biochemie, Abteilung Peptidchemie, 8033 Martinsried bei München, Federal Republic of Germany.

Introduction

Recently, we observed that linear gastrin peptides adopt ordered conformations in trifluoroethanol (TFE) (1). We carried out conformational studies and biological activity tests on a series of gastrin analogs in which the $-(Glu)_5-$ sequence at the N-terminus was systematically elongated from 1 to 5 residues (1). A correlation was observed between increasing conformational order in TFE and biological potency _in vivo_ upon chain elongation. This observation led to the hypothesis that the conformation assumed by gastrin peptides in TFE might be of biological importance.

Little gastrin in aqueous medium containing phospholipids or detergent micelles, assumes a conformation very similar to that observed in TFE (2-4). We have now extended our conformational studies to gastrin peptides in such media. Here, we present some results obtained with little gastrin and des-Trp^1,Nle^{12}-minigastrin in aqueous solution containing increasing amounts of sodium dodecylsulphate (SDS).

Results and Discussion

In aqueous solution at neutral pH the CD spectra of little gastrin and minigastrin are insensitive to the presence of detergent; a

Peptides 1986

random structure is always formed. An interaction of the two hormones with SDS is observed only at pH ≃ 2, i.e. when the carboxylate side chains are fully protonated. Under these conditions the effect of the surfactant on the peptide conformation depends upon its concentration. The far UV CD properties of little gastrin in aqueous solution at pH ≃ 2 and in the presence of increasing amounts of SDS are reported in figure 1. At SDS concentration below CMC the hormone exhibits the typical CD pattern of a β-structure. The same result was obtained with des-Trp^{1},Nle^{12}-minigastrin (not shown). Thus, under these conditions both hormones form β-aggregates stabilized by detergent molecules.

In the presence of micelles of detergent, the behavior of the two gastrin hormones is different. In the case of little gastrin, the increase of detergent concentration above the CMC induces a sharp conformational change. A new CD pattern is formed characterized by two negative bands at 218 and 208 nm. The spectrum is very similar to that observed in TFE except for the intensities of the two negative bands which are higher in the micellar system. These features indicate an increased amount of ordered conformation in the surfactant solution with respect to that observed in TFE. According to our previous conformational hypothesis, the structure is characterized by a β-bend located in the central part of the molecule and by a helical segment at the N-terminus.

In the case of the minigastrin analog, in the presence of SDS micelles, a CD pattern is observed with approximately the same shape as that in TFE but with substantially reduced intensity. Thus, unlike little gastrin, the amount of ordered structure in the micellar system is much lower than in TFE (figure 2).

A possible explanation for the different behavior of the two gastrin peptides is the difference in the N-terminal groups, the free amine (charged at pH 2) in minigastrin, and the uncharged pyroglutamic residue in little gastrin. The conformational ordering of the peptides in SDS solution occurs only at low pH, where the side chains of the -(Glu)$_5$- sequence are uncharged. This strongly suggests that the peptides are solubilized in the interior of the micelles. The hydrophobic environment inside the micelles is res-

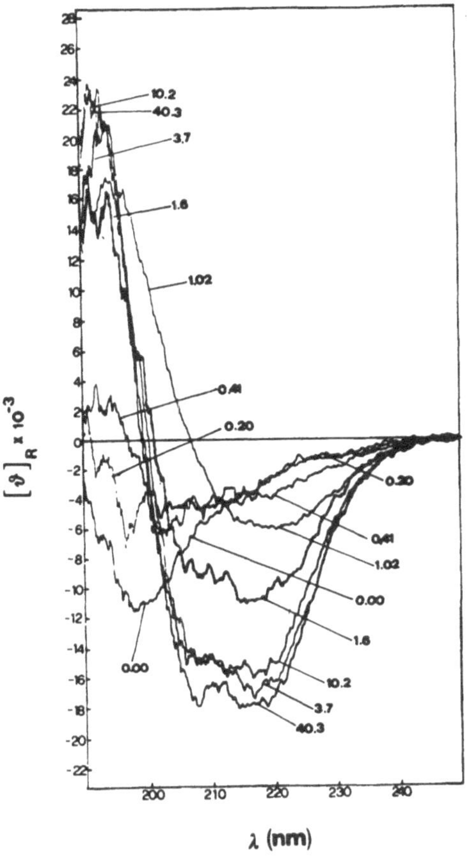

Figure 1: CD spectra of Nle^{15}-little gastrin in aqueous solutions of varying SDS concentration (pH ≅2).

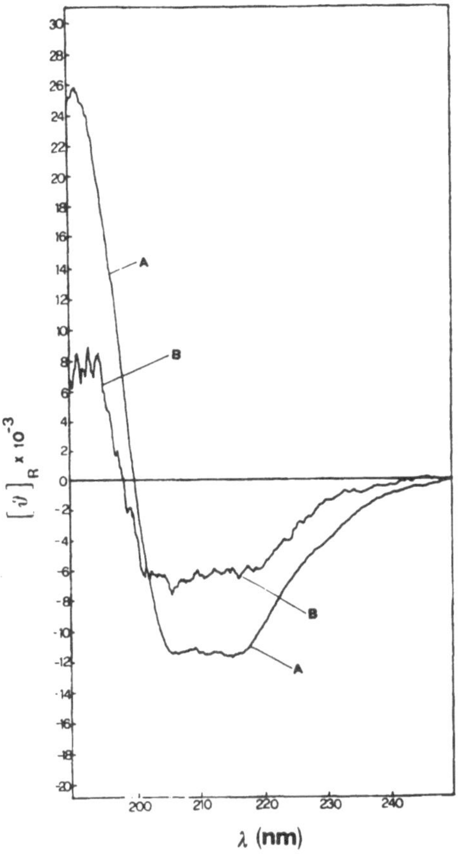

Figura 2: CD spectra of des-Trp^{1},Nle^{12}-minigastrin in 98% TFE (curve A) and 50.0 mM SDS solution (curve B).

ponsible for the enhancement of the structural order of little gastrin. In the case of minigastrin, the positive charge at the N-terminus prevents crossing of the negatively charged micelle surface with consequent incomplete solubilization of the peptide in the interior of the micelles. Considering our previous conformational hypothesis, we conclude that the α-helical segment at the N-terminus is destabilized in the micellar system because of the presence of the positive charge at the N-terminal residue of minigastrin. The β-bend, possibly located in the central sequence Ala-Tyr-Gly-Trp, remains unchanged in the micelles.

Fluorescence emission spectra provide definitive evidence for the solubilization of the two gastrin peptides in the interior of SDS micelles. In water and in aqueous solutions of SDS below the CMC, the maximum of Trp fluorescence emission is always in the range of 352-355 nm. In the presence of micelles, λ_{max} shifts towards the blue by about 10 nm indicating that the Trp moieties have been transfered into a hydrophobic environment. The emission maxima of the two model compounds, Ac-Trp-NH_2 and Ac-Tyr-Trp-NH_2, in H_2O are around 355 nm at any SDS concentration, which indicates that they are unable to penetrate the SDS micelles.

The chiroptical properties in the near UV of the two gastrin peptides in the presence of micelles deserve a special comment. In both cases, the CD signals are very weak, indicative of aromatic moieties without particular conformational constraint. The situation is very different from that observed in TFE where the CD spectra indicate a restricted rotational freedom of the aromatic side chains and in particular, of the indole moiety. According to our working model, in the structure formed in TFE, the restricted rotational freedom of these groups is due to the formation of a hydrophobic cluster involving Nle, Trp and Phe side chains. In the micellar system, the environment is already highly hydrophobic and the non-polar side chains are no longer forced to form a hydrophobic cluster. Thus, in this medium, the aromatic side chains are freely rotating.

References

1. Peggion, E., M.T. Foffani, E. Wünsch, L. Moroder, G. Borin, M. Goodman, and S. Mammi. 1985. *Biopolymers* 24, 647-666.

2. Wu, C.-S.C. and J.T. Yang. 1978. *Biochem. Biophys. Res. Comm.* 82, 85-91.

3. Shirahama, K. and J.T. Yang. 1979. *Int. J. Peptide Protein Res.* 13, 341-345.

4. Wu, C.-S. C., A. Hachimori and J. T. Yang. 1982. *Biochemistry* 21, 4556-4562.

SYNTHESIS AND CONFORMATIONAL STUDY OF CYCLIC HEXAPEPTIDE ANALOGUES OF SOMATOSTATIN : Cyclo(Phe-D-Trp-Lys-Thr-o,m,p-AMPA)

P. Vander Elst, E. van den Berg, D. Gondol, H. Pepermans, L. Van Der Auwera, C. Wynants, D. Tourwé and G. Van Binst

Eenheid Organische Scheikunde (ORGC)
Vrije Universiteit Brussel, Pleinlaan 2
B-1050 Brussels, Belgium

Based on a model for the bio-active conformation of somatostatin, Veber et al. (1,2) and Bauer (3) synthesized cyclic analogues containing the Phe-D-Trp-Lys-Thr sequence. These compounds showed a prolonged duration of action and high biological activity. The highly active analogues cyclo(Xxx-Phe-D-Trp-Lys-Thr-Yyy) all include a cis peptide bond between the Yyy and Xxx residues (4). We now report the synthesis of cyclic somatostatin analogues in which the essential tetrapeptide sequence is cyclized with respectively o,m, and p-(aminomethyl) phenylacetic acid (o-AMPA, m-AMPA and p-AMPA).

The o-AMPA spacer unit mimics a Gly-Gly dipeptide unit with a cis pseudo-peptide bond. The m-AMPA derivative has been studied in order to induce conformational changes in the hydrophobic region as described in the conformation part. The p-AMPA derivative is more similar to cyclo(Phe-D-Trp-Lys-Thr-Aha). (5)

Synthesis

The o-(aminomethyl)-phenylacetic acid derivative, the o-spacer molecule was synthesized starting from 2-indanone, by a Schmidt reaction followed by hydrolyses and t-Boc protection.

The meta and para-(aminomethyl)phenylacetic acid derivatives were synthesized, starting from the corresponding phenylene-diacetic acids, which were converted to their mono-esters and then rearranged to the isocyanates, which were trapped with tBuOH to give the tBoc-protected AMPA esters, followed by hydrolysis of the ester function.

Peptides 1986

The synthetic strategy of the peptide consists of a fragment condensation between the tripeptide mimics x-AMPA-Phe-OMe and the tripeptide tBoc-D-Trp-Lys(Z)-Thr(OBzl)-OH, followed by a cyclization between the Phe and D-Trp residues, using DPPA as cyclization reagent and NMM as base (6,7). Final deprotection was performed by liquid HF treatment.

Conformation analysis

The NMR study of the three analogues has been performed by ^1H NMR at 500 MHz in DMSO and in H_2O/D_2O. The assignments were based on COSY, COSYLR 2D (HH)-INADEQUATE and NOESY experiments as described previously (8,9).

The proposed conformations are based on a detailed analysis of the following parameters : $^3J(C_\alpha H\text{-}NH)$ coupling constants correlations with ϕ dihedral angles, solvent exposure of NH groups releaved by temperature dependence of the NH chemical shift and by the rate of exchange upon addition of D_2O, the intensity of interproton NOE effects measured by 2D NOESY experiments.

The following features are deduced from these measurements : the cyclo(Phe-D-Trp-Lys-Thr-o-AMPA) contains a type II' ß-turn over the D-Trp^8 and Lys^9 residue shown by NOE effects. The rigidity of the ß-turn part of the molecule is confirmed by the invariability of rotamer population of the aromatic side chains with temperature. An hydrogen bond from the pseudo-Gly^{11}NH of the o-AMPA bridge to the pseudo Gly^6-carbonyl is present. It shows a proximity of Thr^{10}NH and Gly^{11}NH by NOE.

The predominance of this C_8 δ-turn, and the presence of the mentioned NOE effect excludes the existence of an hydrogen bond between Phe^7NH and Thr^{10}C=O as in the postulated active conformation of somatostatin. The predominant conformation of this compound is markedley different from the bio-active one.

Concerning the cyclo(Phe-D-Trp-Lys-Thr-m-AMPA) derivative, a type II' ß-turn over the D-Trp^8-Lys residue is shown by NOE effects between $Phe^7\alpha$ and D-Trp^8NH, D-$Trp^8\alpha$ and Lys^9NH and also between Lys^9NH and Thr^{10}NH, stabilized by a H bridge between Thr^{10}NH and Phe^7C=O.

A weak NOE was detected between $Phe^{7}NH$ and $Gly^{6}\alpha$, and between $Gly^{11}\alpha$ and H_2 (arom) of AMPA.
The first NOE cross-peak is only possible when the Phe^{7} amide proton is turned to the outside of the molecule, while the second cross-peak can only exist when the $Gly^{11}NH$ is turned to the inside.
The study of the cyclo(Phe-D-Trp-Lys-Thr-p-AMPA) analogue showed no strong intra molecular hydrogen bonds. The NOE effects indicate a predominant structure with a γ-turn over the D-Trp residue with an hydrogen bond between Lys-NH and Phe-C=O, which interconverts with a type II' ß-turn over D-Trp-Lys.
Only small dihedral angle changes are necessary to obtain this conformational change.
It must be noticed that in the three analogues an upfield shift of the γ-Lys protons is observed. The compounds were not active in in vitro pituitary cell cultures after GM-RF stimulation.
Further work is in progress by quantitative NOE measurements and molecular graphics methods to interpret more deeply these results in order to synthesize new modified analogues.

Acknowledgements

This work has been supported by the Concerted Actions of the Belgian Ministery of Science, the Belgian Science Foundation (FKFO) and the Institute for Scientific Research in Industry and Agriculture (IWONL).

References

1. Veber, D.F., Freidinger, R.M., Perlow, D.S., Paleveda, W.J., Holly, F.W., Strachan, R.G., Nutt, R.F., Arison, B.H., Homnick, C., Randall, W.C., Glitzer, M.S., Saperstein, R. and Hirschmann, R. 1981. Nature 292, 55-58.

2. Freidinger, R.M. and Veber, D.F. 1985. In: "Conformationally Directed Drug Design" (Vida, J.A. and Gordon, M., eds), pp. 169-187, Am. Chem. Soc., Washington D.C.

3. Bauer, W., Briner, U., Doepfner, W., Haller, H., Huguenin, R., Marbach, P., Petcher, T.J. and Pless, J. 1982. Life Sci. 31, 1133-1140.

4. Freidinger, R.M., Perlow, D.S., Randall, W.C., Saperstein, R., Arison, B.H. and Veber, D.F. 1984. Int. J. Pept. Protein Res. 23, 142-150.

5. Fries, J.L., Murphy, W.A., Sueiras-Diaz, J. and Coy, D.H. 1982. Peptides, 3, 811-814.

6. Brady, S.F., Paleveda, W.J., Arison, B.H., Freidinger, R.M., Nutt, R.F. and Veber, D.H. 1983. In: "Peptides Structure and Function", Proc. 8th Am. Pept. Symp. (Hruby, V.J. and Rich, D.H. eds.), pp 127-130, Pierce Chemical Company, Rockford Illinois.

7. Kopple, K.D. and Parameswaram, K.M. 1983. Int. J. Pept. Protein Res. 21, 269-280.

8. Vale, W., Rivier, C. and Brown, M. 1977. Ann. Rev. Physiol. 39, 473-527.

9. Holladay, L.A. and Puett, D. 1976. Proc. Natl. Acad. Sci. US 73, 1199-1202.

10. Wünsch, E. 1974. (Houben Weyl), Synthese von Peptiden, vol. I, pp. 126-128, Georg Thieme Verlag, Stuttgart.

INVESTIGATION OF CARBAZOLCARBONYL-SUBSTITUTED POLY-α-LYSINE, poly[N^6(9_H_-carbazol-9-ylcarbonyl)-lysine].

K. Brunfeldt, K. Kovács
Instituttet for Kemiindustri, The Technical University of Denmark, Building 227, 2800 Lyngby, Denmark.

F. Lauprêtre
Laboratoire de Physico-Chimie Structurale et Macromoléculaire, ESPCI, 10 Rue Vaquelin, 75231 Paris Cedex 05, France.

An increased interest is shown in organic materials with electrical and optical properties. Polymers in the form of liquid crystals can be used as matrix systems for low molecular structures with special physical properties either as substituents or dopants. Substituted poly-α-amino acids may form lyotropic liquid crystals at room T. Reorientation can be brought about by external force and solid films can be prepared by removal of the solvent. For studying photoconductivity as a function of molecular orientation Halstrøm et al. prepared various carbazole-substituted poly-α-amino acids via the corresponding N-carboxy anhydrides using NaOH in MeOH as initiator (1). Lysine was found to be preferable as a 20% solution in THF of poly-[N^6(9_H_-carbazol-9-ylcarbonyl)-L-lysine] readily formed cholesteric lyotropic liquid crystals at room T. IR and CD measurements confirmed the α-helix conformation and also indicated a high degree of orientation of the side chains with respect to the backbone (2). Hydrogen bonding between the urea groups most likely play a role for the orientation. The batch originally prepared has recently been shown to contain N^6-(benzo[b]carbazol-5-ylcarbonyl)-L-lysine residues and N^6-(2-methyl-9_H_-carbazol-9-ylcarbonyl)-L-lysine residues as impurities, the first mentioned in an amount of about 0.5% relative to the N^6-(9_H_-carbazol-9-ylcarbonyl)-L-lysine residues (3). The here investigated compounds have a content of about 0.01% of the benzo[b]carbazole derivative. Higher purity can be obtained by purification of carbazole by preparative HPLC using the results presented in (3).

Peptides 1986

Experiments and results

Carbazole-substituted L and D poly-α-lysine were prepared according to (1). Arbitrary mean molecular weights and narrow molecular distribution can be obtained by selecting the conditions for the polymerization and subsequent preparative fractionation by GPC. In fig. 1 is presented the course of two polymerizations carried out with 0.5g of N-carboxyanhydride in 30ml p-dioxane under stirring at room T. A: Initiated by 30µl of 0.4M NaOH in MeOH and B: 60µl. The reaction was monitored by OR: optical rotation and MW: molecular weight estimation by analytical GPC. MW is the retention time of the maximum of the refractive index signal converted to molecular weight by comparison with polystyrene standards and corrected for the conformational difference between the standards and the substituted poly-α-lysine. Polystyrene is randomly coiled and the product is rigid rods due to the α-helix conformation. A close relation is seen between the OR values and the molecular weights. At a proper reaction rate it is thus possible to obtain a desired mean molecular weight by stopping the reaction by precipitation of the product in abs. EtOH. A narrower molecular weight distribution can be achieved by preparative GPC fractionation by a system as the one described in (4a).

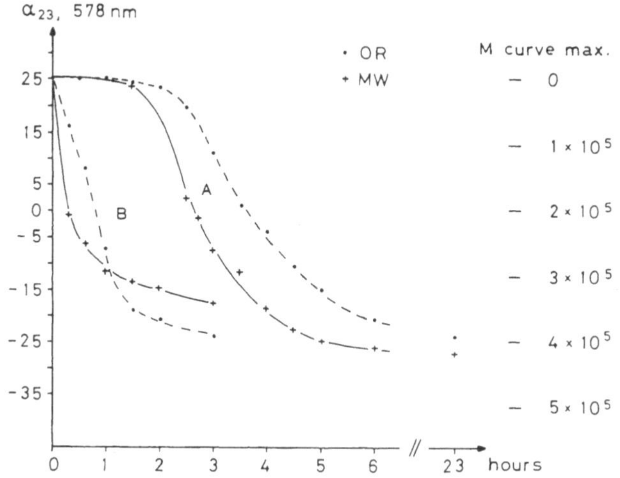

Fig. 1

Monitoring by OR demands a certain difference between the value of the starting material and that of the polymer. Monitoring of liberated CO_2 seems to be preferable. A system under testing is based on a glass tube, ID ~1 cm as reactor and stirring and stripping by passage of an inert gas. If p-dioxane is used as solvent, traces in the gas do not interfere due to negligible absorption at 2200-2500 cm^{-1} (4a).

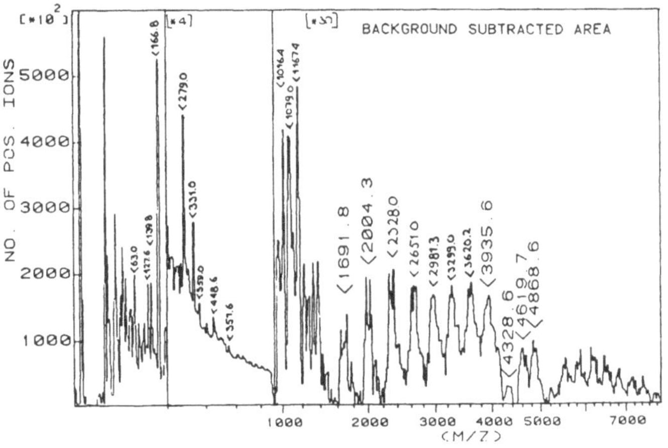

Fig. 2

Characterization of single molecular specimens in a polymer preparation may reveal artifacts. Molecular weight of up to 23.406, bovine trypsin has been recorded by californium-252 plasma mass desorption mass spectrometry (5). Fig. 2 presents a recording of masses on a preparation of the substituted L isomer. Masses 10^3-$8x10^3$ correspond to 3 to 25 repetition units. In the range $1.6x10^3$-$4x10^3$ the difference between neighbouring peaks is 325± 5 corresponding to the molecular weight of the repetition unit. The broadness of the peaks indicates the presence of two specimens, one possibly esterified with methanol.

A prerequisite for tayloring materials for electrical and optical purposes is that the structure found optimal in the liquid crystalline state is preserved when the material is used for practical purposes. Thus regarding the polymer in question in solid form and at moderate heating. C-13 NMR measurements on the polymer as a 2-30% solution in $CDCl_3$ and in amorphous form demonstrated the α-helix conformation and a high stability of the side chains. Heating to 70°C of the solid material caused no difference compared to 27°C. Fig. 3. As expected identical spectra were obtained by measurements on the L and D isomer. The measurements on the solid material D isomer comprised C-13 cross polarisation at 78,47 MHz, proton dipolar decoupling and magic angle spinning using Toss sequence to suppress spinning side bands. For assigning the signals 9-butyl-9<u>H</u>-carbazole-9-carboxamide was used.

Solid films with a nematic structure were prepared by shearing and evaporation of the solvent. Photoconductivity was measured perpendicular to the molecular axis by conventional steady state DC-conductivity technique. An increase in conductivity could be demonstrated by increase in electric field (4b).

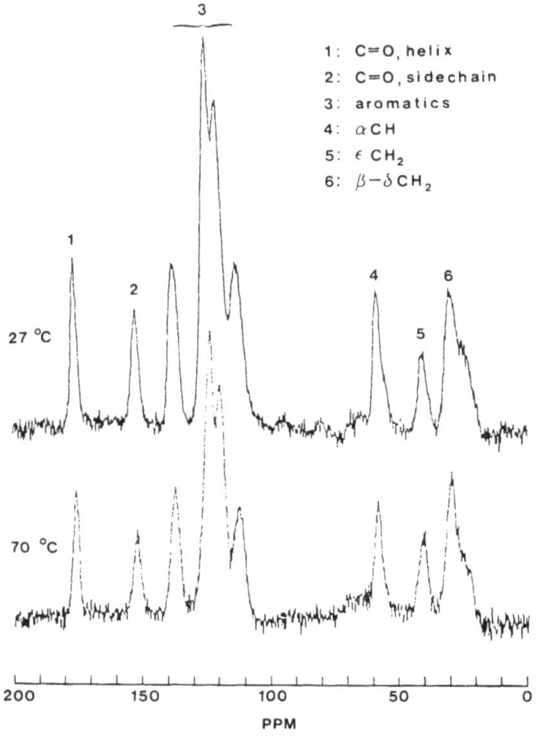

Fig. 3

Acknowledgements. The research was supported by grants from The Danish Council for Technical Research and the EEC Stimulation Program.

References

1. Halstrøm, J., L.L. Chapoy, K. Kovács, K. Brunfeldt, M.A. Qasim. 1981. Peptides 1980. Ed. K. Brunfeldt. Scriptor, Copenhagen p. 759.

2. Chapoy, L.L., D. Biddle, J. Halstrøm, K. Kovács, K. Brunfeldt, M.A. Qasim, T. Christensen. 1983. Macromolecules 16, 181.

3. Clausen, K., K. Brunfeldt. 1986. Macromolecules (in press).

4. Danish report concerning EEC Stimulation contract STI-012-J-C (CD). 1986. Ed. K. Brunfeldt. 4a, K. Brunfeldt, H.J. Styhr Petersen, 4b, B. Marcher.

5. Sundquist, B, R.D. Macpharlane. 1985. Mass Spectrometry Reviews 4, 421.

BIOPHYSICAL PROPERTIES OF SYNTHETIC PRESEQUENCES SUGGEST CONFORMATIONS IMPORTANT FOR PROTEIN LOCALIZATION

D.W. Hoyt, C.J. McKnight, M.S. Briggs, A.N. Stroup, and L.M. Gierasch
Department of Chemistry, University of Delaware, Newark, DE 19716

Nearly every protein that is destined to be translocated across a cytoplasmic or organellar membrane is synthesized with an N-terminal presequence of 15 to 30 residues; for exported proteins the presequence is generally termed a signal sequence. While the presequence is essential for correct localization of synthesized proteins, its roles in the membrane translocation process are still poorly understood (1-3). Surveys of presequences of both exported (4) and mitochondrial (5) proteins indicate that the character and arrangement of the residues are more important than specific sequences and have led to proposals that presequences interact with membranes and/or membrane-resident components.

We have addressed this problem by biophysical studies of isolated, synthetic presequences. Signal peptides studied were derived from genetically-characterized strains of *E. coli* with defects in export of outer membrane proteins. In previous work, we found that a family of signal sequences from the LamB protein had both conformational properties [measured by circular dichroism (CD)] (6) and membrane-interaction properties (measured by surface tensiometry) (7) that correlated with their ability to function *in vivo*. Furthermore, β structure was observed when the wild type signal peptide adsorbed to the charged lipid surface, and α helix was observed when the signal peptide inserted into the hydrophobic region of the lipid (8).

The present work extends these conformational studies to a larger number of signal sequences from *E. coli* outer membrane proteins and to an example of a presequence from a mitochondrial inner membrane protein. CD analysis of presequences in aqueous buffer

Peptides 1986

solution and as a function of concentration of trifluoroethanol or sodium dodecyl sulfate (a membrane-mimetic environment) revealed in all cases strong sensitivity to microenvironment, with helical structure increasing in the more nonpolar solutions [see the Figure for representative data and the Table for results of curve-fitting the observed spectra to combinations of α helix, β structure, and random coil using standard reference curves (9)]. Only one case showed marked secondary structure in aqueous solution: The OmpA wild type signal sequence, synthesized with four residues of the mature protein, adopted predominantly β structure in buffer. Similar results were reported for the signal sequence of parathyroid hormone, which was also synthesized with a segment of the mature region (10). In both these cases the mature region is highly charged; it is possible that interactions of the two charged termini stabilize β structure. Also notable is the fact that the mitochondrial import presequence (F1ATPase β subunit) behaves very similarly to the export presequences despite its different charge

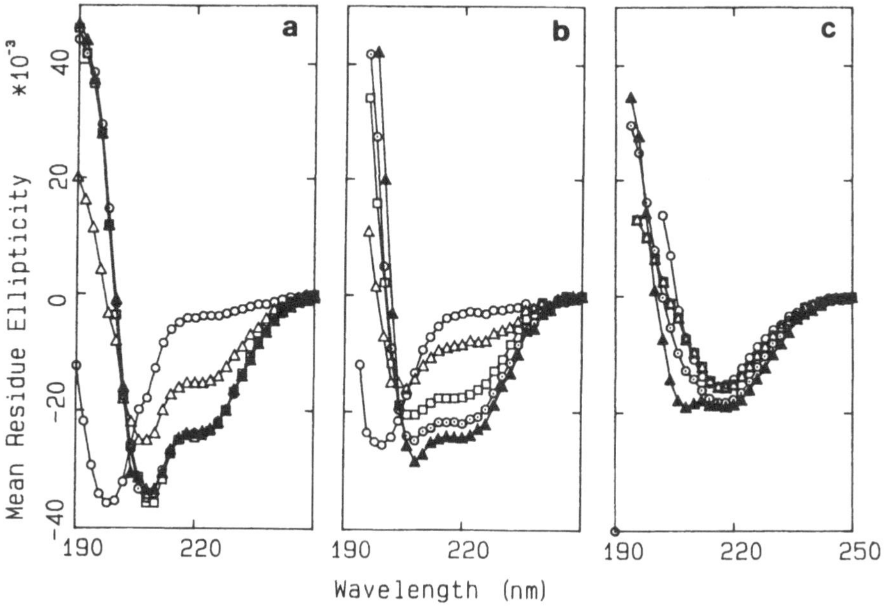

Figure. CD curves of presequences in 5 mM Tris buffer and several concentrations of trifluoroethanol: ○ 0%, △ 20%, □ 40%, ⊙ 60%, and ▲ 80%. a, LamB wild type; b, F1ATPase subunit; c, OmpA wild type plus four residues of mature. Sequences and conditions are given in the Table.

Table 1. Conformations of Synthetic Presequences from CD Analysis

Pre-sequence	Solvent	%Helix	%Beta	%Random
LamB Wild Type	H2O	5	15	80
MMITLRKLPLAVAVAAGVMSAQAMA/	TFE	40	15	45
	SDS	60	10	30
LamB Deletion Mutant	H2O	5	15	80
MMITLRKLP----VAAGVMSAQAMA/	TFE	30	0	70
	SDS	20	15	65
LamB G>C Revertant	H2O	10	10	80
MMITLRKLP----VAACVMSAQAMA/	TFE	35	10	55
	SDS	35	10	55
LamB P>L Revertant	H2O	10	10	80
MMITLRKLL----VAAGVMSAQAMA/	TFE	45	15	40
	SDS	60	25	15
LamB G>R Mutant	H2O	20	0	80
MMITLRKLPLAVAVAARVMSAQAMA/	TFE	40	10	50
	SDS	35	20	45
OmpA Wild Type	H2O	0	100	0
MKKTAIAIAVALAGFATVAQA/APKD	TFE	40	40	20
	SDS	5	80	15
OmpA Deletion Mutant	H2O	0	20	80
MKKTA----VALAGFATVAQA/	TFE	30	20	50
	SDS	35	5	60
OmpA I>N Mutant	H2O	5	15	80
MKKTAIANAVALAGFATVAQA/	TFE	30	10	60
	SDS	35	0	65
F1ATPase ß Subunit	H2O	20	0	80
MVLPRLYTATSRAAFKAAK· · ·	TFE	75	0	25
	SDS	50	0	50

LamB is the *E. coli* lambda phage receptor. OmpA is the *E. coli* outer membrane protein A. F1ATPase is from yeast; the sequence studied represents the first 19 residues of the presequence which has been shown to be sufficient for mitochondrial import. Peptides were studied at approximately 5 X 10^{-5} M. TFE is 50% trifluoroethanol. SDS is 40 mM sodium dodecylsulfate. The slash in the OmpA wild type sequence indicates the break between the presequence and the mature protein.

distribution. Similar results have recently been reported for the presequence of subunit IV of yeast cytochrome oxidase (11).

We conclude that the tendency to adopt α helix in nonpolar environments and the capacity for conformational change are important features of presequences that facilitate protein translocation across membranes. Other studies are in progress on the membrane interactions of these intriguing sequences.

Acknowledgement

We thank Michael Douglas for the sample of yeast F1ATPase presequence. This research was supported by grants from the NIH (GM34962) and the NSF (DCB-8158695).

References

1. Briggs, M.S. and Gierasch, L.M. 1986. Adv. Prot. Chem. 38, 109.
2. Silhavy, T.J., Benson, S.A., and Emr, S.D. 1983. Microbiol. Rev. 47, 313.
3. Rapoport, T.A. 1986. CRC Crit. Rev. Biochem. 20, 73.
4. von Heijne, G. 1985. J. Mol. Biol. 184, 99.
5. von Heijne, G. 1986. EMBO Jour. 5, 1335.
6. Briggs, M.S. and Gierasch, L.M. 1984. Biochemistry 23, 3114.
7. Briggs, M.S., Gierasch, L.M., Zlotnick, A., Lear, J.D., and DeGrado, W.F. 1985. Science 228, 1096.
8. Briggs, M.S., Cornell, D.G., Dluhy, R.A., and Gierasch, L.M. 1986. Science 233, 206.
9. Greenfield, N. and Fasman, G.D. 1969. Biochemistry 8, 4108.
10. Rosenblatt, M., Beaudette, N.V., and Fasman, G.D. 1980. Proc. Natl. Acad. Sci. USA 77, 3983.
11. Roise, D., Horvath, S.J., Tomich, J.M., Richards, J.H., and Schatz, G. 1986. EMBO Jour. 5, 1327.

HIGH AFFINITY AND IMPROVED SELECTIVITY FOR δ OPIOID RECEPTORS EXHIBITED BY NEW LINEAR HEXAPEPTIDES

G. Gacel, J. Belleney, P. Delay-Goyet, C. Seguin, J.-L. Morgat and B.P. Roques
Département de Chimie Organique, U 266 INSERM - UA 498 CNRS, UER des Sciences Pharmaceutiques et Biologiques, 4 avenue de l'Observatoire, 75006 Paris, France.

Introduction

The existence of a plurality of opioid receptors (μ, δ, κ, ...) could explain the multiple pharmacological responses elicited by morphine and opioid peptides (1). Therefore, the biochemical and pharmacological characterization of these different types of receptors remains a very interesting challenge. For this purpose, highly specific probes are required. Among these, the linear hexapeptides DSLET and DTLET were the first really δ selective ligands (2). More recently, two cyclic enkephalins DPDPE and DPLPE have been shown to be even more selective than the linear peptides (3). Unfortunately, this increased selectivity is associated with a decrease of affinity (4) preventing a safe use of $[^3H]$ DPDPE in binding experiments. Therefore, the synthesis of new δ linear probes was undertaken based on a model for δ recognition (5) and from the conformational studies of both DTLET and DPLPE. These studies have led to the synthesis of linear hexapeptides bearing bulky residue(s) such as Ser-OtBu or Thr-OtBu. These peptides are able to fit the δ site without drastic reduction in flexibility.
The synthesis and biochemical properties of some of these new δ agonists are reported here.

Material and methods

The linear hexapeptides of general formula Tyr-D-Ser(OX)-Gly-Phe-Leu-Thr(OY) (X = t.Bu, Y = H, X = H, Y = t.Bu and X = Y = t.Bu) were obtained by liquid phase synthesis as described for DSLET and DTLET (6) , using DCC and HOBt as coupling reagents and Z and benzylesters as protecting groups. The synthesis was achieved by the final condensation of two fragments Z-Tyr-D-Ser(OX) and

Peptides 1986

Gly-Phe-Leu-Thr(OY)-OBzl. Such a strategy was convenient for the preparation of a family of analogues.
[^{3}H] DSTBULET (70 Ci/mmol) was obtained from the 2,5-dibromotyrosyl precursor. [^{3}H] DTLET and [^{3}H] DAGO were from CEA (France). Binding assays were carried out at 37°C under steady-state conditions using a crude rat brain membrane fraction as described (7). K_D and K_I values were obtained from linear regression analysis of experimental data using a desk computer.

Results and Discussion

As shown in table 1, the introduction of a t.butyl group on the OH of Thr^6 in DSLET (3) induces a loss in affinity, slightly higher on μ than δ sites, leading to a better selectivity ratio (0.034) than DTLET. A similar modification leads to compound 4 (DSTBULET) which has the same δ affinity as the parent compound 2 but displays a marked loss of μ affinity, resulting to a selectivity ratio similar to that of DPLPE. The additional effect of simultaneous introduction of two t.bu groups in positions 2 and 6 (5) induces a very high δ selectivity ratio (0.009).

TABLE 1. SELECTIVITIES OF ENKEPHALIN DERIVATIVES ON MU OR DELTA OPIOID RECEPTORS OF RAT BRAIN

COMPOUNDS	μ-site* K_I, nM		δ-site** K_I, nM		$K_I(\delta)/K_I(\mu)$
Tyr-D-Thr-Gly-Phe-Leu-Thr (DTLET)	25.3±2.5	(5)	1.35±0.15	(5)	0.053
Tyr-D-Ser-GLy-Phe-Leu-Thr (DSLET)	31.0±5.0	(4)	4.80±0.80	(4)	0.150
Tyr-D-Ser-Gly-Phe-Leu-Thr(OtBu)	185 ±30	(3)	6.35±0.90	(3)	0.034
Tyr-D-Ser(OtBu)-Gly-Phe-Leu-Thr (DSTBULET)	404 ±50	(6)	4.92±0.60	(6)	0.012
Tyr-D-Ser(OtBu)-Gly-Phe-Leu-Thr(OtBu)	460 ±70	(4)	4.23±0.27	(4)	0.009
Tyr-D-Pen-Gly-Phe-D-Pen (DPDPE)	993 ±151	(3)	10.2 ±1.4	(4)	0.010
Tyr-D-Pen-Gly-Phe-L-Pen (DPLPE)	873 ±210	(4)	10.9 ±1.2	(5)	0.013

* With [^{3}H]DAGO (1 nM) ; ** With [^{3}H]DSLET (2 nM) or [^{3}H]DTLET (1 nM).

Such results might indicate that the presence of one or two bulky t.Bu groups are able to constrain the structure at the N or C terminal part of the peptide defavourising the recognition of the μ-site, as with the Pen containing cyclic peptides. The occurrence of a Gly residue not included in a bridge preserves a higher degree of flexibility in the linear peptides, allowing a less energy-consuming recognition of the δ-site.

In addition, the presence of t.Bu groups is expected to increase the resistance of these derivatives to peptidases.

Taking into account these features, a new δ probe [^{3}H] DSTBULET can be proposed. As shown in fig. 1, this ligand interacts with a single class of sites between 0.5 - 1.5 nM (Bmax = 98.8 $\pm$ 9.4 fmol/mg) with a dissociation constant K_D = 2.90 $\pm$ 0.33 nM. The better δ affinity of [^{3}H] DSTBULET as compared to [^{3}H] DPDPE (K_D = 10.5 nM) is associated with a higher specific binding (55 % at 63 % saturation against 26 % at 67 % saturation for [^{3}H] DPDPE), leading to a greater accuracy for studies of δ receptors. In addition, the high δ-selectivity of [^{3}H] DSTBULET is illustrated by the lack of change of its Bmax after blocking μ-sites by 300 nM DAGO (competitive inhibition).

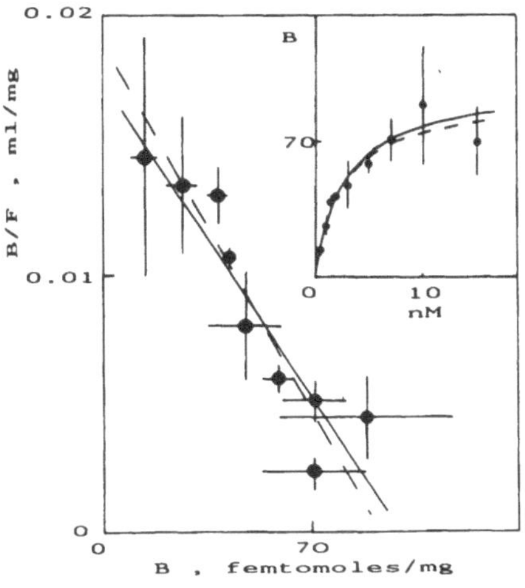

Figure 1. Eadie-Hofstee (– –) and Scatchard (——) analyses of the specific binding of [^{3}H] DStBuLET to rat brain membranes. The corresponding saturation curves are inserted. Each experimental value is the mean of triplicate data of the same assay.

Conclusion

The introduction of bulky residue(s) in linear enkephalin-related hexapeptides represents a new approach in designing δ selective peptides which display the main criteria to investigate both biochemical and pharmacological properties of δ opioid receptors : a good selectivity associated with a high affinity for δ sites and a satisfactory resistance to peptidases.
Moreover, in contrast with DPDPE, their linear structures make them easy to synthesize and stable towards chemical and biological agents during biochemical and pharmacological experiments.
In addition, such analogues can be readily obtained in a tritiated form as shown with [^{3}H] DSTBULET.

References

1. Martin, W.R., C.G. Eades, J.A. Thomson, R.E. Huppler, P.E. Gilbert. 1976. J. Pharmacol. Exp. Ther. 197, 517.

2. Zajac, J.-M., G. Gacel, P. Dodey, J.-L. Morgat, P. Chaillet, J. Costentin, B.P. Roques. 1983. Biochem. Biophys. Res. Commun. 111, 390.

3. Mosberg, H.I., R. Hurst, V.J. Hruby, K. Gee, H.I. Yamamura, J.J. Galligan, T.F. Burks. 1983. Proc. Natl. Acad. Sci. USA 80, 5871.

4. Delay-Goyet, P., J.-M. Zajac, P. Rigaudy, B. Foucaud, B.P. Roques. 1985. FEBS Lett. 183, 439.

5. Fournié-Zaluski, M.-C., G. Gacel, B. Maigret, S. Premilat, B.P. Roques. 1981. Mol. Pharmacol. 20, 484.

6. Gacel, G. M.-C. Fournié-Zaluski, E. Fellion, B.P. Roques. 1981. J. Med. Chem. 24, 1119.

7. Zajac, J.-M., B.P. Roques, 1985. J. Neurochem. 44, 1605.

SYNTHESIS AND BIOLOGY OF CYCLIC ENKEPHALIN PSEUDOPEPTIDES

Arno F. Spatola, Claudio Mapelli, Joseph Humphrey, and J. Vincent Edwards
Department of Chemistry, University of Louisville
Louisville, Kentucky 40292 USA

Thomas F. Burks and Jennifer E. Shook
Department of Pharmacology, School of Medicine, University of Arizona
Tucson, Arizona 85724 USA

Introduction

The concept of combining peptide backbone modifications with conformational constraints represents an example of the continuum between classical peptide structures and synthetic peptide mimetics. Advantages of this approach include the preservation of possibly critical conformational features of the parent structure (1) combined with retention of such properties as altered enzyme resistance, lipophilic balance, and receptor selectivities that may characterize the surrogate structures (2).

Based on parent cyclic enkephalin analogs of varying ring size, we have synthesized several new pseudopeptides which contain the $\psi[CH_2S]$ amide bond replacement. In each case, the modified link is between Phe^4 and Leu^5 of the parent structure as shown in the example in Figure 1.

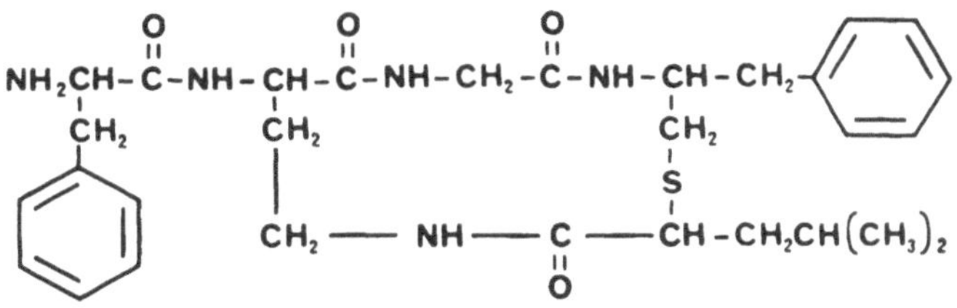

Figure 1. Structure of Tyr-cyclo(D-A_2bu-Gly-Phe$\psi[CH_2S]$Leu).

Peptides 1986

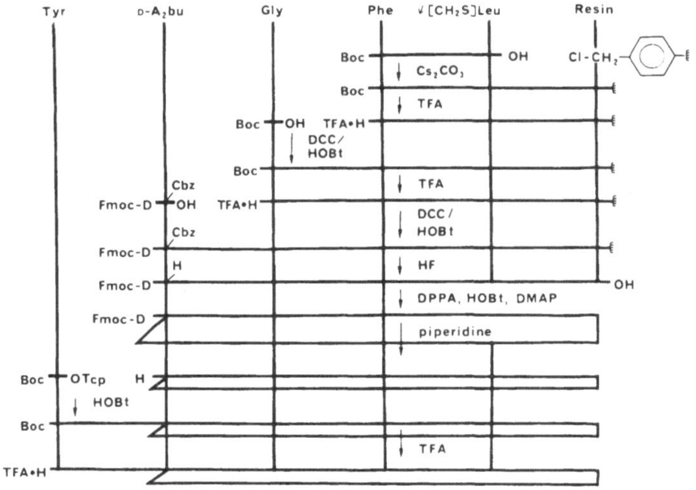

METHOD A - Gisin Coupling

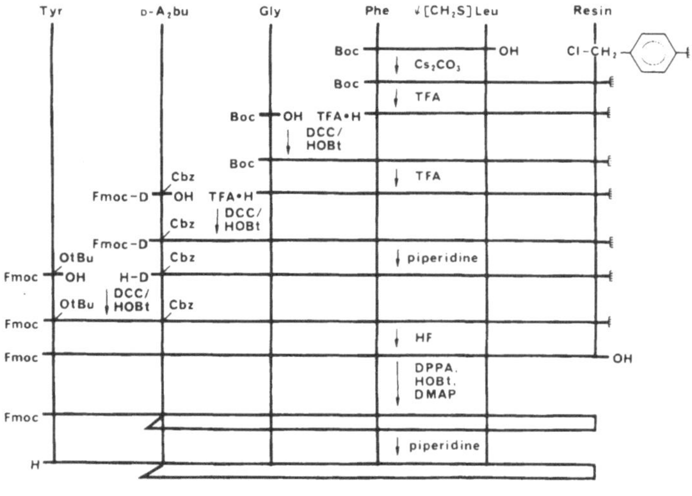

METHOD B

Figure 2. Synthetic schemes for the preparation of cyclic pseudopeptides IIa and IIb using Method A (cyclic 4 + 1 coupling) and Method B (cyclization of linear pentapeptide)

Results and Discussion

The preparation of the 16-membered cyclic pseudopeptide analog Tyr-cyclo(D-Lys-Gly-Pheψ[CH_2S]Leu) has been previously described (4). In those synthetic procedures, two diastereomeric peptides (Ia and Ib) were isolated, corresponding to the L-Leu and D-Leu isomers, respectively. The epimerization during solid phase synthesis was attributed to the use of the Gisin cesium salt procedure (5) for the initial attachment of the pseudodipeptide (Boc-Pheψ[CH_2S]Leu-OH) to the chloromethylated Merrifield resin (6). These compounds proved to be quite potent and displayed prolonged duration of activity when tested in vivo, although both isomers were relatively non-selective toward mu and delta receptors.

Since we felt that the flexibility of the ψ[CH_2S] moiety enabled the molecule to bind equally well to either receptor, it would be of interest to increase the rigidity of the structure by reducing the size of the ring. Accordingly, a smaller 14-membered cyclic pseudopeptide enkephalin containing a D-configuration diaminobutyric acid residue (D-A_2bu) at position 2 instead of D-lysine was prepared by three different routes, including one using the Marglin procedure (Et_3N + DMF) (7) for attachment of the pseudodipeptide to the resin. This confirmed that the cesium procedure caused the epimerization, and also confirmed that the major isomer, possessing the longer retention time on reversed phase HPLC, now corresponded to the L-leucine derivative. Interestingly, instead of the roughly 50/50 mixture of stereoisomers encountered in our previous synthesis of the D-Lys peptides, now, in each case, the major isomer (Pheψ[CH_2S]L-Leu) was predominant in approximately a 70/30 ratio.

Following solid phase synthesis, the linear tetrapeptide was cyclized and coupled with Boc-Tyr-OH (Method A) or the linear Fmoc-protected pentapeptide

TABLE 1. In Vitro Biological Activities of Cyclic Pseudopeptides

Cyclic Peptide	GPI IC_{50} (nM)	MVD IC_{50} (nM)
Pseudo ψ[CH_2S] (IIa and IIb)	2.8 (1.8, 1.5)	2.3 (1.9, 2.9)
Parent (3)	14.1 ± 2.9	81.4 ± 5.8
Pseudo with CTP (1000 nM)	18.2 (9.3, 35.5)	1.7 (1.4, 2.3)
Pseudo with NLX (1000 nM)	339 (162, 1708)	
Pseudo with ICI 174		32.4(20.9, 47.9)

GPI = guinea pig ileum; MVD = mouse vas deferens; NLX = naloxone, and ICI-174 is a sensitive delta-selective antagonist. Values above given in nM (95% CL or ± SEM).

was cyclized and deprotected (Method B). The final purified products were obtained by gel filtration on Sephadex G-15 followed by purification on Vydac C-18 supports using a 0.05% TFA/acetonitrile gradient elution system. The major and minor isomers, IIa and IIb, were subjected to biological testing as a mixture as reported in Table 1.

As can be seen in Table 1, the mixture of cyclic pseudopeptides containing D-A_2bu^2 showed high potency on both the mu and delta receptors. Preliminary results from the testing of the L-isomer yielded an MVD IC_{50} value of 4.7 nM, which suggests that, by inference, the D-isomer apparently has greater δ selectivity. It is apparent that the pseudopeptide analogs display increased *in vitro* activities relative to the all-amide parent, especially with respect to the MVD assay (8). The trends observed upon introduction of the mu-selective antagonists, naloxone and CTP, and the delta selective antagonist, ICI-174, were consistent with the expected behavior. Efforts to separate the diastereomers and to conclusively establish any possible selectivity differences as well as to verify their enhanced stability *in vivo* are currently in progress.

Acknowledgements

This work was supported by grants from the NIH (GM-33376) and the University of Louisville. We are grateful to Dr. Ken Tomer and the NSF Midwest Center for Mass Spectrometry (Grant (CHE 8211164) for FAB-MS determinations.

References

1. Hruby, V. J., Life Sciences (1982) 31, 189-199.

2. Spatola, A. F., Chemistry and Biochemistry of Amino Acids, Peptides, and Proteins (Weinstein, B., ed.), Marcel Dekker, Inc., 1983.

3. DiMaio, J., Nguyen, T. M. D., Lemieux, C., and Schiller, P. W., J. Med. Chem. (1985) 25, 1432-1438.

4. Spatola, A. F., and Edwards, J. V., Biopolymers (1986) in press.

5. Gisin, B., Helv. Chim. Acta (1973) 56, 1476-1482.

6. Edwards, J. V., and Spatola, A. F., J. Liq. Chrom. (1986) 9, 903-919.

7. Marglin, A., Tetrahedron Lett. (1971) 3145-3148.

8. Edwards, J. V., Spatola, A. F., Lemieux, C., and Schiller, P. W., Biochem. Biophys. Res. Commun. (1986), 136, 730-736.

DESIGN AND SYNTHESIS OF RECEPTOR SELECTIVE PEPTIDE NEUROTRANSMITTERS

V.J. Hruby, L.-F. Kao

Department of Chemistry, University of Arizona
Tucson, Arizona 85721

J. E. Shook, K. Gulya, H.I. Yamamura, and T.F. Burks

Department of Pharmacology, University of Arizona
Tucson, Arizona 85724

Introduction

The discovery of numerous peptide neurotransmitters, and of the presence in many cases of multiple receptors for a particular peptide or peptide family of neurotransmitters, raises questions about the physiological roles of these neuropeptides and their receptors. To answer these questions, we have developed an approach for obtaining conformationally constrained peptides with high specificity for particular receptors (1,2). This approach is exemplified by [D-Pen^2, D-Pen^5]enkephalin (DPDPE), a cyclic, highly constrained 14-member ring peptide analog of methionine enkephalin with extraordinary delta opioid receptor specificity (3,4), and by D-Phe-Cys-Tyr-D-Trp-Lys-Thr-Pen-Thr-NH_2 (CTP), a truncated, conformationally constrained analog of somatostatin with greatly reduced somatostatin activity, but potent mu opioid receptor specific antagonist activity (5). We report here further efforts to determine the structural and topological requirements for the high receptor selectivity of DPDPE by use of additional conformational constraints.

Experimental Methods

The cyclic peptide analogs reported here were synthesized by methods similar to those previously reported (3). Purification was accomplished by combinations of partition chromatography, gel filtration and reverse phase-high pressure liquid chromatography (RP-HPLC). Highly pure compounds (>98%) were obtained in all

Peptides 1986

cases as determined by amino acid analysis, thin layer chromatography, RP-HPLC, and fast atom bombardment mass spectrometry. ^{1}H nuclear magnetic resonance (NMR) spectra were examined using a Bruker AM-250 spectrometer. Opioid agonist activity was assessed using the standard mouse vas deferens assay for delta receptors and the guinea pig ileum assay for mu receptors.

Results and Discussion

Utilizing a variety of one and two dimensional NMR methods, including 2D-COSY and 2D-NOESY experiments and conformational calculations (6), we have obtained a low energy conformation consistent with all the data available for DPDPE. The conformation obtained is a highly compact conformation with close transannular interactions between the Tyr^{1} and Phe^{4} aromatic rings and the $D\text{-}Pen^{2}$ β-methyl groups. To further explore the importance of this conformation to the biological potency and receptor specificity of DPDPE, we have begun by examining the conformational consequences and biological changes which accompany further conformation constraints on the Phe^{4} residue of DPDPE. For this purpose, a variety of structural modifications were made which probe specific conformational and structural requirements of the delta opioid receptor for potency and specificity.

Previous investigations have shown that replacement of the peptide amide hydrogen of the Phe^{4} residue in enkephalins with a methyl group leads to an increase in potency at mu receptors. On the other hand, our computer modeling of the DPDPE conformation suggested that N-methylation of the Phe^{4} residue in DPDPE would disrupt its conformation, especially the topological relationships of the Tyr^{1} and Phe^{4} aromatic rings. Consistent with this expectation, $[D\text{-}Pen^{2}, N\text{-}MePhe^{4}, D\text{-}Pen^{5}]$enkephalin (2) (Table I) has greatly reduced potency at the delta receptor and reduced selectivity as well. Furthermore, ^{1}H NMR studies demonstrate that DPDPE and 2 have different conformational properties in solution. On the other hand, examination of the conformation of DPDPE indicated that although the Tyr^{1} and Phe^{4} aromatic rings were in close proximity, para substitution of Phe^{4} residue would be possible

Table 1. Biological Potencies and Receptor Selectivities of DPDPE Analogs[a]

Compound	MVD IC_{50} nM	GPI IC_{50} nM	$[^3H]$DPDPE Ki (nM)	$[^3H]$Nal Ki (nM)	GPI/MVD Ratio
1. $[D\text{-}Pen^2,D\text{-}Pen^5]E$	4.26	16,240	4.0	863.	3810.
2. $[D\text{-}Pen^2,N\text{-}MePhe^4,D\text{-}Pen^5]E$	1016.	186,000	nd	nd	183.
3. $[D\text{-}Pen^2,p\text{-}MePhe^4,D\text{-}Pen^5]E$	10.	15,100	27.	1180.	1510.
4. $[D\text{-}Pen^2,p\text{-}NO_2Phe^4,D\text{-}Pen^5]E$	1.53	1,520	6.6	78.3	993.
5. $[D\text{-}Pen^2,\beta\text{-}MePhe^4,D\text{-}Pen^5]E$	23.4	42,700	13.	2707.	1820.
6. $[D\text{-}Pen^2,\beta,\beta\text{-}Me_2Phe^4,D\text{-}Pen^5]E$	5010.	100,000	nd	nd	20.
7. $[D\text{-}Pen^2,\alpha\text{-}MePhe^4,D\text{-}Pen^5]E$	2200.	16,000	nd	nd	7.2

[a] MVD- mouse vas deferens assay; GPI, guinea pig ileum assay; competitive binding with $[^3H]$DPDPE and $[^3H]$naloxone; E - enkephalin; nd- not determined.

without disruption of the conformation. The effect of such substitutions on interaction of DPDPE with the delta receptor and on receptor selectivity could thus be explored. Interestingly, when 4-p-methylphenylalanine was incorporated into DPDPE (3, Table 1), somewhat reduced potency at delta receptors was observed, and a consequent decrease in selectivity occurred. On the other hand, the 4-p-nitrophenylalanine analog of DPDPE (4, Table I) was essentially equipotent to DPDPE at delta receptors, but its selectivity for delta receptors was greatly reduced due to a large increase in potency at mu receptors for this analog.

An important but largely ignored area of conformational structure-biological activity relationships is that of the relationships of side-chain rotamer population (topology) to biological activity. Among other issues, the question of the relative importance of the proR and proS β-hydrogens in amino acid residues and their importance to peptide-receptor interaction has not been systematically examined. We have begun to examine this problem by substitution of these β-hydrogens with methyl groups. Examination of the low energy conformation proposed for DPDPE suggested that substitution of either β-hydrogen of Phe^4 would not disrupt the overall conformation, but could possibly influence rotamer populations. Thus, the effects of such substitutions on these topological features and on the interaction of DPDPE with delta receptors in regard to potency and selectivity could be explored by this approach. Our results thus far are most interesting. The β-$MePhe^4$ analog 5 is a relatively potent and highly selective analog, while the β,β-dimethylphenylalanine analog 6 has very weak potency at the delta receptor (MVD assay) and little or no activity at the mu receptor (GPI assay). Our present studies indicate that 5 contains the (L,threo)-β-methylphenylalanine residue in the 4 position of DPDPE. Further work is in progress to confirm this and to examine the other diastereoisomeric analogs of DPDPE with a β-$MePhe^4$ substitution. Interestingly, the α-$MePhe^4$ analog 7 (Table 1) has greatly reduced activity at the delta receptor (MVD assay) and is weak but equipotent to DPDPE at the mu receptor (GPI assay), and thus has little opioid receptor selectivity.

Acknowledgements

Supported by grants from the National Science Foundation and the U.S. Public Health Service, NS 19972.

References

1. Hruby, V.J. 1985. Trends Pharm. Sci. 6, 259.
2. Hruby, V.J. 1984. In: Conformationally Directed Drug Design (J.A. Vida and M. Gordon, eds.). ACS Symposium Series No. 251, Washington, D.C., p. 9.
3. Mosberg, H.I., R. Hurst, V.J. Hruby, K. Gee, H.I. Yamamura, J.F. Galligan, and T.F. Burks. 1983. Proc. Natl. Acad. Sci. USA 80, 5871.
4. Akiyama, K., K.W. Gee, H.I. Mosberg, V.J. Hruby, and H.I. Yamamura. 1985. Proc. Natl. Acad. Sci USA 82, 2543.
5. Pelton, J.T., K. Gulya, V.J. Hruby, S.P. Duckles, and H.I. Yamamura. 1985. Proc. Natl. Acad. Sci. USA 82, 236.
6. Hruby, V.J., L.-F. Kao, M. Pettitt, and M. Karplus. 1986. Manuscript in preparation

SYNTHESIS AND STRUCTURE-ACTIVITY RELATIONSHIPS OF SIDE-CHAIN TO SIDE-CHAIN CYCLIZED OPIOID PEPTIDE ANALOGS

P.W. Schiller, T.M.-D. Nguyen, L.A. Maziak, C. Lemieux, B.C. Wilkes

Laboratory of Chemical Biology and Peptide Research
Clinical Research Institute of Montreal
Montreal, Que., Canada H2W 1R7

Introduction

In a further effort to enhance the receptor selectivity of opioid peptides and to manipulate their "efficacy" ("intrinsic activity") we synthesized a series of enkephalin-related analogs which were cyclized through amide bond formation between side-chain amino and carboxyl groups of appropriately substituted Lys (or Orn) and Glu (or Asp) residues. Cyclic lactam analogs of this type with a Phe residue in either position 3 or position 4 of the peptide sequence were prepared by the solid-phase method according to a scheme which permits cyclization with the peptide still attached to the resin, using DCC/HOBt as coupling agents (1). The compounds were tested in the guinea pig ileum (GPI) bioassay (μ-receptor representative) and the mouse vas deferens (MVD) bioassay (δ-receptor representative), and in binding assays based on displacement of [^{3}H]DAGO (μ-selective) and [^{3}H]DSLET (δ-selective) from rat brain membrane preparations.

Results and Discussion

Aside from the cyclic monomers, side-chain-linked antiparallel cyclic dimers were also obtained in various proportions due to intersite reaction on the resin. The extent of dimerization did not depend on the degree of resin substitution (in the range from 0.4 - 1.0 mmol/g titratable amine) but on the solvent used in the cyclization step. Thus, in the case of H-Tyr-D-Orn-Phe-Asp-NH_2 (1), cyclization in DMF produced 32% monomer and 68% dimer, whereas in trifluoroethanol formation of the cyclic monomer was predominant (62%).

Cyclic analog 1 is a highly selective μ-agonist due to its very poor affinity for the δ-receptor, as indicated by the high MVD/GPI IC50-ratio of 107 (Table 1) (2). Inversion of the configuration in position 4 of 1 results in an ana-

Peptides 1986

log (2) with reduced potency in the GPI assay and with only moderate μ-receptor preference. Further conformational restriction of the 13-membered ring in 1 through N-methylation of Phe^3 (analog 3) causes a nearly 200-fold potency drop in the GPI assay, whereas that same modification in the linear opioid peptide morphiceptin (H-Tyr-Pro-Phe-Pro-NH$_2$) increased potency (3). Very low potency in both assay systems is also displayed by analog 4, in contrast to the increased potency observed with correspondingly modified β-casomorphin-5 (H-Tyr-Pro-D-Phe-Pro-Gly-OH) (4). The low potencies shown by 3 and 4 illustrate that structure-activity relationships established with related linear peptides cannot always be extrapolated to cyclic analogs. On the other hand, introduction of a nitro substituent in para position of Phe^3 in 1 (analog 5) produces the same low potency as it does in linear Phe^3 opioid peptides (5). The observation that H-Tyr-L-Orn-Phe-Asp-NH$_2$ (6) has very low potency in both bioassays indicates that as in the case of most linear and cyclic enkephalin analogs the residue in position 2 of the peptide sequence must have D-configuration. The only exception is morphiceptin which requires L-configuration at the Pro^2 residue, as demonstrated by the low potency observed with H-Tyr-D-Pro-Phe-Pro-NH$_2$ (3).

The enkephalin-related cyclic pentapeptide analog 7 is very potent in both assays and, therefore, non-selective (Table 1) (6). Inversion of the configuration of Glu^5 (analog 8) or substitution of (D,L)-homolysine (Hly) for Lys in 7 (analog 9) results in potency losses and in moderate μ-receptor selectivity in the case of 8. C-terminal extension of 7 with Thr (analog 10) lowers potency but has no effect on selectivity, in constrast to the δ-receptor selectivity shown by linear Thr^6 enkephalin analogs (7). Contraction of the 18-membered ring in 7 through elimination of a methylene group in both the 2- and 5-position side-chains slightly lowers potency without affecting selectivity (analog 11). N-methylation of Phe^4 in the latter analog produced a compound (12) with somewhat lower potency in both assays, whereas C-terminal extension according to the sequence of dynorphin A-(1-8) resulted in an octapeptide (13) with an almost identical activity profile. The low K_e-value (1.49) nM) determined for 13 with naloxone as antagonist in the GPI assay indicates that this analog acts via the μ-receptor and lacks the κ-receptor selectivity of dynorphin A-(1-8).

Qualitatively similar potency relationships were observed in the μ- and δ-receptor representative binding assays. However, cyclic analogs 1, 3, 7 and 10 were considerably less potent in the [^{3}H] DAGO binding assay than would have

Table 1. In vitro Activity Profiles of Cyclic Opioid Peptides

No.	Compound	GPI IC50 [nM]	MVD IC50 [nM]	MVD/GPI IC50-ratio	$GPI/[^3H]DAGO$[a] Potency ratio
1	$H-Tyr-D-Orn-Phe-Asp-NH_2$	36.2	3,880	107	7.50
2	$H-Tyr-D-Orn-Phe-D-Asp-NH_2$	632	4,810	7.61	0.894
3	$H-Tyr-D-Orn-Phe(NMe)-Asp-NH_2$	6,460	1,380	0.214	14.4
4	$H-Tyr-D-Orn-D-Phe-Asp-NH_2$	> 30,000	> 20,000	-	-
5	$H-Tyr-D-Orn-Phe(pNO_2)-Asp-NH_2$	2,340	> 100,000	> 43	3.04
6	$H-Tyr-L-Orn-Phe-Asp-NH_2$	16,800	> 100,000	> 6	-
7	$H-Tyr-D-Lys-Gly-Phe-Glu-NH_2$	1.13	0.648	0.573	30.4
8	$H-Tyr-D-Lys-Gly-Phe-D-Glu-NH_2$	12.2	118	9.67	5.14
9	$H-Tyr-(D,L)-Hly-Gly-Phe-Glu-NH_2$	26.5	30.5	1.15	5.36
10	H-Tyr-D-Lys-Gly-Phe-Glu-Thr-OH	21.2	8.03	0.379	20.5
11	$H-Tyr-D-Orn-Gly-Phe-Asp-NH_2$	4.33	5.45	1.26	4.58
12	$H-Tyr-D-Orn-Gly-Phe(NMe)-Asp-NH_2$	8.63	33.3	3.86	-
13	H-Tyr-D-Orn-Gly-Phe-Asp-Arg-Arg-Ile-OH	2.93	8.84	3.02	0.494
14	$[Leu^5]$enkephalin	246	11.4	0.0463	1

[a] Ratio of relative potencies determined with the GPI bioassay and with the $[^3H]$DAGO binding assay.

been expected on the basis of their relative potencies in the GPI assay (Table 1, last column). These discrepancies can be best explained with an increased "efficacy" of these compounds at the μ-receptor and are not due to an additional interaction with κ-receptors in the GPI, since all analogs investigated showed K_e-values for naloxone as antagonist in a range (0.8 - 2.1 nM) typical for μ-receptor interactions. Efficacy enhancement is dependent on the side-chain length of the diamino acid in position 2 (compare analogs 7, 9 and 11) and on the configuration of the dicarboxylic amino acid engaged in ring formation (compare analogs 1 vs. 2 and 7 vs. 8).

In conclusion, cyclization of opioid peptides via side-chains permits the manipulation of the receptor affinity, receptor selectivity and "efficacy" in a more drastic manner than it is possible through peptide analog design based on simple amino acid substitutions. The conformational characterization of these cyclic analogs should permit the design of even more selective opioid receptor ligands and, ultimately, of opioid peptide mimetics.

Acknowledgment

Supported by grants from the Medical Research Council of Canada (MT-5655) and the Quebec Heart Foundation.

References

1. Schiller, P.W., T.M.-D. Nguyen, J. Miller. 1985. Int. J. Peptide Prot. Res. 25, 171.

2. Schiller, P.W., T.M.-D. Nguyen, C. Lemieux, L.A. Maziak. 1985. J. Med. Chem. 28, 1766.

3. Chang, K.-J., E.T. Wei, A. Killian, J.-K. Chang. 1983. J. Pharmacol. Exp. Therap. 227, 403.

4. Matthies, H., H. Stark, B. Hartrodt, H.-L. Ruethrich, H.-T. Spieler, A. Barth, K. Neubert. 1984. Peptides 5, 463.

5. Schiller, P.W., T.M.-D. Nguyen, J. DiMaio, C. Lemieux. 1983. Life Sci. 33, 319.

6. Schiller, P.W., T.M.-D. Nguyen. 1984. Neuropeptides 5, 165.

7. Zajac, J.M., G. Gacel, F. Petit, P. Dodey, P. Rossignol, B.P. Roques. 1983. Biochem. Biophys. Res. Commun. 111, 390.

SYNTHESIS AND BIOLOGICAL ACTIVITY OF ENKEPHALIN RELATED GLYCOPEPTIDES

J.Mª. García Antón, J.L. Torres, G. Valencia, M.A. Busquests, F. Reig,

Laboratory of Peptides. Biological Organic Chemistry Department. C.S.I.C. 08034 - Barcelona.

R.E. Rodriguez

Faculty of Medecine. University of Salamanca

Introduction

The experience gained in the area of the Opioid Peptides should be oriented to the development of highly selective or specific enkephalin analogues of clinical benefit. The design of peptides with antinociceptive efficacy dissociated from its respiratory depression or addictive effects is obviously of the utmost importance. To go further in this concern it would be worthwhile to obtain information on the chemical nature of the different receptor subtypes (μ, δ an K). Despite many attempts for the isolation of opioid binding sites have failed. the presence of glycolipidic material in mammalian surface receptors has been evidenced by many autohors in the last few years (1,2) and it has become widely accepted that exposed sugar residues of glycolipids or glycoproteins serve as determinants for receptor recognition. In this respect, mammalian hepatic receptors are known to be specific for terminal D-galactose, whereas in avian liver the receptors recognize terminal D-glucose.

A posible application based on these facts, could be the use of synthetic specific glycopeptides as ligands for the delivery of pharmaceuticals to target tissues.

Being aware of such a possibility one can assume that the incorporation of sugar moities in an enkephalin sequence could improve the affinity of the molecule for its receptor, exhibiting, as a consequence, a higher biological response.

Peptides 1986

Thus, we decided to synthesize two glycopeptides containing D-glucose and D-galactose, respectively, as sugar moities, being the μ-agonist enkephalin analogue: Tyr-D.Met-Gly-Phe-Pro-NH_2, the parent compound. Sugars have been introduced by means of an amide bond at the C-terminal position of the peptide. The glycopeptides obtained are:

	ß-D-glucopyranosyl amide (I)
Tyr-D.Met-Gly-Phe-Pro	
	ß-D-galactopyranosyl amide (II)

We outline here the synthesis and the "in vitro" and "in vivo" activity of both glycopeptides, which exhibited analgesic effects several times higher than morphine.

Experimental

Glycopeptides (I) and (II) were obtained by the liquid phase procedure. NHS, MA (REMA method), HOBT/DDC were used as coupling agents. Three different strategies have been attempted for the incorporation of the saccharide moiety into the peptide. The one providing better yields for both derivatives is outlined in Fig.1
Removal of tert-butyloxycarbonyl groups was achieved with TFA/CH_2Cl_2 (35:65, v/v) containing a few drops of 2-mercaptoethanol, in 1h at room temperature.
Final glycopeptides were purified by semipreparative reversed-phase HPLC on a 10μm μ-Bondapack C_{18} (30 x 0.78 cm) column, with a 1.1 ml. loop, under isocratic conditions, CH_3CN/H_2O -0.05% TFA (30:70) at a flow rate of 60 mm/min.
Homogeneity of compounds (I) and (II) was checked by analytical HPLC on a 10 μm. (25 x 0.46 cm) reversed phase C_{18}-coated silica gel support, with a programmed gradient elution (λ = 215 mm).
Racemization was checked by GC on a chiral stationary phase: cyano ethyl-siloxane-L-valine-S-α-phenyl ethyl amide.
Characterization of both glycopeptides was ascertained by elemental analysis, amino-acid composition, ^1H n.m.r and FAB-MS. Compounds (I), (II), [$DMet^2$, $ProNH_2^5$] enkephalin and morphine were subjected to the GPI test according to Kosterlitz's description (3).

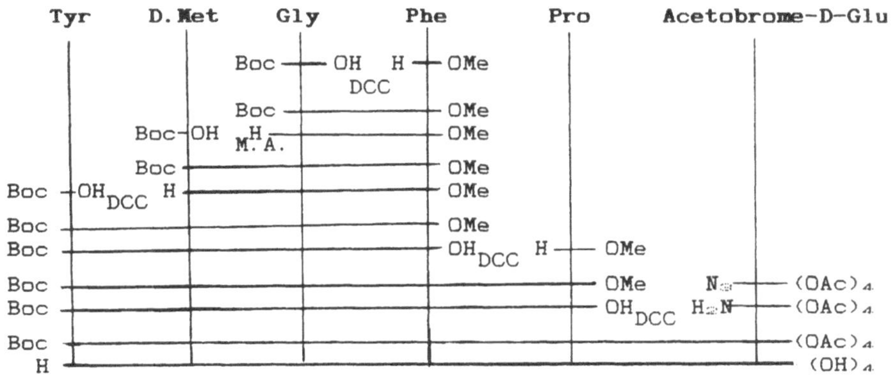

Fig. 1. Strategy of synthesis devised for compounds (I) and (II).

Antinociceptive activity, after intraperitoneal injection of morphine sulphate and glycopeptides (I) and (II), was obtained for the tail inmersion test both in rats and mice (4).
For the tail inmersion test, water at 54.5±0.5°C was used as nociceptive stimulus (4). A cut-off time of 10 sec. was used. Measurements were made 30 min. after the administration of the drugs.

Results and Discussion

Table 1 shows the ED_{50} (95% confident limits) of the antinociceptive activity produced by morphine sulphate and Glycopeptide (I).
No significant differences were observed between compounds (I) and (II). The results obtained show that the antinociceptive activity by intraperitonal administration of the glycopeptides is 2000 times higher, in rats, than the observed with morphine. Besides, the same experiments in mice show that glycopeptides are nearly 200 times more potent than morphine. However, in the GPI test, the IC_{50} of compound (I) was: 64,9 nM, whereas, the corresponding IC_{50} for (II) was: 117,3 nM, whereas, the diference between both derivatives. The IC_{50} for the parent compound under the same conditions is 10,8 nM, which accounts for a slight

Table1. Tail inmersion ED_{50} (μg/animal) after i.p. injection.

Compound	RATS		MICE	
	ED_{50}	nº animal used	ED_{50}	nº animal used
(I)	1.2 (0.75-1.92)	10	5.0 (2.7-9.2)	12
Morphine sulphate	2666 (2050-3465)	6	900 (801-1007)	7

ED_{50} (95% confident limits)

decrease in the affinity for the μ-subtype as a consequence of the saccharide incorporation.

To further implement the above expectancies, other routes of administration and different antinociceptive tests, as well as binding measurements to elucidate receptor selectivity, are under way.

References

1. Ashwell, G. and Harford, J., 1982 Ann. Rev. Biochem 51, 531.

2. Behnan, B.A. and Deber C.M., 1984, J. Biol. Chem. 259, 14.935

3. Kosterlitz, H.W. Lord, J.A., Watt, A.J, 1972. Agonist and antagonist Actions of Narcotic Analgesic Drugs. Univ. Park Press. Baltimore, p. 45.

4. Litchfield, J. T and Wilcoxon, I., 1949. J. Pharmac. 96, 99

DESIGN, SYNTHESIS AND BIOLOGICAL ACTIVITY OF SELECTIVE AGONISTS FOR TACHYKININ RECEPTOR SUBTYPES.

M. Chorev[1], R. Laufer[2], D. Ron[1], J. Frey[3], U. Wormser[2], E. Rubini[1], Z. Selinger[2] and C. Gilon[3].

Departments of Pharmaceutical[1], Biological[2] and Organic[3] Chemistry, The Hebrew University of Jerusalem, Jerusalem 91120, Israel.

Introduction

The newly discovered mammalian tachykinins: neurokinin A (NKA) and neurokinin B (NKB) which together with substance P (SP) constitute the mammalian tachykinin family raised the possibility that each tachykinin acts as a preferred ligand for a particular receptor. The concept of two types of SP receptors designated SP-P and SP-E was first suggested by Iversen and coworkers (1). This classification was based on a different order of relative potencies of series of non mammalian tachykinins when tested by different bioassays. Binding studies using radiolabeled analogs of SP, eledoisin and NKA attempted classification of tachykinin receptors in the CNS (2). However, it should be noted that the natrually occurring mammalian tachykinins are not highly selective and activate all the tachykinin receptors. In addition, many tissues may possess a mixture of different tachykinin receptors and thus deviate from the classical SP-P and SP-E types. It is therefore not surprising that the use of non-selective radioligands for binding studies caused ambiguous results. To overcome some of these difficulties we have developed selective assay for each of the two tachykinin receptors in the guinea-pig ileum (GPI). In this tissue the tachykinins elicit contraction both directly through their action on a muscular tachykinin receptor and indirectly through their action on myenteric neurons causing acetylcholine release which in turn elicits contraction through a muscular muscarinic receptor (3). We have found that neurokinin B is a preferred ligand for the neuronal receptor in the GPI. Since the pharmacological profile of this receptor differed from both the SP-P and the SP-E types we designated this receptor as SP-N type (N for neuronal). To further substantiate the existence of the SP-N receptor and to investigate whether SP-N sites also occur in the CNS we developed ligands with much higher selectivity toward particular receptor subtype than that of the naturally occuring tachykinins.

The rationale for development of selective agonists is based on the assumption that the naturally occurring neuropeptides can exist in several different and interchangeable conformations, each selected for or induced by a particular

Peptides 1986

receptor subtype. From this consideration it follows that conformationally restricted analogs should reveal selectivity toward specific receptor subtype. In this paper we present the synthesis and biological activity of analogs of the hexapeptide $[pGlu^6]SP_{6-11}$ that contain N-alkylated peptide bonds as conformational constraints.

Synthesis and Biological Assay.

Both solution and solid phase methods were employed in the synthesis of analogs. The solution technique used excess mixed anhydride while the solid phase practised either DCC-HOBt or mixed anhydride-HOBt on benzyl resin. Boc N-Me, N-Et, N-allyl, N-Pr and N-Bu amino acids were prepared according to (4,5). Different routes were undertaken to prepare N-iPr and N-iBu derivatives of Phe and for their assembly into the analogs. In both cases we used reductive alkylation of free amine function either at the amino acid stage (6) (Scheme 1) or at the appropriate stage in SPPS (Scheme 2).

$$H_2N\text{-}CH(CH_2\text{-}C_6H_5)\text{-}COOH \xrightarrow[\text{2.NaCNBH}_3]{\text{1.acetone/MeOH}} (CH_3)_2CH\text{-}NH\text{-}CH(CH_2\text{-}C_6H_5)COOH \xrightarrow[\text{MeOH}]{SOCl_2}$$

$$HCl\cdot(CH_3)_2CH\text{-}NH(CH\text{-}C_6H_5)COOMe \xrightarrow[\text{DCC/HOBt}]{\text{BocPheOH/DMF}} \text{BocPhe-N-iPr-PheOMe} \xrightarrow[\text{2. }H^+]{\text{1. NaOH}}$$

BocPhe-N iPr PheOH

Scheme 1. Synthesis of N-iPr PheOH and its incorporation into peptide.

$$(CH_3)_2\text{-}CH\text{-}CHO\text{+}H\cdot\text{Phe-Gly-Leu-Met-}Ⓟ \xrightarrow[\text{2.NaCNBH}_3\text{/Py}]{\text{1.Py/AcOH(pH 7)}} H\cdot\text{N-iBu Phe-Gly-Leu-Met-}Ⓟ$$

Scheme 2. Synthesis of polymer bound N-iBu Phe-peptide.

Difficulties were encountered in coupling of the preceding Boc amino acid to the free N-alkylated amine and in the deprotection of the Boc from either N-alkylated peptides and AA-N-alkylated peptides. Even prolonged reaction times and repetitions resulted in low yields. Analogs and intermediates were purified by RP LPLC and characterized by amino acid analysis FAB-MS sequencing (7) RP-HPLC, TLC and ^{1}H NMR. Biological activity on SP-P and SP-N receptor in the guinea pig ileum was measured according to ref. 3.

Results and discussion

As shown in Table 1 N-methylation of Leu^{10} does not significantly affect activity on either receptor subtype and thus does not improve selectivity (analog 2). Although N-methylation of Phe^7 increases selectivity for the SP-P receptor this analog (3) is a weak SP-P agonist possessing only 5% of the activity of $[pGlu^6]SP_{6-11}$ (1). In contrast substitution of Gly^9 by Sar did not impair SP-P

	Peptide	SP-P receptor	EC_{50}(nM) SP-N receptor	Selectivity SP-P/SP-N EC_{50} ratio
1	$[pGlu^6]SP_{6-11}$	1.6	5	0.32
2	$[pGlu^6,N\text{-}Me\text{-}Leu^{10}]SP_{6-11}$	5	27	0.18
3	$[pGlu^6,N\text{-}Me\ Phe^7]SP_{6-11}$	30	500	0.06
4	$[pGlu^6,Sar^9]SP_{6-11}$	2	33	0.03
5	$[pGlu^6,Pro^9]SP_{6-11}{}^a$ (8)	2.5	1300	0.0019
6	$[pGlu^6,N\text{-}Me\ Phe^8]SP_{6-11}$	500	0.5	1000
7	$[SuccAsp^6,N\text{-}Me\ Phe^8]SP_{6-11}{}^b$	35000	0.5	70000
8	$[pGlu^6,N\text{-}Et\ Phe^8]SP_{6-11}$	4000	0.27	14814
9	$[pGlu^6,N\text{-}Allyl\ Phe^8]SP_{6-11}$	1500	0.27	5555
10	$[pGlu^6,N\text{-}Pr\ Phe^8]SP_{6-11}$	2600	0.36	7222
11	$[pGlu^6,N\text{-}nBu\ Phe^8]SP_{6-11}$	>30000	1800	∿17
12	$[pGlu^6,N\text{-}iPr\ Phe^8]SP_{6-11}$	>10000	>10000	-
13	$[pGlu^6,N\text{-}iBu\ Phe^8]SP_{6-11}$	1000	700	1.4

Table 1. N-alkylation of different peptide bonds in $[pGlu^6SP_{6-11}$ increases agonist selectivity for either the muscular SP-P or for the neuronal SP-N receptors in the guinea pig ileum. a. Designated "Septide" (Selective SP-P receptor peptide). b. Deisgnated "Senktide" (Selective NKB receptor peptide).

potency while it slightly reduces activity on the SP-N receptor (4). These results prompted us to replace Gly^9 by Pro. Indeed this substitution (analog 5) had almost no effect on the potency of the SP-P receptor, but reduced the activity on the SP-N receptor by two orders of magnitude. Most remarkably, an opposite effect on receptor selectivity was obtained by N-methylation of Phe^8. This modification resulted in a 10 fold increase in potency for the SP-N receptor and a concomitant 100-fold decrease in the potency of the SP-P-receptor. Consequently $[pGlu^6,N\text{-}Me\ Phe^8]SP_{6-11}$ (6) is a highly potent and selective SP-N agonist. Further improvement in the selectivity for SP-N receptor was obtained as the result of an effort to increase the water solubility of 6. The water soluble metabolically stable analog $[SuccAsp^6,N\text{-}Me\ Phe^8]SP_{6-11}$ is equipotent to 6 in stimulating the SP-N receptor (EC_{50}=0.5 nM) but considerably less potent on the SP-P receptor (EC_{50}=35 μM) with an SP-N/SP-P potency ratio of 70000. Furthermore analog 7 is devoid of activity in rat vas deferens, a typical SP-E system, even when tested at a concentration 400,000 times higher than its EC_{50} value for the SP-N receptor. A different approach to change selectivity for the SP-N receptor was to alter the size of the group alkylating the Phe^7-Phe^8 bond. Introduction of a straight chain up to three carbons did not impair the potency of the SP-N receptor (as compared to analog 3) but decreased the potency for the SP-P receptor and hence improved selectivity (analogs 8-10). A larger linear alkyl group such as n-Bu (analog 11) causes great reduction in potency for both receptors although it still maintained selectivity for the SP-N receptor. A dramatic effect is caused by introduction of an isopropyl group. This analog (12) was inactive on both receptors even at high concentration. It seems that branching on C^{β} introduces a severe steric hindrance which prevents the formation of conformations compatible with either SP-P or SP-N receptors. Branching on C^{γ} as in analog 13 greatly decreased the potency as well as the selectivity of the analog.

In conclusion selective agonists for tachykinin receptors were obtained by introduction of conformational restrictions attained by N-alkylation. Development of highly potent and selective agonists for tachykinin receptors strongly support the existence of a third tachykinin receptor of the SP-N type, whose endogenous ligand may be neurokinin B.

References

1. Iversen, L.L. 1985 in: Tachykinin Antagonists (R. Hakanson and F. Sundler, eds.) Elsevier Science Publishers, Amsterdam, pp. 291-304.
2. Quirion, R. Trends Neurosci. 1985, 8, 183-185, and references within.
3. Laufer, R., Wormser, U., Friedman, Z.Y., Gilon, C., Chorev, M. and Selinger, Z., 1985, Proc. Natl. Acad. Sci. USA 82, 7444-7448.
4. Cheung, S.T. and N.L. Benoiton, 1977, Can. J. Chem. 55, 906-910.
5. Shuman, R.T., E.L. Smitwick, D.L. Smiley, G.S. Brooke and P.D. Gesellchen, 1983, in: Peptides Structure and Function (J.C. Hruby and D.H. Rich, eds.) Pierce Chem. Co. Ill.
6. Ohfune, Y., N. Kurokawa, N. Higuche, M. Saito, M. Hashimoto and T. Tanaka 1984, Chem. Lett. 441-444.
7. Eckhart, K., H. Schwarz, M. Chorev and C. Gilon, 1986, Eur. J. Biochem. 157, 209-216.
8. Laufer, R., C. Gilon, M. CHorev and Z. Selinger, 1986, J. Med. Chem. 29, 1284-1288.

DISILOXANE-BRIDGED BIOLOGICALLY ACTIVE ANALOGUES OF THE ENKEPHALINS.

J.S. Davies, E.J. Tremeer
Department of Chemistry, University College of Swansea, Wales, U.K.

R.C. Treadgold
Dow Corning Ltd., Barry, S. Glam, Wales, U.K.

Introduction

For over a decade the goal of academia and pharmaceutical industry has been to draft a profile of the pharmacophoric conformation of the enkephalins. In the absence of direct evidence the research effort has relied heavily (1) on circumstantial evidence derived from structure-activity studies using residue substitution and peptide bond surrogacy. Restricting the flexibility of the pentapeptide backbone through bridging across relevant side chains or by insertion of conformationally restricting amino acids, assists in pin-pointing a more limited number of conformations relevant to identifying the pharmacophore. Amongst the key studies can be cited: bridged analogues between an amino group at position 2 and the terminal CO group (2): bridging between a side-chain amino group in 2 and a side chain carboxyl group in position 5 (3); a disulfide bridge between D-Cys^2 or D-Pen^2 and equivalent residues in position 5 (4), including the use of spacer units (5); an azo link between Tyr^1 and Phe^4(6); insertion of Δ-Phe, ∇-Phe(7) and α-substituted residues (8).

Results and Discussion. The present work shows a slightly different bridging technique using a bifunctional reagent to link the side-chains of [Ser^2-Ser^5]-enkephalin. Synthesis of the bridged enkephalin analogues (1) and (2) in the present study was achieved by treating the appropriate Ser^2-diastereoisomers

Peptides 1986

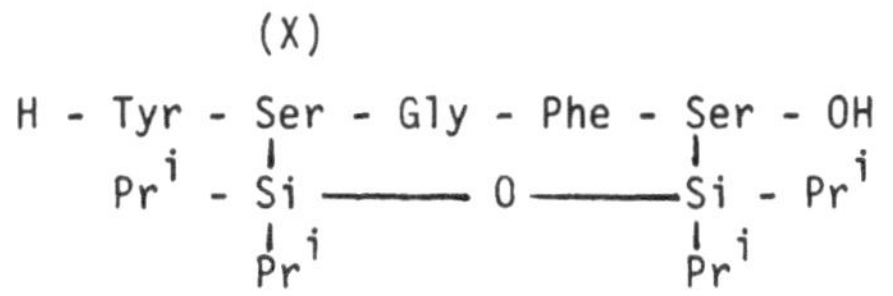

(1) X=L (2) X=D

of Z-Tyr(Bzl)-Ser-Gly-Phe-Ser-OBzl(3) with 1,3-dichloro-1,1,3,3-tetraisopropyldisiloxane [$Pr^i_2Si(Cl)$-O-$Si(Cl)Pr^i_2$] ($TIPDSiCl_2$)(9). Bridging of the two seryl hydroxyls in (3) was effected by the slow addition of $TIPDSiCl_2$ (0.5 eq) every 24 hr until a 2-fold excess of the reagent had been added. Column chromatography on Kieselgel G and FAB-MS analysis of the reaction components gave isolated yields of the cyclic bridged analogues in the 25-30% range without contamination from acyclic analogues.

Scheme 1 summarises the synthetic route which was chosen in the absence of detailed information about the stability of the disiloxane bridge. Acid and base-sensitive protecting groups were therefore avoided after the bridge was in place. Stepwise solution phase synthesis of (3) using the Fmoc protecting

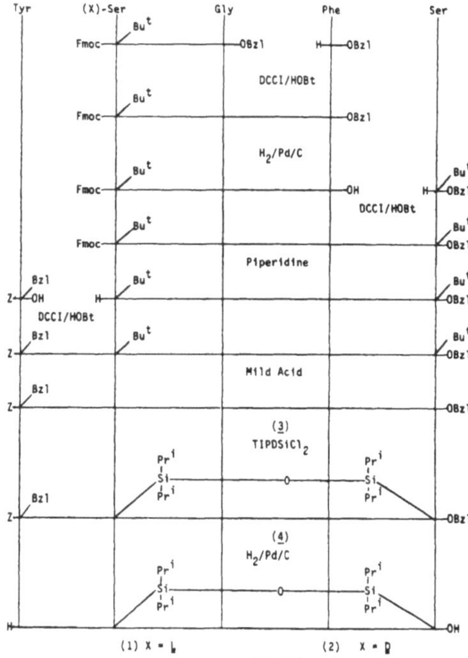

Scheme 1

group initially proved disappointing when a 2+2 fragment approach was used. On release of the Fmoc group from the C-terminal Fmoc-Phe-Ser-OBzl, dioxopiperazine formation became a major competitor to the coupling with Fmoc-Ser-Gly-OH. The 3+1 approach in Scheme 1 gave good coupling yields.

The biological activity of (1) and (2) in the mouse vas deferens assay (carried out for us by ICI Pharmaceuticals Division) have been compared to their linear counterparts:

Compound	Potency relative to [Met] enkephalin
H-Tyr-Ser-Gly-Phe-Ser-OH	Not active
Disiloxane derivative (1)	0.02
H-Tyr-D-Ser-Gly-Phe-Ser-OH	0.02
Disiloxane derivative (2)	1.34

Although relatively low in activity analogue (1) did show a true opiate effect when tested against μ-and δ-selective antagonists. The D-analogue(2) had appreciably more activity with indications that it was δ-selective.

360 MHz N.M.R. analysis on (2) using 2D-techniques for signal correlation and Nuclear Overhauser Effect studies on the amide protons confirmed the existence of a β-bend conformation. Temperature dependence studies on the NH signals revealed the following data (ppm/°C):- D-Ser^2 6.96×10^{-3}, Phe^4 3.36×10^{-3}, Gly^3 4.13×10^{-3}; Ser^5 2.35×10^{-3}. It is only the latter value that comes within the range below 3×10^{-3} usually recognised (10) as evidence for H-bonding. Although the amide NH's all showed linear dependence over most of the temperature range a dramatic shift in Gly^3 and Ser^5 NH signals occurs in the 30-50°C range suggesting a significant conformational change involving these residues.

The n.m.r. results were strengthened by a computer graphics low energy optimisation which predicted the conformation drawn out in the Figure and this together with its biological activity may be considered further support for a similar type of conformation being required for receptor interaction (11).

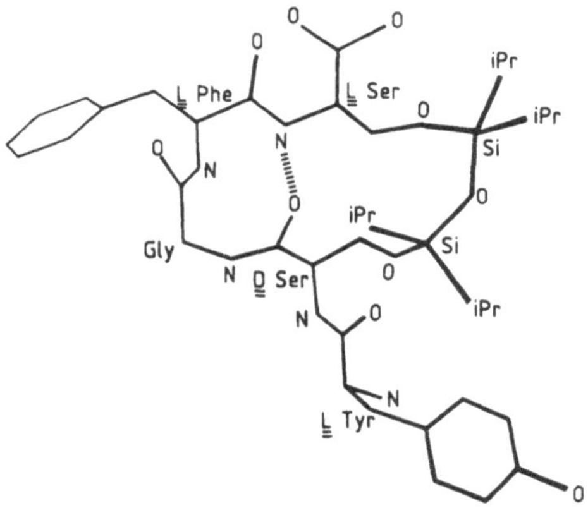

Figure Computer Generated conformation of (2)

Acknowledgements

The work was supported by SERC/Dow Corning Ltd and we thank Dr. W. A. Thomas, Roche Products Ltd. for n.m.r. and computer graphic assistance.

References

1. Davies, J.S., 1986. In "Amino Acids, Peptides and Proteins Vol 17 (J.H. Jones, Senior Reporter), Royal Society of Chemistry, Chapter 3. (in press).
2. J, DiMaio, T.M-D. Nguyen, C. Lemieux, P.W. Schiller, 1982, J.Med.Chem., 25, 1432; H. Kessler, G. Holzemann, C. Zechel, 1985, Int.J.Pept. Protein Res., 25, 267; N.J. Mammi, M. Hassan, M. Goodman, 1985, J.Am.Chem.Soc., 107, 4008.
3. Schiller, P.W., T.M.-D. Nguyen, J. Miller, 1985, Int.J.Pept. Protein Res., 25, 171.
4. Mosberg, H.I., P.W. Schiller, 1984, Int.J.Pept. Protein Res., 23, 462, and refs. cited therein.
5. Mosberg, H.I., J.R. Omnas, 1985, J.Am.Chem.Soc., 107, 2986.
6. Siemion, I.Z., Z. Szewczuk, Z.S. Herman, Z. Stachiva, 1981, Mol.Cell. Biochem., 34, 23.
7. Kimura, H., C.H. Stammer, Y. Shimohigashi, C. Ren-Lin, J. Stewart, 1983, Biochem.Biophys.Res.Commun., 115, 112.
8. Kishore, R., P. Balaram, 1984, Indian J.Chem., 23, 1137.
9. Markiewicz, W.T., 1979, J.Chem.Res., 24,
10. Ravi, A., B.V.V. Prasad, P. Balaram, 1983, J.Am.Chem.Soc., 105, 105.
11. Roques, B.P., C. Garbay-Jaureguiberry, R. Oberlin, M. Anteunis, A.K. Lala, 1976, Nature, 262, C.R. Jones, W.A. Gibbons, V. Garsky, 1976, Nature, 262, 779.

CHARACTERIZATION OF A THIRD TACHYKININ RECEPTOR (SP-N) IN THE MAMMALIAN NERVOUS SYSTEM

R. Laufer[1], Z. Selinger[1], M. Chorev[2], J. Frey[1] and C. Gilon[3]
Departments of Biological[1], Pharmaceutical[2] and Organic[3] Chemistry,
The Hebrew University of Jerusalem, Jerusalem 91904, Israel.

Introduction

Three tachykinin receptors, whose preferred ligands are substance P, neurokinin A and neurokinin B respectively, are recognized in mammalian tissues. These receptors have been designated SP-P, SP-E and SP-N (1-3). We have recently developed highly potent and selective agonists for SP-P and SP-N receptors, which enable a clear distinction between the three tachykinin receptor types (4,5). One of these compounds, succinyl-[Asp^6,N-$MePhe^8$]SP_{6-11} (Senktide) acts on the SP-N receptor of guinea pig ileum myenteric plexus with an EC_{50} value of 0.5 nM, but is a very weak agonist for SP-P receptors in guinea pig ileum muscle (EC_{50}=35 μM) and SP-E receptors of rat vas deferens (no effect at 200 μM). In order to characterize SP-N receptors in the nervous system, we have now prepared a radiolabeled analog of Senktide and used this selective radioligand in binding studies of SP-N sites.

Materials and Methods

Peptides were prepared by conventional methods in solution or by solid phase procedures. N-(desamino-3-iodotyrosyl)-[$Asp^{5,6}$N-$MePhe^8$]SP_{5-11} (I-BH-NH-Senktide) was prepared in both radiolabeled and unlabeled form by coupling iodinated Bolton-Hunter reagent (N-hydroxysuccinimidyl-3-(4-hydroxy,3-iodophenyl)propionate) to [$Asp^{5,6}$,N-$MePhe^8$]SP_{5-11}. Peptides were purified by RP HPLC and characterized by amino acid analysis and FAB-MS sequencing using the method described in ref. 6.

Membranes from rat cerebral cortex were prepared as follows: Tissues were homogenized in ice-cold 10% sucrose. The homogenate was centrifuged at 16,000 g x min and the resulting supernatant was spun down at 200,000 g x min. The crude synaptosomal pellet was lysed by resuspension in 5 mM Hepes, pH 7.5, followed by two washes at 1,000,000 g x min. The final pellet was suspended in 20 mM Hepes, pH 7.4 (1 ml/g of initial tissue wet weight) and stored frozen under liquid N_2. Before use in a binding experiment, membranes were thawed, diluted 2-fold with binding buffer (composition in mM: NaCl, 120; KCl, 4.8;

Peptides 1986

$CaCl_2$, 1; $MgSO_4$, 1; Hepes, 20; pH 7.4) and left to stand at room temperature for 30 min. The membranes were then washed with binding buffer (1,000,000 g x min) and suspended in the same buffer to a protein concentration of 1-2 mg/ml.

Binding assays contained in a final volume of 0.2 ml: binding buffer; 2 mM $MnCl_2$; 50 µl of membrane suspension; and 0.4-0.8 nM ^{125}I-BH-NH-Senktide. Non-specific binding was determined in the presence of 1 µM Senktide. After 30 min at 35°C, the reaction was terminated by addition of 2x2.5 ml of ice-cold binding buffer containing 0.05% BSA, followed by rapid filtration over Whatman GC/B filters that had been presoaked in binding buffer containing 0.5% BSA. The filters were rinsed three times with 5 ml of ice-cold binding buffer containing 0.05% SDS.

Smooth muscle contraction assays were performed as previously described (3,4).

Results and Discussion

I-BH-NH-Senktide is a water-soluble metabolically stable highly potent and selective agonist of the SP-N receptor subtype. This peptide contracts potently the guinea pig ileum with an EC_{50} value of 1.5 nM; the contractile activity is lost in the presence of the muscarinic blocker atropine (no effect at 10 µM), indicating that I-BH-NH-Senktide acts on the neuronal SP-N, but not on the muscular SP-P receptors. The peptide is also devoid of activity in two SP-E systems, the hamster urinary bladder and the rat vas deferens (no effect at 10 µM).

The binding of ^{125}I-BH-NH-Senktide to rat cerebral cortex membranes reached equilibrium after 20 min and was reversible (half-time of dissociation: 22 min). While non-specific binding (20-30% of total) linearly increased with radioligand concentration, specific binding was saturable, reaching a maximal value of 122 fmol/mg protein (Fig. 1).

The equilibrium dissociation constant, K_D, was found to be 1 nM. The linear Scatchard plot (Fig. 1) and the Hill coefficient of 1.05 indicate that ^{125}I-BH-NH-Senktide binds to a single class of non-interacting sites.

The ^{125}I-BH-NH-Senktide binding site seems to be linked to a guanine nucleotide-binding protein. As shown in Fig. 1, the non-hydrolyzable GTP analog 5'-guanylyl imidodiphosphate inhibits specific binding by decreasing both affinity and the number of binding sites. No inhibition was observed with a high concentration (100 µM) of the adenine nucleotide 5'-adenylyl imidodiphosphate.

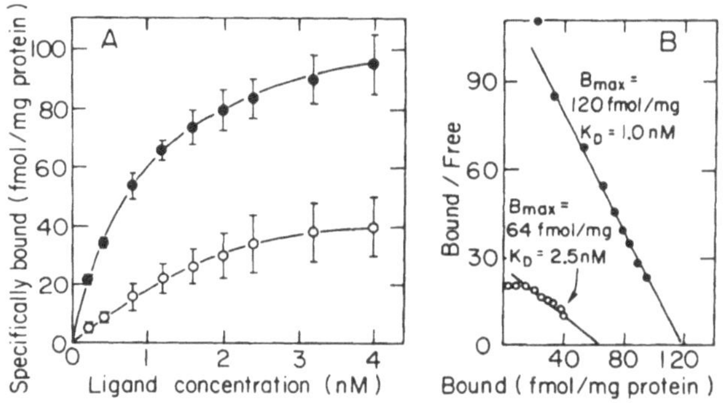

Fig. 1.: Inhibition of specific ^{125}I-BH-NH-Senktide binding to rat cerebral cortex membrane by Gpp(NH)p. A: membranes were incubated with the indicated concentrations of ^{125}I-BH-NH-Senktide in the absence (●) or presence (o) of 50 μM Gpp(NH)p. B: Scatchard analysis.

The potencies of tachykinins and related peptides to compete with the radioligand for binding to rat cortex membranes is described below. The potency order of mammalian tachykinins is characteristic of the SP-N receptor subtype: neurokinin B (K_I=0.5 nM) > neurokinin A (K_I=37 nM) > substance P (K_I=138 nM). The selective SP-N agonists (5) [pGlu^6,N-MePhe^8]SP$_{6-11}$ and Senktide (K_I=1.9 nM) strongly interact with the ^{125}I-BH-NH-Senktide binding site. In contrast, the selective SP-P agonists (1,4) [pGlu^6,Pro^9]SP$_{6-11}$ (K_I=5 μM) and substance P methyl ester (K_I > 25 μM) are only very weakly active, if at all.

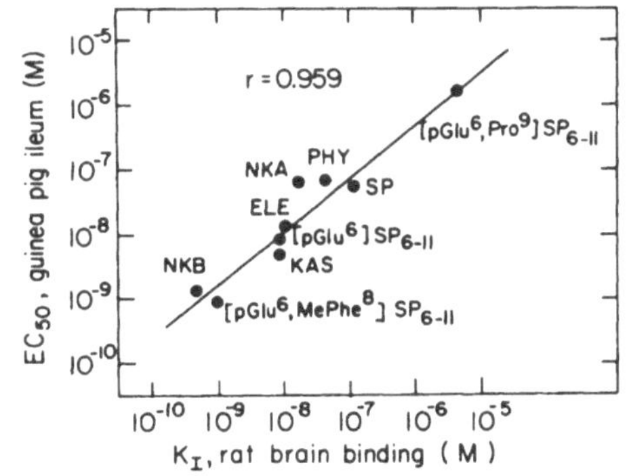

Fig. 2: Correlation between binding affinities of tachykinin-related peptides and their biological potencies in stimulating the SP-N receptor in guinea pig ileum.

A close correlation was obtained between the pharmacological specificity of this binding site and that of the SP-N receptor in guinea pig ileum myenteric plexus which was determined by a functional bioassay (Fig. 2).

These results strongly suggest that the ^{125}I-BH-NH-Senktide binding site represents neurokinin B receptors of the SP-N type. ^{125}I-BH-NH-Senktide combines high receptor affinity and selectivity with metabolic stability and is therefore the radioligand of choice for further studies of the distribution, regulation and function of the SP-N receptor.

References

1 Iversen, L.L. 1985 in: Tachykinin Antagonists, (R. Hakanson and F. Sundler, eds.) Elsevier Science Publishers, Amsterdam, pp. 291-304.

2 Quirion, R. Trends Neurosci. 1985, 8, 183-185.

3 Laufer, R., U. Wormser, Z.Y. Friedman, C. Gilon, M. Chorev and Z. Selinger, 1985, Proc. Natl. Acad. Sci. USA 82, 7444-7448.

4 Laufer, R., C. Gilon, M. Chorev, and Z. Selinger 1986, J. Med. Chem. 29, 1284-1288.

5 Chorev, M., R. Laufer, D. Ron, J. Frey, U. Wormser, E. Rubini, Z. Selinger and C. Gilon. This issue.

6 Eckart, K., H. Schwartz, M. Chorev and C. Gilon, 1986, Eur. J. Biochem. 157, 209.

SYNTHESIS OF PEPTIDES CONTAINING REPEATING SEQUENCES OF ENKEPHALIN

J. Izdebski, J. Bondaruk

Laboratory of Peptides, Department of Chemistry,
University of Warsaw, 02-093 Warsaw, Poland

P. Krząścik

Department of Pharmacodynamics, Warsaw Medical Academy,
00-927 Warsaw, Poland

Introduction

The nucleotide sequence of cloned cDNA for preproenkephalin indicates that the precursor protein containes several copies of Met-enkephalin (1,2).

...Lys-Lys ENK^M Lys-Arg ENK^M Lys-Lys...Lys-Arg ENK^M Lys-Lys...

...Lys-Arg ENK^M Arg-Gly-Leu-Lys-Arg...Lys-Arg ENK^M Arg-Arg...

...Lys-Arg ENK^L Lys-Arg...Lys-Arg ENK^M Arg-Phe

The Met-enkephalin sequences (Tyr-Gly-Gly-Phe-Met and Tyr-Gly-Gly-Phe-Leu, ENK^M and ENK^L respectively) are located between pairs of basic amino acids and could be cleaved by specific enzymes to yield enkephalin molecules. Incomplete processing of the precursor at the dibasic amino acids processing sites generates extended enkephalin-like peptides, such as peptides F, I, B and E (3). These peptides possess enhanced or unique biological activities, e.g. the 25-peptide E containing ENK^M and ENK^L segments at terminal positions exhibits potent opioid activity.

Results

We describe here synthesis and properties of 12-peptide ENK^M-Lys-Arg-ENK^M (peptide BI). The sequence of peptide BI is completely conserved between the bovine, human, rat and toad species in proenkephalin molecules. The peptide itself has not been found in adrenal but formation of this compound cannot be

Peptides 1986

ruled out. Peptide BI may exhibit modyfied activity and also represents a highly efficient precursor of enkephalin. Peptide BI analogs containing two lysine or two arginine residues have also been synthesized and studied.

The peptides were synthesized by the solid-phase method. The protected peptides were assembled on the classical Merrifield resin using Boc group for N-alfa protection. Side chain protecting groups were: Arg(Tos), Lys(2-BrZ) and Met-O.

The important feature of this synthesis was that symmetrical anhydride was formed by the treatment with t-butylmethylcarbodiimide. This reagent we proposed two years ago (4) and recently demonstrated (5) that soluble in CH_2Cl_2 t-butylmethylurea, which is the co-product of the reaction of the formation of the anhydride, interacts with peptide chains. This interaction would permit to overcome the unfavorable consequences of self-aggregation of peptide chains. The reagent is also very convenient from practical point of view: no filtration is needed after the formation of the anhydride.
A possibility of losses of the anhydride due to precipitation or decomposition are eliminated. Chain assembly yields (one coupling with 3 equivalents of preformed symmetrical anhydrides in CH_2Cl_2 except for arginine were 20% of DMF was added) based on quantitative ninhydrin test were (in %):

Tyr -	Gly*-	Gly*-	Phe -	Met -	Arg -	Arg*-	Tyr -	Gly -	Gly -	Phe - Met
99.96	98.82	97.85	99.92	99.85	99.06	97.87	99.90	99.98	99.98	99.98

Tyr -	Gly -	Gly -	Phe -	Met -	Lys -	Arg*-	Tyr -	Gly -	Gly -	Phe - Met
99.90	99.99	99.92	99.33	99.77	99.71	98.56	99.90	99.98	99.98	99.98

In the cases indicated with * second coupling was performed. The yield was increased to above 99.50%.

The peptide-resins were cleaved in HF, 0°, for 1 h with 10% of anisol. The crude peptides were purified by gel filtration (Sephadex G-10) in 0.5 M CH_3COOH, partition chromatography on a Sephadex G-25 column in BuOH:AcOH:H_2O (4:1:5) system, then reduced with NH_4I in TFA and purified on a CM Sephadex column (gradient 0.05 M NH_4OAc, pH 5.5 → 0.5 M NH_4OAc, pH 7.5). Overall yields of purified peptides were about 30%.

In vitro opioid activities of the peptides were determined with assays based on inhibition of electrically evoked contraction of the GPI, MVD, RVD and RaVD. The potency of the peptides in the GPI assay (μ-receptor representative) and the MVD assay (δ-receptor representative) is presented in Table 1.

Table 1. Biological Activity in Smooth Muscule Assays

Peptide	IC_{50}(M)	
	GPI	MVD
ENK^M-Lys-Arg-ENK^M	5.1×10^{-6}	2.2×10^{-7}
ENK^M-Arg-Arg-ENK^M	1.6×10^{-7}	2.0×10^{-8}
ENK^M-Lys-Lys- ENK^M	4.6×10^{-7}	1.7×10^{-8}
Enkephalin	4.2×10^{-7}	5.0×10^{-9}

The effects of all peptides on the GPI and MVD preperations were reversed by naloxone. No activity was observed in the RVD and RaVD assays at concentration lower than 3×10^{-5}. The analgesic activity was determined in mice by the hot-plate method. Doses ranging from 3×10^{-9} to 3×10^{-10} M per mouse, injected i.c.v., produced the effect, which lasted 15-60 min. depending on the dose used.

Acknowledgement

This research was supported by Polish Academy of Sciences grant no CPBP 06.03.4.3.

References

1. Noda, M., Y. Furutani, H. Takahashi, M. Toyosato, T. Hirose, S. Inayama, S. Nakanishi, S. Numa. 1982. Nature 295, 202.

2. Gubler, U., P. Seeburg, B.J. Hoffman, L.P. Gage, S. Udenfriend. 1982. Nature 295, 206.

3. Udenfriend, S., D.L. Kilpatrick. 1984. In: The Peptides, Analysis, Synthesis, Biology (S. Udenfriend and J. Meienhofer, eds.). Academic Press, Inc. Vol. 6, pp. 25-68.

4. Izdebski, J., J. Pełka. 1984. Peptides 1984 (U. Ragnarsson, ed.). Almqvist and Wiksell International, Stockholm. pp. 113-116.

5. Toniolo, C., G.M. Bonora, V. Moretto, C.H. Schneider, H. Rolli, M. Jung, J. Izdebski. 1986. Helv. Chim. Acta 69, 350.

SYNTHESIS AND BIOLOGICAL PROPERTIES OF PERIPLANETIN CC-2 COCKROACH NEUROPEPTIDE FROM CORPORA CARDIACA OF *PERIPLANETA AMERICANA* L.

D. Konopińska

Institute of Chemistry , Wrocław University , Wrocław 50-383 , Poland

G. Rosiński , P.Sujak , A.Lesicki , A.Kasprzyk

Department of Animal Physiology , Poznań University ,Poznań 61-701 , Poland

W. Sobótka

Institute of Organic Chemistry , Polish Academy of Sciences , Warszawa 01-224 , Poland

Key words : periplanetin CC-2 new synthesis , hyperglycemic action on yellow mealworm larvae , cockroach myoactive neuropeptide .

Introduction

Periplanetin CC-2 <Glu-Leu-Thr-Phe-Thr-Pro-Asn-Trp-NH_2 and periplanetin CC-1 (<Glu-Val-Asn-Phe-Ser-Pro-Asn-Trp-NH_2), octapeptides isolated from corpora cardiaca of the American cockroach *Periplaneta americana* L . (1,2), turned out to be identical with myotropic factor M-2 and M-1 (2) , and hyperglycemic hormones HGH-II and HGH-I (3). As in the case of the locust adipokinetic hormone AKH-I (<Glu-Leu-Asn-Phe-Thr-Pro-Asn-Trp-Gly-Thr-NH_2), both neuropeptides elevate the concentration of hemolymph trehalose and diglicerides (4), and accelerate the cockroach heartbeat (1,2). We report here a synthesis of periplanetin CC-2 (1), and its C-terminal tetrapeptide fragment Thr-Pro-Asn-Trp-NH_2 (2), carried out by classical methods (Scheme) . The neuropeptide 1 can be considered as a shortened analogue of AKH-I , where asparagine in position 3 has been replaced by threonine. Bioactivities of 1 and 2 against yellow mealworm *Tenebrio molitor* L. and American cockroach *Periplaneta americana* L . have also been elucidated.

Peptides 1986

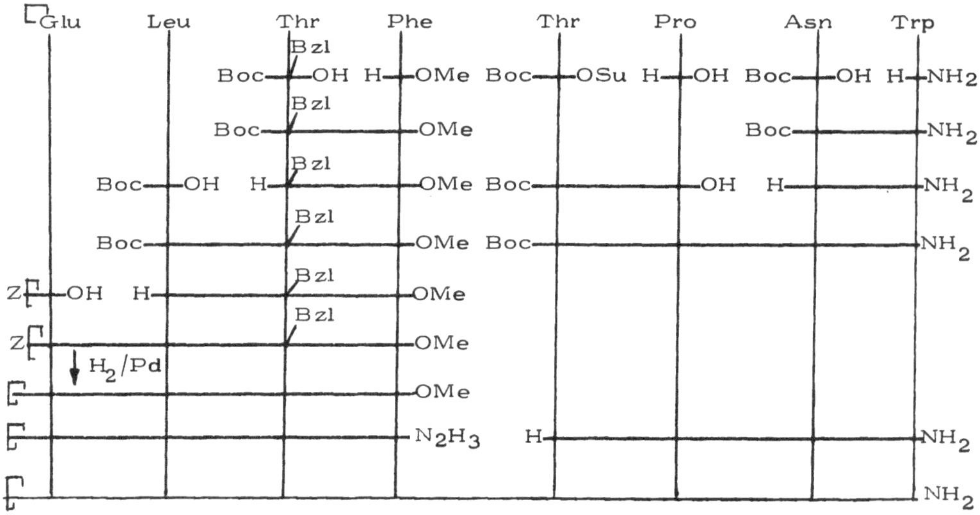

Scheme . A Synthesis of Periplanetin CC-2

Results

In the course of the synthesis of the neuropeptide 1 the N-terminal ⌐Glu-residue was used as Z-⌐Glu-, while the C-terminal tryptophan carboxyl was protected as the amide. All amino groups were blocked with a Boc-residue, which was subsequently removed by HCl in dioxane.For the formation of the peptide bond the following methods were applied: DCC/HOSu, active esters,azide,and mixed anhydrides.Fragment condensation,by the azide method, of two tetrapeptides,applied in the final step of the synthesis afforded octapeptide 1 . When recrystallized from MeOH- ether and AcOH-ether mixtures, and chromatographed on Sephadex LH-20 the product gave 1 as proved by its amino acid composition and molecular weight determination. Biological effects of octapeptide 1 and tetrapeptide 2 included *in vivo* evaluation of their influence on the hemolymph free carbohydrates and the fat body glycogen levels in the yellow mealworm larvae,as well as their cardioacceleratory activity on imagines.The bioassay results were compared with those found for the American cockroach.Determination of the hemolymph free carbohydrate level in the yellow mealworm larvae according to the method of Rosiński (5),was carried out one hour after injection of 20 pmol/insect of peptide 1 or 2 . Values were calculated as percentage increase over 2 μl saline-injected controls.For the peptides 1 and 2 , 40% and 9% increase of the carbohydrate level was observed. More than 80% carbohydrate mobilization by compound 1 on the American cockroach hemolymph was in agreement with reported data (1).After *in vivo* application of a 20 pmole dose of 1 according to the methods of Van Handel (6) and Dubois (7) over 50 % decrease of the yellow mealworm larva fat body glycogen level was observed . The effects of compounds 1 and 2 on yellow mealworm and cockroach heart action were examined , over a concentration range 4.10^{-12} to 4.10^{-11} M , following the method of Miller (8). Both peptides did not affect the yellow mealworm heartbeat whereas the octapeptide 1 , in our bioassays also , led to 30 % increase of the American cockroach heart activity (1). In conclusions drawn from the bioassays of peptides 1 and 2 we wish to emphasize the fact that octapeptide 1 (synthetic periplanetin CC-2) significantly increases hemolymph carbohydrate concentration in yellow mealworm larvae (40 %) , accompanied by decrease (50 %) of the fat body glycogen level. Moreover , species specificity of 1 bioactivity was clearly pronounced when its hyperglycemic and cardioacceleratory action in the American cockroach was compared with lower hyperglycemic effect in the yellow mealworm larvae and complete lack of activity in imagines heartbeat.

Acknowledgement

This research was supported by the Polish Academy of Sciences (grants CPBP-06.03 and CPBR-3.13).

References

1. Scarborough ,R.M., G.C.Jamieson , F.Kalish ,S.J.Kramer ,G.A.McEnroe , C.A.Miller , D.A.Schooley. 1984 . Proc.Natl.Acad.Sci. USA , 81 , 5575 .
2. O'Shea ,M., J.Witten ,M.Schaffer .1984 .Science 213 , 567 .
3. Siegert ,K., P.Morgan , W.Mordue.1985.Biol.Chem.Hoppe-Seyler 366,723.
4. Stone ,J.V., W.Mordue ,K.E.Batley ,H.R.Morris .1976. Nature 263 ,207 .
5. Rosiński ,G., A.Wrzeszcz , L.Obuchowicz .1979. Insect Biochem. 9 ,485.
6. Van Handel , E. 1965. Ann. Rev. Biochem. 11 , 266 .
7. Dubois , M., K.A.Gilles ,J.K.Hamilton ,P.A.Rebers ,F.Smith .1956. Anal . Chem . 28 , 350 .
8. Miller ,T.A..1983.In: Endocrinology of Insects. A.R.Liss Inc. New York, p. 101 .

A NOVEL μ-SPECIFIC REACTIVE LIGAND OF THE OPIATE RECEPTOR

K. Medzihradszky, J. Hepp
Central Research Institute of Chemistry, Hungarian Academy of Sciences, 1025 Budapest

S. Benyhe, A. Borsodi
Biological Research Center, Hungarian Academy of Sciences, 6701 Szeged

Introduction

It is known that in the different organs a number of subtypes of the opiate receptor can be found. These subtypes selectively bind ligands of various constitution, and numerous data on the structure - selectivity relationships are available from the literature. Thus, for example, converting the terminal carboxyl group of the enkephalins into the chloromethyl ketone, a significant shift from the δ to the μ specificity could be observed (1). We synthesized therefore the chloromethyl ketone derivative of DAGO, a μ-specific ligand (2) with the hope to obtain a μ selective compound, capable of covalently reacting with the μ-subtype of the opiate receptor.

Synthesis

As DAGO (H-Tyr-D-Ala-Gly-MePhe-Gly-ol) does not contain free carboxyl group, the glycinol residue had to be replaced by the chloromethyl ketone group. Thus, first the chloromethyl ketone of the BOC-MePhe was synthesized, deblocked and coupled with the BOC-Tyr-D-Ala-Gly-OH protected tripeptide (3) according to the scheme shown in Fig. 1.

Purification of the protected tetrapeptide chloromethyl ketone was performed by column chromatography on silica gel. HCl in formic acid was used for deprotection. The free tetrapeptide chloromethyl ketone (DAMPCK) hydrochloride was sufficiently pure for the bio-

Peptides 1986

logical investigations, it could be stored without significant decomposition in dry state at +4°C for several months.

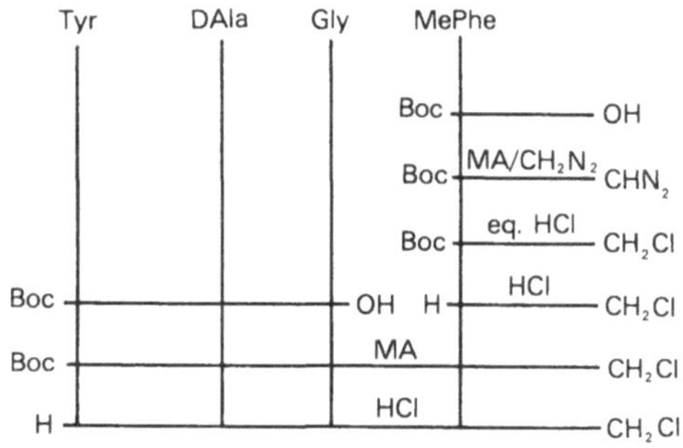

Fig. 1: Synthesis of H-Tyr-D-Ala-Gly-MePhe-CH_2Cl

Receptor binding

Binding was measured on a PVG/C rat brain synaptosome membrane preparate, possessing μ, δ and κ type binding sites, and 1 pmole/mg protein ^{3}H-naloxone binding capacity. IC_{50} values against ^{3}H-naloxone (μ, δ and κ sites), ^{3}H-DALE (D-Ala^2, Leu^5-enkephalin, δ sites) and ^{3}H-DHM (dihydro-morphin, μ sites) for a number of ligands are shown in Table 1. In the last column, the μ/δ ratio, the receptor selectivity index is also given.

Table 1: Receptor Binding of Different Ligands

Ligand	IC_{50}, nmole			
	^{3}H-naloxone μ, δ, κ	^{3}H-DALE δ	^{3}H-DHM μ	μ/δ
DALE	100	3	30	10.0
DALECK	50	15	10	0.66
DHM	10	70	1	0.014
DAGO	8	80	2	0.025
DAMPCK	10	80	2	0.025

DALE: D-Ala^2,Leu^5-enkephalin
DALECK: D-Ala^2,Leu^5-enkephalin chloromethyl ketone

Discussion

The data of Table 1 show that DAMPCK possesses the same affinity toward the receptor as DAGO, the receptor selectivity index indicates a strong preference for the μ subsites, as predicted from our previous observations.

Chloromethyl ketones of the enkephalins proved to be powerful irreversible inhibitors of the opiate receptor (4). This behavior was also shown by DAMPCK. Fig. 2 shows the radioactivity bound to the membrane preparate preincubated with DAMPCK in various concentrations.

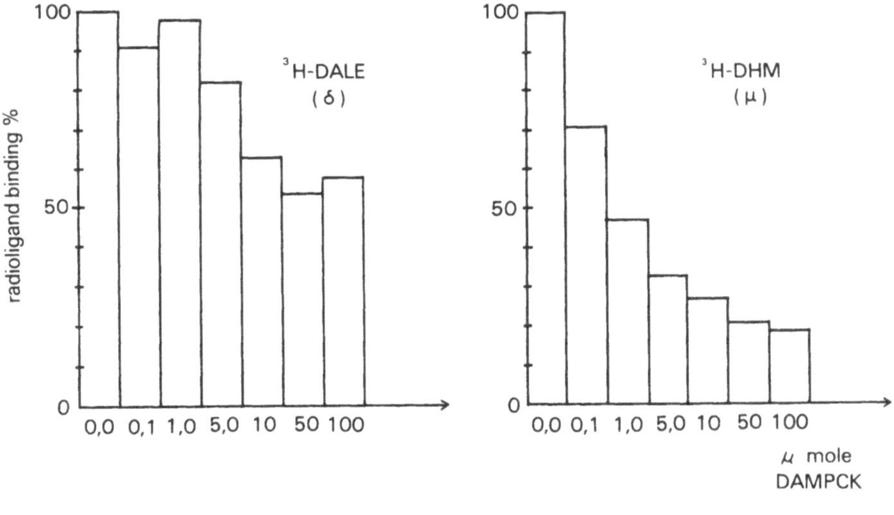

Fig. 2: Irreversible Inhibition of the Opiate Receptor with DAMPCK

^{3}H-DALE binding is inhibited by 50% at a ligand concentration of 50 μmole. Due to the specificity of DAMPCK, the same degree of inhibition of the ^{3}H-DHM binding can be observed at 1.0 μmole concentration, after incubation with 100 μmoles of DAMPCK only 20% of the receptors remained available for the radioligand.

Reaction of the receptor with DAMPCK is a fairly rapid process, maximal inhibition can be reached in ten minutes, as indicated on Fig. 3.

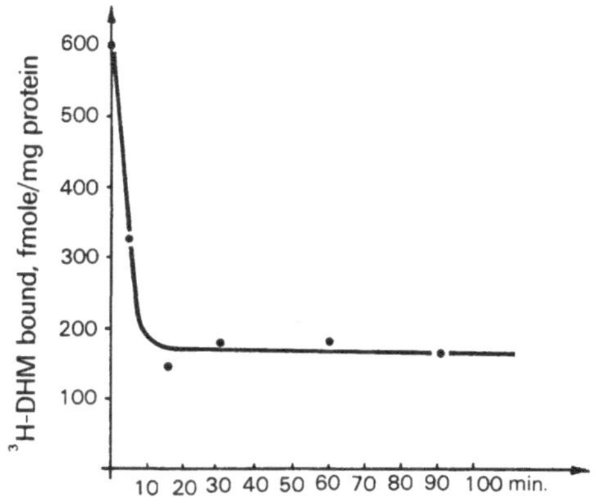

Fig. 3: Time Dependence of the Irreversible Inhibition of the Opiate Receptor by DAMPCK

As a conclusion, DAMPCK proved to be a rather specific affinity label of the μ-receptor. Synthesis of its tritium-labelled analog and application for the characterization of the μ-subsites is in progress.

References

1. Benyhe, S., J. Hepp, M. Szücs, J. Simon, A. Borsodi, K. Medzihradszky, M. Wollemann: in press.

2. Roemer, D., H.H. Buescher, R.C. Hill, J. Pless, W. Bauer, F. Cardinaux, A. Closse, D. Hauser, R. Huguenin. 1977. Nature 268, 547.

3. Szécsi, J., J. Hepp, K. Medzihradszky. 1983. Magy. Kém. Foly. 89, 474.

4. Szücs, M., S. Benyhe, A. Borsodi, M. Wollemann, G. Jancsó, J. Szécsi, K. Medzihradszky. 1983. Life Sci. 32, 2777.

STUDIES ON CARBOXYL TERMINALLY EXTENDED DERMORPHINS

S. Salvadori, M.Marastoni, G.Balboni, R.Tomatis
Department of Pharmaceutical Sciences,University of Ferrara,Italy.

Introduction

In our laboratory we have carried out detailed investigations (1) on the synthesis and pharmacological screening of oligopeptides related to the opioid dermorphin, H-Tyr-D-Ala-Phe-Gly-Tyr-Pro-Ser-$-NH_2$. Moreover it has been reported that a dermorphin-like peptide apparently exists in the rat brain but the amino acid sequence of mammalian dermorphin and its precursor is still unknown (2). On the other hand, biologically active peptides are derived from biosynthetic precursors by enzymatic processing reactions which involve specific endo- and exoproteolytic steps, and often C-terminal amidation. In these precursors the Gly-Xaa-Xbb sequence is generally present at the C-terminus of peptide hormones which comprises two typical signals for post-translational processing: the Gly residue for the formation of C-terminal amide and paired basic residues (Xaa-Xbb) for endoproteolytic processing.
Taking these observations into account, we have prepared dermorphinoyl(DMR)-Gly, DMR-Sar and DMR-Gly-Arg in order to examine i) the effect on opioid activity of C-terminal extension and ii) the ability of amidating enzyme, obtained from a porcine pituitary, to convert DMR-Gly into the corresponding des-glycine amide (dermorphin).

Synthesis

Fig. 1 outlines the synthesis of DMR-Gly-Arg which is representa-

Peptides 1986

tive of the method used for the dermorphinoyl-derivatives prepared. Iodination of DMR-Gly and DMR-Sar was performed with ^{125}iodide and chloramine-T (0.2 µg of peptide with 0.5 mC of ^{125}I) followed by purification on Sephadex G25 in 0.1M acetic acid containing 0.02% NaN_3 at 4°C.

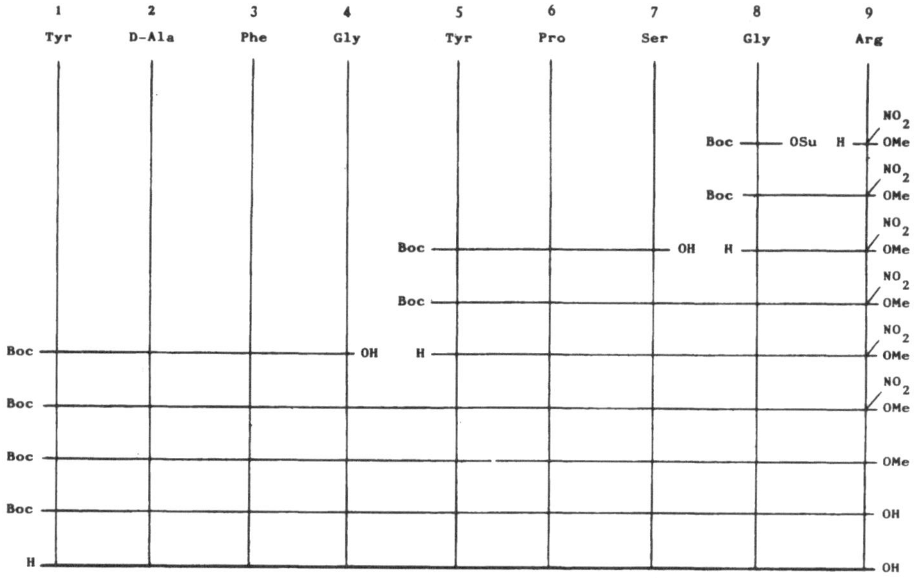

Fig. 1 - Synthetic scheme of DMR-Gly-Arg

Amidation of peptides

An assay system for estimating rate amidation was employed using ^{125}I-labelled peptide substrates incubated with amidating enzyme obtained from a porcine pituitary (3). The reactions were carried out at 37°C, pH 7.0 and samples of the reaction mixtures were removed at intervals for analysis. The formed peptide amide was resolved from unchanged substrate and the extent of amidation was calculated from the yield of the radioactivity emerging in the elution position of the amide compared with the radioactivity emerging in the position of the unchanged substrate (4).

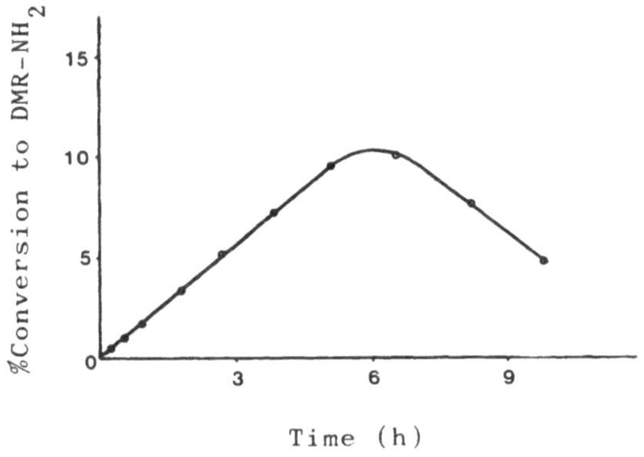

Fig. 2 Formation of dermorphin from DMR-Gly catalysed by an amidating enzyme from a porcine pituitary.

Pharmacological evaluation

The dermorphinoyl-derivatives were tested by the electrically-stimulated guinea pig ileum (GPI) preparation (5) and in vivo, by the mouse tail-flick assay (6) after icv and sc administrations: the results are given in Table 1 and compared with dermorphin, dermorphinoic acid (DMR-OH) and morphine.

Table 1. Biological Activities of Dermorphin Analogues[a]

Compounds	Relative Potencies		
	GPI	Analgesia (icv)	Analgesia (sc)
DMR-OH	11	0.6	0.1
DMR-Gly-OH	3	24	1
DMR-Sar-OH	9	109	3
DMR-Gly-Arg-OH	6	108	11
dermorphin	45	270	4
morphine	1	1	1

[a]All effects observed were completely blocked by naloxone.

Conclusion

In the GPI test and in the tail-flick assay after icv administration, the addition of Gly,Sar or Gly-Arg to the carboxy terminus of dermorphinoic acid (DMR-OH) is relatively tolerated: in fact DMR-Gly, DMR-Sar and DMR-Gly-Arg retain 5-20% of dermorphin activity. In contrast to GPI and icv data, the nonapeptide, containing the basic and hydrophilic Arg residue, displays highest activity following sc administration.

Incubation of DMR-Gly with amidating enzyme led to the progressive formation of dermorphin: under the conditions used, the yield of amide reached a maximum value of 11% in 6h (Fig. 2). Control experiments showed that when synthetic ^{125}I-labelled dermorphin was incubated under the same conditions it was slowly degraded. Thus the formation of the peptide amide by the pituitary enzyme is followed by its partial destruction by other enzymes present in the pituitary extract. No amidation would be expected to take place with DMR--Sar as substrate, and none was observed.

References

1. Salvadori, S., M. Marastoni, G. Balboni, R. Tomatis 1986. J. Med.Chem. 29, 889 and references cited therein.

2. Buffa, R., E. Solcia, G. Rindi, L. Negri, P. Melchiorri 1982. Histochemistry 76, 273.

3. Bradbury, A.F., M.D.A. Finnie, D.G. Smith 1982. Nature 298,686.

4. Bradbury, A.F., D.G. Smith 1983. Biochem.Biophys.Res.Commun. 117, 289.

5. Kosterlitz, H.W., A.J. Watt 1968. J.Br.Pharmacol.Chemother. 33, 266.

6. Janssen, P.A., C.J. Memegeers, J.G. Dony 1958. Arzneim.Forsch. 13, 502.

SYNTHESIS OF ANALOGS OF ENKEPHALIN DERIVATIVES OF 2-D-ALA,5-DES-MET-ENKEPHALIN HYDRAZIDE AT THE FIRST AND FOURTH POSITIONS

G.P.Vlasov,N.Yu.Kozhevnikova,E.N.Krasnikova,N.G.Illarionova
Institute of Macromolecular Compounds of the Academy of Sciences of the USSR,Leningrad, USSR

Introduction

The purpose of the present paper was to study the effect of the modification of the first and fourth positions in the enkephalin series on its biological activity.The compounds of the following general formula were obtained:

X-D-Ala-Gly-Y,

where:

I	X=Tyr	Y=Phe-N_2H_3	V	X=Tyr	Y=NH-NH-Ph
II	X=DOPA	Y=Phe-NH_2	VI	X=DOPA	Y=NH-NH-Ph
III	X=DOPA	Y=Phe-N_2H_3	VII	X=Tyr	Y=NH-NH-Ph(NO_2)
IV	X=DOPA	Y=Phe(NO_2)-N_2H_3	VIII	X=Tyr	Y=NH-NH-Ph$(NO_2)_2$
			IX	X=DOPA	Y=NH-NH-Ph$(NO_2)_2$

and where DOPA is dioxyphenylalanine,NH-NH-Ph is phenylhydrazine, NH-NH-Ph(NO_2) is nitrophenyl hydrazine and NH-NH-Ph$(NO_2)_2$ is dinitrophenyl hydrazine.

Results

Analogs (I-IV) were synthesized by solid-phase method on a polymer support, Sephadex LH-20 (1). The addition of the C-terminated amino acid and the reactions of acylation and deblocking were carried out by the published methods (2,3). Tret-butyloxycarbonyl group was used for the protection of N-amino groups and hydroxyl groups of dioxyphenylalanine were protected by acetyl groups(4). Aminolysis was carried out by simultaneous passing of gaseous ammonia and an inert gas through a suspension of a peptidyl polymer in methanol for two and four hours. The yield of the crude product was 10% and 15%,respectively. The increase in aminoly-

Peptides 1986

sis time led to a complex mixture of products because of fast oxidation of the compound obtained in the course of the reaction. The hydrazinolysis of the peptidyl polymer was carried out with a 20% solution of hydrazine in methanol for seven days in an atmosphere of inert gas. As a result of hydrazinolysis and subsequent deblocking by a 1M solution of p-toluenesulphonic acid in glacial acetic acid, a chromatographically homogeneous product was obtained in 80% yield.

Scheme 1.

Stepwise synthesis of analog (IV) on a polymer support.

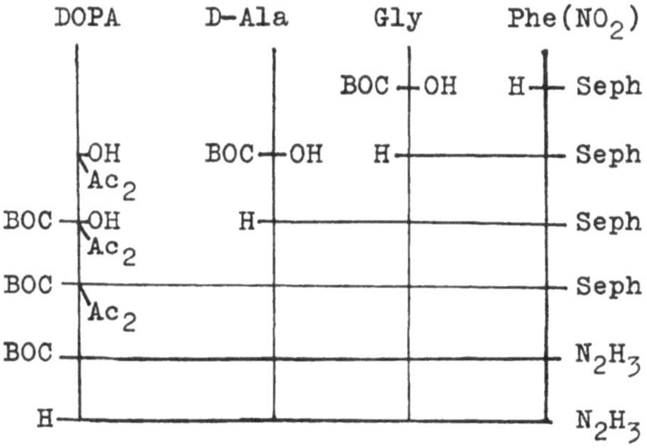

Analogs (I-III) were synthesized by the same procedure as in scheme 1. Analogs (V-IX) were synthesized in solution by using activated pentafluorophenyl esters. .

Scheme 2.

Synthesis of analog (V) in solution.

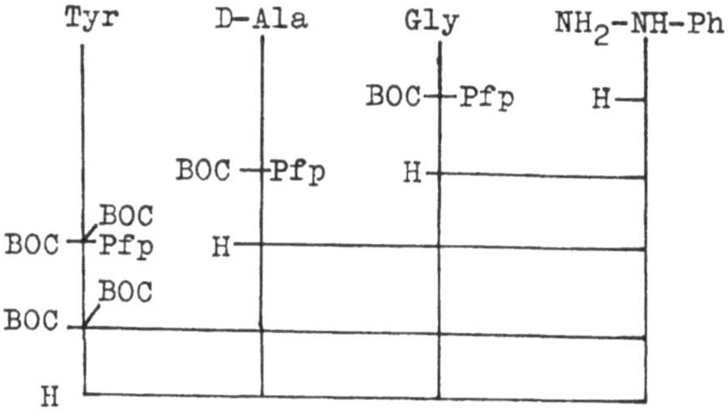

Analogs (VI-IX) were prepared by the same procedure as in scheme 2.

Table 1

Physicochemical characteristics and biological activity of enkephalin analogs

Compound	M.p. °C	$[\alpha]_D^{20}$ c=0,5 CH_3OH	R_f *	E_{Gly}	E_{His}	max of absorption, nm	[θ] of phenol chromophore $\frac{deg\ cm^2}{d\ mole}$	[θ] of phenyl chromophore $\frac{deg\ cm^2}{d\ mole}$	Blocking activity ** $EC_{50} \cdot 10^7$ mole	Analgetic activity *** $ED_{50} \frac{nmole}{mouse}$	Cataleptic activity $ED_{50} \frac{nmole}{mouse}$
I	158	+48	0,74	0,70	0,49	276	-10_{276}	-200_{257}	0,31±0,04	1,3±0,4	1,3±0,4
II	148	+12	0,67	0,74	0,44	280	-	-	4,4±2,9	-	-
III	144	+11	0,71	0,89	0,54	280	$+256_{280}$	$+136_{257}$	2,0±0,7	15,0±3,4	10,7±3,0
IV	154	+12	0,70	0,83	0,51	280	$+1856_{280}$	$+1299_{257}$	5,1±4,9	17,0±3,9	14,2±3,3
V	160	+19	0,70	0,80	0,49	276	-214_{276}	$+172_{257}$	1,7±1,0	20,2±4,1	17,4±3,4
VI	141	+28	0,64	0,64	0,38	280	-643_{280}	$+385_{257}$	77±76	-	-
VII	175	-	0,71	0,80	0,52	276	-	-	21±19	25	25
VIII	150	+15	0,72	0,78	0,56	265 341	-181_{276}	-605_{265}	200	25	25
IX	141	+28	0,67	0,79	0,53	280	-	-	200	-	-

Notes: * values of R_f are given for an n-butanol-acetic acid-pyridine-water mixture (4:1:1:2);

** EC_{50} - effective concentration of a compound leading to a 50% decrease in the amplitude of contractions of a guinea-pig intestine;

*** ED_{50} - dose of compounds leading to an analgetic (cataleptic) effect for 50% of animals upon central introduction.

Dash implies that the compound was not used in this test.

In the synthesis carried out in solution in the case of DOPA-containing analogs the deblocking of hydroxyl groups had been carried out before the α-amino protective group was removed by treatment with a hundredfold excess of hydrazine in methanol in an atmosphere of inert gas for seven days. The physico-chemical constants and biological activity of analogs are given in Table 1. The biological activity of the compounds was studied "in vitro" tests from the determination of the affinity for the peripheral opiate receptors of the μ-type of an isolated segment of the iliac intestine of a guinea-pig and "in vivo" from the capacity of inducing analgesia and catalepsy in mice with central introduction. When the chemical structure of the compounds obtained is correlated with their biological activity, attention is drawn to the fact that the introduction of DOPA instead of Tyr in the first position of 2-D-Ala,5 des-Met-enkephalin hydrazide decreases the biological activity of analogs (III<I, VI<V, IX<VIII). The introduction of the nitro group into phenylalanine of DOPA-containing analog(IV) resulted in an increase in affinity for the μ-receptors by a factor of 40. The replacement of phenylalanine in position four by phenyl hydrazine(V) led to a decrease in activity in the case of tyrosine-containing analog (V I), whereas in the case of dioxyphenyl-containing compound (VI) this modification had little influence upon the biological activity. The introduction of the nitro group into the para-position of the aromatic nucleus of phenyl hydrazine (VII) drastically decreases the activity of this compound and the introduction of two nitro groups results in its complete disappearance (analogs VIII and IX).

The effect of modification of positions 1 and 4 of tetrapeptide hydrazide (I) on the steric structure of the compounds was investigated by the method of circular dichroism. In compounds(II-IV) a strong mutual influence of phenol and phenyl chromophores on their rotational ability was detected when one of them was subjected to modification. This phenomenon may be interpreted by the compact character of the steric structure of analogs as a result of the proximity of the C-and N- ends of their molecules. In compounds (V-IX) a similar mutual influence was not observed. The replacement of phenylalanine in position 4 by an aromatic hydrazine presumably results in a more unfolded state of the molecule.

References

1. Authors certificate 250158 USSR (Vlasov G.P. and Bilibin A.Y) Bull.Izobret.,1969,N 26,p.24.

2. Bilibin A.Y. and Vlasov G.P., Zhur.Odshch.Khim., 1973,v.43, N 8,p.1844-1848.

3. Kozhevnikova N.Yu. and Vlasov G.P., Zhur. Obshch.Khim.,1979, N 4,p.924-936.

4. Harwood H.,Bassid H, J.Am.Chem. Soc., 1957,v 79, p. 4360-4365.

SYNTHESIS AND PRELIMINARY BIOLOGICAL RESULTS OF DIMERIC DERMORPHIN-LIKE PEPTIDES.

R. de Castiglione, G. Perseo, R. Mena

Farmitalia Carlo Erba SpA, Chemical Research & Development
Via dei Gracchi 35, 20146 Milan - Italy

G. Falconieri Erspamer, L. Negri

Institutes of Medical Pharmacology III and of Pharmacology & Pharmacognosy, University of Rome "La Sapienza", 00100 Rome, Italy

Introduction

Dermorphin (H-Tyr-D-Ala-Phe-Gly-Tyr-Pro-Ser-NH_2) is a natural heptapeptide with marked affinity and selectivity for the μ-type opioid receptor (1). A shift towards enhanced selectivity for the δ-type opioid receptor, accompanied by either an increase or decrease in affinity, has been observed in a few synthetic analogues (2).
In the enkephalin series, pentapeptide dimers of formula

H-Tyr-D-Ala-Gly-Phe-Leu-NH \
 $(CH_2)_n$ n= 2,4,6,8,10,12
H-tyr-D-Ala-Gly-Phe-Leu-NH /

have been reported to display enhanced affinity and selectivity for the δ-type opioid receptor (particularly for the shortest connecting chains) (3), whereas dimers of formula

H-Tyr-X-Gly-Phe-NH \
 $(CH_2)_n$ X= D-Ala, D-Thr; n= 0,3
H-Tyr-X-Gly-Phe-NH /

appeared to display increased affinity (n= 0,3) and selectivity (n= 0) for the μ-type receptor (4, 5). Tripeptide dimers of formula

H-Tyr-X-Phe-NH
 |
H-Tyr-X-Phe-NH X= D-Ala, D-Thr, Pro

showed, on the contrary, a general decrease of potency, particularly for the δ-receptor sites (4).
These observations prompted us to try similar modifications in the dermorphin sequence. Accordingly, a series of dimers (n= 0,2,4,12) of dermorphin tri-, tetra-, penta- and heptapeptides, either unsubstituted or substituted in position 2 (D-Arg) and/or in position 4

Peptides 1986

(D-Ala, Sar), have been synthesized and tested.

Results

The connecting bridge $-NH(CH_2)_nNH-$ was introduced essentially either by coupling 2 equivalents of an N-protected amino acid or peptide with a polyethylenediamine via DCC/HOBt ($n \neq 0$), or by condensing the mixed anhydride (M.A.) of an N-protected amino acid or peptide with the corresponding hydrazide ($n = 0$).
Sequence elongation was usually carried out by the M.A. method.
Side-chain functions of the Arg^2 and Tyr^5 residues were protected by the NO_2 and Bzl groups, respectively.
Dermorphin dimers were tested in the guinea pig ileum (GPI) and mouse vas deferens (MVD) bioassays (assumed to be representative of the μ- and δ-type opioid receptor interactions, respectively), and in the hot plate and tail-flick tests for analgesia in the rat by intracerebroventricular (i.c.v.) administration.
The results (expressed as molar potencies relative to dermorphin) are reported in the Table. With the exception of the heptapeptide dimer (No 13), which is as potent as dermorphin in the two *in vitro* bioassays, all other derivatives show a more or less marked increase (2-15 times) of relative selectivity for the δ-type opioid receptor. This behaviour is either the result of a higher affinity for the MVD together with a lower affinity for the GPI organ preparations (dimeric pentapeptides Nos 8-11), or of a greater decrease of potency in the GPI as compared to the MVD bioassays (dimeric tri- and tetrapeptides Nos 1-7, and pentapeptide No 12).
All derivatives, however (except No 5, with the longest connecting chain) still retain an absolute selectivity for the μ opioid receptors.
Regarding central analgesia (which should be preferentially mediated by μ-type receptors) the available data do not allow us to draw any correlation with the *in vitro* bioassays. With one exception, there is a general decrease of potency in the two antinociception tests.
Surprisingly, the dimeric pentapeptide No 12 (which is much less potent than dermorphin in vitro) is as potent as the parent com-

Table. Relative Molar Potencies of Dimeric Dermorphin-like Peptides

N.	COMPOUND	GPI	MVD	HOT PLATE (rat, i.c.v.)	TAIL-FLICK (rat, i.c.v.)
	H-Tyr-D-Ala-Phe-Gly-Tyr-Pro-Ser-NH_2 (dermorphin)	100[a]	100[b]	100[c]	100[d]
1	H-Tyr-D-Ala-Phe-NH H-Tyr-D-Ala-Phe-NH	0.4	2.5	<0.1	<0.1
2	H-Tyr-D-Ala-Phe-Gly-NH H-Tyr-D-Ala-Phe-Gly-NH	1.5	3.0	<0.1	<0.1
3	H-Tyr-D-Ala-Phe-Gly-NH \ $(CH_2)_2$ H-Tyr-D-Ala-Phe-Gly-NH /	1.5	10	≤0.1	<0.1
4	H-Tyr-D-Ala-Phe-Gly-NH \ $(CH_2)_4$ H-Tyr-D-Ala-Phe-Gly-NH /	1.6	4.3	≤0.1	≤0.1
5	H-Tyr-D-Ala-Phe-Gly-NH \ $(CH_2)_{12}$ H-Tyr-D-Ala-Phe-Gly-NH /	0.3	4.5	25	0.3
6	H-Tyr-D-Ala-Phe-Sar-NH \ $(CH_2)_2$ H-Tyr-D-Ala-Phe-Sar-NH /	4	11.5	-	2
7	H-Tyr-D-Arg-Phe-Sar-NH \ $(CH_2)_2$ H-Tyr-D-Arg-Phe-Ser-NH /	5.2	10.5	-	25
8	H-Tyr-D-Ala-Phe-Gly-Tyr-NH H-Tyr-D-Ala-Phe-Gly-Tyr-NH	50	168	5	0.8
9	H-Tyr-D-Ala-Phe-Gly-Tyr-NH \ $(CH_2)_2$ H-Tyr-D-Ala-Phe-Gly-Tyr-NH /	46	137	3	2
10	H-Tyr-D-Ala-Phe-D-Ala-Tyr-NH H-Tyr-D-Ala-Phe-D-Ala-Tyr-NH	23	130	-	-
11	H-Tyr-D-Ala-Phe-Sar-Tyr-NH H-Tyr-D-Ala-Phe-Sar-Tyr-NH	82	230	-	4
12	H-Tyr-D-Arg-Phe-Sar-Tyr-NH H-Tyr-D-Arg-Phe-Sar-Tyr-NH	8.5	20.5	-	100
13	H-Tyr-D-Ala-Phe-Gly-Tyr-Pro-Ser-NH H-Tyr-D-Ala-Phe-Gly-Tyr-Pro-Ser-NH	100	100	20	15

a) IC_{50} 3.30±0.22 nM; b) IC_{50} 29±3.35 nM; c) ED_{50} 13.3 (8.3-40) pmol/rat;

d) ED_{50} 23 (15.3-40) pmol/rat; -, not tested.

pound in the tail-flick test for analgesia.

References

1. de Castiglione, R., A.C. Rossi. 1985. In: VIIIth International Symposium on Medicinal Chemistry (R. Dahlbom and J.L.G. Nilsson, eds.). Swedish Pharmaceutical Press, Stockholm, Vol. 1, pp. 147--159.

2. de Castiglione, R. 1984. In: Highlights in Receptor Chemistry (C. Melchiorre and M. Gianella eds.). Elsevier Science Publishers B.V., pp. 149-168.

3. Shimohigashi, Y., T. Costa, S. Matsuura, H.-C. Chen, D. Rodbard. 1982. Molecular Pharmacol. 21, 558-563.

4. Lipowski, A.W., M. Konopka, B. Osipiak, W.S. Gomulka. 1983. In: Peptides 1982 (K. Blaha and P. Malon eds.). W. de Gruyter & Co., Berlin, pp. 481-486.

5. Lipowski, A.W., A.M. Konecka, J. Sroczynska. 1982. Peptides 3, 697-700.

ENKEPHALIN ANALOGUES CONTAINING POLAR AMINO ACID RESIDUES

G. W. Hardy and L. A. Lowe,
Department of Medicinal Chemistry,

T. W. Smith,
Department of Pharmacology,

The Wellcome Research Laboratories,
Langley Court, Beckenham,
Kent, U. K.

INTRODUCTION

Since the discovery of the endogenous opioid peptides, a vast number of analogues have been prepared in the search for novel analgesic agents. As part of a study of structure-activity relationships we have developed several series of modified enkephalin peptides. In particular, high opiate activity was found in analogues bearing a D-Met^2 residue (1). As a further development in this area we have investigated the properties of tetrapeptides in which this residue is replaced by a polar amino acid residue, namely D-Arg or D-Glu. The synthesis and biological activities of a number of such derivatives, having further modifications at the N-terminus and/or the phenylalanyl residue, are described.

METHODS

SYNTHESIS

The peptides were assembled using classical solution methodology as illustrated for one example in the Scheme below. D-Arginine was incorporated without protection of the guanidino group; the side chain carboxyl group of D-glutamic acid was protected as the methyl ester. The final 2+2 fragment coupling was mediated by DCCI-HOBt. For D-Arg^2-peptides, extraction into citric acid, neutralisation to pH 6 and extraction into EtOAc/n-BuOH gave the crude protected tetrapeptide. Deprotection with 1M HCl-HOAc followed by ion-exchange chromatography on CM52 cellulose provided the pure peptide for testing. Peptides containing D-Glu^2(OMe) were saponified prior to removal of the Boc group.
Phenylalanines bearing substituents on the ring were either purchased or prepared by standard methods. Resolution was achieved by enzymic hydrolysis of

Peptides 1986

a suitable derivative with either hog renal acylase or carboxypeptidase A (2). The N-terminus was amidinated, in selected cases, by reaction with 1-amidino-3,5-dimethylpyrazole (3).

SCHEME

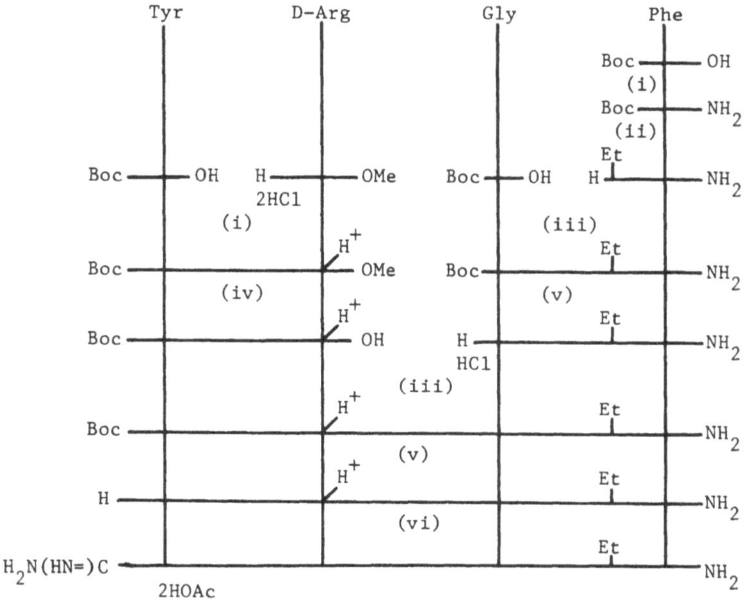

BW A368C

Reagents: (i) mixed anhydride; (ii) a: HCl-HOAc, b: EtI, $NaHCO_3$, MeCN; (iii) DCCI, HOBt, DMF; (iv) a: NaOH aq, b: HCl aq; (v) HCl-HOAc-anisole; (vi) 1-amidino-3,5-dimethylpyrazole acetate, NEt_3, EtOH, DMF.

BIOLOGICAL TESTING

Three assays were used to assess the activity of the compounds prepared:

a) Opiate activity was checked in vitro by suppression of the co-axially stimulated contractions of the Guinea-pig ileum (4).

b) Antinociceptive effects in vivo were measured in the peripheral phenylbenzoquinone-induced writhing test in mice (5). Both the subcutaneous and the oral routes of administration were investigated

c) Activity in vivo was also assessed in the hot-plate test in mice after subcutaneous dosing (6).

RESULTS

The peptides synthesised and their biological activities are summarised in the Table. A comparison of the activities of (1) and (2) shows that the former peptide has higher activity. As a result of this observation, further investigation was carried out in the D-Arg[2] series.
We next chose to investigate the influence of changing the aryl substituent on the phenylalanine residue; much previous work has demonstrated that small electron withdrawing substituents have a beneficial effect on opiate activity. The activities of (3)-(7), particularly (6) and (7), support these general observations. The 4-chloro analogue (4) has particularly high activity in the hot plate test.
The effect of varying the position of substitution was investigated with the fluoro group, (5), (8) and (9). Broadly, the results show little variation, apart from the lower activity of (9) in the Guinea-pig ileum test.
Finally we wished to increase the polarity of the molecules still further and chose to do so by N-amidination of (1), (2) and (5) to give compounds (10)-(12). As can be seen, such a modification increases the in vivo activity of the compounds and, in particular, gives a derivative (10) with potent oral activity in the P.B.Q. test.

In conclusion, we have examined structure activity relationships in a short series of enkephalin analogues containing polar amino acid residues and have discovered an analogue (10, BW A368C) with high opiate activity and which is effective after oral administration.

ACKNOWLEDGEMENTS

We are grateful to Gail Mills and Pang Yi Sang for expert assistance with the synthetic work and to Clare Shankley and Rhonda Follenfant for the biological assays.

Table. Structure and Biological Activities of the Analogues Prepared

No.	Structure[@]	G.P.I.* ED50, nM	P.B.Q.[#] ED50 mg/kg, 30min s.c.	p.o.	H.P.[+] ED50, mg/kg, 30 min s.c.
1	Tyr-D-Arg-Gly-(Et)Phe(4-NO_2).NH_2	5.2	0.06	13.9	0.52
2	Tyr-D-Glu-Gly-(Et)Phe(4-NO_2).NH_2	42.2	3.0	19.4	NT
3	Tyr-D-Arg-Gly-(Et)Phe(4-CF_3).NH_2	6.3	NT	NT	16.9
4	Tyr-D-Arg-Gly-(Et)Phe(4-Cl).NH_2	0.64	NT	NT	0.03
5	Tyr-D-Arg-Gly-(Et)Phe(4-F).NH_2	2.2	0.13	7.9	1.6
6	Tyr-D-Arg-Gly-(Et)Phe(4-SO_2Me).NH_2	>6500	NT	NT	NT
7	Tyr-D-Arg-Gly-(Et)Phe(4-NHSO_2Me).NH_2	>3200	NT	NT	NT
8	Tyr-D-Arg-Gly-(Et)Phe(2-F).NH_2	0.4	0.15	5.9	1.95
9	Tyr-D-Arg-Gly-(Et)Phe(3-F).NH_2	25.0	0.08	NT	1.7
10	Am-Tyr-D-Arg-Gly-(Et)Phe(4-NO_2).NH_2	3.9	0.005	1.9	0.34
11	Am-Tyr-D-Glu-Gly-(Et)Phe(4-NO_2).NH_2	42.0	0.54	34.0	NT
12	Am-Tyr-D-Arg-Gly-(Et)Phe(4-F).NH_2	0.09	0.04	25.6	0.9

[@] The symbol (Et)Phe is used to represent an N^{α}-ethyl-phenylalanine residue; Am = -C(=NH)NH_2

[*] Inhibition of electrically induced contractions of the Guinea-pig Ileum longitudinal muscle *in vitro*

[#] Phenylbenzoquinone induced writhing test in mice

[+] Hot plate analgesic test in mice

REFERENCES

1. Smith, T. W. and Wilkinson, S. 1982. In "The Chemical Regulation of Biological Mechanisms" (A. M. Creighton and S. Turner, eds.) R.S.C. Special Pub. No. 42, London, pp 231-254.

2. "Chemistry and Biochemistry of the Amino Acids" Ed. G. C. Barrett, Chapman & Hall, London, 1985.

3. U. K. Patent No. 2,053,925.

4. Paton, W. M. D. , 1957, Brit. J. Pharmacol., 11, 119-127.

5. Hendershot, L. C. and Forsaith, J., 1959, J. Pharmac. Exp. Ther., 125, 237.

6. Eddy, N. B. and Leimbach, D., 1953, J. Pharmacol. Exp. Ther.,107, 385.

CONFORMATIONS OF TACHYKININ PEPTIDES IN MEMBRANE ENVIRONMENTS

G. Andrew Woolley and Charles M. Deber
Research Institute, Hospital for Sick Children, Toronto M5G 1X8; and Department of Biochemistry, University of Toronto, Toronto M5S 1A8, Ontario, Canada

Introduction

Substance P - an undecapeptide neuromodulator involved in the transmission of pain information (1) - exerts its biological effects by binding to membrane-embedded protein receptors. The influence of membrane lipids on conformations of peptide ligands in the receptor recognition process has been emphasized recently by several authors (2-4). In continuing studies to define peptide/lipid interactions in detail, as well as protein/membrane interactions generally, we have been using spectroscopic and physical techniques to characterize the conformation(s) of substance P in model membrane environments. Substance P is a member of a family of peptides called tachykinins which share a common C-terminus (Phe-X-Gly-Leu-Met-NH_2) (Figure 1) and a common range of biological effects (5).

We have previously characterized the complexes formed between substance P and lipid micelles (6). Circular dichroism (CD) spectra of substance P showed that the peptide was largely unordered in aqueous media, but displayed significant induced secondary structure upon addition of sodium dodecylsulfate (SDS), lysophosphatidylglycerol (LPG), or lysophosphatidylcholine (LPC). On the basis of these CD data, as well as data from ultraviolet absorbance, pH titration, sedimentation velocity and high resolution 1-D and 2-D COSY NMR experiments, substance P was proposed to interact with the micellar membrane as a monomer, with its N-terminal residues in contact with the head group region of the membrane. As well, a

Peptides 1986

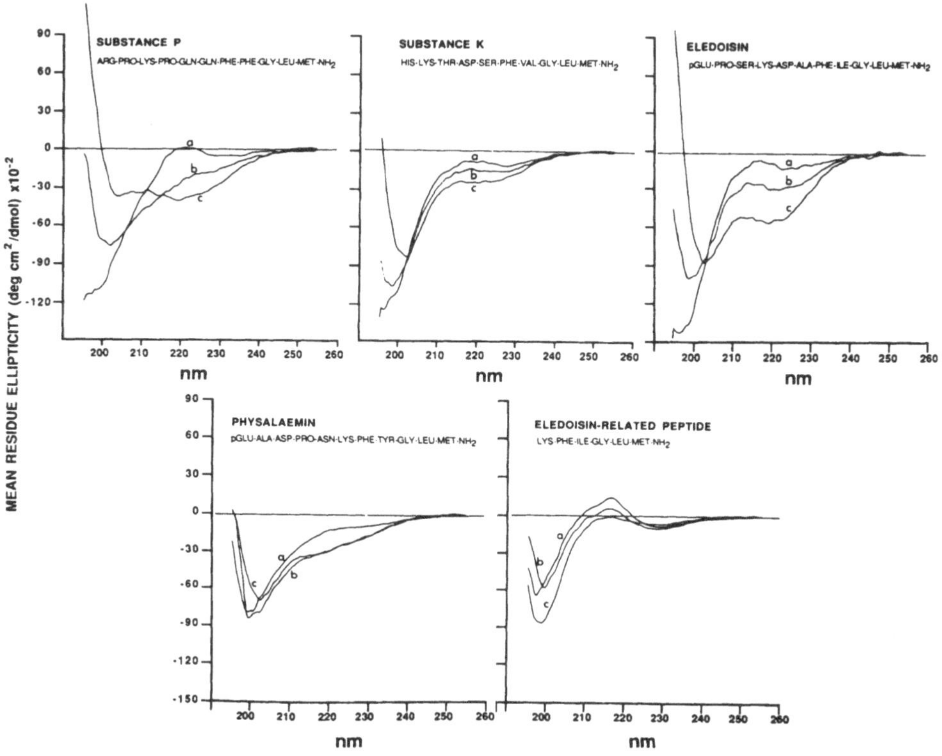

Figure 1: Circular dichroism spectra of tachykinin peptides in 14.3 mM phosphate buffer, pH 7.3. Spectra were recorded on a Jasco J-41A spectropolarimeter. In each panel, curve (a) is free peptide, curve (b) is in the presence of 18 mM lysophosphatidylcholine (LPC), and curve (c) is in the presence of 18 mM sodium dodecylsulfate (SDS). Lipids are above their critical micelle concentration in all experiments. Peptide concentrations were: substance P, 0.12 mM; substance K, 0.15 mM; eledoisin, 0.075 mM; physalaemin, 0.24 mM; and eledoisin-related peptide, 0.50 mM. Although peptide concentrations varied for technical reasons, CD spectra were generally insensitive to further additions of SDS.

partial helix is likely in the region Pro^4-Gln^5-Gln^6-Phe^7-Phe^8 with both Phe residues in contact with the apolar interior of the micelle. In the present work, we report that peptides of the tachykinin family display a sequence-dependent propensity toward membrane-induced secondary structure, a diversity which may ultimately correlate with aspects of their biological activity.

Results and Discussion

CD spectra of substance P and several related peptides are presented in Figure 1. In each instance, the free peptide in aqueous buffer [curve (a) in each panel] displays an essentially random spectrum. Addition of either excess neutral lipid LPC [curves labelled (b)] or excess negative lipid SDS [curves labelled (c)] induces secondary structure - presumably of an α-helical type - in substance P, eledoisin, and to some extent, substance K, while spectra of physalaemin and eledoisin-related peptide remain relatively unaffected. Induction of secondary structure was greater with the negatively-charged SDS (vs. equal concentrations of LPC), but this phenomenon was independent of net charge on the peptide: substance P has a charge of +3; eledoisin-related peptide +2; substance K +1; and eledoisin and physalaemin both 0.

These findings emphasize the notion that various solvents/ environments can significantly alter the conformational "ensemble" of peptides. Thus, for example, the finding of induced secondary structure for substance P both in isotropic [methanol (7)] and anisotropic media [micelles (6)] (vs. aqueous media) suggests a possible requirement for conformational flexibility of biologically-active peptides in the microenvironment of their receptor proteins. Indeed, a lipid-induced conformation may be the one actually recognized by the receptor. Preliminary studies we performed by Chou-Fasman predictive criteria (8) of the sequences of the five peptides in Figure 1 indicate that each has a stretch of 4-7 (partially or completely overlapping) residues with dual propensity for α-helix and β-sheet structure; note that peptide/receptor interactions are by nature intermolecular, as is β-structure. Indeed, some of the peptides studied herein also displayed CD spectra compatible with β-structure at intermediate lipid/peptide ratios (data not shown). However, in view of the complexity of the overall series of biological events governing activity, it remains unlikely that any one *in vitro* parameter, such as ellipticity at 222 nm in CD spectra, can totally auger peptide activity or indicate its mechanism of action. Nevertheless, the present results suggest that conformational flexibility is a basic phenomenon underlying peptide biology.

Acknowledgement

This work was supported, in part, by a grant to C.M.D. (MT-5810) from the Medical Research Council of Canada. G.A.W. holds a Connaught Foundation studentship.

References

1. Pernow, B. 1983. Pharmacological Rev. 35, 85-141.
2. Deber, C.M. and B.A. Behnam 1984. Proc. Natl. Acad. Sci. U.S.A. 81, 61-65.
3. (a) Gremlich, H., U. Fringeli and R. Schwyzer. 1984. Biochemistry 23, 1808-1810; (b) Gysin, B. and Schwyzer, R. 1984. Ibid., 1811-1818.
4. Zetta, L., A. De Marco, M. Bourdenneau and C. Brevard. 1985. In: Peptides: Structure and Function. Proc. 9th Amer. Pep. Symp. (C.M. Deber, V.J. Hruby and K.D. Kopple, eds.). Pierce Chem. Co., Rockford, ILL. pp. 890-894.
5. Erspamer, V. 1981. Trends in Neurosciences; pp. 267-269.
6. Woolley, G.A. and C.M. Deber. 1987. Biopolymers 26, 0000 (in press).
7. Chassaing, G., O. Convert and S. Lavielle. 1986. Eur. J. Biochem. 154, 77-85.
8. Chou, P.Y. and G.D. Fasman. 1978. Ann. Rev. Biochem. 47, 215-276.

CONFORMATION ACTIVITY RELATIONSHIP OF A SERIES OF PEPTIDES RELATED TO DERMORPHIN. A 500 MHz NMR STUDY.

M.A. Castiglione-Morelli, T. Tancredi, E. Trivellone
I.C.M.I.B. Del C.N.R., Arco felice, Napoli, Italy

S. Salvadori, R. Tomatis
Istituto di Chimica Farmaceutica, University of Ferrara via Scandiana, Ferrara, Italy

P.A. Temussi*
Dipartimento di Chimica, University of Naples, via Mezzocannone 4, 80134 Napoli, Italy

Introduction

The majority of studies of structure-activity relationship (SAR) of opioid peptides is based on changes of chemical constitution (1,2) that affect both the interaction of the individual side chains with parts of the receptor and the global peptide conformation. It would be desiderable to isolate these two effects by establishing direct conformation-activity relationships (**CAR**). A possible way is to use the biological activity as a sieve: in fact, the changes in biological activity are used in conventional SAR studies as a means of discriminating among favorable and unfavorable substitutions. A strategic choice of analogs may allow the selection of specific conformational features, *i.e.* it is possible to reduce the mesh of the sieve, thus trapping selected conformational features related to spectroscopic parameters.

It is the purpose of this paper to look for conformation-activity relationships (**CAR**) in a series of short dermorphin analogs in which the orientation of the ring of the third residue has been systematically altered, on the basis of a receptor model previously developed by us (3). We report a 500 MHz ^{1}H nmr study of the following peptides : Tyr-D-Ala-Phe-Gly-NH_2, the N-tetrapeptide fragment of dermorphin, (henceforth called **P1**, with an arbitrary relative activity of 100) Tyr-D-Ala-Ph-Gly-Gly-NH_2 (**P2**, with relative actvity 0.5), Tyr-D-Ala-Ph-Abu-Gly-NH_2 (**P3**, rel. activity 2.3) , Tyr-D-Ala-Cha-Gly-NH_2 (**P4**, rel. activity 36.5), Tyr-D-Ala-dehydrophe-Gly-NH_2 (**P5**, rel. activity 1.0) and Tyr-D-Ala-D-Phe-Gly-NH_2 (**P6** , rel. activity 4.0) . A further simple analog was also investigated for comparison *i.e.* Phe-D-Ala-Phe-Gly-NH_2 (**P7**, rel. activity 3.7).

Peptides 1986

Results

Assignments of all resonances of the seven peptides studied were achieved by means of 2D techniques. A combination of COSY and J-resolved experiments was sufficient in all cases to assign unambiguously the resonances of each residue . Table 1 and table 2 summarize, respectively, the chemical shifts at 300 K and the NH temperature coefficients, measured in the range 295-335 K, of the peptides in DMSO-d_6.

TABLE 1 - ^{1}H Chemical shifts (ppm from TMS) measured at 300 K in DMSO$_{d6}$

	Tyr **Phe** (P7)				**D-Ala**			**Xxx**				**Gly**	
	H_α	H_β	H_N	Ar	H_α	H_β	H_N	H_α	H_β	H_N	Ar	H_α	H_N
P1	-	2.77 2.75	-	6.98 6.64	4.22	0.90	7.97	4.49	3.09 2.79	8.34	7.25	3.68 3.62	8.26
P2	3.39	2.83 2.44	-	6.98 6.65	4.42	1.14	8.08	5.47	-	8.64	7.37	3.65	8.51
P3	3.99	2.88	-	7.00 6.69	4.40	1.12	8.49	4.22	1.89	8.30	7.26	3.65	7.97
P4	3.99	2.88	8.09	7.02 6.70	4.41	1.06	8.52	4.37	-	8.30	-	3.57	8.16
P5	3.36	2.83 2.47	-	6.99 6.65	4.28	1.28	8.26	-	7.19	9.15	7.40 7.34	3.70	8.20
P6	3.80	2.85 2.73	-	6.98 6.65	4.26	1.05	8.30	4.47	3.05 2.79	8.24	7.27	3.69 3.59	8.11
P7	3.66	2.77 2.53	-	7.24	4.23	0.90	8.02	4.50	3.10 2.91	8.38	7.24	3.37	8.02

The NH chemical shifts do not show any outstanding value, with the obvious exception of the resonance of the NH of dehydrophenylalanine in Tyr-D-Ala-dehydrophe-Gly-NH_2 that experiences the direct paramagnetic shift of the adjacent olefinic bond. All other NH shifts are very close to the corresponding values previously measured for dermorphin and its fragments and to the average literature values of small linear peptides in random conformation (4).

TABLE 2 - NH Temperature coefficients (ppb/K) measured in $DMSO_{d6}$

	Tyr Phe (P7)	D-Ala	Xxx	Gly
P1		-6.50	-5.30	-4.00
P2	-	-5.04	-8.74	-7.89
P3	-	-3.09	-3.68	-3.41
P4	-3.10	-4.20	-4.70	-5.90
P5	-	-5.78	-6.79	-5.36
P6	-	-4.40	-5.60	-6.50
P7	-	-2.67	-4.97	-4.86

The NH's temperature coefficients are spread over a quite wide range, indicating that the accessibility to the solvent is influenced by the primary structure. Most of the chemical shifts of the non-exchangeable protons have also values close to literature values for random conformations (4).The only spectral parameter of high diagnostic value is the chemical shift of the methyl group of D-Ala. It had already been noticed in dermorphin and its fragments (5) that the methyl chemical shift is unusually low with respect to the standard value for alanine (4) and the observed ring-current shift had been considered as an indication that the alanine beta protons are located between the aromatic rings of Tyr^1 and Phe^3. Such an arrangement points to the relevance of a well characterized family of conformations if not to the existence of a small number of conformers. It seems significant that Tyr-D-Ala-Phe-Gly-NH_2, in the *beta* type II conformation consistent with the proposed model receptor (3), has both the methyl group of D-Ala^2 and the ring of Phe^3 in equatorial positions, at a short enough distance to induce a large diamagnetic shift in the resonance of the methyl group of D-Ala.

In all N-fragments of dermorphin the high-field shift of the resonance of this group is paralleled by a high *mu*activity; the same correlation holds for the peptides of table1, with the obvious exception of Phe-D-Ala-Phe-Gly-NH_2. This result can be interpreted on the basis of the quoted model of *mu* receptor (3).

According to this model the activity of a *mu* opioid depends on the presence of five features : a **T** skeleton (that in the case of these peptides is assured by the *beta* turn), a hard base, a soft base, a *cis* arrangement of two aromatic rings, and the presence of a phenolic OH on the stem of the **T**.

Although in Phe-D-Ala-Phe-Gly-NH_2 the methyl group of D-Ala is between the two aromatic rings, the substitution of Tyr with Phe makes the peptide ca. 27 times less active than Tyr-D-Ala-Phe-Gly-NH_2, showing the great importance for the activity of Tyr in the first position of the peptide sequence.

In those peptides in which Phe^3 was substituted with other aromatic residues or has different configuration (Tyr-D-Ala-D-Phe-Gly-NH_2) the methyl chemical shift has "normal values", *i.e.* around 1.1 p.p.m., because the aromatic ring is too far away to influence it or is not properly aligned even when the side-chain is in the equatorial position of the C_{10} ring. An obvious, partial, exception in this respect is shown by the chemical shift of Tyr-D-Ala-Cha-Gly-NH_2. The almost normal value (1.05 p.p.m.) of the resonance of D-Ala in this compound indicates that the greater part of ring current effect observed in the other peptides can be attributed to the side chain of the third residue. On the other hand, the *mu* activity, although lower than that of the parent tetrapeptide (Tyr-D-Ala-Phe-Gly-NH_2), is of the same order of magnitude, and definitely higher than those of peptides in which the aromatic ring of the third residue is misoriented. This finding can be interpreted by saying that a a side chain of bulkiness and hydrophobicity comparable to that of Phe can still be accomodated by the receptor provided it can assume the proper orientation with respect to the other side chains. The NMR study of these analogs of dermorphin shows that a conformation activity relationship (**CAR**) can be established even for small linear peptides in a strongly solvating medium, provided one studies a carefully chosen serie of agonists. The interpretation of the nmr parameters in terms of *mu* activities confirms the relevance of folded conformations of the family of C_{10} rings, and the crucial role played, for *mu* agonists, by the side-chain of the third residue.

References

1. Hansen, P.E.,B.A. Morgan .1984. ***The Peptides* 6**, 269-315.

2. Schiller, P.W. .1984. ***The Peptides* 6**, 219-268.

3.Pastore, A.,T. Tancredi, P.A. Temussi, S.Salvadori, R. Tomatis.1985. In : ***Peptides: Structure and Function - Proceedings of the Ninth American Peptide Symposium, Toronto, Ont.*,** (Deber, C.M., Hruby, V.J. & Kopple, K.D., Eds.) Pierce Chemical Company, Rockford, Illinois, pp. 529-532.

4.Wuthrich, K. 1976. ***NMR in Biological Research: Peptides and Proteins.*** North-Holland Publ. Co., Amsterdam.

5.Pastore,A.,P.A.Temussi,T.Tancredi, S.Salvadori, R.Tomatis .1984. ***Biopolymers* 23**, 2349-2360.

SYNTHESIS OF CHOLECYSTOKININ PEPTIDES

B. Penke, M. Zarándi, J. Zsigó, G.K. Tóth, K. Kovács

Institute of Medical Chemistry, Szeged University Medical School, Szeged, H-672o Hungary

G. Telegdy

Institute of Pathophysiology, Szeged University Medical School, Szeged, H-672o Hungary

Introduction

The synthesis of Tyr-O-sulfate containing cholecystokinin (CCK) peptides involves several problems. Although many sulfating agents are known (1) and the in vivo sulfate ester (SE) formation of Tyr in peptides might be part of an important regulatory process (2), the chemical synthesis of peptide SEs has been a challenge. Selective sulfation of Tyr in the presence of other hydroxyamino acids and the acid lability of the SE group requires specific side chain protection and a unique strategy. We wanted to resolve the synthetic problems of the cholecystokinin peptides in order to a.) work out an optimal synthesis of short peptides, e.g. CCK-8, CCK-8 analogs, caerulein, etc. b.) find a synthesis strategy (either solid phase or classical solution method) for the preparation of O-sulfated CCK-33 and longer CCK-peptides c.) work out a mild and simple method for reducing of methionine-sulfoxide content of peptides formed during the synthesis, storage, etc.

Results

We have found that pyridinium-acetyl sulfate (PAS,3) is an optimal sulfating agent if tryptophan and methionine are present in the peptide. Sulfation with pyridine-SO_3 complex causes a series of side reactions (slow sulfonation of Trp and His, N-sulfonation of the free groups, etc.) PAS proved to be the optimal reagent for the introduction of sulfate ester group onto the Tyr residue,

Peptides 1986

only the side chain of serine and threonine had to be protected with acetyl group, and the free α- and ε-amino functions with Fmoc group. The guanidino function of arginine can be unprotected during sulfation. We have worked out an optimal solid phase strategy for the synthesis of short CCK-peptides. The cholecystokinin-octapeptide (CCK-8) was synthesized with Boc-amino acids as follows:

Fmoc-Asp(OcHex)-Tyr(2-Br-Z)-Met-Gly-Trp-Met-Asp-(OcHex)-Phe-MBHA

↓ HF

Fmoc-Asp-Tyr-Met-Gly-Trp-Met-Asp-Phe-NH_2

↓ 1. PAS(5 equ., 2 days in DMF-Py 1:1
2. 2o% piperidine in DMF, 1o min

H-Asp-Tyr(SO_3Na)-Met-Gly-Trp-Met-Asp-Phe-NH_2

We have optimized the solid phase synthesis of caerulein using N^{α} - Boc protection:

Z-Pyr-Gln-Asp—Tyr—Thr-Gly-Trp-Met-Asp—Phe-MBHA
Xan OcHex 2BrZ Ac OcHex

↓ HF

Pyr-Gln-Asp-Tyr-Thr(Ac)-Gly-Trp-Met-Asp-Phe-NH_2

↓ 1. PAS (5 equ.), 2 days in DMF-Py 1:1
2. Hydrolysis of the acetyl group with NaOH

Pyr-Gln-Asp-Tyr(SO_3Na)-Thr-Gly-Trp-Met-Asp-Phe-NH_2

Both the crude CCK-8 and the caerulein was purified on preparative HPLC column /Vydac C_{18},15-2oμ) using o.o5 M CH_3COONH_4 (pH 6.5) and an acetonitrile gradient. The motivation for the previous synthetic studies was to optimize conditions and strategies that would lead to a scheme amenable to yield longer CCK-peptides (CCK-33, CCK-39, etc.). We took in account all of side reactions known by working out the synthesis strategy for CCK-33. We could not find any advantages to the previous fragment condensation strategies (4,5), therefore we used the stepwise building-up of the whole CCK-33 on solid phase. The problems of the synthesis can be summarized as follows:

1. The present status of the Fmoc-strategy does not allow the

synthesis of CCK-33 or longer cholecystokinins. During the acidolytic cleavage of the N^G -Mtr group of arginine complete desulfation of Tyr-O-sulfate occurs.

2. We have not found any selective chemical sulfation agent for the tyrosine if other hydroxyamino acids (Ser, Thr) are present in the peptide molecule. Therefore, the alcoholic hydroxyl groups must be protected (e.g. with O-acetyl groups) during sulfation.

3. The CCK-33 peptide chain is labile in HF. Both under low-high and high HF-conditions, the acid labile Asp-Pro bond was partially cleaved.

Despite these problems we succeeded in synthesizing sulfated CCK-33 in solid phase using the Boc-strategy. The synthesis was performed on methylbenzhydrylamino polymer using the following side-chain protecting groups: - OcHex for Asp, acetyl for Ser, Fmoc for Lys, Tos for Arg and His, as well as 2-Br-Z for Tyr. Boc-asparagine and glutamine were coupled simply with the DCC/HOBt method. The partially protected CCK-33 was cleaved from the resin with a special HF procedure and successfully sulfated with pyridinium-acetyl sulfate.

It is well known that the methionine residues in CCK-peptides are very sensitive to oxidation. Even under mild conditions (e.g. isolation, storage in solution) the methionine may be oxidized to Met(O) giving a mixture of side-products. There are a series of reducing agents but their preparative use is limited. The most frequently used mercaptoacetic acid method suffers of two disadvantages: the reagent has low reactivity and can acylate the reactive side chain groups. N-methylamido-mercaptoacetic acid (MMA) was reported to be an excellent reducing agent in analytical experiments (6). We have found that in nonaqueous medium in the presence of acetic acid MMA forms an active ester (CH_3-CO-CH_2-CO--NH-CH_3) and causes acetylation as side reaction during the Met(O) $\longrightarrow$ Met reduction. If water is present, no acetylation occurs. On the basis of these results, we have worked out a preparative method which enables the large scale reduction of the methionine-sulfoxide content to Met with the MMA-reagent without side reactions. The free or protected peptide was dissolved in a mixture of DMF-H_2O 1:1 or CH_3OH-H_2O (1:1) containing 25-33% MMA and reacted at 2o ^oC. After 3 days the reduction was complete.

This method was successfully used for the reduction of partially oxidized CCK-8, tetragastrin and ACTH 1 - 24 and proved to be very efficient also in large scale. It can be used also for the reduction of water-insoluble peptides.

References

1. Houben-Weyl " Methoden der Organischen Chemie", 1963, VI/2, Thieme p. 452.

2. Unsworth, C.C., F. Hughes, J.S. Morley 1982. Nature 295,519.

3. Penke,B., F.Hajnal, J. Lonovics, G. Holzinger, T. Kadar, G. Telegdy, J. Rivier 1984. J. Med. Chem. 27, 845.

4. Bodanszky, M., N.Chaturvedi, D. Hudson, M. Itoh, 1972. J. Org. Chem. 37, 23o3.

5. Mori,Y., H. Yajima 1976, Chem. Pharm. Bull (Japan) 24, 2781.

6. Houghten, A., C.H. Li 1979. Anal. Biochem. 98, 36.

CYCLIC HEXAPEPTIDE CCK ANTAGONISTS

R. M. Freidinger, W. L. Whitter, R. S. L. Chang, V. J. Lotti and D. F. Veber
Merck Sharp & Dohme Research Laboratories, West Point, Pennsylvania 19486

The development of orally effective long acting CCK antagonists which bind to receptors with affinity as great as CCK itself has been recently reported from our laboratory. These 3-substituted benzodiazepines characterized by structure I were developed using asperlicin (II) as a lead.[1,2] Other highly rigid, related structures have also been found to be receptor antagonists of CCK. X-ray crystal structure data is available for some of these, offering a unique opportunity for retrospective evaluation of possible relationships of these structures to CCK. A successful evaluation might offer insight into the process of design of analogs of other important peptides. In general, the large choice of possible conformations available to a peptide such as CCK results in multiple solutions and ambiguous conclusions. If hypotheses can be tested through design of peptidal conformationally constrained CCK antagonists, then it may prove possible to overcome the ambiguities.

Hypothetical problems relating to the possibility that the non-peptide antagonist binds at a different site overlapping that of the peptide agonist[3] may be overcome in this instance due to the existence of a peptide antagonist CBZ-Tyr(SO_3H)-Met-Gly-Trp-Met-Asp-NH_2.[4] Thus modeling studies can ultimately involve evaluation also of the conformational relationship of the peptide agonist and antagonist.

The possibility of a turn in the solution conformation of CCK was first proposed on the basis of fluorescence studies of CCK-7.[5] More recently, two proton NMR studies have reported evidence for folded conformations of CCK-8[6] and CCK-7[7] in solution. The latter authors have postulated a specific β-turn involving Gly-Trp-Met-Asp and a Met-Gly-Trp γ-turn in the solution conformation. The high degree of flexibility normally associated with short linear peptide sequences makes this type of conclusion rather tenuous and in any case may bear no relation to the bioactive conformation. The possibility of a turn in the bioactive conformation of CCK is supported by the reported high activity of CCK-7 analogs with a D-Ala substitution for Gly.[8,9] Such modification tends to stabilize turns.

We have found that replacement of L-Trp by D-Trp in the antagonist structure III gives an antagonist (IV) of similar potency (see table) also consistent with a turn in the bioactive conformation of antagonists.

Peptides 1986

This positive result with a partially constrained linear sequence encouraged us to prepare more constrained, cyclic hexapeptide sequences in which turn conformations must exist. The analogs were designed to achieve different types of β-turns at differing points in the active sequence. A central importance of the indole of tryptophan has been assumed based on reported structure-activity studies[10] as well as the importance of the indole nucleus in the nonpeptidal antagonists.[1] Tyrosine sulfate has been deemed less important since we have found that the dessulfates of III and IV have receptor affinity comparable to the parent compounds.

The cyclic hexapeptides of Table I were prepared using standard methodology with cyclization via diphenyl phosphoryl azide.[11] Purification was accomplished by silica gel chromatography and/or preparative reverse phase HPLC. All analogs are >96% pure and were characterized by amino acid analysis and NMR spectroscopy and gave the correct molecular ion by FAB mass spectroscopy.

Table I contains data for binding of peptides III-VII to rat pancreas and guinea pig brain CCK receptors. Most of the analogs show a modest preference for the former receptor. Compounds V and VII behave as pure antagonists analogous to I and in contrast to III showing no change in pancreas CCK receptor affinity in the presence of guanyl nucleotides or in the absence of $MgCl_2$.[12] It is significant that cyclic analogs V-VII are comparable in potency to linear peptides III and IV, thus further supporting a turn in the bioactive antagonist conformation. The turn suggested by these results would have Gly and Trp at positions i+1 and i+2, and thus differs from the solution conformation proposed for CCK-8.[6] NMR studies indicate multiple conformations in solution for these active cyclic peptides casting further doubts on the validity of conclusions related to the linear sequence. Although the analogs prepared so far do not give a definitive picture of the active conformation, they reduce the number of variables to be considered in modeling the relationship of non-peptide antagonists such as I with peptidal CCK antagonists.

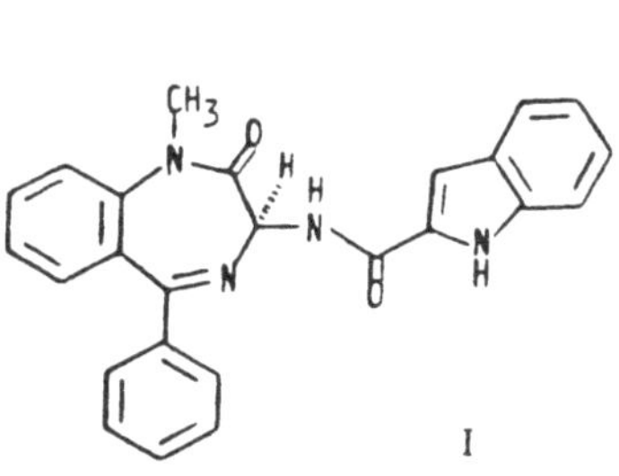

I

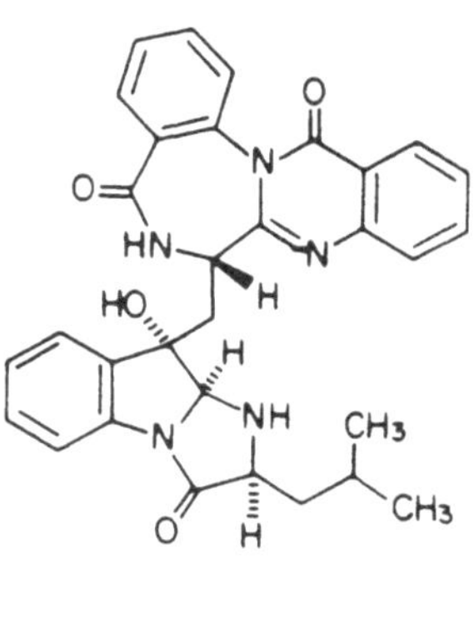

II

TABLE I

		IC50 (M), ^{125}I-CCK Pancreas	 Brain
SO_3H | Cbz-Tyr-Met-Gly-Trp-Met-Asp-NH_2	**III**	3.3	7.6
III DesSulfate		16	13.3
SO_3H | Cbz-Tyr-Met-Gly-D-Trp-Met-Asp-NH_2	**IV**	14	17.5
IV DesSulfate		13	17
Phe-Met-Gly | | Pro-Met-Trp	**V**	29	14
Phe-Met-D-Ala | | Pro-Met-Trp	**VI**	30	11
Phe-Met---Gly | | Pro-Met-D-Trp	**VII**	5.3	17

REFERENCES

1. B.E. Evans, M.G. Bock, K.E. Rittle, R.M. DiPardo, W.L. Whitter, D.F. Veber, P.S. Anderson and R. M. Freidinger. 1986. Proc. Nat. Acad. Sci. 83, 4918-4922.

2. R.S.L. Chang and V.J. Lotti. 1986. Proc. Nat. Acad. Sci. 83, 4923-4926.

3. E.J. Ariens, A.J. Beld, J.F. Rodrigues de Miranda and A.M. Simonis in "The Receptors" Vol. 1 (R.D. O'Brien, Ed.) Plenum Press, New York, p.33.

4. M. Spanarkel, J. Martinez, C. Briet, R.T. Jensen, and J.D. Gardner. 1983. J. Biol. Chem. 258, 6746-6749.

5. P.W. Schiller, S. Natarajan and M. Bodanszky. 1978. Int. J. Pept. Prot. Res. 12, 139-142.

6. I. Koizuka, H. Watari, N. Yanaihara, Y. Nishina, K. Shiga and K. Nagayama. 1984. Biomed. Res. 5(Suppl.), 161-168.

7. a) C. Durieux, J. Belleney, J.Y. Lallemand, B.P. Roques, and M.C. Fournie-Zaluski. 1983. Biochem. Biophys. Res. Comm. 114, 705-712; b) M.C. Fournie-Zaluski, J. Belleney, B. Lux, C. Durieux, D. Gerard, G. Gacel, B. Maigret and B.P. Roques. 1986. Biochem. 25, 3778-3787.

8. B. Penke, F. Hajnol, J. Lonovias, G. Holzinger, T. Kadar, G. Telegdy and J. Rivier. 1984. J. Med. Chem. 27, 845-849.

9. C. Yanaihara, N. Suguira, K. Kashimoto, M. Kondo, M. Kawamura, S. Naruse, A. Yasui and N. Yanaihara. 1985. Biomed. Res. 6, 111-115.

10. H. Adachi, H.M. Rajh, G.I. Tesser, J.J.H.H.M. DePont, R.T. Jensen and J.D. Gardner. 1981. Biochem. Biophys. Acta. 678, 358-363.

11. P.N. Maton, R.T. Jensen and J.D. Gardner. 1986. Horm. Metabol. Res., 18, 2-9.

12. S.F. Brady, W.J. Paleveda, B.H. Arison, R.M. Freidinger, R.F. Nutt and D.F. Veber. 1983. In: "PEPTIDES: Structure and Function", Proceedings of the 8th American Peptide Symposium, (V. Hruby and D. Rich, Eds.) pp. 127-130.

13. R.S.L. Chang, V.J. Lotti, T.B. Chen and K.A. Kunkel. Mol. Pharmacol. (in press).

SYNTHESIS OF PSEUDO-PEPTIDE AND PEPTIDE ANALOGUES OF THE C-TERMINAL HEPTAPEPTIDE OF CHOLECYSTOKININ AND PRELIMINARY RESULTS ON THEIR IN VITRO ACTIVITY ON RAT PANCREATIC ACINI

M. Rodriguez, M.F. Lignon, M.C. Galas[§], C. Mendre, & J. Martinez[*]

Centre CNRS-INSERM de Pharmacologie-Endocrinologie, Rue de la Cardonille, 34094, Montpellier, France.

Introduction

Gastrin and cholecystokinin have in their sequence the same C-terminal pentapeptide : -Gly-Trp-Met-Asp-Phe-NH_2. Our recent results on gastrin-related peptides allowed us to point out the significance of the peptide bonds of the C-terminal tetrapeptide, particularly the bond between Met and Asp, for the biological gastrin-like activity on acid secretion (1). The importance of the phenylalanyl C-terminal residue was also demonstrated, and its deletion or modification led to gastrin receptor antagonists (2). According to these results and because gastrin and cholecystokinin overlap in their activities, we investigated the role of the peptide bonds of the CCK-7 C-terminal tetrapeptide and of the replacement of the phenylalanyl C-terminal residue.

Results

The syntheses of the pseudo-peptides Z-Tyr(SO_3^-)-Met-Gly-Trpψ(CH_2NH) Nle-Asp-Phe-NH_2 1, Z-Tyr(SO_3^-)-Met-Gly-Trp-Nleψ(CH_2NH)Asp-Phe-NH_2 2, Z-Tyr(SO_3^-)-Met-Gly-Trp-Nle-Aspψ(CH_2NH)Phe-NH_2 3, were carried out by the conventional technique in solution. The CH_2NH bond was prepared from the N-O-dimethyl hydroxamate of the N-protected aminoacid or peptide derivatives, which were converted into the corresponding aldehydes in the presence of $AlLiH_4$ (3). After reaction with the trifluoroacetate salt of the aminoacid or peptide derivatives in a methanol/acetic acid mixture in the presence of $NaBH_4CN$ (4) as reducing reagent, they yielded after further coupling and deprotection the corresponding pseudo-peptide compounds 1 - 3. In these syntheses, Z-Tyr(SO_3^-) hemi-barium salt was introduced in the form of its N-hydroxysuccinimide ester obtained in situ. The peptide derivatives Boc-Tyr(SO_3^-)-Met-Gly-Trp-Nle-Asp-α-2-Phenylethyl Ester 4, and Boc-Tyr(SO_3^-)-Met-Gly-Trp-Leu-Asp-α-2-Phenylethyl Ester 5 were prepared stepwise by the conventional technique of peptide synthesis in solution starting from the α-2-Phenylethyl Ester of Boc-Asp(Bzl) (2). Peptides 1 - 5 were purified by HPLC on a Beckmann Ultrasphere ODS (5µm) column using a mixture of ammonium acetate 0.01M (pH 6.5) / methanol (62:38) and identified by aminoacid analysis, elemental analysis (C,H,N), and by their 360MHz ^1H NMR spectroscopy.

Compounds 1 - 5 were tested for their ability to stimulate amylase release from isolated rat pancreatic acini and to inhibit the binding of ^{125}I-BH-CCK-9 (5) to rat pancreatic acini. Each pseudo-peptide tested had a biphasic dose-response curve for stimulating amylase release, in that, as the pseudo-peptide concentration increased, amylase release increased, became maximal and then decreased by approximately 50% with supramaximal concentrations. Replacement of the peptide bond between Nle and Asp by a CH_2NH bond caused a 20-fold reduction in potency, whereas replacement of the bond between Trp and Nle or Asp and Phe-NH_2 by a CH_2NH bond caused respectively a 100-fold and a 1000-fold reduction (Table 1).

Peptide Derivatives	Potency (M)	IC50 (M)
Boc-[Nle^{28}, Nle^{31}]CCK-8	$3 \cdot 10^{-10}$	$3 \cdot 10^{-9}$
1	$3 \cdot 10^{-8}$	$2 \cdot 10^{-7}$
2	$5 \cdot 10^{-9}$	$3 \cdot 10^{-8}$
3	$3 \cdot 10^{-7}$	$3 \cdot 10^{-7}$
4	10^{-7}	$1.5 \cdot 10^{-8}$
5	10^{-7}	$5 \cdot 10^{-8}$

Table 1 : Ability of pseudo-peptide and peptide derivatives 1 - 5 to stimulate amylase release and inhibition of binding of ^{125}I-BH-CCK-9 to isolated rat pancreatic acini. In each experiment, each value was determined in duplicate, and results given are means of at least 5 separate experiments.

The peptide derivatives 4 and 5 stimulated amylase release from isolated rat pancreatic acini (Table 1), but we did not observe a decrease of amylase release at supramaximal concentrations (Fig. 1). To examine directly the interaction of Boc-[Nle[28], Nle[31]]-CCK-8 and its derivatives with CCK-receptors, we measured their ability to inhibit binding of ^{125}I-BH-CCK-9 to rat pancreatic acini. Each tested derivative was able to inhibit binding of ^{125}I-BH-CCK-9, compounds 2, 4 and 5 being the most potent. There was a close correlation between the ability of these derivatives to inhibit the binding of ^{125}I-CCK-9 and to the derivative's ability to stimulate enzyme secretion.

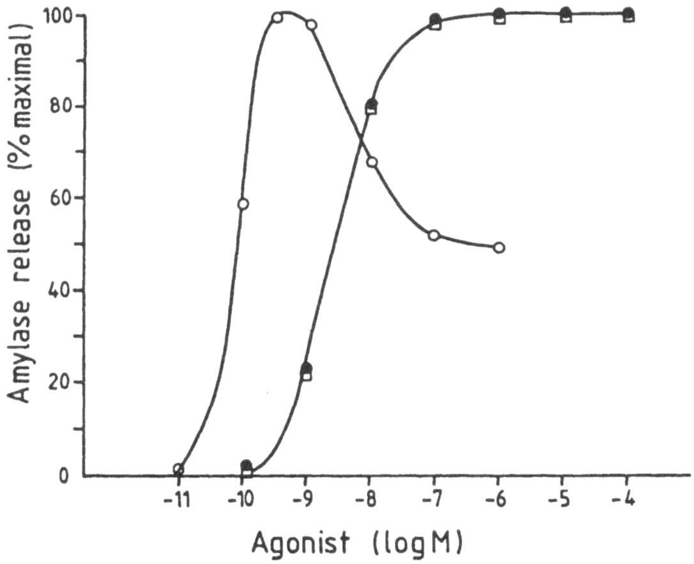

Fig 1.: Stimulation of amylase release from rat pancreatic acini caused by Boc-[Nle[28], Nle[31]]CCK-8 (○) and peptide derivatives 4 (●) and 5 (□). Amylase release was determined during a 30 min. incubation at 37°C. In each experiment, each value was determined in duplicate and results given are means of 6 separate experiments.

Conclusion

From this work, it appears that the peptide bonds of the C-terminal tetrapeptide of CCK-7 are not crucial, neither for the pancreozymin activity, nor for the binding to CCK-receptors from rat isolated pancreatic acini. In contrast to the results observed for gastrin (1). The pseudo-peptides 1 - 3 are able to stimulate amylase release from pancreatic acini and in terms of the concentration that caused maximal stimulation, the relative potencies were Z-Tyr(SO_3^-)-Met-Gly-Trp-Nleψ(CH_2NH)Asp-Phe-NH_2 2 > Z-Tyr(SO_3^-)-Met-Gly-Trp-ψ(CH_2NH)Nle-Asp-Phe-NH_2 1 > Z-Tyr(SO_3^-)-Met-Gly-Trp-Nle-Aspψ(CH_2NH)Phe-NH_2 3. The in vitro results of these pseudo-peptides in inhibiting the binding of ^{125}I-BH-CCK-9 to pancreatic acini occur over the same concentration range, suggesting that they act at the CCK receptor. Peptides 4 and 5 are very potent in inhibiting the binding of ^{125}I-BH-CCK-9 to pancreatic acini (Table 1) and interestingly behave as partial CCK agonists. They exhibit the same maximal response on amylase release from pancreatic acini but with no decrease of this response at supramaximal concentrations (Fig 1). These derivatives may represent interesting "tools" for pharmacological evaluation of CCK action.

Acknowledgement

§ M.C.G. is a recipient of ARI "Chimie Biologie" (F2RCB) which is gratefully acknowledged.

References

1. Martinez, J., J.P. Bali, R. Magous, J. Laur, M.F. Lignon, M. Rodriguez & B. Castro. 1985. C.R. Acad. Sci. Paris, 300, 437.

2. Martinez, J., M. Rodriguez, M., J.P. Bali & J. Laur. 1986. J. Med. Chem. (in press).

3. Fehrentz, J.A. & B. Castro. 1983. Synthesis, 676.

4. Jacobson, K.A., D. Marr-Leisy, R.P. Rosenkranz, M. Verlander, K.L. Melmon & M. Goodman. 1983. J. Med. Chem. 26, 492.

5. ^{125}I-BH-CCK-9 was kindly provided by L. Pradayrolle and D. Fourmi, Toulouse, France and had the same binding characteristic on pancreatic acini as BH-CCK-8 from Amersham.

SURPRISING FINDINGS WITH SOME DELETION-MODIFIED SELECTIVE V_2 ANTAGONISTS OF ARGININE VASOPRESSIN (AVP)

In memory of I. Photaki

M. Manning, J. Przybylski, B. Lammek, Z. Grzonka, A. Olma, K. Bankowski, A. Misicka, M. Kruszynski and W.H. Sawyer

Medical College of Ohio, Toledo, Ohio and Columbia University, New York, NY, USA

Introduction

Tail deletions and substitutions in AVP antagonists have hitherto yielded highly consistent effects (1-5). Deletion or substitution of the C-terminal glycine or glycinamide from potent and non-selective antagonists of the antidiuretic (V_2) and the vascular (V_1) responses to AVP led in most instances to full or almost full retention of antagonistic potencies (2, 3). While exploring the effects of deleting the proline residue from AVP V_2 antagonists (1), we encountered surprisingly inconsistent findings which are reported here. We also report on the effects of tail deletion and substitution modifications in our most potent V_1 antagonist [1-(β-mercapto-β,β-cyclopentamethylenepropionic acid), 2-O-methyltyrosine]arginine vasopressin ($d(CH_2)_5[Tyr(Me)^2]AVP$) (6). All analogs were synthesized by the solid phase method (7). Their properties are given in Table 1 - 4.

TABLE 1: Deletions or Substitutions in the C-terminal Position in the Potent AVP (V_1) Antagonist $d(CH_2)_5Tyr(Me)AVP$ Lead to Potent and Selective Antagonists

	Antivasopressor		Antidiuretic	
	ED[a]	pA_2[b]	Activity (u/mg)	pA_2[b]
$d(CH_2)_5[Tyr(Me)^2]AVP$[c]	0.16 ± 0.01	8.62 ± 0.03	0.31 ± 0.07	
1) $desGly^9d(CH_2)_5[Tyr(Me)^2]AVP$[d]	0.27 ± 0.06	8.44 ± 0.10	0.010 (mixed)	~6.1
2) $desGly(NH_2)^9d(CH_2)_5[Tyr(Me)^2]AVP$[d]	0.23 ± 0.02	8.46 ± 0.05	0.0005 (mixed)	~5.4
3) $desGly(NH_2)^9d(CH_2)_5[Tyr(Me)^2ArgNHMe^8]VP$[d]	0.12 ± 0.01	8.73 ± 0.02	~0.002 (mixed)	~5.9
4) $desGly(NH_2)^9d(CH_2)_5[Tyr(Me)^2ArgEDA^8]VP$[d]	0.12 ± 0.02	8.75 ± 0.06	0.004 (mixed)	~6.3
5) $d(CH_2)_5[Tyr(Me)^2Ala(NH_2)^9]AVP$[d]	0.13 ± 0.03	8.74 ± 0.08	~0.002 (mixed)	~6.0

[a]The effective dose (ED) is the dose (in nanomoles/Kg) of antagonist which reduces the response to a given dose of AVP to equal the response to 1/2 of this dose. [b]Estimated *in vivo* pA_2 values represent the negative logarithms of the "effective doses" divided by the estimated volume of distribution in the rat (67 ml/Kg). [c]Reference 6. [d]This communication. Analogs 1 and 2 represent deletions of the C-terminal glycine and glycinamide, respectively. In Analogs 3-5 the glycinamide is replaced by methylamine, ethylenediamine and alaninamide, respectively.

Peptides 1986

C-Terminal Deletions and Substitutions in $d(CH_2)[Tyr(Me)^2]AVP$ (Table 1)

The desGly and desGly(NH_2) analogs of $d(CH_2)_5[Tyr(Me)^2]AVP$ retain almost the full V_1 antagonistic potency of the parent analog. Remarkably, replacement of the glycinamide residue by methylamine ($-NH-CH_3$), ethylenediamine ($-NHCH_2CH_2NH_2$) and alaninamide led to V_1/V_2 antagonists which are equipotent but much more selective than $d(CH_2)_5[Tyr(Me)^2]AVP$.

Effects of Pro^7 Deletion From V_1 and V_2/V_1 Antagonists (Table 2)

Deletion of Pro^7 from AVP leads to drastic losses of V_1 and V_2 agonistic potencies. We (1) and others (4, 5) have previously found that the deletion of Pro^7 from the potent V_2/V_1 antagonist $d(CH_2)_5[D\text{-}Tyr(Et)^2]VAVP$ (8) led to either 50% retention (4, 5) or full retention (1) of V_2 antagonistic potency depending on the assay used. We also reported an almost full retention of V_1 antagonism for this $desPro^7$ analog (1). Consistent with these findings are those reported here for the $desPro^7$ analogs of the potent V_1 antagonists $d(CH_2)_5AVP$ and $d(CH_2)_5[Tyr(Me)^2]AVP$ (6), which retain between 20 and 50% of the antagonistic activity of their respective parents.

TABLE 2: Effects of Pro^7 Deletion from AVP and AVP V_1 and V_2/V_1 Antagonists

	Peptide	Antiantidiuretic		Antivasopressor	
		ED[a]	pA_2[b]	ED[a]	pA_2[b]
	AVP[c]	Agonist (330 u/mg)		Agonist (382 u/mg)	
1)	$desPro^7AVP$[d]	Agonist (1.2 ± 0.2 u/mg)		100 ± 8	5.83 ± 0.03
	$d(CH_2)_5AVP$[c]	Agonist (0.03 ± 0.01 u/mg)		0.56 ± 0.11	8.16 ± 0.09
2)	$desPro^7d(CH_2)_5AVP$[d]	Mixed (approx. 0.001 u/mg)		2.4 ± 0.4	7.48 ± 0.07
	$d(CH_2)_5[Tyr(Me)^2]AVP$[c]	Agonist (0.31 ± 0.07)		0.16 ± 0.01	8.62 ± 0.03
3)	$desPro^7d(CH_2)_5[Tyr(Me)^2]AVP$[d]	Mixed ~94	~5.86	0.31 ± 0.06	8.42 ± 0.05
	$d(CH_2)_5[DTyr(Et)^2]VAVP$[c]	1.1 ± 0.2	7.81 ± 0.07	0.45 ± 0.11	8.22 ± 0.12
4)	$desProd(CH_2)_5[DTyr(Et)^2]VAVP$[e, f]	1.1 ± 0.1	7.82 ± 0.07	0.56 ± 0.09	8.09 ± 0.07

[a,b]For definitions of ED and pA_2 see corresponding footnotes in Table I. [c]From Reference 11. [d]This communication. [e]First reported by Ali et al., Reference 4. [f]From Reference 1.

$desPro^7$ Analogs of Selective V_2/V_1 Antagonists: Surprising Findings (Table 3)

Deletion of Pro^7 from the highly potent and selective V_2 antagonists $d(CH_2)_5[D\text{-}Ile^2,Ile^4]AVP$ (9) and $d(CH_2)_5[D\text{-}Leu^2,Ile^4]AVP$ (1) brought about dramatic losses in V_2 antagonism. The resulting $desPro^7$ analogs reported here exhibit less than 2% of the potency of their respective parents. Likewise the $desPro^7$ analog of $d(CH_2)_5[D\text{-}Phe^2,Ile^4]AVP$ (10) exhibits only approximately 6% of the V_2 antagonism of its parent.

TABLE 3: Deletion of Pro^7 from Selective AVP V_2/V_1 Antagonists Leads to Drastic Losses of Antagonistic Potencies

Peptide	Antiantidiuretic		Antiantivasopressor	
	ED[a]	pA_2[b]	ED[a]	pA_2[b]
$d(CH_2)_5[DIle^2,Ile^4]AVP$[c]	0.7 ± 0.2	8.04 ± 0.10	26 ± 3	6.42 ± 0.06
1. $desPro^7d(CH_2)_5[D\text{-}Ile^2,Ile^4]AVP$[d]	~90 (mixed)	~5.9	~97 (mixed)	~5.9
$d(CH_2)_5[D\text{-}Leu^2,Ile^4]AVP$[e]	2.3 ± 0.47	7.50 ± 0.10	67 ± 6	6.01
2. $desPro^7d(CH_2)_5[D\text{-}Leu^2,Ile^4]AVP$[d]	~165 (mixed)	~5.6	~70 (mixed)	~6.0
$d(CH_2)_5[D\text{-}Phe^2,Ile^4]AVP$[f]	0.5 ± 0.1	8.24 ± 0.08	0.99 ± 0.12	7.86 ± 0.05
3. $desPro^7d(CH_2)_5[D\text{-}Phe^2,Ile^4]AVP$[d]	8.4 ± 1.6	6.92 ± 0.07	10.6 ± 1.6	6.83 ± 0.07

[a,b]For definitions of ED and pA_2 see corresponding footnotes in Table I. [c]From Reference 9. [d]This communication. [e]From Reference 1. [f]From Reference 10.

Contrasting Effects of desGly/desGly(NH_2) and $desPro^7$ Modifications in a Selective V_2 Antagonist (Table 4)

The data in this Table illustrate very clearly the contrasting effects of deletions at positions 9 and 7 in the potent and selective V_2 antagonist $d(CH_2)_5[D\text{-}Ile^2,Ile^4]AVP$ (9). Whereas the deletion of Gly^9 leads to a significant enhancement of V_2 antagonism, the deletion of Pro^7 leads to a drastic loss of antagonism. At present we have no logical explanation for these extraordinary findings.

TABLE 4: Contrasting Effects of Pro^7 and Gly/Gly(NH_2) Deletions in Selective AVP V_2 Antagonist

Peptide	Antiantidiuretic		Antivasopressor	
	ED[a]	pA_2[b]	ED[a]	pA_2[b]
$d(CH_2)_5[D\text{-}Ile^2,Ile^4]AVP$[c]	0.7 ± 0.2	8.04 ± 0.10	26 ± 3	6.42 ± 0.06
$desGly^9$[d,e]	0.37 ± 0.03	8.26 ± 0.04	~17 (mixed)	~6.7
desGly-(NH_2)^9[d,e]	0.9 ± 0.1	7.88 ± 0.05	~400 (mixed)	~5.2
$desPro^7$[d]	~90 (mixed)	~5.9	~97 (mixed)	~5.9

[a,b]For definitions of ED and pA_2 see footnotes to Table I. [c]From Reference 9. [d]This communication. [e]Most selective V_2/V_1 antagonists reported to date; ED ratios (V_1/V_2) = ~46 and ~440.

SUMMARY AND CONCLUSION

The properties of the desGly and desGly(NH_2) and 9-substituted analogs of the potent AVP V_1 antagonist $d(CH_2)_5[Tyr(Me^2)]AVP$ (6) are consistent with those previously reported for potent V_2/V_1 antagonists posessing similar modifications (1-3, 11). Their high potency and selectivity make the 9-substituted analogs in Table 1 attractive new pharmacological tools for studies on the cardiovascular role(s) of AVP in normal and pathophysiological states (11). The desGly and desGly(NH_2) analogs of d(CH_2)$_5$[D-Ile^{2}Ile^4]AVP (Table 4) are the most selective V_2/V_1 antagonists reported to date.

We have shown that while the effects of deleting Pro^7 from the two potent and selective V_1 antagonists d(CH_2)$_5$AVP and d(CH_2)$_5$[Tyr(Me)^2]AVP (6) are highly consistent with those previously observed for the desPro^7 analog of the potent but non-selective V_2/V_1 antagonist d(CH_2)$_5$[D-Tyr(Et)^2]VAVP, i.e., substantial retention of antagonistic potencies, this is in striking contrast to the effect of the Pro^7 deletion from selective V_2/V_1 antagonists which contain D-Ile, D-Leu or D-Phe at position two, which resulted in dramatic losses of V_2 antagonism. These inconsistent findings with desPro^7 analogs are unprecedented in our experience (11). This is the first instance we have encountered where a single modification, in this case the deletion of the Pro^7 residue from AVP antagonists, has had such contrasting effects in similar type molecules. These highly puzzling findings point to the complexity inherent in the design of receptor antagonists and illustrate the need for extreme caution in attempting to generalise about the biological effects of any given molecular modification based on data obtained from a single parent analog, e.g., in this case, either of the parent analogs d(CH_2)$_5$[D-Tyr(Et)^2]VAVP (8) or d(CH_2)$_5$[D-Ile^2,Ile^4]AVP (9) if looked at in isolation, could provide very misleading information on the generalised effects of desPro^7 modifications in AVP antagonists (1, 4, 5). Nonetheless, these very exciting findings (a) beg for an explanation and (b) may offer very valuable clues to the key roles of the position 2 and 7 substituents in modulating AVP agonist and antagonist peptide/receptor interactions.

ACKNOWLEDGMENT

We thank Ms. Valerie Murphy for her help in manuscript preparation and Ms. N.C. Wo, Mr. W.V. Greim and Mr. G.H. Pelton for performing most of the bioassays. This work was supported in part by research grants from the National Institute of General Medical Sciences (GM25280) and the National Institute of Arthritis, Diabetes and Digestive and Renal Diseases (AM-01940). The following authors are visiting investigators to MCO; J.P., B.L., Z.G. and M.K. from the University of Gdansk Poland; A.O. from the Technical University of Lodz, Poland; K.B. and A.M. from the University of Warsaw, Poland.

REFERENCES

1. Manning, M., A. Misicka, S. Stoev, E. Nawrocka, W.A. Klis, A. Olma, K. Bankowski, B. Lammek, W.H. Sawyer. 1985. In: V. Hruby, C.M. Deber, K.D. Kopple, eds. Proc. 9th Amer. Pep. Symp. Rockford, Illinois, Pierce Chem. Co. 599-602.

2. Manning, M., A. Olma, W.A. Klis, A. Kolodziejczyk, E. Nawrocka, A. Misicka, J. Seto, W.H. Sawyer. 1984. Nature 308, 652-653.

3. Ali, F.E. and W.F. Huffman. 1984. U.S. Patent: 4,481,193.

4. Huffman, W.F. and M.L. Moore. (a) 1984. U.S. Patent: 4,469,679. (b) 1985. U.S. Patent: 4,542,124.

5. Ali, F.E., W. Bryan, H.L. Chang, W.F. Huffman, M.L. Moore, G. Heckmann, L.B. Kinter, J. McDonald, D. Schmidt, D. Schue, and F.L. Stassen. 1986. J. Med. Chem. 29, 984-988.

6. Kruszynski, M., B. Lammek, M. Manning, J. Seto, J. Haldar, and W.H. Sawyer. 1980. J. Med. Chem. 23, 364-368.

7. Merrifield R.B. 1963. J. Am. Chem. Soc. 85, 2149-2154.

8. Manning, M., A. Olma, W.A. Klis, A.M. Kolodziejczyk, J. Seto, and W.H. Sawyer. 1982. J. Med. Chem. 25, 45-50.

9. Manning, M., E. Nawrocka, A. Misicka, A. Olma, W.A. Klis, J. Seto, and W.H. Sawyer. 1984. J. Med. Chem. 27, 423-429.

10. Manning, M., A. Olma, W.A. Klis, J. Seto, and W.H. Sawyer. 1983. J. Med. Chem. 26, 1607-1613.

11. Manning, M. and W.H. Sawyer. 1985. In: R.W. Schrier, ed. Vasopressin. New York: Raven Press 131-144.

VASOPRESSIN ANALOGS WITH STRONG AGONISTIC - ANTAGONISTIC PROPERTIES

M. Zaoral, I. Bláha, M. Lebl and T. Barth
Institute of Organic Chemistry and Biochemistry, Czechoslovak Academy of Sciences, 166 10 Praha 6, Czechoslovakia

Introduction

Some 20 years ago we demonstrated that both the quantity and quality of the biological effects of vasopressins were strongly dependent on the properties of the amino acid in position 8 of the C-terminal peptide side chain of these hormones. By simple changes of its properties such as the change of configuration, homologous change, functionalization and derivatization, it is possible to modify the biological activities of vasopressins almost at will. We have demonstrated the value of this piece of information on the preparation of the first "superactive" vasopressin analog, 1-desamino-8-D-arginine vasopressin (DDAVP). DDAVP has rapidly found its way into medical practice and it has been designated as belonging among the 15 most important drugs produced so far in this century. Our previour work has recently been summarized (1).

The change of configuration in position 2 has been frequently used for the preparation of potent vasopressin inhibitors (2). However, our previous studies indicated that the structural changes formed a kind of hierarchical succession, according to their influence on the biological activities of vasopressins (1). The change of configuration in position 8 seems to play a dominant role in this succession. The idea on the hierarchy of structural changes encouraged us to an attempt at introducing two changes of configuration into a vasopressin molecule simultaneously, namely into position 8 and 2. We assumed that the change of configuration in the first mentioned position would outweigh that in the latter, and by combining these changes we hoped to obtain a favorable combination of the relevant biological properties.

Peptides 1986

Results

The solid phase synthesis, purification and assay for biological activities of the double-D-substituted vasopressin analogs was carried out as usual (1). The biological activities of two representative compounds are shown in Table 1. The data in the table

Table 1. Biological Activities of Double-D-Substituted Vasopressin Analogs

[X,Y] VP	AD	BP	UT	G
[Mpr^1, D-Tyr^2, D-Arg^8]	120-150%	0.1	inh.[a]	0.1 part.agon.
[Mpr^1, D-Tyr^2, Abu^4, D-Arg^8]	40-60%	inh.[b]	inh.[c]	inh.[d]
DDAVP	100%	0.96	5.1	

VP vasopressin, AD antidiuretic activity, BP pressor activity, UT uterotonic activity, G galactogogous activity.
[a]pA_2 6.3, 7.9 (in situ, in vitro), [b]pA_2 6.8, [c]pA_2 7.1, 8.2 (in situ, in vitro), [d]pA_2 6.6.

are in accordance with the working idea. The compounds combine the properties of 2-D- and 8-D-substituted analogs in the desired way. The first compound has markedly higher antidiuretic activity than DDAVP, it has very low pressor effect and is a potent inhibitor of the uterotonic activity. The second compound has only approximately 50% of the antidiuretic activity of DDAVP; however, it is a potent inhibitor of all of the side effects studied. The compounds represent a new type of vasopressin analogs with strong agonistic - antagonistic properties

After completing the study of a group of compounds of the type shown in Table 1 we were faced with a similar problem as two decades ago, during the development of DDAVP. The compounds in Table 1 are highly and selectively active as antidiuretics but they are not active enough to compete effectively with DDAVP. To improve their competitive ability we made an attempt to increase their antidiuretic potency.

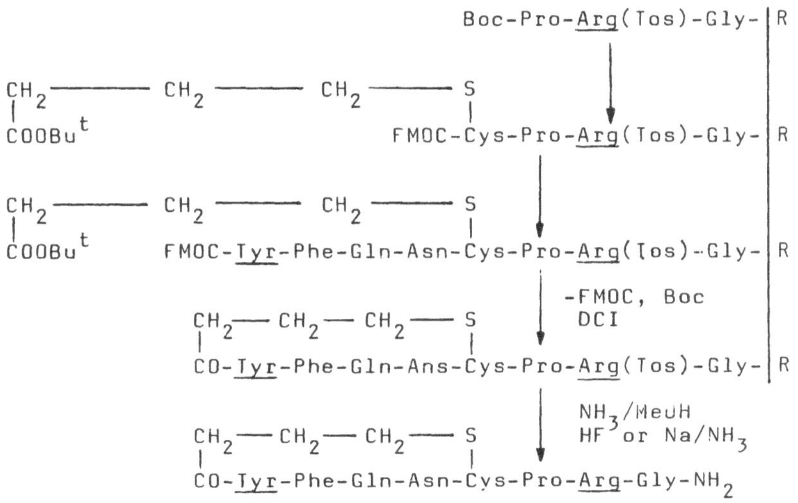

Scheme 1. Synthesis of double-D-substituted carba-vasopressin analogs. The underlined amino acids are of D-configuration.

As has been shown by Jošt et al. (3), oxytocin and vasopressin carba-analogs frequently have higher biological activities than their disulfide bridge counterparts. We took advantage of this valuable piece of knowledge and we prepared such analogs of selected members of our double-D-substituted group of vasopressins. The compounds were obtained by a somewhat modified solid phase procedure as compared to that which is usually employed for this purpose in our laboratories. Its principle is shown in Scheme 1. The 7-9 sequence is prepared by means of Boc-amino acids. The bridge amino acid is attached as FMOC-tert-butyl ester. The synthesis is then continued with FMOC-amino acids. In the penultimate step the Boc and FMOC groups are removed, the cycle is closed and the peptide is split off from the resin by ammonolysis. The Tos residue, protecting the guanido group of D-Arg^8, is removed with liquid HF or with Na/NH_3. The synthesis is then completed in the usual manner. The biological properties of the compounds prepared are shown in Table 2. From the data in the table it is evident that the carba-substitution caused an unexpectedly high increase of the antidiuretic effect. Two of the compounds have a 7x and 10x higher antidiuretic effect than DDAVP. The carba-substitution does not have a negative effect on the inhibitory properties of the analogs. It is evident that the $carba^6$-analog has much lower antidiuretic activity.

Table 2. Biological Activities of Double-D-Substituted Carba-Vasopressin Analogs

[X,Y]VP	AD	BP	UT[x]	G
[dCar[1], D-Tyr[2], D-Arg[8]]	7.9	pA_2 7.2	pA_2 7.5	3.6
[dCar[1], D-Tyr[2], Val[4], D-Arg[8]]	10.0	pA_2 6.8	pA_2 7.9	0.4
[Mpr[1], D-Tyr[2], Car[6], D-Arg[8]]	1-2	pA_2 6.8	pA_2 7.8	1.0
DDAVP	1	0.96	5.1	

[x]in vitro. dCar desamino-carba, Mpr 3-mercaptopropionic acid. Other abbreviations have the same meaning as in Table 1.

The carba analogs under consideration are very stable compounds. Their N- and C-terminal part is protected against the attack of proteolytic enzymes and they do not contain the labile disulfide bridge. Evidently, they are suitable for application per os. Indeed, in tests on experimental animals (rats) they showed a strong antidiuretic effect when applied per os. The double-D-substituted carba-vasopressin analogs certainly deserve serious consideration as possible successors to DDAVP. Perhaps it should also be mentioned that these compounds are not easily obtainable by methods of genetical engineering.

References

1. Zaoral,M. 1985. Int. J. Peptide Protein Res. 25, 561.
2. Manning,M., Sawyer,W.H. 1985. In: The Vasopressins (R.W. Schrier, ed.). Raven Press, New York, p. 131.
3. Jošt,K. 1987. In: CRC Handbook of Neurohypophyseal Hormone Analogs, Vol. I (K. Jošt, M. Lebl and F. Brtník, eds.). CRC Press, Inc. Chapter III. Boca Raton, Florida.

PHOTOREACTIVE AGONISTS AND ANTAGONISTS OF NEUROHYPOPHYSIAL HORMONES FOR RECEPTOR STUDIES

F. Fahrenholz, E. Kojro and R. Boer

Max-Planck-Institut für Biophysik, Kennedyallee 70,
D-6000 Frankfurt am Mainz, F.R.G.

Introduction

Photoreactive analogues of vasopressin have been synthesized and used to identify vasopressin receptor proteins in plasma membranes from different target organs (1,2):V_1 vasopressin receptors (coupled to Ca^{2+}-mobilization) in rat liver and V_2 vasopressin receptors (coupled to adenylate cyclase) in kidney medulla from different mammalian species. The question as to whether vasopressin agonists and antagonists interact with the same subunits of vasopressin receptor proteins remains open. Furthermore, at present we have no information which amino acid residues in the vasopressin molecule are close to the hormone binding site in the hormone-receptor complex. In this report we describe the synthesis of potent photoreactive vasopressin agonists and antagonists an their employment in receptor studies to answer these questions.

Material and Methods

The parent cyclic nonapeptides were prepared by solid phase synthesis following described methods (2). The photoreactive residue was introduced into these vasopressin analogues at the 6-NH_2 group of Lys^8 by reaction with methyl-4-azidobenzoimidate (2). Radioiodination of the photoreactive antagonist was carried out with a 3-fold molar excess of peptide over 125-iodine. Tritium labelling of the agonist was performed by iodination of the parent peptide at Tyr^2 and catalytic dehalogenation in the presence of tritium with 10 % Pd/Al_2O_3 as catalyst. The labelled photoreactive ligands were purified by HPLC. Photoaffinity labelling experiments were performed as described (1,2).

Peptides 1986

Results and Discussion

Fig. 1 shows the structure of the photoreactive vasopressin antagonist with 1-ß-mercapto-ß-,ß-cyclopentamethylenepropionic acid (Mca) in position 1, N-methylalanine (MeAla) in position 7 and N^6-(4-azidophenylamidino)-lysine (Lys(N_3PhCNH)) in position 8 of vasopressin (VP).

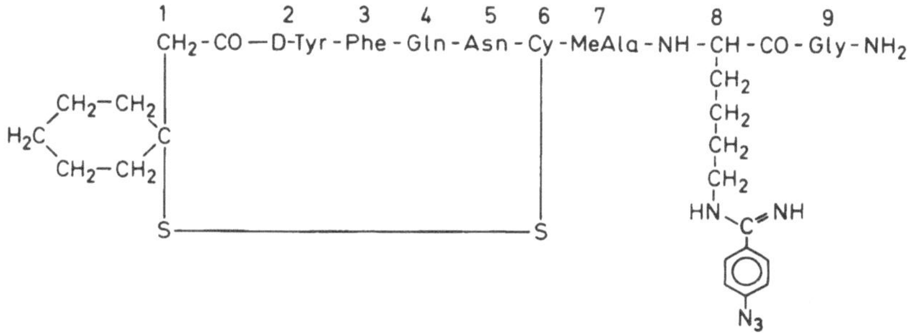

Fig. 1 Structure of [Mca[1], D-Tyr[2], MeAla[7], Lys(N_3PhCNH)[8]] VP.

This analogue retains the positive charge in position 8 of the vasopressin sequence and is a potent V_1 antagonist of the in vitro glycogenolytic activity on isolated hepatocytes (pA_2 = 9.35 ± 0.31 n = 6) and of the in vivo vasopressor activity of vasopressin (pA_2 = 7.69 ± 0.20 n = 5). Monoiodination with ^{125}I at D-Tyr[2] results in a photoreactive antagonist of high specific radioactivity (2000 Ci/mmol) which is roughly as potent as vasopressin in binding to the hepatic V_1 receptor (K_D = (1.8 ± 0.3) x 10^{-9}M).

In photoaffinity labelling experiments with rat liver membranes the ^{125}I-antagonist bound covalently to both vasopressin receptor subunits with the relative molecular masses (M_r) of ~ 30 000 (α-subunit) and 38000 (ß-subunit). The specific labelling of the same two subunits has been demonstrated using the tritium labelled analogue with ß-mercaptopropionic acid in position 1 and N^6-(4-azidobenzoyl)-lysine in position 8 [Mpa[1], Lys(N_3PhCO)[8]] VP (1). This analogue retains full agonistic properties on isolated hepatocytes (1). Our studies provide evidence that V_1 antagonists can interact with both subunits of the V_1 receptor in rat liver membranes. However, due to the ß,ß-dialkyl-substitution at position 1 and the conformational restraint of antagonists, they can not adopt a biologically active conformation in the receptor bound state.

In the series of photoreactive vasopressin analogues retaining antidiuretic and vasopressor activity properties (3), the N^6-azidophenylamidino analogue $[Mpa^1, Lys(N_3PhCNH)^8]$ VP (peptide 6, Table 1) showed the highest binding affinity for the hepatic V_1 receptor (1) and the renal V_2 receptor (ref. 2 and Table 1). Tritium labelling of the parent peptide $[Mpa^1, Lys^8]$ VP resulted in a specific radioactivity of 52 Ci/mmol. The photoreactive tritium labelled ligand incorporated specifically into a M_r= 30 000 protein from bovine kidney which has been identified as the renal V_2 receptor binding subunit (2). A comparison of diffenrent ^3H-labelled photoreactive agonists (Table 1) demonstrates that the highest yield between the ligands and the renal V_2 binding subunit is obtained with analogues containing an aryl azido group in position 8. Roughly the same yield is observed for the N^6-azidobenzoyl analogue - without a positive charge in position 8 (peptide 5, Table 1) - and for the N^6-phenylamidino analogue which retains a positive charge in positio 8 (peptide 6, Table 1).

Table 1. Specific labelling of the V_2 receptor binding subunit in bovine kidney by different agonists a)

Peptide b)	Position of reactive group	$\frac{K_D}{K_{D\,(AVP)}}$ c)	Yield d)
1 $[Asu^{1,6}, Phe(p\text{-}N_3)^2]$ AVP	2	7	-
2 $[Asu^{1,6}, Phe(p\text{-}N_3)^3]$ AVP	3	18	+
3 $[Mpa^1, Dab(N_3PhCO)^4]$ AVP	4	11	-
4 $[Mpa^1, Dab(N_3NO_2Ph\text{-}\beta\text{-}Ala)^4]$ AVP	4	6	++
5 $[Mpa^1, Lys(N_3PhCO)^8]$ VP	8	15	++++
6 $[Mpa^1, Lys(N_3PhCNH)^8]$ VP	8	1,6	++++

a) from refs. 2,4 and this communication

b) Asu = L-2-aminosuberic acid, AVP = arginine vaopressin,
Dab = 2,4-diaminobutyric acid , $Dab(N_3NO_2Ph\text{-}\beta\text{-}Ala)$ =
N^4(4-azido-2-nitrophenyl-ß-alanine)-2,4-diaminobutyric acid,
$Phe(p\text{-}N_3)$ = 4-azido-L-phenylalanine

c) apparent dissociation constant for arginine vasopressin
$K_D \sim 1 \times 10^{-9}$M.

d) ++++: corresponds to a specific radioactivity of the labelled binding subunit of 0.4 - 0.6 Ci/mmol V_2 receptor;
\- : no specific labelling.

These results suggest a close contact between the side chain of the basic amino acid in position 8 in the vasopressin molecule on the one hand and the hormone binding site of the renal vasopressin receptor on the other hand.

Acknowledgements

We thank Dr. S. Keppens and Professor H. De Wulf, Afdeling Biochemie, Katholieke Universiteit te Leuven, for determining the antiglycogenolytic potency of vasopressin analogues, Dr. J. Slaninová and Dr. T. Barth, Institute of Organic Chemistry and Biochemistry, Czechoslowak Academy of Science, Prague, for performing the vasopressor tests. We wish to thank Dr. J.L. Morgat from the Service de Biochemie, Département de Biologie CEN Saclay, Gif-sur-Yvette, for his help in the catalytic dehalogenation experiments. This work was supported by grant Fa 48/23-6 from the Deutsche Forschungsgemeinschaft.

References

1. R. Boer and F. Fahrenholz (1985) J. Biol Chem. 260, 15051 - 15054.
2. F. Fahrenholz, R. Boer, P. Crause and M. Toth (1985) Eur. J. Biochem. 152, 589 - 595.
3. F. Fahrenholz, P. Eggena, D. Gazis, M.V. Toth and I.L. Schwartz (1986) Endocrinology 118(3), 1026 - 1032.
4. F. Fahrenholz and P. Crause (1984) Biochem. Biophys. Res. Commun. 122, 974 - 982

ACTH-PHOTOAFFINITY LABELS, SYNTHESIS AND APPLICATION OF ANALOGUES WITH AGONISTIC AND PARTIAL AGONISTIC PROPERTIES

E. Escher, Z. Mackiewicz, S. Sircar, D. Bourque, F. Bouthillier, J.-G. Lehoux
Laboratories of Pharmacology and Biochemistry, Medical School, University of Sherbrooke, Sherbrooke, QC, Canada J1H 5N4

Introduction

Receptor labeling experiments have shown in several cases that the pharmacological properties (agonists or antagonists) of a label influence the observed size of the native receptor (1,2,3). On ACTH some work with photoaffinity-labels has already been accomplished (4,5). The label utilized was [(azido-nitrophenylsulfenyl)-Trp^9]ACTH, a partial agonist, which at lower temperatures is reported to be an antagonist (5). In order to obtain purely agonistic and, if possible, purely antagonistic labels, a series of [Phe^2,Nle^2]ACTH (1-24) analogues were prepared with the photoaffinity labeling groups in several positions.

Preparation of Labels

Five different labeling peptides were prepared for this study: in a preliminary experiment, ACTH (1-24) was modified with 4'-azidobenzoic acid, separated by HPLC, and the fraction with best yield and potency was retained. (Peptide 2) In the second attempt, four labeling sequences were built up by the solid-phase procedure and labeling precursors placed either in pos. 2, 7, 11 or 16 of the sequence [Phe^2,Nle^4]ACTH (1-24). As labeling amino acid precursors were utilized (4'-NO_2)Phe for the pos. 2 and 7 analogue and Nε-(4'-NO_2)Benzoyl-Lys for the pos. 11 and 16 analogue respectively (Peptides 3 - 6).

Peptide synthesis was carried out on Merrifield resin with the Boc-TFA-HF Scheme and symmetrical anhydride couplings. The side chain protecting groups were Tos for His and Arg, Cl_2-Z for Lys and Tyr, Bzl for Ser and Glu. 40% TFA containing 0.1% of N -acetyl-Trp was utilized to prevent t-butylation of Trp (6). The cleaved peptides were purified by filtration over LH20 with DMF

Peptides 1986

and preparative HPLC on C-18 reversed phase material (Macherey-Nagel Nucleosil, 30u,) with gradients of 10 to 60% CH_3CN at pH2 (0.05% TFA) or pH5 (0.1 N $AcONH_4$). The peptides were converted to the photolabels by hydrogenation, diazotation at -25°C and azide formation according to (7) and the scheme (Peptides 7 - 10).

Scheme:

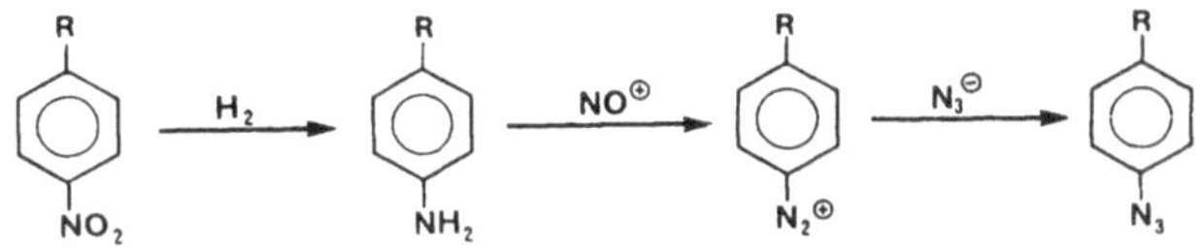

Table. Biological activities of ACTH-(1-24) analogues

No Structure		R_f	a^E	R.A.
1. ACTH (1-24)		.45	1	100
2. [$(N_3$-Bzo)Lys^x]	ACTH (1-24)	.46	.4	56
3. [NO_2Phe^2,Nle^4]	"	.46	1.4	45
4. [Phe^2,Nle^4,NO_2Phe^7]	"	.43	.8	10
5 [$Phe^2,Nle^4,(NO_2Bzo)Lys^{11}$]	"	.46	1	0.7
6. [$Phe^2,Nle^4,(NO_2Bzo)Lys^{11}$]	"	.41	1	5
7. [N_3Phe^2,Nle^4]	"	.45	1	260
8. [$Phe^2,Nle^4,(N_3Phe^7$]	"	.46	1	14
9. [$Phe^2,Nle^4,(N_3Bzo)Lys^{11}$]	"	.46	1^*	0.1
10.[$Phe^2,Nle^4,(N_3Bzo)Lys^{16}$]	"	.43	1^*	2.5

R_f: values on TLC (butanol: acetic acid: water = 5:2:3). a^E is the intrinsic activity, i.e. the capacity of maximal stimulation. R.A.: Relative affinity compared to ACTH = 100%.NO_2 in 4'-nitro, N_3 is 4'-azido, Bzo is benzoyl. ACTH (1-24) induced half maximal corticosteroid production at $4.1 \cdot 10^{-10}$ [M] on adrenocortical cells. * completely antagonized at 10^{-8}[M] a dose of 10^{-9}[M] 1.

All peptides had satisfactory analytical data (HPLC, TLC, amino acid analysis); some are presented together with the steroidogenic properties in the Table. The labeling peptides were iodinated and purified with HPLC (8). For the labeling peptide 2 the iodine position has been established to be simultaneously ($3'$-^{125}I)Tyr in pos. 2 and 23 by Edman degradation.

Results

In order to prove a correlation of protein labeling and receptor identity, irreversible biological effects had to be shown. Rat glomerulosa cells were incubated with 2, irradiated and washed. A continuous maximal steroid production for both, Aldosterone and Corticosterone was observed. Radioactive 2 incubated in the dark with glomerulosa cell membranes was competitively displaced by 1 with an IC_{50} of $1.18 \cdot 10^{-10}$ [M]. Photolabeling experiments were carried out on glomerulosa membranes as described for angiotensin II (9). Radioactive 2 ($1.1 \cdot 10^{7}$ cpm) was incubated with membranes (3.5 mg protein, Lowry) and photolyzed with four 100W black ligh lamps for 2 min. After photolysis, the membranes were washed three times by centrifugation and resuspension and divided in several aliquots. Control experiments with membranes non-photolyzed or photolyzed in presence of 10^{-5}[M] ACTH were treated similarily. For SDS-PAGE, samples containing 100 ug of protein were solubilized and applied to 7.5% polyacrylamide gels. After migration, the gels were stained, de-stained, exposed to autoradiography or cut into 2 mm slices (Fig. 1,2). In every case, a clear strong band at 60 + 4 kD was seen and a second, variably strong but weaker at 99 + 5 kD. For gel filtration, samples were solubilized in Triton X and eluted from a Biogel A-5 column. A single radioactive peak was observed which obeyed competitivity and corresponded to a Stokes radius of 47 + 3.5 A, (Fig. 3). Sucrose density gradient centrifugation of the combined fractions from the gel filtrations produced a sedimentation constant of 5.1 + 0.5 Svedberg units (see Fig. 4), which together with the Stokes radius produces a calculated effective molecular weight of 96.5 + 9.5 kD.

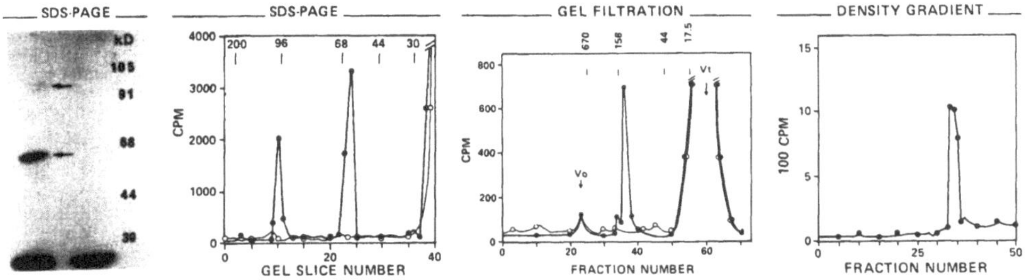

Fig. 1-4: Elution patterns for SDS-PAGE 1) autoradiography, 2) sliced gel, 3) Biogel A-5, 4) density centrifugation. The left lane in 1) and the solid lines in 2) - 4) are the labeling experiments, the middle lane in 1) and the fine lines in 2) and 3) are the competitive protection experiments.

Discussion

From the preliminary labeling studies with photolabel 2 the ACTH receptor seems to consist of a 100 kD protein that under strong denaturation dissociates into a 60 kD protein containing the hormone binding site and into others, not yet known subunits. Due to the partial agonistic properties of 2 and in analogy to the results obtained on the angiotensin II receptor (1), we propose that the 100 kD protein is the native ACTH-receptor without guanylnucleotide sensitive N- or G-protein attached to it and thus similar to the antagonist occupied angiotensin II receptor (1). The studies with peptides 7 and 8 (full agonists), and 9 and 10, which have antagonistic properties at lower doses, will produce this information in the very near future.

Acknowledgements

We are gratefully indebted to Dr. S. Lemaire for the Edmann degradation, to Mrs. C. Theberge for typing this report and for financial support from the Canadian Kidney Foundation, the Canadian Heart Foundation (grants to EE) and from the Medical Research Council (grant of JGL). EE and JGL are scholars of the Fonds de la Recherche en Santé du Québec.

References

1. Guillemette, G., G. Guillon, J. Marie, C. Pantaloni, M.N. Balestre, E. Escher, S. Jard. 1984. J.Receptor Res. 4, 267.
2. De Léan, H. Ong, J. Gutkowska, P.W. Schiller, N. McNicoll. 1984. Mol.Pharmacol. 26, 498.
3. Limbird, L.E., D.M. Gill, J.M. Stadel, A.R. Hickley, R.J. Lefkowitz. 1980. J.Biol.Chem. 255, 1854.
3a. Kilpatrick, B.F., M.G. Caron. 1983. J.Biol.Chem. 258, 13528.
4. Ramachandran, J., J. Hagman, K. Muramoto. 1981. J.Biol.Chem. 256, 11424.
5. Ramachandran, J., K. Muramoto, M. Kenez-Keri, G. Keri, D.I. Buckley. 1980. Proc.Natl.Acad.Sci. 77, 3967.
6. Escher, E., P. Parent, J. Mizrahi, P. D'Orléans-Juste, S. Dion, G. Drapeau, D. Regoli. 1985. In: Tachykinin Antagonists (R. Hakanson and F. Sundler, eds.). Elsevier Amsterdam, p. 267.
7. Laczko, E., E. Escher. 1981. Helv.Chim.Acta 64, 621.
8. Gallo-Payet, N., E. Escher. 1985. Endocrinology 117, 38.
9. Guillemette, G., E. Escher. 1983. Biochemistry 22, 5591.

THE INFLUENCE OF PEPTIDE-RECEPTOR DISSOCIATION RATE CONSTANTS ON RESPONSES TO AND APPARENT POTENCY OF PEPTIDE HORMONES

Diana Gazis

Department of Physiology and Biophysics and Center for Polypeptide and Membrane Research, Mount Sinai School of Medicine, New York, NY 10029

Two parameters, potency and plasma half-life, determine whether peptides are effective as therapeutic and physiological agents and whether they are long acting. However, a third parameter, the peptide-receptor dissociation rate constant (k_d), strongly influences both the apparent potency of an analog and the time course of the response *in vivo* and should be considered during peptide design and during physiological investigation of peptide action. This article summarizes an investigation of the role of the peptide-receptor dissociation rate constant in determining binding and response which was performed by comparing the results of graphic simulations of receptor kinetics with the results of biological experiments in which dissociation rate was varied. The biological experiments were carried out using the pressor response to arginine vasopressin as a model response and a series of vasopressin analogs which differed in pressor potency as a model series of probes which varied in dissociation rate constant.

In *in vitro* assays, k_d, by itself, does not influence binding, response or potency because these parameters are measured at equilibrium. The following equations describe binding and response (if first order kinetics apply) at equilibrium:

$$D+R \underset{k_d}{\overset{k_a}{\rightleftharpoons}} DR; \quad K_D=k_d/k_a \sim \text{Potency}; \quad \text{Response} \sim DR/R_T=1/[1+(K_D/D)]$$

Binding and response do not depend on individual association and dissociation rates but only on the ratio of these rates.

Peptides 1986

In *in vivo* assays (if injections are used), dissociation rate constant does affect binding, response, and potency measurement, because equilibrium cannot occur. When equilibrium is not present, the appropriate binding equation is a rate equation, such as the following:

$$\text{Response} \sim DR/R_T = [1-e^{-(k_a D+k_d)t}][1/(1+(K_D/D))].$$

(The particular form of the rate equation used depends on the algorithm which describes the time course of peptide concentrations - - for example, peptide could be infused or injected).

It is apparent from the above rate equation that dissociation rate does affect time to equilibrium, but two other parameters, dose and association rate also have an influence. Although mathematically, association rate affects time to equilibrium, this effect is undetectable *in vivo* because in order to compare peptides with differing association rates, we must give doses which produce similar responses. To produce the same response, as we increase association rate (which makes equilibrium come faster), we must decrease dose (which makes equilibrium come more slowly).The result is that the two effects cancel each other out and association rate appears not to affect time to equilibrium binding. Under actual *in vivo* experimental conditions, then, only dissociation rate and one additional parameter, dose, visibly affect time to equilibrium binding. As dose or dissociation rate increase, equilibrium binding is reached more quickly. When dissociation rates increase, peptides leave their receptors more quickly and, association rates remaining the same, are weaker peptides. Under non-equilibrium conditions, weak peptides are closer to equilibrium than strong peptides. This difference between weak and strong peptides is more pronounced at low doses.

A third factor, plasma half-life, does not affect time to equilibrium directly, but determines how strongly the effects of dose and dissociation rate affect responses to injections. After an injection, peptide is swept away through degradation and elimination before equilibrium binding at the initial concentration can occur and response,

Table 1. Comparison of Mathematical Predictions about the Effects of Altering Dissociation Rate Constant, Dose, and Plasma Half-Life on *In Vivo* Biological Responses with Pressor Responses to Vasopressin Analogs which Vary in Dissociation Rate Constant[a] and Plasma Half-life[b].

Mathematical Prediction #1	Experimental Result #1	Analog (10 mU injected): Arginine vasopressin	Oxytocin	Oxypressin
The slower the dissociation rate, the longer the time to maximum response.	k_d	0.5	2.2	2.1
	Time to maximum response (min)	1.43	0.54	0.58
	Potency (I.U./mg)	435	5	3
	Plasma half-life (min)	2.5	2.5	2.5

Mathematical Prediction #2	Experimental Result #2: Dose of Arginine Vasotocin injected (mU)	Time to maximum response (min)
The lower the dose the longer the time to maximum response	20	1.62 ± 0.2
	40	1.50 ± 0.1
	80	1.18 ± 0.12
	160	1.05 ± 0.19

Mathematical Prediction #3	Experimental Result #3	Analog (10 mU injected): Arginine Vasotocin	Arginine Vasopressin
The longer the plasma half-life, the longer the time to maximum response.	Plasma half-life (min)	5	2.5
	Time to maximum response (min)	1.85	1.43
	k_d	0.5	0.4

Mathematical Prediction #4	Experimental Result #4: Analog	k_d	Response to 15 mU Dose: Injected	Infused (10 min)
The slower the dissociation rate (i.e., the smaller the k_d), the more the potency will be underestimated by injection results.	Oxytocin	2.2	27.2 mm Hg	17.2 mm Hg
	Arginine-Vasopressin	0.5	27.6	22.8

[a] Dissociation rate constants were determined from the decay of pressor responses after injection of supramaximal amounts of inhibitor (Ref.1).

[b] Plasma half-lives were determined by the method of Gazis and Sawyer (Ref. 2).

accordingly, is smaller than it would have been if it could have gone to equilibrium. Peptides which take a long time to reach equilibrium binding (that is, peptides with slow dissociation rates) are most affected.

Table 1 summarizes both the mathematical predictions described above and experimental data showing that these predictions are fulfilled if one uses analogs of vasopressin for probes which vary in potency, half-life, and dissociation rate constant, and the pressor response as the *in vivo* biological indicator of peptide-receptor binding. The effect of dissociation rate on response is not only observable but is more pronounced than the effect of either dose or plasma half-life.

Only slow dissociation rates affect *in vivo* responses, for only when dissociation is slow do responses lag behind those appropriate (were equilibrium to apply) for the plasma peptide concentration present. When dissociation is fast, binding is close to the equilibrium value throughout the response. However, in a series of analogs of a given peptide, weak analogs will dissociate fast and potent analogs will dissociate slowly. The potency of weak analogs will be estimated correctly from *in vivo* results, but the potencies of strong analogs, which are likely to be of therepeutic and physiological interest, will be underestimated because of the lag in reaching equilibrium binding. The underestimation will be greatest at low doses and in peptides with short half-lives.

In summary, peptide-receptor dissociation rates, if slow enough, can affect *in vivo* responses as strongly as plasma half-life or the equilibrium dissociation constant. Since this effect is most likely to occur with potent peptides, it should be considered when designing and assaying potential therapeutic agents.

References

1. Gazis, D., Roy, J., and Schwartz, I. L.: Fed. Proc. 41, 1353 (1982).

2. Gazis, D., and Sawyer, W. H.: Proc. Soc. Exptl. Biol. Med. 157, 584-589(1978).

SYNTHESIS AND BIOLOGICAL ACTIVITY OF NOVEL GROWTH HORMONE RELEASING FACTOR ANALOGS

A.M. Felix, E.P. Heimer, T.F. Mowles, H. Eisenbeis, P. Leung, T.J. Lambros, M. Ahmad, C.-T. Wang

Peptide Research Department and Animal Science Research Department, Hoffmann-La Roche Inc., Nutley, N.J. 07110

P. Brazeau

Montreal University, Quebec, Canada

Introduction

Peptides with growth hormone releasing activity were first isolated from human sources and characterized independently by Vale et. al. (1) and Guillemin et al. (2), Fig. 1. Hypothalamic growth hormone releasing peptides have subsequently been isolated from several animal species and shown to have significant homology to human GRF. Structure-activity studies have established the importance of the NH_2-terminus and that GRF(1-29)-NH_2 possesses the full intrinsic activity <u>in vitro</u> (1). GRF has a short half life ($t_{1/2}$ 6.8 min) (3) and frequent injections are required to mimic the effects of single daily injections of growth hormone. Therefore a program has been initiated to synthesize analogs with enhanced or prolonged biological activity.

H-Tyr-Ala-Asp-Ala-Ile-Phe-Thr-Asn-Ser-Tyr-Arg-Lys-Val-
Leu-Gly-Gln-Leu-Ser-Ala-Arg-Lys-Leu-Leu-Gln-Asp-Ile-
Met-Ser-Arg-Gln-Gln-Gly-Glu-Ser-Asn-Gln-Glu-Arg-Gly-
Ala-Arg-Ala-Arg-Leu-NH_2

Figure 1. Primary Sequence of Human Growth Hormone Releasing Factor (Somatocrinin)

Peptides 1986

Strategy

Chou-Fasman empirical probability calculations predict regions of beta-turn (residues 6-10) and alpha-helical conformation (residues 16-26). Circular dichroism measurements of GRF(1-44)-NH_2 reveals significant (∿62%) alpha-helical conformation. Circular dichroism of a series of GRF fragments in 75% MeOH, Table 1, shows the onset of secondary structure for GRF(22-44)-NH_2 and substantial alpha-helical conformation for GRF(13-44)-NH_2 (∿35%).

The NH_2-terminal region of GRF also has the potential to form an amphiphilic alpha-helix (4). Analogs of GRF were designed in which Gly at position 15 is replaced by other hydrophobic amino acids which can (a) extend the alpha-helical region and (b) maximize the potential of this region to form an amphiphilic alpha-helical structure. GRF(1-44)-NH_2, fragments and analogs were synthesized by the solid phase procedure and purified to homogeneity by high performance liquid chromatography. Characterization included in vitro biological activity, FAB-mass spectroscopy and tryptic digestion followed by hplc mapping.

Biological Results

In vivo studies (3-10 μg/kg s.c.) in cattle has established the equivalency of GRF(1-44)-NH_2 and GRF(1-29)-NH_2. Subcutaneous administration of GRF(1-44)-NH_2 and GRF(1-29)-NH_2 (3-30 μg/kg) in porcine animals were equipotent (weight basis) and demonstrated a good dose-response relationship. Treatment of swine with GRF(1-44)-NH_2 (30 μg/kg s.c.) [once-a-day for 30 days] enhanced body weight (5.5%), improved feed utilization (8.9%) and increased lean/fat ratio favoring muscle gain at the cost of fat (16.7%) (5). Milk production was enhanced in lactating cattle (∿15%) given either GRF(1-44)-NH_2 or GRF(1-29)-NH_2 (1 μg/kg s.c.) [6 times/day for 10 days] (6).

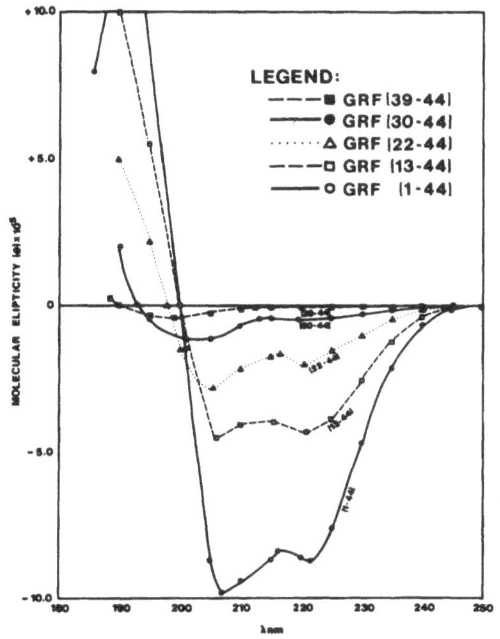

Figure 2. Circular Dichroism of Growth Hormone-Releasing Factor and Fragments (in 75% Methanol)

	Relative Potency (in vitro bioassay)
GRF(1-44)-NH_2	1.0
GRF(1-29)-NH_2	0.8
[Val^{15}]-GRF(1-29)-NH_2	1.02
[Leu^{15}]-GRF(1-29)-NH_2	3.5
[Ala^{15}]-GRF(1-29)-NH_2	4.0
[Aib^{15}]-GRF(1-29)-NH_2	3.7
[Sar^{15}]-GRF(1-29)-NH_2	0.04
[Val^{15}]-GRF(1-29)-OH	0.38
[Leu^{15}]-GRF(1-29)-OH	0.65
[Ala^{15}]-GRF(1-29)-OH	1.5
[Leu^{15}]-GRF(1-44)-NH_2	1.5
[Ala^{15}]-GRF(1-44)-NH_2	1.5
[D-Ala^{2}]-GRF(1-29)-NH_2	1.6
[Met$(O)^{27}$]-GRF(1-44)-NH_2	0.30
[Leu^{27}]-GRF(1-44)-NH_2	0.63
[Nle^{27}]-GRF(1-29)-NH_2	2.1

Table 1. GRF Analogs

Analogs of GRF(1-29)-NH_2 with increasing biological potency (rat pituitary bioassay) were obtained when Gly^{15} was replaced by Val, Leu or, preferably, Ala (Fig. 3). Biological potency also increased in the same order (Ala > Leu > Val) for the GRF(1-29)-OH series. This potency may be attributable to enhancement of alpha-helical conformation since replacement of Gly^{15} by Sar^{15}, reported to lead to decreased intramolecular hydrogen-bonding, results in a loss of biological activity. These observations have been confirmed using an *in vitro* superfusion of rat pituitary cell bioassays. The most potent GRF analogs have also been evaluated and growth hormone levels in blood plasma have shown excellent correspondence with *in vitro* bioassays. Limited field trials are being conducted with selected analogs to improve performance in pigs (growth, feed utilization and improved lean to fat ratios) and in lactating cattle (feed utilization and enhanced milk production).

Conclusions

Analytical conditions have been established for the characterization of GRF and analogs including the use of analytical hplc, FAB-mass spectroscopy and tryptic digestion/hplc mapping. Circular dichroism measurements are in agreement with calculations predicting significant alpha-helical conformation between residues 16-26.

Analogs of GRF with increased biological potency have been synthesized in which Gly^{15} was replaced by other hydrophobic amino acids (increasing order of biological potencies: Ala > Leu > Val). Rationale for enhanced biological potency is attributable to increased alpha-helicity and maximization of amphiphilic structure.

Limited *in vivo* field trials in lactating cows have demonstrated a good dose-response relationship and equivalency (weight basis) of GRF(1-44)-NH_2 with the shortened analog, GRF(1-29)-NH_2. Pig growth trials using GRF(1-44)-NH_2 have been successfully completed showing positive weight gain, improved feed utilization and increased lean to fat ratio. Ala^{15}-substituted analogs of GRF demonstrating increased potency in releasing growth hormone *in vitro* and *in vivo* are now being tested in performance enhancement trials.

References

1. Rivier, J., J. Spiess, M. Thorner, W. Vale. 1982. Nature 300, 276.

2. Guillemin, R., P. Brazeau, P. Böhlen, F. Esch, N. Ling, W.B. Wehrenberg. 1982. Science 218, 585.

3. Frohman, L.A., T.R. Downs, T.C. Williams, E.P. Heimer, Y.-C.E. Pan, A.M. Felix: J.Clin.Inv. (in press).

4. Kaiser, E.T. and F.J. Kezdy. 1984. Science 223, 249.

5. Etherton, T., J.P. Wiggins, C.S. Chung, J.F. Rebhun and P.E. Walton. 1985. Endocrine Soc. Abst. #475.

6. Pelletier, G., T. Petitclerc, H. Lapierre, J. Morisset, P. Gaudreau, Y. Couture and P. Brazeau: (in prep.).

SYNTHESIS AND BIOLOGICAL ACTIVITIES OF THREE ARGININE VASOPRESSIN ANALOGS WITH 4-HYDROXYPROLINE IN POSITION 7

Diana Gazis, Angeliki Buku, John D. Glass, Irving L. Schwartz, Herman R. Wyssbrod

Department of Physiology and Biophysics and Center for Polypeptide and Membrane Research, Mount Sinai School of Medicine, New York, NY 10029 USA

Introduction

Substitutions for the normally occurring proline residue in position 7 of arginine vasopressin drastically alter the specificity of the molecule and also, in some cases, produce analogs with activities greater than those of the parent compound. To investigate the structural basis for these changes in biological activity, we synthesized and bioassayed the following three analogs with hydroxy groups on carbon 4 of proline: [7-(trans-4-hydroxy)-L-proline] arginine vasopressin (trans-OHPro^{7}AVP, [1-desamino,7-(trans-4-hydroxy)-L-proline] arginine vasopressin (d-trans-OHPro^{7}AVP) and [1-desamino,7-(cis-4-hydroxy)-L-proline) arginine vasopressin (d-cis-OHPro^{7}AVP).

Methods

Analogs were prepared by stepwise solid phase peptide synthesis[1] on a benzhydrylamine hydrochloride (BHA) resin using a Vega 96 Automatic peptide synthesizer. The following bioassay preparations were used: antidiuretic activity on water loaded anesthetized rats[2], pressor activity on urethane-anesthetized, phenoxybenzamine-treated rats[3], and uterine activity *in vitro*[4] in solutions with or without 0.5 mM Mg^{++}.

Peptides 1986

Results and Discussion

Biological activities of the three hydroxyproline compounds and of other relevant analogs are shown in Table 1. Introduction of a hydroxy group on proline enhances antidiuretic and uterine activities but depresses pressor activity. The cis enantiomer is somewhat less active than the trans enantioner, but it is still very active. Deamination increases antidiuretic activity as it does in all the other analog pairs shown except one.

Table 1. Biological Activities of 7-Hydroxyproline-Substituted and Other 7-Substituted Arginine Vasopressin Analogs

Analog	Biological Activity in U/mg			
	Antidiuretic	Pressor	Uterus in Vitro	
			no Mg^{++}	Mg^{++}
Trans-OH-Pro[7] AVP	712	4	73	268
d-Trans-OH-Pro[7] AVP	780	1	225	285
d-Cis-OHPro[7] AVP	614	8	63	176
[7-DehydroProline]AVP[5]	1260	255	41	
d[7-DehydroProline]AVP[5]	4134	240	69	
AVP[6]	323	369	12	
dAVP[7]	1745	346	28	
[7-N-Methylalanine]AVP[8]	343	11	3	4
d[7-N-Methylalanine]AVP[8]	113	1	2	8
[7-Sarcosine]AVP[8]	188	4	9	36
d[7-Sarcosine]AVP[8]	580	1	38	49
[7-Glycine]AVP[5]		0.01	0.1	

In Table 1 the activities of the hydroxyproline-substituted analogs are compared with two types of 7-substituted analogs - - - those (N-methylalanine, sarcosine, glycine) in which proline is replaced with a residue which has only part of a proline ring and those (dehydroproline) in which a double bond has been introduced into the proline ring. All 7-substituted analogs have antidiuretic activity, but those with some electronegativity on the proline ring (hydroxyproline or dehydroproline) are extremely active - - - more active than vasopressin. For pressor activity, the critical requirement seems to be an intact proline ring with no added bulk. Uterine activity is greatest in the hydroxyproline analogs, which have strikingly higher activities than vasopressin.

Acknowledgements

This research was supported by Grant AM-10080 from the National Institute of Arthritis, Metabolic Diseases and Diabetes and from the Life Sciences Foundation.

References

1. Merrifield, R.B., 1962, J. Am. Chem. Soc, 85, 2149.

2. Sawyer, W. H., 1958, Endocrinology 63, 694.

3. Dekanski, J., 1952, Brit. J. Pharmacol. 7, 567.

4. Munsick, R. A., 1960, Endocrinology 66, 451.

5. Ross, C. R., Smith, C. W., Chan, Y. L., and Walter, R. W., 1979, J. Med. Chem. 22, 926.

6. Manning, M., Coy, E., Sawyer, W. H., Acosta, M., 1873, J. Med. 16, 463.

7. Manning, M., Balaspiri, L., Moehring, J., Haldar, J., Sawyer, W. H., 1976, J. Med. Chem. 19, 842.

8. Grzonka, Z., Lammek, B., Kasprzykowski, K., Gazis, D., and Schwartz, I. L., 1983, J. Med. Chem. 26, 555.

STUDIES WITH PEPTIDES RELATED TO THE SEQUENCE 4-9 OF ACTH

R. Geiger, H.J. Gerhards, G. Wiemer, F.J. Hock
HOECHST Aktiengesellschaft, D-6230 Frankfurt/Main 80, FRG

Introduction

Behavioural effects of ACTH and short-chain analogues have been established for a long time (1,2,3). QSAR studies revealed that the hexapeptide $ACTH_{4-9}$ contains the essentialelements for the avoidance behaviour effect. Amino acid exchanges in several positions as well as C-terminal elongation of the peptide chain increase the potency of this essentialcore dramatically. Compared with $ACTH_{4-10}$, a peptide corresponding to $ACTH_{4-16}$-NH_2 with replacements by $Met(0)^4$, $D\text{-}Lys^8$, and Phe^9 exhibited a 100.000 fold increase of the effect on the delay of extinction in the pole-jumping test (4).

We observed that the turnover of acetylcholine (ACH) in the midbrain area is influenced by the dipeptides Lys-Phe and D-Lys-Phe if their amino group is protected by benzyloxycarbonyl (Z) and the carboxyl group is esterified. The corresponding dioxopiperazines are inactive but the effect is increased by substitution with a basic amide.

It has been shown, on the other hand, that the main metabolite Phe-D-Lys-Phe of the behaviourally potent hexapeptide H-Met(0)-Glu-His-Phe-D-Lys-Phe-OH still possesses behavioural activity, albeit much less than $ACTH_{4-10}$ (5). Thus, the substitution of the carboxyl group with a basic amide may have potentiated the weak activity in this structure to the extent that it became manifest.

This reminds one of the role of the basic amino group in Lys^{11} of ACTH (6). The hormone is known to increase ACH turnover in

Peptides 1986

the brain (7). We therefore focussed on the contribution of this basic amino group to the biological activity of short-chain analogues of ACTH in the central nervous system (CNS).

Results and discussion

A series of basic amides of peptides related to the essential amino acid sequence of ACTH has been synthesized by stepwise N-terminal elongation of an amino acid unit which bears an amino group in a position corresponding to that of Lys^{11} in ACTH. To assess the differences between the carboxylic form and the basic amide, we mainly refer to compounds VII and IX in Table 1 since most information is available for the former peptide whose structure has been optimized with regard to behavioural potency by Greven, De Wied and their co-workers (4,5).

Table 1: Influence of Peptides Related to $ACTH_{4-9}$ on Acetylcholine Turnover in the Striatum

Compound		Acetylcholine level in the striatum (rat)	
		Dose with max. effect (µg/kg s.c.)	% control
I	Z — Lys-Phe-OMe	10	118
II	Z-D-Lys-Phe-OMe	10	121
III	Z-D-Lys-Phe-NH-$(CH_2)_8$-NH_2	1	130
IV	ACTH-(1-24)	0.01	78
V	H-Met —— Glu-His-Phe — Arg-Trp-OH	1000	152
VI	H-Met(O)—Glu-His-Phe-D-Lys-Phe-OH	1	71
VII	H-Met(O_2)-Glu-His-Phe-D-Lys-Phe-OH	1	80
VIII	H-Met(O)—Glu-His-Phe-D-Lys-Phe-NH-$(CH_2)_8$-NH_2	0.01	74
IX	H-Met(O_2)-Glu-His-Phe-D-Lys-Phe-NH-$(CH_2)_8$-NH_2	0.01	81
X	H-Met(O_2)-Glu-Ala-Phe-D-Lys-Phe-NH-$(CH_2)_8$-NH_2	0.1	143
XI	Glutaroyl-Glu-Ala-Phe-D-Lys-Phe-NH-$(CH_2)_8$-NH_2	0.01	82
XII	H-Met(O_2)-Ala-Ala-Phe-D-Lys-Phe-NH-$(CH_2)_8$-NH_2	0.1	135
XIII	Z-Ala-Ala-Phe-D-Lys-Phe-NH-$(CH_2)_8$-NH_2	0.01	177
XIV	H-Met(O)—Cys-His-Phe-D-Lys-Phe-Cys-NH-$(CH_2)_4$-NH_2	10	65
XV	H-Cys —— Glu-His-Phe-D-Lys-Phe-Cys-NH-$(CH_2)_4$-NH_2	10	61

Decrease in ACH should reflect increased utilization of ACH since the co-administration of the choline uptake inhibitor hemicholinium-3 and IX potentiates the decrease of ACH induced by hemicholinium-3 (8).

The potency of IX in decreasing the ACH level corresponds to that of $ACTH_{1-24}$, whereas VII is about 100 times less potent. In contrast, $ACTH_{4-9}$ augments ACH, but only at a high dose. Modifications in position 4-6 have different consequences. Full potency is retained after exchange of His with Ala and $Met(O_2)$ with glutaric acid (XI), replacement of Ala for His alone (X), however, inverses the effect at a tenfold higher dose. Additional substitution of $Met(O_2)$ by benzyloxycarbonyl and Glu by Ala (XIII) leads to the most potent ACH increasing compound.

More detailed experiments showed that ACH is decreased by IX in the striatum, hippocampus and hypothalamus. In contrast to the different potency of VII and IX in lowering ACH in the striatum, both compounds are equipotent in the stimulation of choline acetyltransferase (CAT). This indicates that differences are a con-consequence of utilization rather than of formation of ACH.

The influence of IX on ACH turnover is part of a complex process since the effects are still seen after 24 hrs, whereas the half-life of the peptide is short. Increase of cGMP parallels ACH decrease in the hippocampus, striatum and cerebellum. Enhanced glucose uptake as a general phenomenon accompanying stimulation of brain functions has been seen in every area tested so far at 0.1 mcg/kg IX s.c. in contrast to VII which is inactive after a single dose of 10 mcg/kg (9) and becomes only effective after chronic treatment with 100 mcg/kg i.p. for 10 days (10).

It is not fully clear which variable can be correlated with cognitive and memory functions. In the 8-arm radial maze performance test (11), compound IX in doses of 10 and 3 mcg/kg s.c., inhibits the amnesic effect of scopolamine in rats.

The biological significance of the elevation of ACH levels in the CNS after administration of e.g. XIII has still to be evaluated.

We conclude that by substitution of the free carboxyl group with a basic amide (IX), effects of VII and related compounds in the CNS are increased to a degree similar to that provided by Lys^{11} to the hormonal action of ACTH. Conformational restraints by cyclization (XIV and XV) do not lead to increased potency.

References

1. Mirsky, I.A., R. Miller, M. Stein. 1953. Psychosom. Med. 15, 574.
2. De Wied, D. 1969. In: Frontiers in Neuroendocrinology (W.F. Ganong and L. Martini, eds.). Oxford University Press London/New York, p.97.
3. De Wied, D. 1977. Ann.N.Y.Acad.Sci. 297, 263.
4. van Nispen, J.W., H.M. Greven. 1982. Pharmac.Ther. 16, 67.
5. De Wied, D., W.H. Gispen. 1977. In: Peptides in Neurobiology (H. Gainer ed.). Plenum Press New York/London, p.397.
6. Geiger, R., H.G. Schröder. 1973. Hoppe-Seyler's Z. Physiol. Chem. 345, 156.
7. Wood, P.L., D. Malthe-Sorenssen, D.L. Cheney, E. Costa. 1978. Life Sci. 22, 673.
8. Wiemer, G., H. Gerhards, R. Geiger. 1985. Soc.Neurosci. Abstr. 11, Part 1, p.709
9. Delanoy, R.L., A.J. Dunn. 1978. Pharmac.Biochem.Behav. 9, 21
10. McCulloch, J., P.A.T. Kelly, A.M.L. van Delft. 1982. Eur. J. Pharmac. 78, 151
11. Stäubli, U., M. Baudry, G. Lynch. 1984. Behav.Neural Biol. 40, 58

STRUCTURE-ACTIVITY RELATIONSHIPS OF VASOPRESSIN: SYNTHESES OF NEW VASOPRESSIN ANALOGUES

Zbigniew Grzonka, Leszek Łankiewicz, Franciszek Kasprzykowski, Adam Liwo

Institute of Chemistry, University of Gdańsk, 80-952 Gdańsk, Poland

Per Melin

Ferring Pharmaceuticals, 200 62 Malmö, Sweden

Introduction

As a part of our continuing program of investigation into structure-activity relationships of vasopressin we have synthesized six analogues containing modifications at positions 1,2,4 and 7. These peptides are as follows: [7-glycine]AVP, (1); [1-β-mercaptopropionic acid,7-glycine]AVP (2); succinyl-bis[AVP], (3); [1-thiosalicylic acid]AVP, (4); [1-thiosalicylic acid,2-D-phenylalanine]AVP, (5); [1-thiosalicylic acid,2-D-phenylalanine, 4-isoleucine]AVP, (6).

Results

Arginine-vasopressin and its 1,2,4,5 and 6 analogues were prepared by solid-phase method. Succinyl-bis[AVP] (3), was synthesized by acylation of AVP with di-p-nitrophenyl succinate. The results of syntheses are summarized in Table 1.

7-Glycine substituted analogues

[Gly^7]arginine- and lysine-vasopressins since their syntheses (1,2) were considered as almost inactive analogues. This is in contrast to highly active and very selective [7-sarcosine]AVP (3). Therefore, we decided to resynthesize [Gly^7]AVP and to recheck its potency. As can be seen from the data presented in Table 2,

Peptides 1986

[Gly^7]AVP, as well as its new deamino derivative (2), possesses relatively high antidiuretic activity , whereas vasopressor potency is diminished. Thus, the A/P selectivity of both analogues is 3-4 orders higher than that of the parent hormone.

Table 1

Analogues of arginine-vasopressin (AVP)

Peptide	Total yield of pure analogue %	$[\alpha]_D^{24}$ c=0.5 1N AcOH deg.	TLC R_f in solvent systems BAW	BAWP
[Gly^7]AVP	41.2	-3.7	0.23	0.39
[Mpa^1,Gly^7]AVP	16.4	-36.8	0.18	0.27
Suc-bis [AVP]	20.5	-90.9	0.07	0.11
[Ths^1]AVP	12.4	-121.9	0.21	0.40
[Ths^1,D-Phe^2]AVP	5.7	-94.1	0.20	0.36
[Ths^1,D-Phe^2,Ile^4]AVP	10.1	-200.5	0.31	0.65

Symbols: Mpa - β-mercaptopropionic acid; Suc - succinic acid; Ths - thiosalicylic acid; BAW - n-butanol-acetic acid-water (4:1:5, v/v, upper phase); BAWP - n-butanol-acetic acid-water-pyridine (15:3:3:10, v/v).

Table 2

Biological activity of arginine-vasopressin analogues

Analogue	Biological activity: Antidiuretic A IU/mg	Vasopressor P IU/mg	Selectivity[a] A/P
[Gly^7]AVP	80	0.11	730
[Mpa^1,Gly^7]AVP	172	0.019	9050
Suc-bis [AVP]	0.20	0.6	
[Ths^1]AVP	0.12	0.04	
[Ths^1, D-Phe^2]AVP	0.009	0.03	

[a] The ratio A/P for AVP is 0.9.

Succinyl-bis[AVP]

Dimerization of AVP using succinic acid as a divalent carrier leads to a compound possessing very low agonistic activity (Table 2). This analogue was devoid of antagonistic properties.

1-Thiosalicylic acid substituted analogues

It is well known that the introduction of β-mercapto-β,β-cyclopentamethylenepropionic acid residue (Mca) in position 1 gave analogues with antagonistic potencies. Here, we would like to report a new type of conformationally constrained analogues with thiosalicylic acid (Ths) in position 1. As opposed to the Mca^1 substituted analogues ("perpendicularly" situated cyclohexyl ring in relation to hexapeptide ring), Ths^1 containing analogues ("parallel" orientation of aromatic ring; C_α and C_β of residue 1 are part of aromatic nucleus)are devoid of antagonistic properties. Their antidiuretic and vasopressor activities are very low (Table 2).
Molecular mechanics calculations (energy minimization) have been carried out for the cyclic moiety of deamino-vasopressin, [Mpa^1]-AVP, and its [Mca^1]- and [Ths^1]-analogues. The force field of Weiner et al. (4), which uses the united atom approach has been employed. The MM2 program (5) has been adapted to this force field. In the calculation all degrees of freedom were relaxed. The starting backbone conformation of each molecule was the same as the lowest energy conformation of oxytocin (6). The side chains were attached as not to coincide with each other. To generate the initial structures of the analogues, cyclohexane or benzene fragment was properly attached to the cyclic moiety, regardless of initial strains in the case of thiosalicylic acid analogue. Table 3 lists some of the calculated dihedral angles of AVP analogues. As can be observed the substitution of β-mercaptopropionic acid by its cyclohexyl containing analogue (Mca) does not change the conformation of cyclic hexapeptide. On the other hand, the introduction of Ths residue in position 1 leads to the dramatic change of dihedral angles. This in our opinion, can explain different biological behaviour of Ths substituted analogues.

Table 3

Calculated dihedral angles of the first three residues in AVP analogues

Residue	Φ	Ψ	ω	χ^1	χ^2	χ^3
[Mpa[1]]AVP:						
Mpa[1]		(-133)	(-178)	(58)	(80)	-97
Tyr[2]	-118	155	-179	-173	-119	-90
Phe[3]	-54	-37	179	-173	66	
[Mca[1]]AVP:						
Mca[1]		(-130)	(-178)	(57)	(80)	-101
Tyr[2]	-119	153	180	-171	-119	-90
Phe[3]	-54	-37	178	-173	66	
[Ths[1]]AVP:						
Ths[1]		(-159)	(-171)	(10)	(105)	87
Tyr[2]	-61	152	178	-163	-122	-90
Phe[3]	-55	-39	178	-177	64	

Acknowledgment

This research was supported by Polish Academy of Sciences (grants Nos. CPBP-01.13.2.9 and CPBR-3.13.4.3.2).

References

1. Papsuevich, O.S., Krikis, A., Chipens, G. 1972. Zh. Obshch. Khim. 42, 224.
2. Kolc, J., Zaoral, M., Šorm, F. 1967. Collect. Czech. Chem. Commun. 32, 2667.
3. Grzonka, Z., Lammek, B., Kasprzykowski, F., Gazis, D., Schwartz, I.L. 1983. J. Med. Chem. 26, 555.
4. Weiner, S.J., Kollman, P.A., Case, D.A., Singh, U.Ch., Ghio, C., Alagona, G., Profeta, S., Weiner, P. 1984. J. Am. Chem. Soc. 106, 765.
5. Allinger, N.L., Yuh, Y.H. 1980. QCPE, program No. 395.
6. Spasov, V.Z., Popov, E.M. 1981. Bioorg. Khim. 7, 25.

OXYTOCIN ANALOGS CONTAINING TWO DISULFIDE BONDS

J. Hlaváček, J. Konvalinka, J. Slaninová, I. Frič

Institute of Organic Chemistry and Biochemistry, Czechoslovak Academy of Sciences, 166 10 Prague 6, Czechoslovakia

Introduction

Recently, a series of oxytocin (Ia)* analogs with sterically hindered carboxyterminal tripeptide part was described. Some of the analogs exhibited interesting changes in biological activities of oxytocin and increased stability towards chymotrypsin cleavage (2,3).

Now we describe the synthesis and some properties of the analogs Ib and Ic containing an additional disulfide bond in the carboxyterminus of the oxytocin molecule.

H - Cys - Tyr - Ile - Gln - Asn - Cys - R

Ia, R = Pro - Leu - $GlyNH_2$

Ib, R = Cys - Leu - $CysNH_2$ Ic, R = Pro - Cys - $CysNH_2$

Results

The analogs Ib and Ic were prepared by the coupling of two cyclic disulfide fragments, Boc - tocinoic acid (4,5) and tripeptides H-Cys-Leu-$CysNH_2$ or H-Pro-Cys-$CysNH_2$, using dicyclohexylcarbodiimide and 1-hydroxybenzotriazole as condensation reagents in dimethylformamide. The Boc group was removed by treatment with TFA and the analogs purified on Biogel P-4 in 3M AcOH and by RP HPLC in 50% MeOH

---

*Nomenclature of amino acids and peptides follows published suggestions (1). TFA, trifluoroacetic acid; AcOH, acetic acid. Amino acids used in this study were all in the L - configuration.

Peptides 1986

-0.05%TFA. Similar reaction conditions were used in the synthesis of protected precursors Z-Cys(Bzl)-Leu-Cys(Bzl)NH_2 and Z-Pro-Cys(Bzl)-Cys(Bzl)NH_2 which were built up from the carboxyterminus of the peptide chain on coupling of the corresponding Boc- and Z-amino acids. The Bzl and Z groups were cleaved by means of Na/NH_3 and the disulfide bridge closed by oxidation with $K_3Fe(CN)_6$ in high dilution. The cyclic tripeptides were desalted on Dowex 50x2, monomers separated on Sephadex G-10 in 0.2M AcOH and further purified by RP HPLC in 40% MeOH - 0.05% TFA or by gradient elution using the same solvents.

The sterically hindered analogs Ib and Ic are nearly devoid of the oxytocin - like activity (6-9). Analog Ib exhibits a weak antagonist activity (pA_2 = 6.5) on isolated rat uterus. Its intrinsic uterotonic activity has not reached the peak oxytocin activity at any concentration of the analog. The galactogogic and pressor activity is about three orders of magnitude lowered similarly to the uterotonic activity of the analog Ic. The increased rigidity of the cyclic tripeptide part in these analogs leads probably to the change of overall molecule conformation resulting in a strong decrease of affinity to receptor. On the other hand, the weak inhibitory effect suggests that the analog Ib interacts with a uterine receptor without possibility to accept a biologically active conformation.

The increased stability in vitro of the analog Ib against chymotrypsin cleavage was observed (Fig.1). The Edman degradation of the product isolated by HPLC after its digestion with chymotrypsin

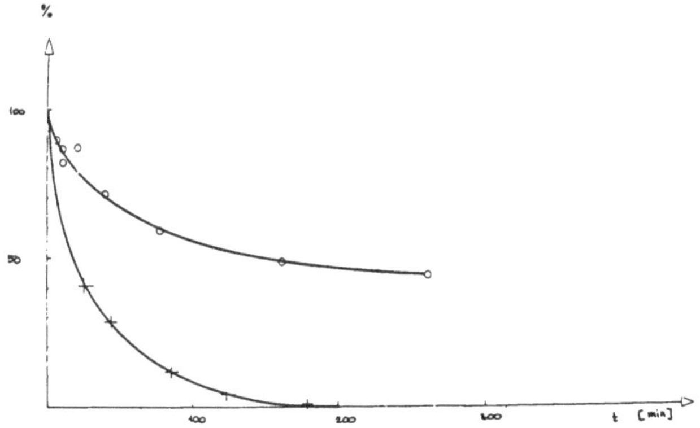

Fig.1. Time course of cleavage of oxytocin Ia (x---x) and the analog Ib (o---o) with chymotrypsin. The amount of the non-cleaved compound was determined by HPLC. The results are expressed as percent of non-cleaved peptide in minutes of incubation time (4).

revealed the cleavage of the peptide chain between Cys and Tyr residues in positions 1 and 2. Thus, the Leu-Gly peptide bond (positions 8 and 9) normally cleaved in oxytocin, is probably protected against chymotrypsin degradation in this analog due to the presence of disulfide bond spanning positions 7 and 9.

Circular dichroism spectra of the cyclic tripeptides are interesting above all as revealing spectral and conformational properties of the disulfide group closing an 11- and 8- membered ring. The tripeptide H-Cys-Leu-CysNH_2 does not show any significant disulfide bands in the neutral buffer solution (curve 1 in Fig. 2) whereas in acid solution and in trifluoroethanol (curve 2 in Fig. 2), it exhibits weak negative bands in the 250 and 270 nm region. (For the CD spectra of the protected tripeptide see Ottnad et al, 10). The disulfide group in this tripeptide thus appears to remain still relatively flexible and exist as a mixture of conformers in which the strained transoid conformers tend to prefer. The tripeptide H-Pro-Cys-CysNH_2 exhibits, on the other band, only positive bands in both solvents (a maximum at 250 and a shoulder at 270 nm in neutral buffer, curve 3, and a relatively strong maximum in trifluoroethanol, curve 4 in Fig. 2). In this smaller ring the disulfide group appears to be more rigid and prefer the rectangular conformation.

The replacement of the linear tripeptide with a cyclic tripep-

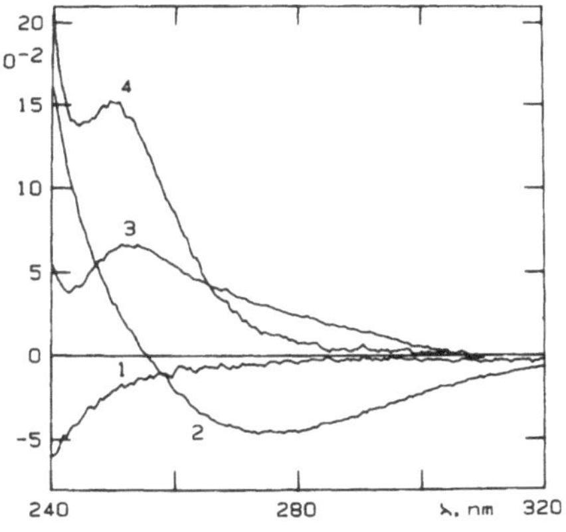

Fig.2 H-Cys-Leu-CysNH_2 (1,2)
H-Pro-Cys-CysNH_2 (3,4)
Buffer pH 7.5 (1,3)
Trifluoroethanol (2,4)

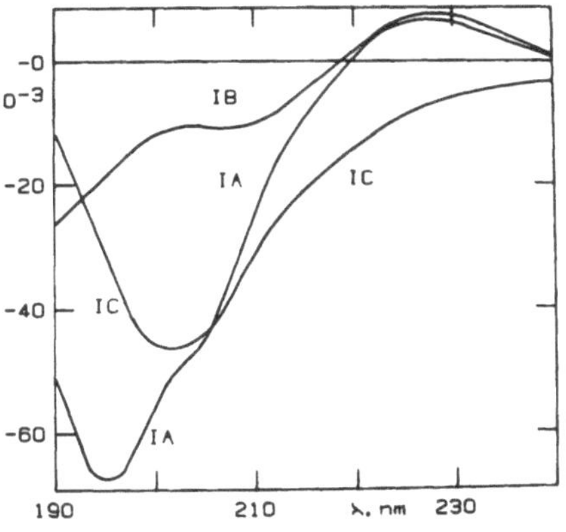

Fig.3 Oxytocin (IA)
/Cys^7, Cys^9/OXT (IB)
/Cys^8, Cys^9/OXT (IC)
in Buffer pH 7.5

tide affects strongly the CD behavior and conformation of oxytocin as it is apparent from Fig.3, where the CD spectra of oxytocin (IA) and analogs IB and IC in neutral buffer solution are compared. In IB only the shortwavelength part of the spectrum is changed. This indicates that there is no interaction between the cyclic tripeptide and the disulfide and Tyr region of the 20- membered ring moiety but the peptide backbone conformation is substantially changed. Spectrum of IC differs from that of oxytocin both in the longwavelength and shortwavelength spectral region. It appears that there exists an interaction between the bulky tripeptide chain and the region of the Tyr residue. This corresponds with the more flexible connection of the two ring moieties in the analog IC. The above interpretation of CD spectra is based on previous studies; see for example Frič (11).

References

1. IUPAC - IUB: Nomenclature and Symbolism for Amino Acids and Peptides, Recommendations 1983. 1984. Eur.J.Biochem. 138, 9.
2. Hlaváček, J., K. Jošt. 1985. In: Proc. F.E.C.S. 3rd Int.Conf. Chem. and Biotechn. of Biol.Act.Nat.Prod. (R. Vlahov, ed.). Publ. House Bulg. Acad. Sci., Vol.3, p. 244.
3. Hlaváček, J. In: Handbook of Neurohypophyseal Hormone Analogs. (K. Jošt, M. Lebl, F. Brtník, eds.). CRC Press, Boca Raton, in press.
4. Hlaváček, J., J. Pospíšek, J. Slaninová, T. Barth, K. Jošt. 1984. In: Peptides 1984 (U. Ragnarsson, ed.). Almqvist and Wiksell Int., p. 415.
5. Mühlemann, M., M.I. Titov, R. Schwyzer, J. Rudinger. 1972. Helv. Chim.Acta 55, 2854.
6. Holton, P. 1948. Brit.J.Pharmacol. 3, 328.
7. Munsick, R.A. 1960. Endocrinology 66, 451.
8. Bisset, G.W., B.J. Clark, J. Haldar, M. Harris, G. Lewis, M. Rocha e Silva. 1967. Brit.J.Pharmacol.Chemotherap. 31, 537.
9. Krejčí, I., B. Kupková, I. Vávra. 1967. Brit.J.Pharmacol. Chemotherap. 30, 497.
10. Ottnad, M., P. Hartter, G. Jung. 1976. Eur.J.Biochem. 66, 115
11. Frič, I. In: Handbook of Neurohypophyseal Hormone Analogs. (K. Jošt, M. Lebl, F. Brtník, eds.). CRC Press, Boca Raton, in press.

SYNTHESIS OF ALA-SUBSTITUTED ANALOGS OF ACTH/αMSH-(4-10) AND THEIR EFFECTS ON STEROIDOGENESIS AND PIGMENT DISPERSION

M. Lőw, L. Kisfaludy

Chemical Works of G. Richter Ltd., P.O.B. 27, H-1475 Budapest, Hungary

K.Sz. Szalay

Institute of Experimental Medicine, Hungarian Academy of Sciences, P.O.B. 67, H-1450 Budapest, Hungary

H. Medzihradszky-Schweiger

Research Group for Peptide Chemistry, Hungarian Academy of Sciences H-1088 Budapest, Hungary

Introduction

ACTH/αMSH-(4-10) ($\overset{4}{\text{Met}}$-Glu-His-Phe-Arg-Trp-$\overset{10}{\text{Gly}}$) dose-dependently stimulates corticosterone production in zona fasciculata cells and aldosterone production in zona glomerulosa cells of rat adrenals (1). It also shows a high melanotropic activity in the frog skin test (2). To get an insight into the role of the amino acid side chains in the above effects, analogs substituted with L-alanine in one of the positions 4, 5, 6, 7, 8 and 9 were synthesized and tested.

Peptide synthesis

All peptides were synthesized in solution. The scheme of the synthesis of ACTH-(4-10) is shown in Fig. 1. The analogs were synthesized essentially in the same manner as used in the synthesis of ACTH-(4-10), except that intermediates containing L-alanine in the given positions were employed. The crude products were purified by reversed phase column chromatography on silica gel, or by crystallization. All peptides were characterized by melting point, optical rotation, TLC, RP-HPLC, UV-spectrum and amino acid analysis (Table 1).

Peptides 1986

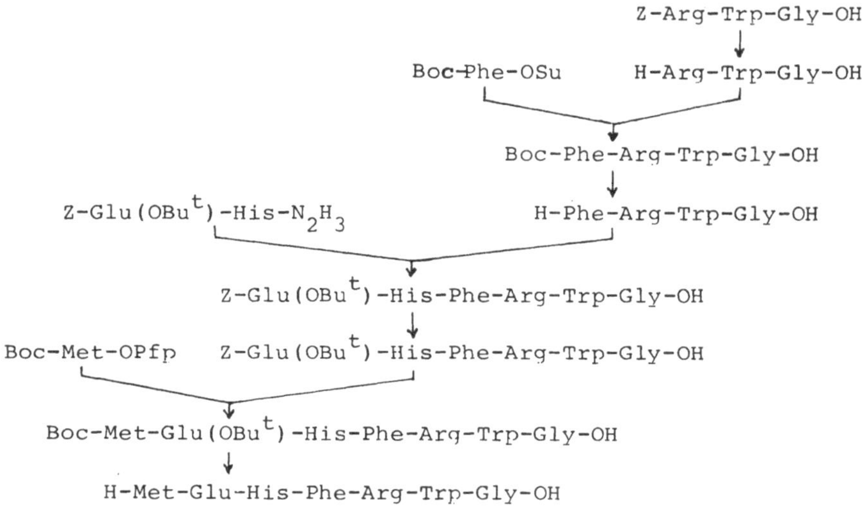

Fig. 1. Synthesis of ACTH-(4-10)

Table 1. Physical Constants of Protected and Free Heptapeptides

	protected heptapeptide		free heptapeptide			
	m.p.[a] °C(dec.)	R_f[b]	purification	m.p. °C(dec.)	$[\alpha]_D^g$	R_f^c
ACTH-(4-10)	209-211	0.20	d	216-218	-13.2°	0.30
Ala^4-ACTH-(4-10)	205	0.17	e	207	-23.3°	0.19
Ala^5-ACTH-(4-10)	199	0.14	e	202	-14.4°	0.37
Ala^6-ACTH-(4-10)	212	0.32	f	213	-24.9°	0.47
Ala^7-ACTH-(4-10)	196	0.16	e	180	-15.5°	0.18
Ala^8-ACTH-(4-10)	217	0.42	e	190	-17.9°	0.47
Ala^9-ACTH-(4-10)	217	0.13	e	222	-21.9°	0.16

a crystallized from ethanol-water (1:1)
b and c Kieselgel 60 (Merck) /
ethyl acetate-(pyridine-acetic acid-water=20:6:11)=3:2 and 2:3
d crystallization from ethanol-water
e reversed phase chromatography in methanol-water(0,1 % TFA)
f crystallization from dimethylformamide-water
g in acetic acid-water (4:1), c=0.4

Biological assay

The zona glomerulosa and zona fasciculata cell suspensions were prepared by collagenase digestion of rat adrenal glands. The peptides were added to the cell suspensions in amounts to produce from 10^{-9} to 10^{-3} mol/l concentrations. The corticosterone content of the incubation media was determined by fluorometry after chloroform extraction. The chloroform extract was assayed for aldosterone by radioimmunoassay; the whole process was described previously (1). The skin darkening effect was measured on skin of the laughing frog (Rana ridibunda), by the method of Shizume et al. (3), with the aid of a Specol reflectometer.

Results

The steroidogenic effect of the analogs was different in the two adrenal cell systems. The Ala^4, Ala^6-, Ala^7- and Ala^9-analogs had no stimulatory effect on zona fasciculata corticosterone production even at 1 mmol/l concentration; the Ala^8-analog had a

Table 2. Relative Potencies of Analogs of ACTH-(4-10) on Steroid Release from Adrenal Cells and in Frog Skin Darkening Assay

	zona[a] fasciculata corticosterone	zona[a] glomerulosa aldosterone	frog skin[b] darkening	melanotropic activity U/mmol
ACTH-(4-10)	1	1	1	$1.3 \cdot 10^6$
Ala^4-ACTH-(4-10)	0	0.13	0.08	$1.0 \cdot 10^5$
Ala^5-ACTH-(4-10)	1.19	0.91	0.09	$1.1 \cdot 10^5$
Ala^6-ACTH-(4-10)	0	0	0.42	$5.5 \cdot 10^5$
Ala^7-ACTH-(4-10)	0	0.11	0.37	$4.8 \cdot 10^5$
Ala^8-ACTH-(4-10)	0.31	0.08	0.29	$3.8 \cdot 10^5$
Ala^9-ACTH-(4-10)	0	0.05	0.05	$7.0 \cdot 10^4$

a relative effect at 1 mmol/l concentration to ACTH-(4-10)
b calculated from concentrations producing equal darkening

distinct stimulatory effect, and the effect of Ala^5-ACTH-(4-10) actually exceeded that of the parent compound (Table 2).

In zona glomerulosa cells Ala^5-ACTH-(4-10) exhibited about the same effect as ACTH-(4-10) did. The effect of the Ala^4-, Ala^7-, Ala^8- and Ala^9-analogs (1 mmol/l) was only 5-10% (Table 2). In contrast to these results, Ala^6-ACTH-(4-10) (1mmol/l) inhibited aldosterone production, induced by ACTH-(4-10) (100 nmol/l), by 30%.

The skin darkening effect of the analogs showed slight variations. None of them produced an effect similar to that of the parent compound. The potency of Ala^6-, Ala^7- and Ala^8- analogs was 30-40% compared with ACTH-(4-10), whereas the Ala^4-, Ala^5- and Ala^9-analogs had 5-8% relative potencies (Table 2).

Conclusions

The ACTH-(4-10) receptors seem to be different in the three types of cells examined. For triggering corticosterone production in zona fasciculata cells, the presence of Met^4, His^6, Phe^7 and Trp^9 in the molecule seems to be essential, in contrast to the role of Glu^5. For triggering aldosterone production of zona glomerulosa cells only His^6 seems to play a dominant role. By contrast with these, for triggering melanin dispersion in melanofores the presence of the each amino acid side chain appears to be nearly equally important and absence of one of them does not totally inactivate the peptide.

References

1. Szalay, K.Sz., D. DeWied, E. Stark and G. Folly. 1985. Regulatory Peptides 11, 187-192.

2. Li, C.H., E. Schnabel, D. Chung and T.B. Lo. 1961. Nature 189, 143.

3. Shizume, K., A.B. Lerner and T.B. Fitzpatrick. 1954. Endocrinology 54, 553-560.

SYNTHESIS AND BIOLOGICAL ACTION OF ACTH-MSH FRAGMENTS CONTAINING MELPHALAN

H. Medzihradszky-Schweiger, H. Süli-Vargha, J. Botyánszki and K. Medzihradszky

Research Group for Peptide Chemistry, Hungarian Academy of Sciences, 1088 Budapest, Hungary

Introduction

Peptide hormones or biologically active fragments containing melphalan can be applied for the covalent labelling of the hormone receptor. They offer a possibility for the chemical characterization of the receptor and may also be used for therapeutical purposes, as receptor blockers, hormone preparates of long lasting activity or antitumor agents. With the same purposes in mind earlier we prepared and investigated some enkephalin derivatives (1), now we present the synthesis and the biological evaluation of the melphalan analogs of some ACTH-MSH fragments.

Synthesis

The following melphalan analogs of the MSH-ACTH fragments were prepared by the classical solution methods of peptide synthesis:

I H-Mel-Lys-Pro-Val-NH_2
II H-Mel-Trp-Gly-Lys-Pro-Val-NH_2
III H-Mel-Glu-His-Phe-Arg-Trp-Gly-OMe
IV H-Mel-Glu-His-D-Phe-Arg-Trp-Gly-OH

In general, the hormone fragments (H-Lys(BOC)-Pro-Val-NH_2, H-Trp-Gly-Lys(BOC)-Pro-Val-NH_2, H-Glu(OtBu)-His-Phe-Arg-Trp-Gly-OMe, H-Glu(OtBu)-His-D-Phe-Arg-Trp-Gly-OH) were first synthesized by methods described earlier, and were acylated thereafter with BOC-Mel-OSu. The protected peptides were purified either by precipitation from solvent mixtures (in the case of peptide I and II), or by column chromatography on silica gel in cold room (the protected

Peptides 1986

derivatives of III and IV). TFA was used for the removal of the protecting groups. Purification of the free peptides I and II was performed by precipitation from ethanol-ether. Compounds III and IV proved to be sufficiently pure after deprotection. The free peptides were homogeneous on Sephadex G 10 chromatography, electrophoresis and TLC.

Stability

Melphalan, like all nitrogen mustard derivatives, is a reactive compound, which shows a certain degree of lability under different conditions. According to literature data (2) the half life time of melphalan is 10 hours in 1 N NH_4OAc, 20 hours in 10% AcOH and 42 hours in 1 N NaCl solution. Enhanced lability is expected when this compound is built into peptides containing basic amino acid residues. Therefore we tested the stability of our compounds in buffer solutions of different pH values at 37°C. As an example, decomposition of compound II, measured by the alkylation of nitrobenzyl-pyridine is shown in Fig. 1.

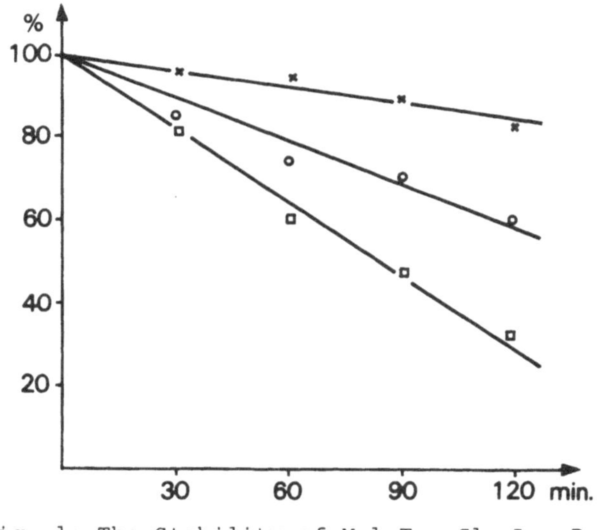

Fig. 1: The Stability of Mel-Trp-Gly-Lys-Pro-Val-NH_2
x = 0.05M Tris buffer, pH 6.0, 37°C
o = 0.05M Tris buffer, pH 7.0, 37°C
□ = 0.05M Tris buffer, pH 8.0, 37°C

Melanocyte stimulating activity

Activities were determined by the frog skin reflectance method of Shizume et al. (3) and expressed as units/mmole, the melanocyte stimulating activity of α-melanotropin taken as 4.10^{10} units/mmole. As a comparison, the activity of a heptapeptide analog, H-Phe-Glu-His-Phe-Arg-Trp-Gly-OMe was also measured. The values obtained are summarized in Table 1.

Table 1. Melanotropic Activity

Peptides	U/mmole
α-Melanotropin	4.10^{10}
Mel-Lys-Pro-Val-NH_2	1.10^{4}
Mel-Trp-Gly-Lys-Pro-Val-NH_2	2.10^{4}
Mel-Glu-His-Phe-Arg-Trp-Gly-OMe	3.10^{6}
Mel-Glu-His-D-Phe-Arg-Trp-Gly-OH	6.10^{7}
Phe-Glu-His-Phe-Arg-Trp-Gly-OMe	2.10^{6}

Discussion

It was interesting to see, whether substitution of the N-terminal amino acids by melphalan in MSH-ACTH fragments of different length would change the biological activity. From Table 1. it is clear that there is no dramatic change in the biological action in any direction, Phe-Glu-His-Phe-Arg-Trp-Gly-OMe is of about the same activity as compound III and similar activity is shown by the natural heptapeptide sequence, H-Met-Glu-His-Phe-Arg-Trp-Gly-OH as well. It is remarkable, however, that in contrast to the natural sequences, peptides containing melphalan behave another way, they bind more or less irreversibly to the melanocytes, as shown in the washing out step of the biological measurement. For example, after treatment with compound II only 30 per cent regeneration of the skin could be observed even after two hours.

It is very likely that this irreversible binding involves the MSH-receptors. Fig. 2 shows the biological effect of 0.1 ng/ml MSH (taken as 100%), the activity of the melphalan-analogs, finally the repeated measurement of the effect of 0.1 ng/ml MSH on the same skin piece. It is clearly seen that compound I has no inhibitory effect, compound II, III and IV however show 50-75% inhibition of the melanotropin action. It might be concluded that there is no reactive nucleophile in the neighborhood of the Lys-Pro-Val binding site of the melanotropin receptor.

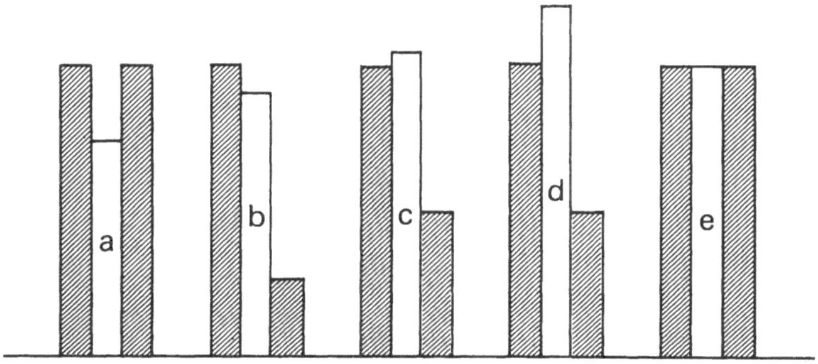

Fig. 2: The Inhibitory Effect of the Melphalan Analogs
▨ = 0.1 ng/ml MSH before and after the treatment with the analogs
a = 62 µg/ml Mel-Lys-Pro-Val-NH_2
b = 60 µg/ml Mel-Trp-Gly-Lys-Pro-Val-NH_2
c = 1.1 µg/ml Mel-Glu-His-Phe-Arg-Trp-Gly-OMe
d = 40 ng/ml Mel-Glu-His-D-Phe-Arg-Trp-Gly-OH
e = 1.1 µg/ml Phe-Glu-His-Phe-Arg-Trp-Gly-OMe

References

1. Medzihradszky, K., K. DiGlèria. 1984. In: Peptides, 1984. Proc. 18th European Peptide Symposium, Djurönäset. (U. Ragnarsson, ed.). Almquist & Wiksell, p. 297.

2. Hsieh K.-h., G.R. Marshall. 1981. J. Med. Chem. 24, 1304.

3. Shizume, K., A.B. Lerner, T.B. Fitzpatrick. 1954. Endocrinology 54 553.

NOVEL GnRH AGONISTS CONTAINING L-β-ASPARTIC ACID DERIVATIVES IN POSITION 6

J. Seprődi, J. Érchegyi, Zs. Vadász, I. Teplán
1st Institute of Biochemistry, Semmelweis University Medical School,
Budapest 8, P.O.B. 260, H-1444

B. Kanyicska
Institute of Experimental Medicine, Budapest, Hungary

Introduction

It has been shown that an L amino acid substitution of the glycine residue in position 6 of the gonadotropin-releasing hormone (GnRH) results in biologically almost inactive peptides (1). This explains why almost all potent agonistic and antagonistic GnRH analogues contain either D configurational or some optically inactive amino acids in the centre of the molecule (2). We assumed that an L-β-aspartyl residue in position 6 of GnRH might favour a similar conformation to the D amino acid-6 substituted superactive analogues. Here we describe a series of L-β-aspartyl-6 GnRH analogues with considerable potency yet only containing only naturally occuring L amino acids.

Materials and Methods

The IV-VIII peptides were synthesized by solid phase techniques using a Beckman 990 peptide synthesizer. Benzhydrylamine resin support and tert-butyloxicarbonyl protected amino acids were used to prepare the peptides. BOC-Asp-N-RR' derivatives were prepared by the condensation of BOC-Asp(OBzl)-OH with the amine component in the presence of dicyclohexyl-carbodiimide, followed by catalytic hydrogenation to give the free acid derivative. The peptides I-III and IX-XI were synthesized by solution phase methods following the (1-5)+(6-9) fragment condensation strategy. Final peptides were characterized by HPLC, TLC, amino acid analysis and optical rotation (3).
For the determination of in vivo LH releasing activity, analogues were injected subcutaneously into mature male mice. The serum LH level was determined by RIA (4).

Peptides 1986

Table 1. Structure and Biological Properties of GnRH Analogues

	Compounds	k' [x]	Relative Potency at 15 min [xx]	Relative Potency during 4-6 hours [xxx]
	D-Phe[6], desGly[10]-GnRH-EA	15.1	135.4	15
	pGlu-His-Trp-Ser-Tyr-Gly—Leu-Arg-Pro-Gly-NH_2	1.9	100	1
I	— Asp-OBzl — EA	17.3	74.3	12
II	— Asp-OH — EA	2.9	64.1	1
III	— Asp-OMe — EA	4.3	57.0	20
IV	— Asp-NH-$C(CH_3)_2$-CH_2-C_6H_5 — Gly-NH_2	16.8	30.6	1.7
V	— Asp-NH-C_6H_4 —	4.4	8.7	1.2
VI	— Asp-N<(CH_2-CH_2)(CH_2-CH_2)] —	3.4	7.0	0.4
VII	— Asp-NH-CH<(CH_2)(CH_2)>CH_2 —	2.8	6.5	0.4
VIII	— Asp-NH_2 —	1.4	3.8	0.4
IX	— Asp-$N(Et)_2$ — EA	6.0	3.2	0.8
X	— Asp-OMe — NMeLeu — EA	5.8	2.9	0.9
XI	— Asp-OtBu — Leu — EA	13.5	2.1	1.1

x relative retention times were determined by HPLC on 25x0.4 cm column packed with C18 bonded silica (10 µm, made by BST) in methanol/0.1 M ammonium-acetate buffer 47:53 mixture at pH 4

xx relative potencies were normalized to the ED_{50} of nativ GnRH (=22.5 ng/animal). ED_{50} values were determined in adult male mice. After the s.c. injection of 0.00125-125 µg/kg analogue at 15 min animals in group of 6 were decapitated and LH level was determined from blood by RIA (4).

xxx calculated from the total amount of LH released (see Figure) after a single s.c. injection of 1 µg analogue during 4-6 hours period.

Results and Discussion

The structure of the analogues with their relative LH releasing potency are shown in Table 1. A close correlation between the relative retention time (k') and the biological potency at 15 minutes after the injection can be established for the compounds IV-VIII, but the rest of the analogues listed do not show any tendency like this. For instance the analogue II with a polar -COOH group is about as effective as the much less polar benzyl-ester (I) or methyl-ester (III) derivatives. Furthermore the tert-butyl-ester XI and diethyl-amide IX derivatives exert only a slight activity in spite of their relatively high lypophylic character. These data may lead us to the conclusion, that the contribution of the side chain character of the residue in position 6 to the biological activity, if there is any of it at all, can not be significantly expressed after 15 minutes.

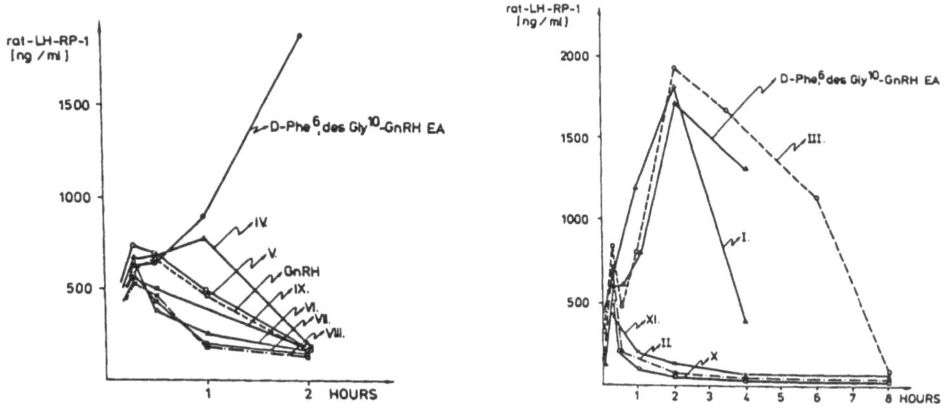

Figure. LH releasing effect of GnRH analogues.
Structures of analogues and the total amount of LH released are in Table 1.

The time dependent LH release of the analogues can be seen in the Figure. On the basis of the total amount of the LH released, the analogues can be separated into two groups: compounds II and IV-XI are more or less as active as the native GnRH and they show no significant LH releasing potency after 60 minutes. On the other hand, compounds I and III can release 12-20 times more LH than the GnRH and they have a strong long acting effect similar to that of the superactive D-Phe^6,desGly^{10}-GnRH EA analogue. It is interesting to notice that the LH release profile of the superactive analogues has two separated peaks, one of them at 30 minutes and the other at 120 minutes, which may suggest a two step mechanism of the release under these experimental conditions.

Why the other aspartyl-ester derivatives, compounds X and XI, show no superagonistic properties still remains to be elucidated. A possible explanation may be the following: while I and III GnRH derivatives can easily be hydrolyzed in vivo (decomposition rate about 50 % per hour monitored by HPLC at 37 ^{o}C pH=7.35), all the others remain definitely stabil at neutral pH. Thus it can be well assumed that the intermediate compound formed from unstable ester derivatives may be the only one having as much affinity to the receptor and/or antimetabolic properties as the D amino acid substituted GnRH analogues.

References

1. Monahan M.W., M.S. Amoss, H.A. Anderson and W. Vale. 1973. Biochemistry 12, 4616.

2. Dutta A.S., and B.J.A. Furr. 1985. Annual Reports in Medicinal Chemistry 20, 203.

3. Seprődi J., J. Érchegyi, I. Teplán and I. Mező. Hung. Pat. No. 187 503.

4. Beamer W.G., S.M. Murr and I.I. Geshwind. 1972. Endocrinology 90, 823.

CHARACTERIZATION OF NOVEL SPECIES SPECIFIC GnRH ANALOGS WHICH CAN BE USED FOR THE INDUCED ARTIFICIAL PROPAGATION OF FISHES

Gy. Kéri, A. Horváth, T.Gulyás, I. Teplán
1st Institute of Biochemistry, Semmelweis University Medical School, 1444 Budapest, P.O.Box 260, Hungary

The existence of structural variation in gonadotropin hormone releasing hormone /GnRH/ in vertebrates has been known only from the last few years /1,2/. The first non-mammalian GnRH was isolated from chicken hypothalamus and its chemical structure was established as /Gln^8/-GnRH /3,4/. Later a second form of GnRH /His^5, Trp^7, Tyr^8/-GnRH was found in chicken hypothalamus /5/. Another type of GnRH was isolated from salmon hypothalamic tissue. Its structure was characterized as /Trp^7, Leu^8/-GnRH /6/.

In order to develop GnRH agonists which can be used for the artificial propagation of fishes mammalian, salmon and chicken gonadotropin-releasing hormones and their analogs were synthesized and tested in vitro in cultured and superfused rat pituitary cells and in vivo in various fishes. /Table 1./

Table 1

SPECIES SPECIFIC GnRH ANALOGS AND THEY DERIVATIVES

	1.	2.	3.	4.	5.	6.	7.	8.	9.	10.	
	Glp	His	Trp	Ser	Tyr	Gly	Leu	Arg	Pro	Gly-NH_2	mammalian
	Glp	His	Trp	Ser	Tyr	Gly	Leu	<u>Gln</u>	Pro	Gly-NH_2	chicken
	Glp	His	Trp	Ser	Tyr	Gly	<u>Trp</u>	<u>Leu</u>	Pro	Gly-NH_2	salmon
1./	Glp	His	Trp	Ser	Tyr	<u>DPhe</u>	Leu	<u>Gln</u>	Pro	Gly-NH_2	
2./	Glp	His	Trp	Ser	Tyr	<u>DPhe</u>	<u>Trp</u>	<u>Leu</u>	Pro	Gly-NH_2	
3./	Glp	His	Trp	Ser	Tyr	Gly	<u>Trp</u>	<u>Gln</u>	Pro	Gly-NH_2	
4./	Glp	His	Trp	Ser	Tyr	Gly	<u>Phe</u>	<u>Gln</u>	Pro	Gly-NH_2	
5./	Glp	His	Trp	Ser	Tyr	Gly	<u>Phe</u>	<u>Leu</u>	Pro	Gly-NH_2	
6./	Glp	His	Trp	Ser	Tyr	Gly	<u>Trp</u>	<u>Leu</u>	Pro	NH-C_2H_5	
7./	Glp	His	Trp	Ser	Tyr	<u>DPhe</u>	Leu	<u>Gln</u>	Pro	NH-C_2H_5	
8./	Glp	His	Trp	Ser	Tyr	Gly	<u>Trp</u>	<u>Gln</u>	Pro	NH-C_2H_5	

Peptides 1986

The artificial propagation of cultured fishes could be an inexhaustible source to satisfy mankind's increasing demand for dietary proteins. A lot of scientific efforts has been directed toward propagation techniques of cultured fishes in the last years. One of the greatest problems to be solved is that several kind of fishes having great economical value do not reproduce in captivity.

Several hormones like steroids, gonadotropins and GnRH analogs were tried in various kinds of fishes for intervention in the hypothalamic-pituitary-ovarian axis to induce maturation and ovulation /2/, without achieving the final goal to fulfill induced artificial propagation of fishes with a well defined know how, which is applicable for commercial purposes, i.e. for food production.

We have investigated the effect of these GnRH agonists in vivo on fishes, measuring their ability to induce ovulation and in vitro in cultured and in superfused rat adenohypophyseal cells.

The in vitro results using cultured rat adenohypophyseal cells demonstrated that these agonists of the "species specific" GnRH analogs are less potent than GnRH in stimulating LH and FSH release in the cultured rat pituitary cells but their mechanism of action is similar. These analogs are also capable to sensitize or desensitize LH and FSH production in the cultured rat pituitary cells on a dose and/or time dependent way similarly to GnRH /7/.

The analogs were more effective in the superfused cells where their FSH releasing activity were more pronounced.

In a series of in vivo experiments using special technologies what we have developed for propagation of fishes we tested our analogs on sterlets and perches and we found that the best analogs are the D-Phe^6-Gln^8-GnRH, D-Phe^6-Gln^8-GnRH-EA and D-Phe^6-Trp^7-Leu^8--GnRH-EA. From this point we have concentrated our efforts on these analogs.

Table 2. shows the effect of D-Phe^6-Gln^8-GnRH on various fishes, in the spawning season and the effect of D-Phe^6-Gln^8-GnRH out of season.

Table 2

Effect of D-Phe^{6}-Gln^{8}-GnRH and D-Phe^{6}-Gln^{8}-GnRH-EA on various fishes

Fish	Hormone	Treatment	Apl.	Treated	Reacted (Ovulated)[+]	%
Tench	D-Phe^{6}-Gln^{8}-GnRH	10 µg-8 hours-100 µg	im.	9	7	77,7
Carp	D-Phe^{6}-Gln^{8}-GnRH	50 µg-8 hours-300 µg	im.	5	2	40,0
Pike-pearch	D-Phe^{6}-Gln^{8}-GnRH	100 µg in natural environment	im.	84	84	100
Pike	D-Phe^{6}-Gln^{8}-GnRH	10 µg-4 hours-100 µg	ip.	300	300	100
Pike	D-Phe^{6}-Gln^{8}-GnRH-EA	10 µg-4 hours-100 µg	ip.	100	100	100
Hung. Cat fish	D-Phe^{6}-Gln^{8}-GnRH	10 µg/kg bw.-6 hours-50 µg/kg bw.	ip.	31	24	77
Hung. Cat fish	D-Phe^{6}-Gln^{8}-GnRH-EA	10 µg/kg bw.-6 hours-50 µg/kg bw.	ip.	17	15	88
Sterlet	D-Phe^{6}-Gln^{8}-GnRH	10 µg-6 hours-100 µg	ip.	258	201	78
Sterlet	D-Phe^{6}-Gln^{8}-GnRH-EA	10 µg-6 hours-100 µg	ip.	32	28	87

[+] The fertility rate of the ovulated eggs was 80-90 %

These results clearly demonstrate that using these "species specific" GnRH analogs we were able to propagate artificially several kinds of fishes having great economical value which were not possible to propagate artificially before.

References

1. King, J.A. and Millar, R.P. /1979/ Science 206, 67-69.
2. King, J.A. and Millar, R.P. /1980/ Endocrinology 106,707-717.
3. King, J.A. and Millar, R.P. /1982/ J.Biol.Chem. 257, 10729-10732.
4. Miyamoto, K., Hasegawa, Y., Minegishi, T., Nomura, M., Takahashi, Y., Igarashi, M., Kangawa, K. and Matsuo H. /1982/ Biochem. Biophys. Res. Commun. 107, 820-827.
5. Miyamoto, K., Hasegawa, Y., Igarashi, M., Kangawa, K. and Matsuo, H. /1983/ In: Peptide Chemistry E. Munekata, ed. The Protein Research Foundation, pp. 94-104.
6. Sherwood, N., Eiden, L., Brownstein, M., Spiess, J., Rivier, J. and Vale, W. /1983/ Proc. Natl. Acad. Sci. USA 80,2794-2798.
7. Kéri, Gy., Nikolics, K., Teplán, I. and Molnár, J. /1983/ Mol. and Cell Endocrinology 30, 109-120.

SYNTHETIC PEPTIDE ANTAGONISTS OF GLUCAGON

R.B. Merrifield, C. Unson, E.M. Gurzenda and D. Andreu
The Rockefeller University
New York, N.Y. 10021, USA

Introduction

The design and synthesis of glucagon analogs that reversibly inhibit the hormone would be useful for studies on its mode of action and its role in diabetes. Toward that end we have developed a solid phase synthesis of glucagon (1) and its analogs that results in high yields of homogeneous peptides. Our initial search for inhibitors began with a series of replacement analogs at residues 22 and 23 and an attempt to correlate peptide conformation with receptor binding and adenyl cyclase activation. Analogs that favored a β sheet structure were active but of reduced potency while those that favored a helical structure were completely inactive but retained low levels of binding. However, they were not sufficiently effective competitive inhibitors. We then tested the idea that secretin, which does not bind to the glucagon receptor, may have evolved from glucagon or a common precursor by a series of steps such that intermediates lost their ability to activate adenyl cyclase while retaining some affinity for the glucagon receptor.

Results

Two of the synthesized secretin-glucagon hybrids, [Asp^3,Glu^9]-glucagon and [Asp^3,Glu^9,Arg^{12}]glucagon, fit this postulate. They were < 0.001% active but retained 2% binding capacity (2). In competitive cyclase activation assays they were able to inhibit glucagon activity completely. The pA_2 values were 5.6 and 5.5 and the inhibition indices, $(I/A)_{50}$, were 35 and 75 (Table 1).

Peptides 1986

Table 1. Competitive Inhibition of Glucagon by Synthetic Analogs

Analog	Type of Replacement					Membrane binding	Adenyl Cyclase Assays			
No.	Delete Res 1	Secretin Residues	Superagonist		Amide	Relative %	Max act. %	Rel. potency %	Inhib. index [I/A]50	pA2
1		Asp^{3}Glu^{9}Arg^{12}				2.2	0	0.0002	75	5.5
2		Asp^{3}Glu^{9}				1.9	0	0.001	35	5.6
3		Asp^{3}				0.74	17	2.2	38	
4		Glu^{9}				7.0	24	0.2		
5			DPhe^{4}			100 (69)	100 (86)	100 (700)		
6		Asp^{3}Glu^{9}	DPhe^{4}			2.2	14	0.005	100	6.0
7		Asp^{3}	DPhe^{4}			3.5	29	0.02	105	6.7
8				Lys^{17,18}Glu^{21}		100 (500)	100	300-(700)		
9		Asp^{3}		Lys^{17,18}Glu^{21}		1	37	2.4	1000	
10		Asp^{3}		Glu^{21}		1	0	0.005	670	5.1
11	desHis^{1}					8.1	36	0.1-1		
12	desHis^{1}	Asp^{3}Glu^{9}Arg^{12}				0.8	0	0.002	59	5.9
13	desHis^{1}	Asp^{3}				0.45	16	0.006	174	6.8
14	desHis^{1}	Glu^{9}				11	0	0.0008	36	7.6
15	desHis^{1}		DPhe^{4}			13	48	0.1-3		
16	desHis^{1}	Asp^{3}Glu^{9}	DPhe^{4}			2.4	0	0.0001	23	7.4
17	desHis^{1}	Asp^{3}	DPhe^{4}			7.4	0	0.0007	40	7.1
18	desHis^{1}			Lys^{17,18}Glu^{21}		15	28	0.63		
19	desHis^{1}	Glu^{9}		Lys^{17,18}Glu^{21}		10	0	0.0001	12.6	7.5
20	desHis^{1}	Asp^{3}	DPhe^{4}	Lys^{17,18}Glu^{21}		2.8	0	0.005		
21	desHis^{1}	Glu9			NH_2		0	0.0001	12	7.2

The single-replacement analogs, [Asp^{3}- and [Glu^{9}]-glucagon were weak partial agonists. Based on this lead we decided to combine the secretin-glucagon hybrids with four other kinds of structural changes.

One way to improve a peptide analog inhibitor is to introduce additional structural changes that are already known to produce superagonists (3). We have, therefore, combined the secretin-glucagon changes with either D-Phe^{4}, shown by Sueiras-Diaz et al. (4) to be 700% as potent as glucagon in raising blood glucose in vivo, or Lys^{17,18},Glu^{21}, shown by Krstenansky et al. (5) to be of 700% potency in the in vitro cyclase assay and to have 500% relative affinity in the [^{125}I]glucagon membrane binding assay. None of these analogs had the improved antagonist properties we were seeking. Most of them retained considerable agonist activity but reduced potency and binding affinity.

Removal of His^{1} was known to give a weak partial agonist and to reduce binding and potency of glucagon (6). DesHis^{1} glucagon in combination with [Asp^3,Glu^9,Arg^{12}]glucagon or [Asp^3]glucagon made little change, but [$desHis^1$,Glu^9]glucagon retained no cyclase activity while showing substantial binding affinity (11%) and giving a good pA_2 value of 7.6 (analog 14, Table 1).

When the Asp^3 or Glu^9 changes were combined with both $desHis^1$ and the superagonist changes, some further improvements were obtained. $DesHis^1$ reduced the binding and activity of D-Phe^4 to the same extent as it did natural glucagon, but in combination with D-Phe^4 and Asp^3-Glu^9 it completely eliminated cyclase activity while the inhibition index was reduced to 23 and pA_2 increased to 7.4. The result with [$desHis^1$,Asp^3,D-Phe^4]glucagon was similar. The best inhibitor to date combines Glu^9 with $desHis^1$ and $Lys^{17,18}$,Glu^{21}. It has the same membrane binding affinity and pA_2 value as $desHisGlu^9$, but the additional replacements with $Lys^{17,18}$,Glu^{21} appear to have improved $(I/A)_{50}$; 12 vs 36. The

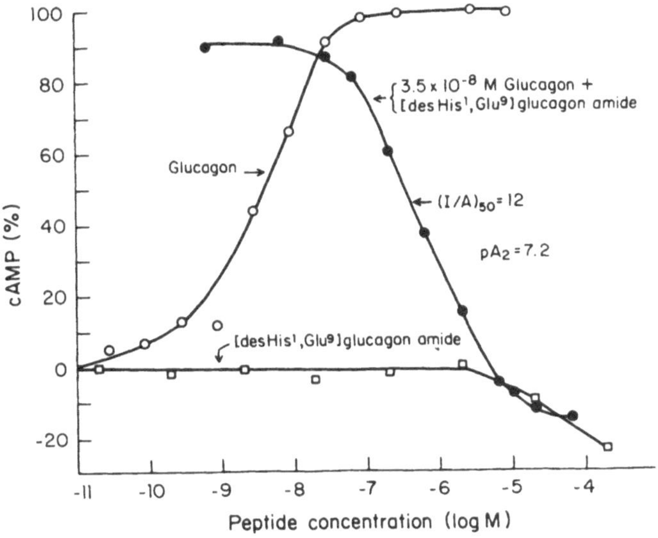

Fig 1 Adenyl cyclase assays of [$desHis^1Glu^9$]glucagon amide in the absence □ and presence ● of glucagon

C-terminal amide made a similar improvement in the inhibition index and retained a good pA_2 (Fig 1). Although none of these totally synthetic glucagon analogs is as good an inhibitor as N^{α}-Tnp [Har^{12}]glucagon reported by the Hruby laboratory (7), several other obvious changes in structure are expected to lead to further improvements. The effective analogs must next be tested *in vivo* for their effect on blood glucose and as inhibitors of glucagon in diabetic animals.

Acknowledgement

This work was supported by NIH Grant No. AM 24039.

References

1. Mojsov, S. and Merrifield, R.B. (1984) Eur. J. Biochem., 145, 601-605.

2. Andreu, D. and Merrifield, R.B. (1985) in Peptides: Structure and Function (C.M. Deber, V.J. Hruby and K.D. Kopple, eds.) Pierce Chem. Co., pp 595-598.

3. Rosenblatt, M., Mahaffey, G.H. and Potts, G.T. Jr. (1977) in Peptides (M. Goodman and J. Meienhofer, eds.) Wiley, N.Y. pp. 581-583.

4. Sueiras-Diaz, J., Lance, V.A., Murphy, W.A. and Coy, D.H. (1984) J. Med. Chem., 27, 310-315.

5. Krstenansky, J.L., Trivedi, D., Johnson, D. and Hruby, V.J. (1985). J. Am. Chem. Soc., 108, 1696-1698.

6. Lin, M.C., Wright, D.E., Hruby, V.J. and Rodbell, M., (1975) Biochemistry, 14, 1559-1563.

7. Bregman, M.D., Trivedi, D. and Hruby, V.J. (1980) J. Biol. Chem. 255, 11725-11731.

NEW PHOTO-REACTIVE DERIVATIVES OF INSULIN FOR AFFINITY-LABELLING OF THE INSULIN RECEPTOR

D. Ambrosius, S. Bala-Mohan, C. Behrendt, K. Schäfer, A. Schüttler and D. Brandenburg

Deutsches Wollforschungsinstitut, Veltmanplatz 8, D-5100 Aachen
Federal Republic of Germany

Introduction

While most studies on hormone binding and action are carried out under reversible conditions, irreversible crosslinking of hormones and receptors, particularly by photoaffinity labelling (1), has opened new pathways. Photo-activatable insulin derivatives have already given valuable information on the subunit structure of the insulin receptor as well as on the biological activity and internalisation of the covalent hormone-receptor complex (2). Labelling insulin and structurally related molecules at different sites allows further mapping of surfaces involved in hormone-receptor interactions, correlations between formation, activity and fate of the complex, as well as studies on receptor topography.

Results

Des-(B26-B30)-insulin-B25-(4-Nap-iminoethyleneamide)

Our recent finding, that des-(B26-B30)-insulin-B25-amide is fully active (3) has prompted the enzyme-catalyzed semisynthesis of a corresponding photoinsulin as the first light-activatable insulin derivative with the label in a C-terminal position.
Nap-ethylenediamine, prepared by an improved procedure, was coupled to Boc-Gly-Phe-Phe-OH with DCC/HOBT. After deblocking with HCOOH, the labelled tripeptide (8 eq.), which was homogeneous in HPLC, was coupled to A1,B1-$(Msc)_2$-des-(B23-B30)-insulin (porcine) in DMF/glycerol/water (5:5:1) at pH 6.5 in the presence of trypsin (3) at room temp. for 38 h. After gel filtration, deprotection and purification on DEAE-Trisacryl M (pH 8.0, 40% isopropanol,

Peptides 1986

NaCl gradient elution) the shortened photoinsulin was isolated in 6% overall yield (coupling yield was ca. 50%). It was homogeneous according to electrophoreses at 3 pH values and RP-HPLC.
IR-spectrum: Azide band at 2118 cm^{-1}
Amino acid analysis: Gly +0.87, Phe +2.06

The biological activity in vitro under reversible conditions is 45% and equals that of the B25-benzylamide (M. Casaretto, to be publ.). This indicates that the larger residue does not interfere with receptor binding. Preliminary irradiation experiments have shown that covalent binding to viable adipocytes causes persistent activation of lipogenesis (see ref. 2).

B29,B29'-Bis(Nap-Ahe)-B1,B1'-azelaoyl-insulin-dimer
The insulin receptor is a symmetrical hetero-tetramer (β-α-α'-β') (4). For studies of intramolecular relationships of the insulin-binding α-subunits and receptor valency, as well as interactions between different receptors, which are thought to be essential for biological action, a bivalent dimeric photoinsulin was designed. First, A1,B29-$(Msc)_2$-insulin (40 μmol/ml) was specifically cross-linked between the B1-groups (5) by reaction with $Aze(ONp)_2$ (0.55 eq., in DMSO, 2 h, room temp.) in the presence of 5.5 eq. HOBT. After gel filtration (Sephadex G-25 f, 10% acetic acid), deblocking and gel filtration (as before), the dimer (yield 52%) was selectively protected at A1 by reaction with Msc-ONSu at low pH (6.0). After removal (SP-Trisacryl M, pH 3.0) of mono- and tri-substituted dimers (5 and 18%, resp.), 42% of homogeneous A1,A1'-$(Msc)_2$-B1,B1'-Aze-$(insulin)_2$ were obtained. Acylation with 8 eq. of Nap-Ahe-ONSu gave 85% disubstitution after 6 h. The product was deprotected and subjected to gel filtration. Purification was effected by semipreparative HPLC on Nucleosil $7C_{18}$ (column: 5 x 250 mm) by elution with 0.07% TFA/acetonitrile (gradient: 30 - 70% B in 60 min). The yield of labelled dimer was 5% (based on insulin).
All products were homogeneous according to HPLC and electrophoreses (pH 2.6 and 3.6) and gave the expected bands before and after cleavage of disulfide bonds.
The biological potency of the photoinsulin dimer under reversible conditions, determined as the stimulation of lipogenesis in isolated rat adipocytes, was 2%, and binding affinity to insulin

receptors on cultured human IM-9 lymphocytes 4% (molar basis, on insulin).

This dimer contains both photo groups in one chain (Nap-Ahe-B-Aze-B'-Ahe-Nap), which is not cleavable under the conditions used for SDS-polyacrylamide gel electrophoresis of solubilized receptors. Although binding and biopotency are unexpectedly low, it should be useful for crosslinking studies, which are currently under way.

Acknowledgement

We wish to thank Cornelia Diaconescu and Dr. P. De Meyts (Bruxelles) for bioassays, as well as Dr. Monika Casaretto and Marie-Therese Sommer for HPLC analyses.

This work was supported by SFB 113 Diabetesforschung Düsseldorf.

References

1. Eberle, A.N., P.N.E. De Graan. 1985. In: Methods in Enzymology, Vol. 109 (L. Birnbaumer and B.W. O'Malley, eds.). Academic Press, Inc., Orlando, Florida. pp. 129-156

2. Brandenburg, D., C. Diaconescu, G. Klotz, P. Mucke, J. Neffe, D. Saunders and A. Schüttler. 1985. Biochimie 67, 1111-1117

3. Fischer, W.H., D. Saunders, D. Brandenburg, A. Wollmer and H. Zahn. 1985. Biol. Chem. Hoppe-Seyler 366, 521-525

4. Gammeltoft, S.. 1984. Physiol. Reviews 64, 1321-1378

5. Schüttler, A. and D. Brandenburg. 1982. Hoppe-Seyler's Z. Physiol. Chem. 363, 317-330

Abbreviations

Ahe = 6-aminohexanoyl-

Aze = azelaoyl-

Nap = 2-nitro-4-azidophenyl-

SELECTIVE N- AND C-TERMINAL ELONGATION AT A1 AND B30 OF INSULIN: NEW APPROACHES TO PROINSULIN RELATED CONVERSION INTERMEDIATES

V. K. Naithani, H.-G. Gattner and H. Zahn
Deutsches Wollforschungsinstitut an der TH Aachen, Veltmanplatz 8, 5100 Aachen, Federal Republic of Germany

Introduction

The biosynthesis of insulin proceeds by the proteolytic cleavage of proinsulin at the paired basic residues (Lys-Arg and Arg-Arg), which link insulin to the C-peptide. Recent studies (1,2) have indicated the occurence of the single site mutation at the cleavage sites (Lys-X instead of Lys-Arg or Arg-X instead of Arg-Arg). These molecular abnormalities impair the complete conversion of pro-insulin to insulin. In these cases the proinsulin conversion intermediates are formed in which the C-peptide is still attached to insulin either at the N-terminus of the A-chain (AC-type) or at the C-terminus of the B-chain (BC-type). The aim of the present investigation is to develop semisynthetic procedures suitable for proinsulin related conversion intermediates with or without mutation. We selected Leu to substitute Arg^{32} to prepare the BC-type of conversion intermediate as mutational model. The reason being that the single nucleotide change in the codon for Arg (see ref. 3 for the gene sequence) could also result in replacement by Leu.

Results

N-terminal elongation at Gly-A1 of insulin: The following procedures were applied to achieve this aim (Scheme):

1. Using B1-protected insulin: a) B1-citraconyl insulin (4) was reacted with the mixed anhydride of Boc-Arg (3 fold molar excess) for 1 min. The protecting groups were removed and the product was

Peptides 1986

purified by ion exchange chromatography on SP-Sephadex. Yield Arg^{65}-insulin 25% and B29 Arg-insulin 2-3%.

Boc-Lys(Boc)-Arg and Trt-Gly-Ser(Bu^t)-Leu-Gln-Lys(Boc)-Arg were coupled essentially in the similar manner as described above. The derivatives $(Lys\text{-}Arg)^{64-65}$- and $(Gly\text{-}Ser\text{-}Leu\text{-}Gln\text{-}Lys\text{-}Arg)^{60-65}$-insulin were also contaminated with traces of corresponding εB29 substituted isomer.

b) B1-Boc insulin was prepared in 25% yield by a single step reaction of t-butylaminocarbonate (5) with insulin in 50% dioxane-water mixture containing 0.1 M NaOAc, pH 6.5. The product was purified by ion exchange chromatography on SP-Sephadex.

B1-Boc insulin was enzymatically reacted with a large excess of Boc-Arg-OMe (100 fold molar excess added in three lots at an interval of 3.5 min) using trypsin-TPCK at pH 9.5. The protecting groups were removed and Arg^{65}-insulin was purified as described above. Yield 20-25%.

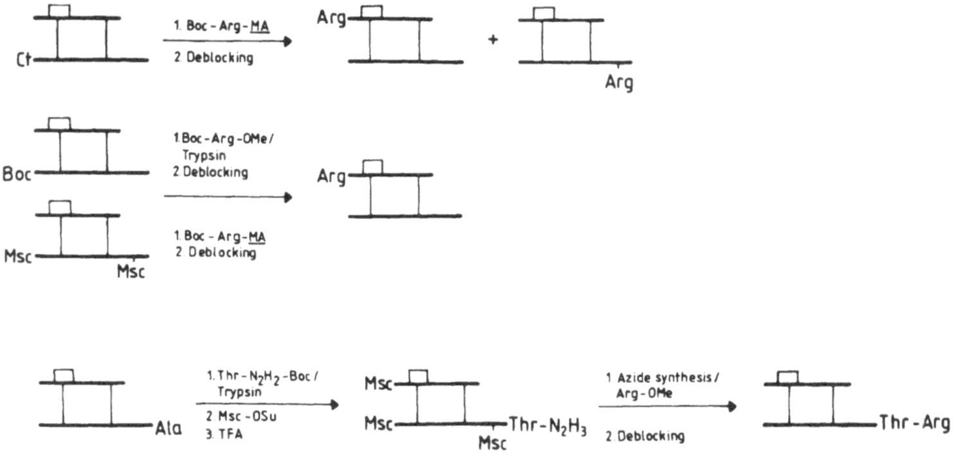

Scheme. Semisynthesis of N- and C-terminal elongated insulin derivatives.

2. Using B1,B29 Msc_2-insulin: The free A1 amino function was acylated with Boc-Arg using the mixed anhydride procedure. The protecting groups were removed and Arg^{65}-insulin was purified as described earlier. Yield 30%.

C-Terminal elongation at Thr-B30 of insulin: Porcine insulin was converted into the Msc_3 human insulin hydrazide derivative (Scheme). This was then coupled with Arg-OMe, Arg-Arg and Arg-Leu-OBu^t by the azide procedure. The protecting groups were removed and the desired insulin-Arg^{31}, insulin-Arg^{31}-Arg^{32} and insulin-Arg^{31}-Leu^{32} were purified by ion exchange chromatography on SP-Sephadex.

From the isoelectrofocusing and HPLC data, it could be shown that the elongated insulin derivatives contained some mono des-amido derivative (5-10%). The HPLC data also revealed the presence of traces of other unknown impurities (2-5%). These derivatives were also characterized by the amino acid analysis and end group determination.

Discussion

A series of proinsulin conversion intermediates have been isolated from the commercial crystalline bovine (6) and porcine insulin (7). The N-terminal elongated insulin derivatives bearing 1-3 arginines have been prepared by semisynthesis (8). We have developed semi-synthetic procedures suitable for the N- as well as the C-terminal elongatior of insulin. The derivatives prepared are: Arg^{65}-, (Lys-$Arg)^{64-65}$-, (Gly-Ser-Leu-Gln-Lys-$Arg)^{60-65}$-insulin, insulin-Arg^{31}, insulin-Arg^{31}-Arg^{32} and insulin-Arg^{31}-Leu^{32}. The numbering of the extended residues on insulin refers to the corresponding human proinsulin sequence. The N-terminal extended derivatives contain porcine insulin while the C-terminal contain human.

For the N-terminal elongation we first applied B1-citraconyl insulin because of the significant difference in the pKs (9) of the Gly-A1 (8.6) and Lys-B29 (11.2). However, the mixed anhydride acylation showed that besides Arg^{64}-insulin (25%), traces of εB29 Arg-insulin (2-3%) was also formed.

When t-butylaminocarbonate was allowed to react with insulin at pH 6.5 besides A1,B1 Boc_2-insulin (40%) a reasonable amount of B1-Boc insulin (25%) was also isolated. The B1-protected insulin was selectively acylated at the A1-amino function using trypsin. Inspite of the high excess of Boc-Arg-OMe used, the yield of Arg-insulin was low (25%).

The acylation using B1,B29 blocked insulin proceeded in higher yield and the product obtained was devoid of contaminating isomer. Thus this is the method of choice to prepare N-terminal elongated insulin derivatives.

In the C-terminal elongation the preparation of Msc_3-insulin (B30) hydrazide was straight forward. The last step involving the azide synthesis proceeded in 30-35% yield.

References

1. Robbins, D.C., P.M. Blix, A.H. Rubenstein, Y. Kanazawa, K. Kosaka, H.S. Tager. 1981 Nature 291, 679.

2. Gabby, K.H.,R.M Bergenstal, J. Wolff, M.E. Mako, A.H. Rubenstein. 1971. Proc. Natl. Acad. Sci. USA. 76, 2882.

3. Naithani, V.K., H.-G. Gattner. 1982. Hoppe-Seyler's Z.Physiol. Chem. 363, 1443.

4. Bell, G.F., R.L. Pictet, W.J. Rutter, B. Cordell, E. Tischer, H.M. Goodman. 1980. Nature. 284, 26.

5. Harris, R.B., I.B. Wilson. 1984. Int.J.Peptide Protein Res. 23, 55.

6. Steiner, D.F., O. Hallund, A.H. Rubenstein, S. Cho, C. Bayliss. 1968 Diabetes. 17, 725.

7. Chance, R.E., R.M. Ellis, W.W. Bromer. 1968. Science 161, 165.

8. Friesen, J. Doctoral Thesis RWTH Aachen. 1976.

9. Bradbury, J.H., L.R. Brown. 1977. Eur. J. Biochem. 76, 573.

SYNTHESIS OF GLUCAGON-LIKE PEPTIDES: DEVELOPMENT OF SPECIFIC ANTISERA AND ANALYSIS OF PROGLUCAGON PROCESSING.

S. Mojsov, J.F. Habener

Laboratory of Molecular Endocrinology, Howard Hughes Medical Institute, Harvard Medical School
Boston, Massachusetts 02114

Introduction

Glucagon, a 29-residue peptide that regulates glycogenolysis and gluconeogenesis in the liver, is synthesized in the pancreas in the form of a precursor of 180-residues (1,2,3) (Fig. 1). Throughout its sequence the precursor, preproglucagon, contains multiple pairs of basic amino acids (underlined in Fig. 1) that are characteristics of the sites that are cleaved during the post-translational processing of prohormones. Predicted cleavages at these basic residues can potentially produce two other peptides 37- and 33-residues long, whose structures are similar to that of glucagon, termed glucagon-like peptides (GLP's) I and

```
-20                                          1                 10
 M K T V Y I V A G L F V M L V Q G S T Q H A P Q D T E E N A
                                                 33            40
 R S F P A S Q T E P L E D P D Q I N E D K R H S Q G T F T S
                                                               70
 D Y S K Y L D S R R A Q D F V Q W L M N T K R N R N N I A K
                                                              100
 R H D E F E R H A E G T F T S D V S S K L E G Q A A K E F I
                                                              130
 A W L V K G R G R R D F P E E V A I A E E L*G R R H A D G S
                                                              160
 F S D E M N T I L D N L A T R D F I N W L I Q T K I T D K K
```

Figure 1. Amino acid sequence of pancreatic rat preproglucagon. Overlines designate regions of sequences that were synthesized for immunization. (*)denotes leucine amide.

II. Moreover, the presence of a typical amidation sequence GlyArgArg at the carboxy terminal of the intervening peptide II (IPII) (residues 111-123, Fig. 1) suggests that it can also be liberated during processing. The sequence of the amino terminal part of proglucagon together with glucagon and the first

Peptides 1986

intervening peptide (residues 1-69, Fig. 1) is closely related to the sequence of glicentin (4). To determine whether (i) pancreatic glucagon and intestinal glicentin arise from the same precursor and (ii) glucagon-like peptides are liberated in the pancreas and/or intestine, we prepared antisera to synthetic glucagon-like peptides and used them for analyses of proteins in extracts of rat pancreas and intestine.

Results and Discussion

Six peptides varying in length from 12 to 37 residues were synthesized by the stepwise solid phase method in a manual shaker (5). The synthetic strategy followed the one described for glucagon (6). Most residues were incorporated in a single coupling step with dicyclohexylcarbodiimide in dichloromethane (CH_2Cl_2) (2.5-fold excess), except when indicated for a given synthesis. The subsequent couplings were as preformed symmetrical anhydrides in CH_2Cl_2 (2.5-fold excess). Peptides were purified by preparative C-18 hydrophobic chromatography; homogeneity was assessed by (i) HPLC on a) µBondapak C-18, and b) Protein-Pak DE 52 ion-exchange, (ii) amino acid analysis and (iii) quantitative preview sequencing, and was found to be satisfactory.

For immunizations we used glucagon (33-61), GLPI (87-101), GLPII (134-149) and IPII amide (111-122) where numbers in parentheses refer to sequence positions in Fig. 1. The subpeptides of GLPI and GLPII were chosen because they contain the least sequence similarities between each other and glucagon. IPII amide was synthesized using p-methylbenzhydrylamine resin (4g, 0.4 mmol nitrogen/g, U.S. Biochem. Corp.). Isoleucine (position 118, Fig. 1) required two symmetrical anhydride couplings. Chromatographic analysis on µBondapak C-18 HPLC of the crude peptide directly after low/high HF cleavage showed that under these conditions the major side-reactions at the four glutamic acid residues were avoided (Fig. 2A). Similar results to those obtained with IP-II were obtained with the 15-residues long GLPI and 16-residues long GLPII subpeptides using chloromethyl resin as solid support (data not shown).

We also synthesized the 37-residue GLPI, amino acids (72-108) and 33-residue GLPII (126-158) for use as synthetic standards in the radioimmunoassays. The synthesis of the GLPI started with Boc-Gly-OCH_2-Pam-resin (3.3g, 0.82 mmol/g, Applied Biosystems). Multiple couplings were required for: arginine at position 107 (two), isoleucine at 100 (three) and histidine at 78 (four); 18.1g of dry protected peptide-resin were recovered (98% of the theoretical yield). GLPII was synthesized starting with Boc-Asp(OBzl)-OCH_2-Pam-resin (3.5g, 0.73 mmol/g, Applied Biosystems). Multiple couplings were required for isoleucine at position 156 (four), arginine at 145 (two), threonine at 144 (two), aspargine at 140 (two) and histidine at 126 (two). The yield was 12.7g of dry protected peptide-resin (94%). Crude peptides directly after low-high HF cleavages were 80% to 90% homogeneous as assessed by HPLC on C-18 columns (Fig. 2B and 2C). Quantitative preview sequencing showed that the by-products in both peptides eluting in front of the main peak on HPLC contained peptides lacking amino terminal histidine.

The agreement between results obtained with the synthesis of GLPI and GLPII clearly indicates that these peptides of considerable complexity can be prepared in both high yield and homogeneity by the use of highly substituted Pam resin supports in combination with conditions of low-high HF deprotection and cleavage.

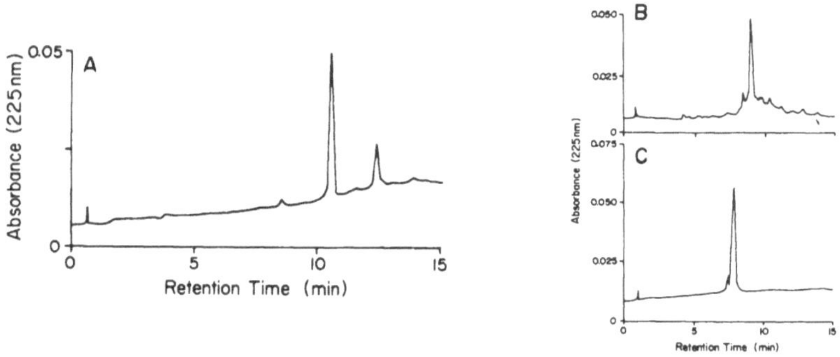

Figure 2. Analytical μBondapak C-18 HPLC of crude (A) IPII amide, (B) GLPI and (C) GLPII directly after low-high HF. Solvent system: buffer A (90% water, 0.1% TFA and 10% CH_3CN), buffer B (10% water, 0.1% TFA and 90% CH_3CN). Linear gradient for (A) 5-95% buffer B in 15 min. and for (B) and (C) 10-70% buffer B in 15 min.

Antisera raised against synthetic peptides were specific for the detection of their homologous antigens. In a competitive-binding radioimmunoassay no cross-reactivity was detected among glucagon and glucagon-like peptides despite their structural similarities (Fig. 3).

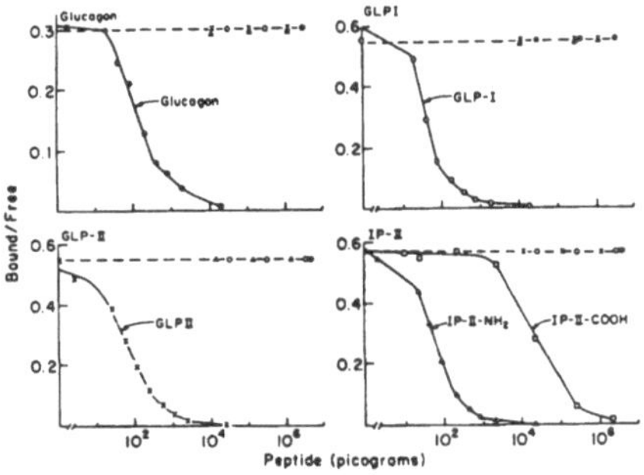

Figure 3. Competitive-binding radioimmunoassay curves for antisera against glucagon, glucagon-like peptides and IPII amide.

The specificity was further confirmed in the assays of gel filtration fractions of extracts of pancreas and intestine (Fig. 4). The major glucagon immunoreactivity was present as glucagon in the pancreas and higher molecular weight molecule, glicentin in the intestine. In the extracts of intestine but not in pancreas immunoreactivities coeluting with GLPII (1-33) and IPII amide standards were detected. Both intestine and pancreas contained GLPI (1-37) immunodeterminants. In addition, both tissues contain higher molecular weight fragments representing incompletely processed proglucagon. IPII immunoreactivity coeluted on ion-exchange DE 52 HPLC with the synthetic amidated IPII (111-122) and not carboxylated IPII (111-123) (data not shown).

The detection of the same immunodeterminants of glucagon and glucagon-like peptides in the pancreas and intestine implies that pancreatic glucagon and intestinal glycentin arise from a tissue-specific post-translational processing of a similar if not identical precursor. Moreover the specific liberation of GLPI, GLPII and IPII amide in intestine suggests that they are new biologically active peptides.

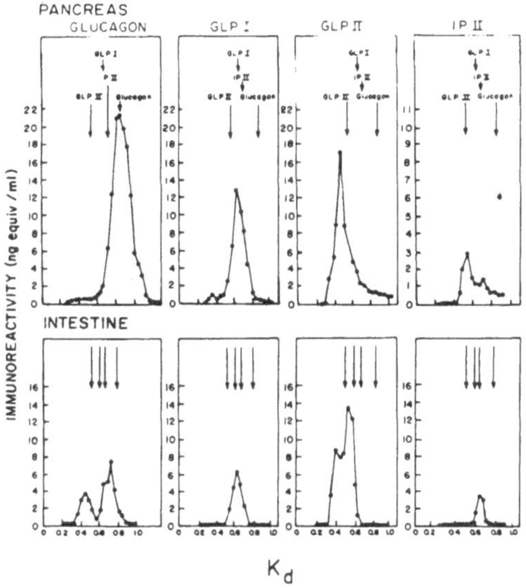

Figure 4. Gel filtration chromatography and radioimmunoassay on extracts of rat pancreas and intestine. Arrows indicate elution positions of synthetic standards.

Acknowledgment

This work was supported in part by Grant AM 30834 from the National Institute of Health.

References

1. Heinrich, G., Gros, P., Habener, J.F. 1984. J. Biol. Chem. 259, 14082.

2. Bell, G.I., Santerre, R.F., Mullenbach, G.T. 1983. Nature 302, 716.

3. Lopez, L.C., Frazier, M.L., Su, C.J., Humar, A., Saunders, G.F. 1983. Proc. Natl. Acad. Sci. U.S.A. 80, 5485.

4. Thim, L., and Moody, A.J. 1981. Regul. Pept. 2, 139.

5. Merrifield, R.B. 1963. J. Am. Chem. Soc. 85, 2149.

6. Mojsov, S., Merrifield, R.B. 1984. Eur. J. Biochem. 145,601.

STRUCTURE-ACTIVITY STUDIES OF SOMATOSTATIN ANALOGS, SUBSTITUTED AT POSITIONS 4 AND 5.

D. Sarantakis
Wyeth Laboratories, Research Division, Medicinal Chemistry
P.O. Box 8299, Philadelphia, PA 19101, USA.

Introduction

Until now a large number of somatostatin analogs have been prepared for both structure-activity studies and practical goals, such as treatment of diabetes. In this communication we report structural requirements for the aminoacids at position 4 and 5 of the somatostatin sequence, for the expression of relative specificity towards the suppression of the release of Growth Hormone (GH) Insulin (INS) and Glucagon (GLUN).

Results

The chemical synthesis and biological testing methodology of the analogs has been reported previously (1).

A structure-activity study of the peptides shown in Table 1 allows certain conclusions to be drawn. The presence of an L-basic aminoacid at position 4 leads to some selectivity towards GLUN release inhibition. Replacement of the Asn^5 by Glu or His as in analogs 2 and 3 provides compounds with definite selectivity towards GLUN. Analogs 4 and 5 with neutral or basic aminoacids at position 5 possess very low GLUN lowering activity.

A combination of the two modes of substitution, that is, Lys, Arg or His at position 4 and Glu, His or Tyr of position 5 results in analogs 6 to 11 which show excellent relative specificity towards GLUN. It is interesting to compare the biological profile of the analogs 11 to 14 which incorporate the D-isomers of the aminoacids promoting GLUN specificity. In this case the compounds possess potent GLUN and INS lowering activity and at the same time prolonged biological activity. It seems that the altered orientation in space of the side chains of the aminoacids, stimulates the INS lowering potency. The extended biological life of the analogs containing D-aminoacids most likely is due to their resistance to enzymatic degradation. Analogs 16 to 20 with an L-aminoacid at position 4 and a D-aminoacid at position 5 display a striking shift towards GH and INS suppressing activity, while analog 15 with a D-aminoacid residue at position 4 and a typtophan at position 5 is very specific for GH lowering.

Peptides 1986

TABLE 1

INHIBITION OF GH, INS and GLUN. RELEASE BY SOMATOSTATIN ANALOGS (RAT)

H-Cys-X^4-X^5-Phe-Phe-D-Trp-Lys-Thr-Phe-Thr-Ser-Cys-OH

Analog	X^4, X^5	Dose µg/kg rat	Percent inhibition GH	INS	GLUN	Duration GH Inhibition (Hours)
	Somatostatin	200	85*	39*	47*	0.25
1	Arg^4 Asn^5	500	66*	50*	67*	0.25
		200	80*	23	70*	
		50	52*	42+	51*	
2	Lys^4,His^5	2000	84*	45*	94*	0.25
		500	87*	49*	83*	
		100	ND	30*	60*	
3	Lys^4,Glu^5	3000	92*	32	55+	0.25
4	Lys^4,Arg^5	1800	81*	47*	31	ND
5	Lys^4,Ser^5	2000	78*	32+	26	ND
6	Arg^4,His^5	100	76*	71*	100*	2
		10	67*	32*	62*	
		5	52*	36+	64*	
7	His^4,His^5	500	72*	52*	94*	2-4, GLUN 2
		100	88*	27	71+	
		20	33	9	52*	
		5	0	8	0	
8	Arg^4,Glu^5	400	87*	45	78*	4
		200	85*	62*	97*	
9	His^4,Glu^5	200	77*	24	94*	ND
		50	64*	6	78*	
10	His^4,Tyr^5	200	78*	21	85+	0.25
11	Arg^4,Tyr^5	200	75*	78*	72*	0.25
		50	61*	65*	86*	
		10	38*	18	51	
12	D-His^4,D-His^5	200	82*	39+	69*	2
13	D-His^4,D-Tyr^5	200	72*	22	78+	8
		50	50*	43*	79*	
		10	59*	34	51*	
		2	61*	11	38	
14	D-Arg^4,D-Tyr^5	2000	88*	55*	79*	4-5
		100	ND	41*	76*	
		10	87*	32+	51+	
		2	88*	13	30	
15	D-Lys^4,Trp^5	250	78*	14	22	0.25
		100	80*	3	9	
		50	76*	2	1	
16	D-Arg^5	2000	85*	85*	16	4
17	D-Tyr^5	100	80*	50*	12	2-4
18	Arg^4,D-Ser^5	2000	83*	77*	32	ND
19	Arg^4,D-Arg^5	1000	97*	60*	28	2
20	Arg^4,D-His^5	5000	95*	55*	34	ND

* = $p < 0.01$ + = $p < 0.05$ ND = not determined

The above structure-activity relations suggest that relative specificity towards GLUN suppression in our series of somatostatin analogs, requires the presence of basic aminoacids at position 4. It is worth noting that all three basic aminoacids, Lys, Arg, and His have an N-atom in the same relative position (Figure 1). The requirements for position 5 comprise the presence of an aminoacid with a side chain which can supply an exchangable proton. The interaction with the receptor(s) must involve a carboxylate group and perhaps an important hydrogen bond close to position 5. The alteration of the orientation of the side chains of the aminoacids at positions 4 and 5 by the incorporation of D-isomers brings about the emergence of the INS lowering activity, while a D-aminoacid at position 4 only provides GH lowering specific analogs.

Figure 1

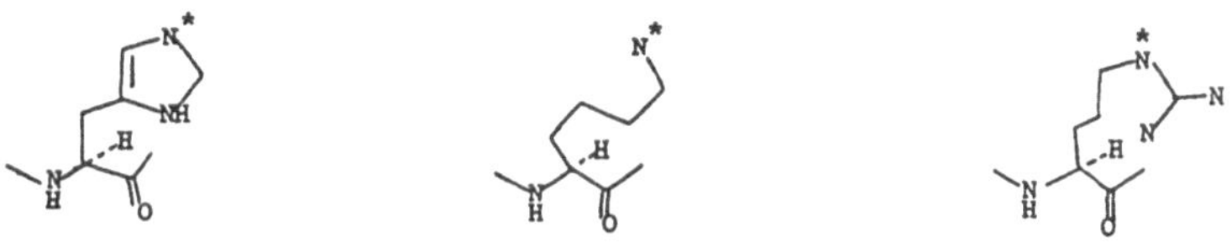

L-Histidine L-Lysine L-Arginine

There is strong evidence (2) that the residues Phe-(D or L)-Trp-Lys-Thr form a β-turn and it has been suggested that Phe^{6} and Phe^{11} interact through hydrophobic forces. Our analog 21 (Figure 2) which was first reported (3) in 1981 and similar ones described (4) later by the Sandoz group in 1982 were designed to take advantage of the stabilization of the β-turn resulting from the presence of the disulfide bridge in the place of the two Phe^{6} and Phe^{11}. The hexapeptide 22 (Figure 2) also reported (5) by the Merck group contains a stabilized β-turn. The octapeptide 21 suppresses the release of GH in rats at doses as low as 20 mg/kg while not altering INS or GLUN release at a dose of 100 mg/kg. At the same time 21 suppresses GH for five hours at a dose of 1 mg/kg. The hexapeptide 22 is a potent inhibitor of the release of GH, INS and GLUN.

Figure 2

```
Lys-Thr- Cys-D-Phe-OH          Lys-Thr—Phe
 |        |                     |       |
 |        S2                    |       |
 |        |                     |       |
D-Trp-Phe-Cys- Phe-H          D-Trp-Phe-Pro
```

In comparing our analogs 1 to 20 and the smaller compounds 21 and 22 it can be seen that the small size of the last compounds precludes the complete occupation of the space of the receptor site for GH, INS, and GLUN occupied by our larger analogs 1 to 20. The residues at positions 4 and 5 of our analogs must interact with additional portions of the receptor site. The pharmacophore for peptides with relative specificity towards GLUN release inhibition requires a positively charged N-atom (position 4) and an exchangable proton (position 5).

References

1. Sarantakis, D., Teichman, J. Fenichel, R. Lien, E. 1981. Second International Symposium on Somatostatin, Serono Symposia, Athens, Greece.

2. Veber, D.F., Holly, F.W., Palaveda, W.J., Nutt, R.F., Bergstrand, J.S., Torchiana, M., Glitger, M., Saparstein, M.S., Hirschmann, R. 1978. Proc. Natl. Acad. Sci. USA 75 2636.

3. Sarantakis, D. 1981. US Patent 4,282,143.

4. Bauer, W., Briner, U., Doepfner, W., Haller, R., Huguenin, R., Marbach, P., Petcher, T.J., Pless, J. 1982. Life Sci. 31 1133.

5. Veber, D.F., Freidinger, R.M., Perlow, D.S., Palaveda, W.J., Holly, F.W., Strachan, R.G., Nutt, R.F., Arison, B.H., Homnick, C., Randall, W.C., Glitger, M., Saperstein, R., Hirschmann, R. 1981. Nature 292 55.

SYNTHESIS OF SOMATOSTATIN ANALOGUES OF REDUCED SUSCEPTIBILITY TOWARDS INTESTINAL ENZYMES

Mark C Allen and Roy Wade

Ciba-Geigy Pharmaceuticals, HORSHAM, West Sussex RH12 4AB UK

Studies of the mechanisms of degradation of the somatostatin analogue cyclo(-Asn-Phe-Phe-D-Trp-Lys-Thr-Phe-Gaba-) in rat intestinal juice in vitro (1) with analogs containing tritium (2) had established a single primary enzymatic cleavage site. It was probable therefore, that modifications of the peptide in the region of the scissile Lys^{9}-Thr^{10} bond would provide analogues with increased stability to enzymic degradation. The modifications were restricted to those that would create only minimal changes in physicochemical properties.

All the peptides described were synthesised by solution-phase procedures. As in earlier work (2,5) 4-aminobutanoic $acid^{12}$ was chosen as the carboxyl-terminal residue of the linear precursor since cyclisation could be affected without fear of racemisation. Since all the modifications to the structure were to be made in the D-Trp-Lys-Thr (8-10) region of the molecule it seemed prudent to include this region of variability in a single fragment. It was intended that the build up of the target peptide would then proceed through the 8-12 pentapeptide (by a 3+2 coupling) to the final 5-12 linear octapeptide. The tactics of functional group protection were as described previously (2).

α-Aminoadipamic acid (Aad) was prepared as the N^{ω}-benzyloxycarbonyl derivative by permanganate oxidation of N^{ε}-benzyloxycarbonyl-L-lysine by the method of Liberek (6) and was incorporated into the 8-10 tripeptide fragment as the N^{α}-benzyloxycarbonyl derivative. Removal of the N^{ω}-benzyloxycarbonyl group was found to be necessary due to

Peptides 1986

the apparent readiness of the molecule to eliminate benzyl urethane under the action of base (6), and the subsequent discovery that warming the N^{ω}-protected product in glacial acetic acid caused decarbobenzoxylation and formation of the 2,5-dioxopiperazine. The stepwise build up of the 8-10 fragment was mediated by DCC-HOBt (7) couplings as was the coupling of tripeptide with the C-terminal dipeptide to provide the fully protected C-terminal pentapeptide. Removal of the α-amino protecting group of this fragment followed by a coupling with the N-terminal tripeptide (2) by the azide method (8) gave the fully protected linear octapeptide in reasonable yield. Terminal group deprotection was effected by standard methods and the product was cyclised using a 10-fold excess of both DCC and HOBt in DMF at 50°C for 16 hours in 16% overall yield. Side-chain deprotection by 90% aqueous TFA gave the required product (8) following crystallisation from acetonitrile.

The synthesis of analogues possessing the $D\text{-}Lys^{9}$ (16) $L\text{-}Trp^{8}$, $D\text{-}Lys^{9}$ (23) and $L\text{-}Orn^{9}$ (31) substitutions were prepared by routes and methods exactly analogous to those described for the synthesis of the $L\text{-}Aad^{9}$ (8) derivative and proceeded without difficulty.

D-Allothreonine was prepared from D-threonine by the method of Elliott (9) and converted smoothly to the fully α-protected derivative Z-D-aThr-OTmse. Butylation by standard procedures however, did not proceed smoothly and only very small yields of product were obtained - possibly as a result of acid catalysed premature de-esterification. The need for carboxyl protection of this residue was circumvented by incorporating Z-D-aThr(But)-OH into the C-terminal (10-12) tripeptide and building up the octapeptide by a 2+3 followed by a 3+5 coupling. This was effected without serious difficulties, save for the almost complete insolubility of the final product.

Amino acid analyses of the five octapeptides are displayed in Table 1 as are the stabilities following their incubation in diluted rat intestinal juice <u>in vitro</u> (1). The results support the premise of a primary initial cleavage site following lysine in the original octapeptide since all the analogs tested displayed a considerable enhancement of stability.

TABLE 1

Amino Acid Analysis[a] and Stabilities[b] of Somatostatin Analogs cyclo(-Asn-Phe-Phe-X-Y-Z-Phe-Gaba-)

	Compounds			Amino Acids Analysis								Stability
	X	Y	Z	Aad	Asp	Gaba	Lys	Orn	Phe	Thr/ aThr	Trp	% Intact
	D-Trp	Lys	Thr									0[c]
(8)	D-Trp	Aap	Thr	1.00	1.00	1.01	-	-	3.00	0.95	0.49	82.5
(16)	D-Trp	D-Lys	Thr	-	0.97	0.98	1.00	-	3.00	0.89	0.93	52.5
(23)	L-Trp	D-Lys	Thr	-	0.97	0.97	1.01	-	3.00	1.02	0	100.0
(31)	D-Trp	Orn	Thr	-	0.98	1.02	-	1.02	3.80	0.91	0.92	76.5
(42)	D-Trp	Lys	D-aThr	-	1.01	1.01	0.98	-	3.00	0.92	0.89	-[d]

Footnotes

a Following acid hydrolysis for 24h in 6M-HCl containing phenol (1% w/v) at 110°C

b After 60 min following incubation in rat intestinal juice *in vitro*

c Only 10% intact after 10 min

d Insoluble

References

1. Allen, M.C., McMartin, C., Peters, G.E., and Wade, R., (1984) Regulatory Peptides, 10, 29-35.

2. Allen, M.C., Brundish, D.E., Fullerton, J.D., and Wade, R., (1986), J.C.S. Perkin Trans. I, 989-1003.

3. Kitigawa, K., and Izumiya, N., (1959) J. Biochem. (Japan), 46, 1159-1165.

4. Bennett, H.P.J., Elliott, D.F., Evans, B.E., Lowry, P.J. and McMartin, C., (1972) Biochem. J., 129, 695-701.

5. Rink, H., Kamber, B. and Sieber, P., (1978), Eur. Pat., 1295.

6. Liberek, B. and Jablonka, R., (1981) in Peptides, 1980 (K. Brunfeldt ed.) pp. 242, Scriptor, Copenhagen.

7. König, W. and Geiger, R., (1970) Chem. Ber., 103, 788-798.

8. Honzl, J. and Rudinger, J., (1961) Coll. Czech. Chem.Commun., 26, 2333-2344.

9. Elliott, D.F., (1950) J. Chem. Soc. (C), 62-68.

10. Sieber, P., Riniker, B., Brugger, M., Kamber, B., and Rittel, W., (1970) Helv. Chim. Acta, 53, 2135-2150; Farrault, J.L. and Mercer, E.H., (1952) Biochem. Biophys. Acta, 8, 355-364; Staub, A., Sinn, L. and Behrens, O.K., (1955), J. Biol. Chem., 214, 619-626.

PEPTIDE CONJUGATES AS DIAGNOSTIC REAGENTS IN CLINICAL IMMUNOLOGY

H. Rolli, C.H. Schneider
University of Bern, Institute for Clinical Immunology, Inselspital
CH-3010 Bern, Switzerland

The diagnosis of antibody-mediated allergies to micromolecular antigens including drugs and chemicals of the environment by means of skin tests is not straightforward. This is in contrast to the situation with macromolecules where skin testing is an efficient, fast and reliable procedure. The reason for the difference is well understood in immunochemical terms: Whereas the macromolecule exhibits a number of antigenic determinants allowing bridging of antibodies on the membranes of vasoactive cells, the low-molecular weight compounds are generally monovalent and incapable of bridging antibodies (Fig. 1). It is well established that antibody bridging constitutes the initiating event for anaphylactic reactions, including the immediate-type, local skin reactions evoked for diagnostic purposes. Such reactions may only be expected to occur when micromolecules can be applied in a bivalent or oligovalent form or when they are sufficiently reactive in the skin to form oligovalent conjugates with autologous carriers.

We have prepared penicilloylated oligolysines via random polymerization techniques and used these as skin test reagents for the detection of penicillin allergy in patients over many years. More

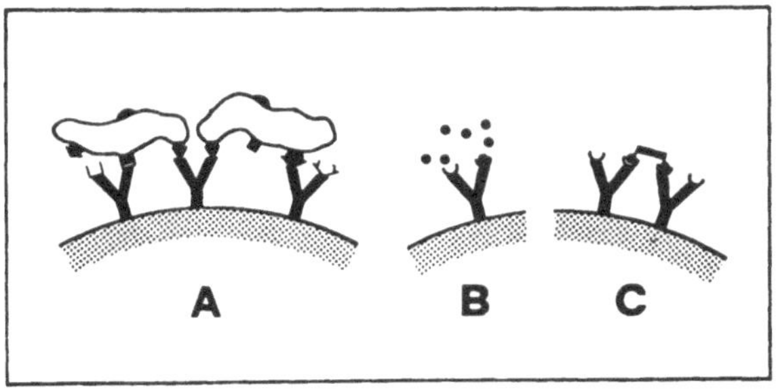

Fig. 1 Bridging of antibodies on the membrane of vasoactive cells. A: Possible bridging by macromolecules with 4 different antigenic sites. B: No bridging with micromolecules. C: Bridging by dihaptenic conjugate.

Peptides 1986

recently, fully defined reagents, obtained by stepwise synthesis in solution, were developed and a set of penicillin test reagents has been made generally available to the medical profession. Other reagents which include haptens related to cephalosporins, pyrazolines, pyrazolidines and aspirin are at various stages of development. All reagents used clinically, are characterized by carriers which comprise two penta-L-lysine chains bound together by aliphatic dicarboxylic acids. It seems worthwile to review some aspects of this work at the present stage.

A variety of BPO-conjugates were evaluated in order to arrive at optimal test reagents and some are listed in Fig. 2. It was found that the efficacy per BPO group in eliciting guinea-pig PCA depends to some extent on the carrier, oligolysines being frequently better than peptide carriers containing Leu, Ala and Gly in addition to Lys. A very important parameter is the ability of the conjugates to elicit systemic reactions, i.e. dangerous generalized anaphylaxis after local application to the skin. In guinea-pig PCA tests, where the elicitor was administered i.d., it was found that within a series of BPO-oligolysines, only the small congeners up to the hexalysine conjugate produced such reactions. BPO_3-LAG_6 and BPO_2-LAG_{12} were also able to evoke generalized reactions, but penicilloylated octalysine and higher homologs did not show this capacity.

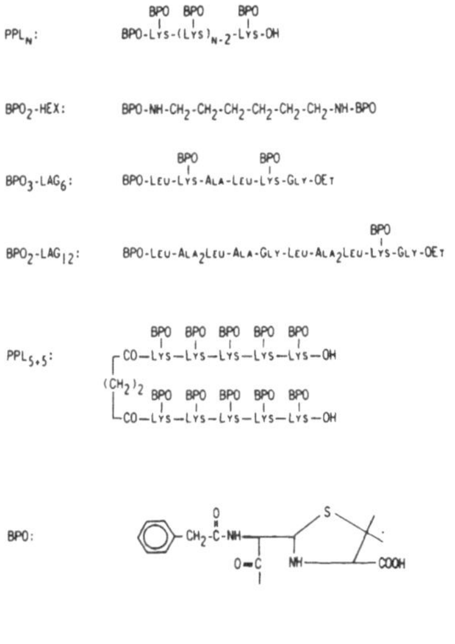

Fig. 2 Some benzylpenicilloyl-conjugates with their abbreviations.
The BPO-substituent is a derivative of D-benzylpenicilloic acid. The four possible diastereoisomers (α - δ) are not distinguished.

Further abbreviation:

PCA : passive cutaneous anaphylaxis.

Other considerations were related to the possible helical content of certain conjugates after it was found that a PPL_{18} conjugate showed 30% helical content in neutral aqueous solution (1). It was thought desirable to generally use conjugates with minimal conformational fixation. Therefore, the bis-pentalysine type of carrier was finally selected and the diagnostically used PPL_{5+5} (Fig. 2) indeed showed no helix formation, neither in aqueous solution nor in the helix-supporting solvent CF_3CH_2OH. It also showed no evidence for β-structure formation, when assessed by IR. absorption of solid samples (1).

The succinic acid bridge of PPL_{5+5} blocks the two α-amino termini of the pentalysines, which is of advantage because free α-termini of oligolysines are difficult to penicilloylate completely. The bis-pentalysine conjugates do not produce systemic anaphylactic reactions after local application to the skin for diagnostic purposes; they are on the other hand small enough to become rapidly eliminated by the kidneys and to be non-immunogenic.
For a variety of haptenic groups, a slightly modified carrier was selected (carrier II or PAL_{5+5}, Fig. 3). It allows six haptens to be attached instead of ten as in PPL_{5+5}. The six carboxylates present provide adequate water solubility even for conjugates with highly hydrophobic haptenic groups. The synthesis, which is carried out on a ten gram scale, is outlined in Fig. 4. The intermediate penta-L-lysines are made by TPM-synthesis (2,3).

The efficacy of PPL_{5+5} (BPO_{10}-carrier I) and of BPO_6-PAL_{5+5} was compared in guinea-pig PCA. It was found that the eliciting potency, when defined per haptenic group concentration but not, when defined per conjugate molar concentration, was virtually identical for these two very similar reagents. On a molar basis therefore, the potency of the BPO_{10}-conjugate was 1.66 times that of the BPO_6-conjugate. Since for steric reasons, it is difficult to visualize

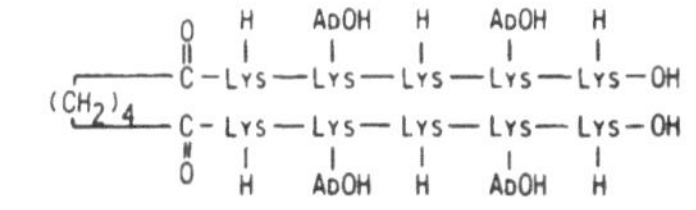

Fig. 3 Structure of carrier II or PAL_{5+5}.
$Ad(OH)_2$: adipic acid

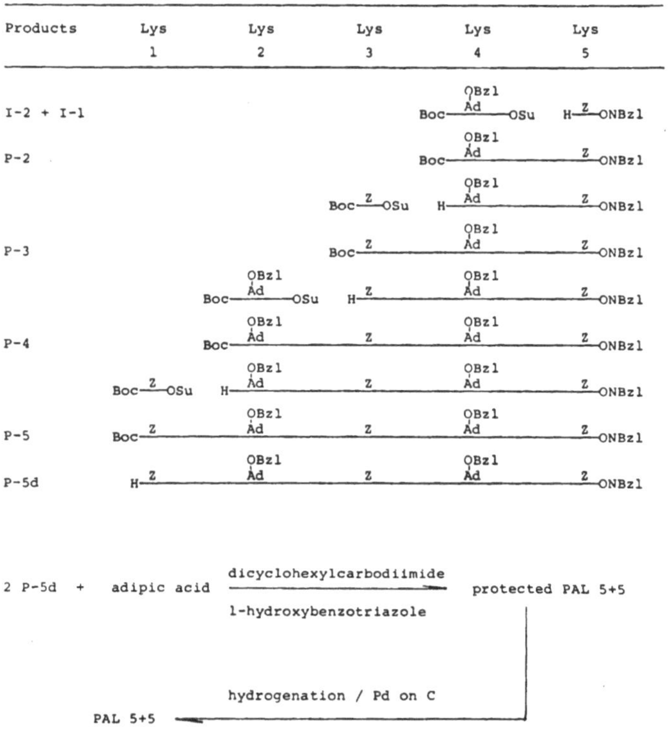

Fig. 4 Synthetic scheme for carrier II (PAL_{5+5}).

that more than two links per conjugate are made with cell-bound antibody, the greater efficacy per mole of the BPO_{10}-conjugate cannot be explained by formation of more antigen-antibody cross-links. Our interpretation is, that clustering of haptenic groups enhances affinity and therefore efficacy. The term "cluster affinity" was applied and it was shown that cluster affinity can be readily integrated into the established concepts of antigen-antibody association (4).

The diagnostic value of PPL_{5+5} is well established. When used in conjunction with the chemically reactive compounds penicillin and penicilloic acid, it will detect almost all allergic patients who might undergo pronounced adverse reactions after penicillin administration. Similar evidence with less well-defined reagents was previously reported (5, 6).

The situation with regard to other drugs is less clear. It has to be taken into account that antibodies are generally directed against some protein-attached or cell-membrane-linked structure of the allergenic micromolecular compound. Such structures may differ considerably from the original ones, particularly when metabolic steps are involved for their establishment. For the test reagents, one has therefore to pick the "right" haptenic structures which can be a burdensome task because the immunochemical specificities involved may be quite strict. This has been recently shown for the 1,2-diphenyl-pyrazolidinedione series (7).

References

1. Rolli, H., I.F. Lüscher, C.H. Schneider, C. Toniolo, G.M. Bonora. 1982. Helv. Chim. Acta 65, 1965.
2. Schneider, C.H., H. Rolli, K. Blaser. 1980. Int. J. Peptide Protein Res. 15, 411.
3. Rolli, H., C.H. Schneider. 1981. Chimia 35, 403.
4. Schneider, C.H., H. Rolli: to be published.
5. Solley, G.O., G.J. Gleich, R.G. van Dellen. 1982. J. Allergy Clin. Immunol. 69, 238.
6. Mendelson, L.M., Ch. Ressler, J.P. Rosen, J.E. Selcow. 1984. J. Allergy Clin. Immunol. 73, 76.
7. Kasper, M.F., C.H. Schneider, H. Rolli, B.D. Angst, A.L. de Weck: Immunobiology (in press).

PREDICTION OF HUMAN RENIN EPITOPES USING A NEW COMPUTER ALGORITHM

Genevieve Evin, William D. Carlson, Mark Handschumacher, Jiri Novotny, Edgar Haber

Massachusetts General Hospital and Harvard Medical School
Boston, Massachusetts 02114, USA

Francois-Xavier Galen, Jacob Bouhnik, Joel Menard, Pierre Corvol

Institut National de la Sante et de la Recherche Medicale, Unite 36, 17 rue du Fer-a-Moulin, 75005-Paris, FRANCE

A new method for predicting the antigenic determinants of proteins has recently been reported (1). Epitopes are predicted as those surface regions that are accessible to a large spherical probe of 1 nm radius (the size of an antibody binding domain), and are defined by calculating the contacts between the large probe and the surface of the protein. We have applied the Large Probe Accesibility Program to a model for the three-dimensional structure of human renin that had been constructed from the atomic coordinates of three homologous aspartyl proteases by maximizing sequence homology and refining the structure for energy minimization (2). Seventeen segments were proposed as epitopes, or parts of epitopes, of human renin. Peptides corresponding to 5 of these segments were synthesized and studied for their reactivity toward antibodies raised against human renin.

Peptide Synthesis

The sequences of the peptides are shown in Table 1. All peptides were prepared by the Merrifield solid phase method. p-Methylbenzhydrylamine resin was used as solid support and benzotriazolyl-oxy-tris(dimethylamino)phosphonium hexafluorophosphate (BOP) (3) as a coupling reagent, using a standard Boc-protection strategy. The syntheses were carried out manually, with a 4-fold excess of Boc amino acid and reagent. The pH was checked with moist indicator paper and maintained between 7 and 8 with diiso-

Peptides 1986

Table 1. Sequences of Synthetic Peptides Proposed as Renin Epitopes

Y-133-144	Tyr-Ile-Glu-Gln-Ala-Ile-Gly-Arg-Val-Thr-Pro-Ile-Phe-amide
C-180-188	Cys-Gly-Ser-Asp-Pro-Gln-His-Tyr-Glu-Gly-amide
Y-211-224	Tyr-Gly-Ser-Ser-Thr-Leu-Leu-Cys-Glu-Asp-Gly-Cys-Leu-Ala-Leu-amide
C-290-296-G	Cys-Tyr-Ser-Ser-Lys-Lys-Leu-Cys-Gly-amide
Y-300-310	Tyr-Ile-His-Ala-Met-Asp-Ile-Pro-Pro-Pro-Thr-Gly-amide

propylethylamine. Under these conditions, maximum coupling efficiency could be obtained, recouplings were avoided, and fast Boc-amino acid incorporation was observed (20-30 min). The peptides were cleaved from the resin and totally deprotected by treatment with HF in the presence of 10% anisole. For cyclization, disulfide bonds were formed by air-oxidizing the crude material obtained from HF and stirring for 24 hours at room temperature in 0.02 M ammonium acetate (pH 8.7) at a peptide concentration of 0.1 mg/ml. Each peptide was purified by filtration on Sephadex G-25, followed by ion-exchange chromatography (CM or DEAE cellulose, depending on peptide ionic charges). Purity was assessed by reverse-phase HPLC, TLC and amino acid analysis.

Tests of Antigenicity

All peptides were tested for their binding to 11 polyclonal and 7 monoclonal antibodies prepared against pure human renin (4.5). Two assays were performed: a radioimmunoassay (RIA), in which the peptide was labeled with ^{125}I and incubated with various dilutions of antibody, and an enzyme-linked immunosorbent assay (ELISA), in which the peptide was first adsorbed on a plate and then incubated with the antibody. Antibody retained on the plate was then quantitated by a colorimetric assay. The results are summarized in Table 2.

Table 2. Recognition of the Synthetic Peptides by 11 Polyclonal Antirenin Antibodies

Peptide	Number of Antisera Binding to Peptide	
	RIA	ELISA
Y-133-144	5	1
C-180-188	4	3
Y-211-224	11	8
C-290-296-G	0	3
Y-300-310	5	2

Even though none of the monoclonal antibodies bound to any peptide in either assay, all peptides were recognized by at least a few of the polyclonal antibodies and were thus characterized as antigenic. Among these peptides, Y-211-224 seemed to correspond to a dominant epitope, since it was recognized by all antirenin antisera tested. Reduction of the disulfide bond followed by carboxymethylation of the cysteines resulted in a complete loss of antigenicity. This result indicates that the epitope is conformationally dependent and confirms the presence of a disulfide between cysteines 217 and 221 in the native structure of human renin. Two peptides, Y-133-144 and Y-300-310, were recognized by a greater number of antibodies in the RIA than in the ELISA. One interpretation of this finding is that these peptides adopt a secondary structure in solution that more closely resembles their secondary structure in native renin; the epitopes associated with these peptides may also be conformationally dependent.

Thus, five epitopes of human renin have been successfully identified by computer prediction. A dominant epitope has been found and defined as conformationally dependent.

References

1. Novotny, J., M. Handschumacher, E. Haber, R. Bruccoleri, W. Carlson, D. Fanning, J.A. Smith, G. Rose. 1986. Proc. Natl. Acad. Sci. USA 83, 226.

2. Carlson, W., M. Karplus, E. Haber. 1985. Hypertension 7, 13.

3. Castro, B., J.R. Dormoy, G. Evin, C. Selve. 1975. Tetrahedron Lett., 1219-1222.

4. Galen, F.X., T. Guyenne, C. Devaux, C. Auzan, P. Corvol, J. Menard. 1979. J. Clin. Endocrinol. Metab. 48, 1041.

5. Galen, F.X., C. Devaux, S. Atlas, T. Guyenne, J. Menard, P. Corvol, D. Simon, P. Richer, G. Badouaille, J.P. Richaud, P. Gros, B. Pau. 1984. J. Clin. Invest. 74, 723.

DESIGN AND SYNTHESIS OF IMMUNOGENIC COMPLEXES BUILT UP OF A PEPTIDE ANTIGEN, GLYCOPEPTIDE ADJUVANT AND POLYMER CARRIER

T.M. Andronova, V.A. Rar, E.A. Makarov, V.T. Ivanov

Shemyakin Institute of Bioorganic Chemistry, USSR Academy of Sciences, ul. Miklukho-Maklaya, 16/10, 117871 Moscow V-437, USSR

Introduction

Rational design of synthetic vaccines implies construction of immunogenic complexes containing the three title components, viz. the peptide antigenic determinant (epitope), low molecular immunoadjuvant and a polymeric carrier, also participating in presentation of the peptide to the immune system. The present work provides an example of constructing such conjugates.

Results

As a peptide moiety of title complexes we have chosen the 205-213 segment of the surface VP_1 protein of foot and mouth disease virus (O_1K strain). According (1,2) the C-terminal sequences of that protein to which our peptide belongs are able to induce virus neitralizing antibodies. The peptide was synthesized in solution by classical method as shown in Fig. 1. The protected intermediate peptides were obtained in 75-90% yields by active ester stepwise elongation. The carbodiimide 4+5 coupling afforded after DMF/Ether precipitation the protected nonapeptide subjected to two consequtive deprotection steps. The final product VP_1(205-213) was isolated by preparative HPLC on Silasorb C18 (16x250 mm) column in 50 MeOH - 0.02 M Et_3N-acetate (pH 6.0). Polycationic poly(DL-Ala)--poly(Lys) (AL) and polyanionic copolymer of maleic anhydride and vinylpyrrolidone (MAVP) were employed as carriers. The former was obtained by the method (3) from the 10,000 (TSK HW 40 column) molecular mass fraction of poly(Lys), the Ala:Lys ratio being 8.9:1. The latter, with vinylpyrrolidone:maleic anhydride was prepared as described in (4) and generously provided by L.A. Fonina. The adjuvant was N-acetyl-

Peptides 1986

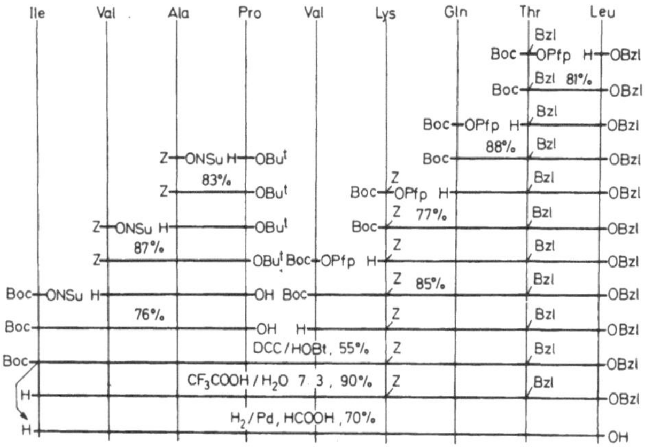

Fig. 1. Synthesis of the VP_1(205-213) segment.

glucosaminyl-N-acetylmuramyl-alanyl-D-isoglutamine (GMDP) (5). A variety of relative arrangements of the three components can be envisaged in the synthetic vaccine. Some of them, schematically presented in Fig. 2 and in the Table 1 were materialized in this work. Samples 4-7 were obtained by 1-ethyl-3(3'-dimethylaminopropyl)carbodiimide hydrochloride (WSCD/HOBt) coupling of variable amounts of GMDP followed by unprotected nonapeptide with the AL polymer. Compounds 12-14, also belonging to type I in Fig. 2 were prepared by simultaneous coupling of the MAVP with GlcNAc-MurNAc-

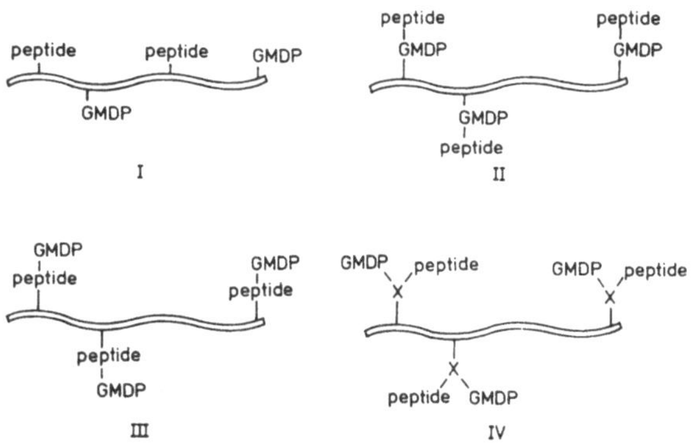

Fig. 2. Arrangements of the peptide, GMDP and the polymer in the triple conjugates.

Table 1. Composition of Immunogenic Complexes

No	Structure	Content (%, w/w)	
		of VP_1(205-213)	of GMDP
1	VP_1(205-213)~AL	3.5	-
2	VP_1(205-213)~AL	7.0	-
3	VP_1(205-213)~AL	10.2	-
4	VP_1(205-213)~AL~GMDP	5.2	5.2
5	VP_1(205-213)~AL~GMDP	6.1	5.4
6	VP_1(205-213)~AL~GMDP	5.0	12.3
7	VP_1(205-213)~AL~GMDP	2.0	31.0
8	$(GMDP)_2$-VP_1(205-213)~AL	2.0	3.5
9	$(GMDP)_2$-VP_1(205-213)~AL	11.1	16.6
10	MAVP~VP_1(205-213)	3.0	-
11	MAVP~VP_1(205-213)	11.2	-
12	VP_1(205-213)~MAVP~$HNCH_2CH_2NHGMDP$	2.9	1.5
13	VP_1(205-213)~MAVP~$HNCH_2CH_2NHGMDP$	6.1	1.9
14	VP_1(205-213)~MAVP~$HNCH_2CH_2NHGMDP$	13.5	3.8
15	MAVP(Aca)~VP_1(205-213)	6.1	-
16	MAVP(Aca)~VP_1(205-213)	2.9	-
17	VP_1(205-213)~MAVP(Aca)~$HNCH_2CH_2NHGMDP$	1.1	1.0
18	VP_1(205-213)~MAVP(Aca)~$HNCH_2CH_2NHGMDP$	2.1	2.0
19	VP_1(205-213)~MAVP(Aca)~$HNCH_2CH_2NHGMDP$	4.3	1.5
20	VP_1(205-213)~MAVP(Aca)~$HNCH_2CH_2NHGMDP$	1.0	2.5

Ala-D-iGln- $NHCH_2CH_2NH_2$ and the nonapeptide. The GMDP ethylendiiamine was obtained from GMDP pentafluorophenylate after treatment with tenfold excess of the diamine. A similar series with the ε-aminocaproyl spacer (compounds 17-20) was obtained from the activated polymer was prepared as described in (6). The respective two component conjugates 1-3, 10, 11, 15 and 16 were prepared. The above mentioned pentafluorophenylate was used to acylate the aminoterminal and the ε-Lys amino moiety of the nonapeptide, giving rise to $(GMDP)_2$-VP_1(205-213). The latter was coupled by WSCD/HOBt with AL to give 8 and 9 (type III conjugate in Fig. 2). Conjugates 1-20 are purified by gel-filtration on Sephadex G-50, characterized by amino acid and carbohydrate analysis and are presently subjected

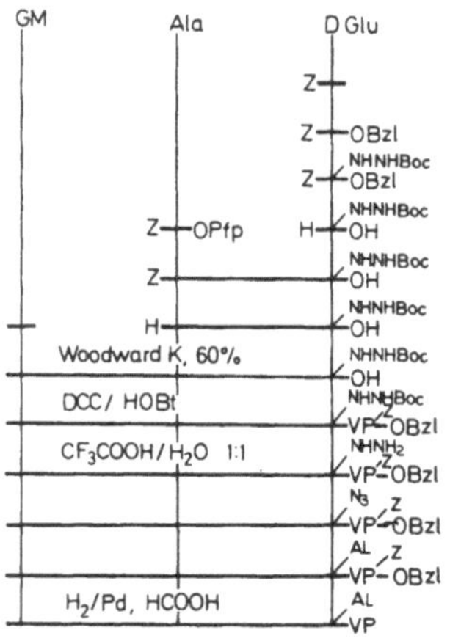

Fig. 3. Synthesis of the type II conjugate.

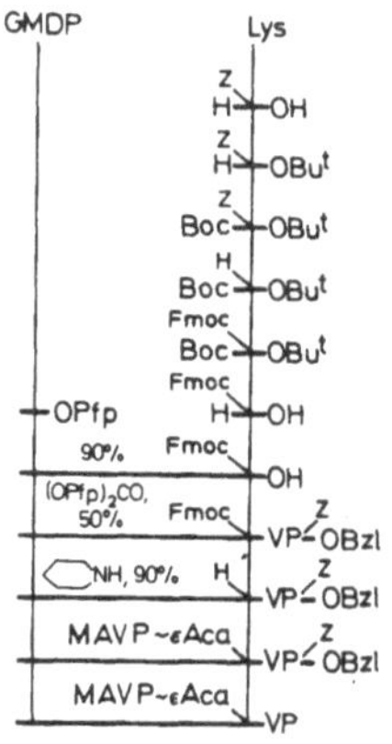

Fig. 4. Synthesis of the type IV conjugate.

to biological studies.

Type II and IV conjugates are being synthesized according to schemes in Fig. 3 and 4. Details will be presented when the work is completed.

References

1. Bittle, J.L., R.A. Houghten, H. Alexander, T.M. Shinnick, J.G. Sutcliffe, R.A. Lerner, D.J. Rowlands, F. Brown. 1982. Nature (London) 298, 30.

2. Pfaff, E., M. Mussgay, H.O. Bohm, G.E. Schulz, H. Schaller. 1982. EMBO Journal 1, 869.

3. Sela, M., E. Katchalski, M. Gehatia. 1956. J.Amer.Chem.Soc. 78, 746.

4. Conix, A., G. Smets. 1955. J.Polym.Sci. 15, No.79, 221.

5. Rostovtseva, L.I., T.M. Andronova, V.P. Mal'kova, I.B. Sorokina, V.T. Ivanov. 1981. Bioorg.Khim. (Russian) 7, 1843.

6. Andreev, S.M., M.V. Sidorova, O.A. Rakova, D.E. Tsvetkov, L.A. Fonina. 1978. Bioorg.Khim. (Russian) 13 (in press).

THYMULIN ANALOGUES FOR PROBING SPECIFIC BINDING SITES ON LYMPHOBLASTOID CELLS

Didier Blanot*, Louis-Noël Gastinel+, Mireille Dardenne+ and Jean-François Bach+
*UA 04/1131 CNRS, Université de Paris-Sud, 91405 Orsay, France ; + INSERM U.25, Hôpital Necker, 75015 Paris, France

Introduction

Thymulin (or facteur thymique sérique, FTS) is a thymic nonapeptide, the sequence of which is <Glu-Ala-Lys-Ser-Gln-Gly-Gly-Ser-Asn-OH. It was isolated from pig serum on its ability to induce the appearance of Thy-1 antigen on Thy-1-negative rosette-forming spleen cells. The presence of zinc ions has been shown to be necessary for the expression of its activity in the rosette assay (for a recent review, see ref. 1).

The presence of specific binding sites for thymulin on four lymphoblastoid T-cell lines (HSB_2, 1301, CEM and JM) has been established by using two different tritiated probes : a methylated analogue, with a 5-Ci/mmol specific activity (2,3), and a radioactive thymulin obtained by catalytic tritiation of an acetylenic precursor, with a 120-Ci/mmol specific activity (4-7). In order to further characterize these receptors, other kinds of probes are necessary. In this communication, we wish to describe the synthesis of seven new thymulin analogues carrying either fluorescent groups, a biotinyl moiety, or affinity labellers.

Syntheses

Thymulin (I) was a gift from Dr. Lefrancier (8). The syntheses of [Glu^1]thymulin (II) and H-Gln(Mbh)-Ala-Lys(Boc)-Ser(Bu^t)-Gln(Mbh)Gly-Gly-Ser(Bu^t)-Asn-OBu^t.AcOH (III) have already been described (9,10). Dansyl chloride was purchased from Merck, fluorescein isothiocyanate (FITC) and biotin N-hydroxysuccinimide ester from Sigma, [cysteamine-^{35}S]N-succinimidyl-3-[(2-nitro-4-azidophenyl)-2-aminoethyldithio]propionate (SNAP) from Amersham and N-succinimidyl-4-azidobenzoate from Pierce. The purity of the synthesized compounds was checked by TLC and HPTLC in several solvent systems and by amino acid analysis after acid hydrolysis.

<u>[N^{ϵ}-Dansyl-Lys^3]thymulin (IV)</u> was obtained by reaction of dansyl chloride on I in 0.1 M sodium bicarbonate, pH 9.5, for 3 hrs. It was purified by gel filtra-

Peptides 1986

tion on Bio-Gel P_2, then by HPLC on μBondapak C_{18}. Yield: 65%.

[N^{α}-Dansyl-Gln^1]thymulin (V). Dansyl chloride was allowed to react on III in DMF for 24 hrs, in the presence of N-methylmorpholine (NMM). The isolated nonapeptide derivative was deprotected by CF_3CO_2H-anisole 10:1 for 3 hrs (11). The product was purified by preparative TLC on cellulose in n-propanol-water 2:1. Overall yield: 23%.

[N^{ϵ}-Fluorescein-5 -thiocarbamoyl-Lys^3]thymulin (VI) was synthesized and purified as for IV, using FITC as a reagent. Yield: 35%

[N^{α}-Fluorescein-5 -thiocarbamoyl-Gln^1]thymulin (VII). FITC was condensed with III in DMF in the presence of NMM. After 24 hrs, the reaction was complete and resulted in only one new TLC spot. However, after deprotection as for V, the product was heterogeneous and mainly contained a polar, non-fluorescent compound. A minor fluorescent compound could be purified by preparative TLC on cellulose, then by HPLC on μBondapak C_{18} in 50 mM ammonium formate pH 3.5-methanol (65:35). Overall yield: 1.2%.

[N^{α}-Biotinyl-Gln^1]thymulin (VIII). Biotin N-hydroxysuccinimide ester was condensed with III in DMF in the presence of NMM. The isolated nonapeptide derivative was deprotected as for V. The product was purified by preparative TLC on cellulose, then by HPLC on Ultrasphere-ODS (30 mM ammonium formate pH 3.5, gradient from 3% to 30% of methanol for 30 min). Overall yield: 23%.

[^{35}S](3-[(2-Nitro-4-azidophenyl)-2-aminoethyldithio]-propionyl-Glu^1)thymulin (IX). [^{35}S]SNAP (12) (1 Ci/mmol), dissolved in DMF, was added to II, dissolved in 0.1M sodium phosphate, pH 7.5. After 24 hrs in the dark, the product was purified by HPLC on Lichrosorb RP-18 (A: 30 mM ammonium formate pH 3.5-methanol 97:3 ; B: 30 mM ammonium formate pH 3.5-methanol 50:50 ; gradient from 0 to 100% of B for 50 min). Yield: 19%.

[4-Azidobenzoyl-Gln^1]thymulin (X). N-Succinimidyl-4-azidobenzoate was condensed with III in DMF in the presence of NMM. After 24 hrs in the dark, the reaction mixture was evaporated and deprotected as for V. The product was purified by HPLC on Lichrosorb RP-18 in 2% acetic acid (gradient from 3 to 80% of acetonitrile for 50 min). Yield: 41%.

Results and Discussion

Syntheses. They were carried out by reaction of appropriate reagents with a protected [Gln^1]thymulin derivative followed by deprotection (analogues V, VII, VIII and X), with thymulin itself (analogues IV and VI) or with [Glu^1]thymulin (analogue IX). In the first case, a problem arose only during the synthesis of

VII. Whereas the coupling of FITC with III had proceeded correctly, the yield of the desired compound after deprotection was very low. This was undoubtly due to a reaction of the fluorescein-thiocarbamoyl-peptide in acidic medium, reaction which is analogous to that occurring in the Edman degradation (13).

In the third case, we tried to couple the reagent with a free peptide directly. [Glu^1]thymulin (III) was chosen in order to avoid the cyclization of the glutaminyl residue in [Gln^1]thymulin. It had been demonstrated that Glu in position 1 did not modify neither the biological activity nor the binding to the receptors (unpublished results). pH conditions were studied in order to favor the coupling on the α-NH_2 of the peptide. This procedure allowed us to obtain IX with a moderate yield. However, it was absolutely ineffective for the coupling of N-succinimidyl-4-azidobenzoate with III.

Biological activity and binding experiments. The biological activity of the analogues in the rosette assay and their effect of inhibition of the binding of [^{3}H]thymulin on 1301 cells are shown on Table 1. It can be seen that the N^{ϵ}-substituted derivatives are poorer competitors than the N^{α}-substituted ones, the IC_{50} of which are of the same order of magnitude as that of cold thymulin. This is in good correlation with the biological activity in the rosette assay.

Table 1. Biological Activity and Inhibition of [^{3}H]Thymulin Binding.

Analogue	Biological activity[a]	Inhibition of binding IC_{50} (nM)[b]	R.P.[c]
Thymulin (I)	+	30	1
IV	-	6000	0.005
V	+	60	0.5
VI	-	12000	0.0025
VII	+	240	0.125
VIII	N.D.[d]	20	1.5
X	N.D.[d]	60	0.5

[a]in the rosette assay; [b]concentration of peptide inhibiting 50% of [^{3}H]thymulin binding on 1301 cells; [c]relative potency (thymulin = 1); [d]not determined.

Microscopic visualization experiments. Attempts to visualize the receptors on 1301 cells either directly (with V or VII) or indirectly (with VIII and fluorescent avidine) were unsuccessful. Possible explanations are as follows. In the first case, the density of sites per cell might be too weak for a direct optical detection. In the second case, the receptor of thymulin might render the probe inaccessible to avidin ; a spacer between biotin and the nonapeptide would then be necessary.

Photoaffinity experiments. HSB_2 cells were incubated with radioactive analogue IX either in the dark or under a UV lamp. In both cases, the binding was very low. Moreover, in the dark, it was inhibited only by a large excess of cold thymulin (IC_{50} > 100 µM). The affinity of IX for the receptors as well as its specific radioactivity are too weak for such experiments.

On the other hand, the good results obtained with non-radioactive analogue X (relative potency on 1301 cells : 0.5 ; on CEM cells : 1) seem to indicate that radiolabelled X could be a suitable tool for affinity labelling of the thymulin receptors on lymphoblastoid cells.

References

1. Pléau, J.M., Gastinel, L.N., Bach, J.F. 1985. Methods Enzymol. 116, 269.
2. Pléau, J.M., Fuentes, V., Morgat, J.L., Bach, J.F. 1979. C.R.Acad. Sci. Paris Sér. D 288, 445.
3. Pléau, J.M., Fuentes, V., Morgat, J.L., Bach, J.F. 1980. Proc. Natl. Acad. Sci. USA 77, 2861.
4. Gastinel, L.N., Pléau, J.M., Dardenne, M., Sasaki, A., Bricas, E., Morgat, J.L., Bach, J.F. 1982. Biochim. Biophys. Acta 684, 117.
5. Gastinel, L.N., Pléau, J.M., Goldstein, G., Bach, J.F. 1983. Thymus 5, 79.
6. Auger, G., Blanot, D., Magnin, M., Gastinel, L.N., Dardenne, M., Bach, J.F. 1984. In: Peptides 1984 (Ragnarsson, U., ed.). Almqvist & Wiksell International, Stockholm, p. 435.
7. Gastinel, L.N. 1986. Doctoral Thesis, University of Paris VII.
8. Lefrancier, P., Derrien, M., Amiot, J.L., Choay, J. 1984. In: Peptides 1984 (Ragnarsson, U., ed.). Almqvist & Wiksell International, Stockholm, p.251.
9. Martinez, J., Blanot, D., Auger, G., Sasaki, A.N., Bricas, E. 1980. Int. J. Peptide Protein Res. 16, 267.
10. Auger, G., Blanot, D., Bricas, E., Pléau, J.M., Dardenne, M., Bach, J.F. 1982. Hoppe-Seyler's Z. Physiol. Chem. 363, 331.
11. Blanot, D., Martinez, J., Auger, G., Bricas, E. 1979. Int. J. Peptide Protein Res. 14, 41.
12. Schwartz, M.A., Prem Das, O., Hynes, R.O. 1982. J. Biol. Chem. 257, 2343.
13. Muramoto, K., Kamiya, H., Kawauchi, H. 1984. Anal. Biochem. 141, 446.

SYNTHESIS AND PHARMACOLOGICAL PROPERTIES OF ADAMANTYLAMIDE ANALOGS OF MURAMYL-DIPEPTIDE (MDP).

M. Flegel, J. Seifert, H. Farghali, K. Masek
Institute of Pharmacology, Czechoslovak Academy of Sciences
Albertov 4, 128 00 Prague 2, Czechoslovakia

M. Krojidlo
Laboratory of Peptides, Leciva-Pharmaceuticals, 143 10 Prague 4

Introduction

The structure - activity relationship studies of muramyl peptides resulted in the synthesis of compounds, where a number of biological activities could be dissociated (1). At present the search for new analogs with increased immunostimulatory activity and less side effects still continues. It was shown previously (1,2) that muramic acid presence is not essential for the immunostimulatory activity and that the introduction of lipophilic features into the structure could be important. The aim of this work was to link 1-aminoadamantane to the carboxyl group in the side chain of D-isoglutamine of MDP and also to its peptide part only.

Results and Discussion

Peptides were prepared according to scheme 1. The protected dipeptide acid (I) was coupled with 1-aminoadamantane either by DCC-HOBT or mixed anhydride method. After splitting BOC group (40% TFA-DCM) and desalting (IRA-400) product (III) was purified by HPLC (Separon SIX C18, mobile phase: 55% MeOH, 45% 0.1% TFA). AdDP was then used for condensation with 1-α-O-benzyl-4,6-O-benzyliden-N-acetylmuramic acid (pivaloylchloride or DCC-HOBT). Protected adamantylamide of muramyldipeptide (IV) was treated with sodium in liquid ammonia (4) and purified by HPLC. The purity of the compounds was

Peptides 1986

evaluated by TLC, HPLC and mass spectra.

Scheme 1: Synthesis of Adamantylamide analogs

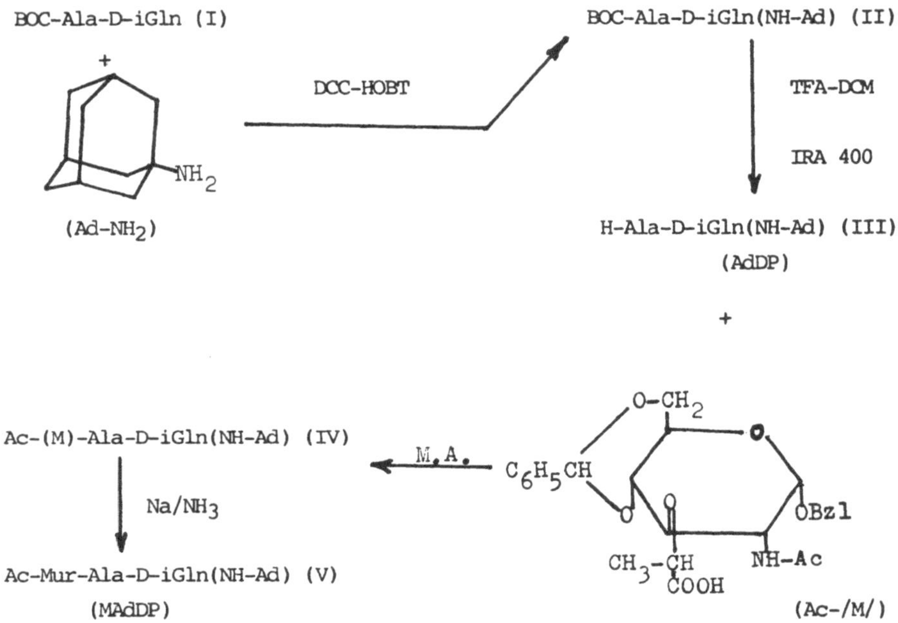

The synthesis of stereochemically homogenous muramic acid can be a rather difficult task (3). The protected product contains usually 10-25% of (S)-1-carboxyethyl- e.g. isomuramic diastereomer, the presence of which can influence pharmacological results. On the other hand the synthesis can be carried out with diastereomeric mixture if HPLC is used for the purification after the last reaction step (Separon SIX C18, 40% MeOH, 60% 0.1% TFA).

The synthesis of disaccharide containing glycopeptides, modification of muramyl residue by increasing its lipophilicity or substitution of Ala by more hydrophobic amino acids (1) are important factors that influence the activity profile leading to compounds with protracted and enhanced immunostimulatory properties more or less without significant side effects. The most important structure feature is however the configuration and functionality of D-glutamic acid (5). Using of 1-aminoadamantane, which is known to interact with the cell membrane and is used clinically (6) for the synthesis of analogs of MDP led among the other findings also to the conclusion, that for the immunomodulatory activity the presence of muramyl residue is not essential. The interesting profile of biological activity of AdDP and the simplicity of its synthesis could later influence the decision for its development as a drug.

From the results presented in table 1 is clear that both adamantylamide analogs have the adjuvant activity comparable either to MDP or FCA, the muramyl peptide derivative being more active.

Table 1. Induction of delayed hypersensitivity to ovalbumin in guinea pig

compound	dose(μg/kg)	skin response (24 hours)
FIA	-	3.39 ± 0.22
FCA	1000	5.73 ± 0.33*
MDP	100	6.07 ± 0.21*
AdDP	100	5.82 ± 0.27*
MAdDP	100	7.54 ± 0.81*

The effects on the synthesis of DNA (7) are summarised in table 2.

Table 2. Utilisation of ^{3}H thymidine for the synthesis of DNA thymine in rat organs after administration of MDP and AdDP

compound	dose (μg/kg)	cpm/μmol of DNA thymine			
		Liver	Kidney	Thymus	Spleen
control	-	1.750	1.750	940	11.150
MDP	1000	4.100	2.350	1400	10.900
AdDP	740	2.950	2.200	1550	16.500

As can be seen, the effect of AdDP was comparable to the effect of either MDP or FCA. The activity was observed even in the spleen, where MDP and FCA are without the effect. It is noteworthy, that AdDP in relevant doses is nonpyrogenic and in addition to the activities mentioned above a hepatoprotective effect was observed against CCl_4 liver injury as summarised in table 3. It is noteworthy that to observe a significant hepatoprotective activity of MDP (9) one has to increase the dose several hundred folds. The mechanism of hepatoprotective effect is so far not completely clear, but it does not seem to be related simply to immunological activity.

Table 3. Hepatoprotective Effect of AdDP in rats poisoned with CCl_4

Enzyme	Activity (μKat/l) in		
	controls	0.29 ug/kg	2.9 ug/kg
AST	38 ± 18	28.7 ± 17	24.8 ± 16.5*
ALT	57 ± 25	44 ± 37	39 ± 25*
LD	182 ± 134	157 ± 116	118 ± 100*
n	(50)	(32)	(20)

AST,ALT = aminotransferases, LD = Lactate dehydrogenase, n = number of rats per experiment, * Statistically significant results; $p < 0.05$

References

1. Arlette, A., Lederer, E. 1984. Med.Res.Rev. 4, 111.
2. Masek, K., Flegel, M. 1983. Zbl.Pharm. 122, 127.
3. Gross, P.H., Rimpler, M. 1986. Liebigs Ann.Chem. 37.
4. Zaoral, M., Jezek, J., Straka, R., Masek, K. 1978. Coll.Czech.Chem. Commun. 43, 1977.
5. Lefrancier, P., Derrien, M., Jamet, X., Choay, J., Lederer, E., Audibert, F., Parant, M., Parant, F., Chedid, L. 1982. J.Med.Chem. 25, 87.
6. Induen, M.K. 1981. Mechanismy antivirusnogo dejstvija proizvodnych adamantana. Riga - Zinatne.
7. Masek, K., Zaoral, M., Jezek, J., Straka, R. 1978.Experientia, 34,1363.
8. Seifert, J. 1982. Eur.J.Biochem. 128, 277.
9. Farghali, H., Machková, Z., Kamenîková, L., Janku, I., Masek, K. 1984. Meth. and Find.Exptl.Clin.Pharmacol. 6, 449.

IMMUNOGENICITY OF HEPATITIS B SURFACE ANTIGEN SYNTHETIC PEPTIDE ANALOGUES

B. Lammek, G. Kupryszewski
Institute of Chemistry, University of Gdańsk, 80-952 Gdańsk, Poland

U. Ragnarsson
Institute of Biochemistry, Biomedical Center, University of Uppsala
Box 576, S-75123 Uppsala, Sweden

H. Świderska, M. Jastrzębska, A. Nowosławski
National Institute of Hygiene, Institute of Immunopathology,
00-971 Warsaw, Poland

Introduction

Hepatitis B surface antigen HBsAg is the major constituent of the envelope of the hepatitis B virus HBV and of the 22 nm noninfective particles produced in excess during infection and released to the host circulation. The HBV cannot be grown in the tissue culture. HBsAg purified from human blood has been employed as a vaccine. The major polipeptide of antigen contains 226 amino acid residues; its 121-161 amino acid sequence is postulated to be a hydrophilic region exposed on the surface of the intact HBV or HBsAg particles (1). The presence of β-turns in the regions 138-149 and 122-137 suggested that they may be responsible for the antigenicity of determinant "a" of HBsAg, common for all subtypes of HBV. Therefore, this two amino acid sequences were selected as a model for design of analogues with expected immunogenicity (2,3).

Materials and methods

Seven modified peptide analogues were synthesized by the solid phase method on chloromethylated resin. The obtained peptides have a following structure:

Abu-Abu-Thr-Lys-Pro-Thr-Asp-Gly-Asn-Abu-Thr-Abu-Gly-Gly (1)

Peptides 1986

Ala-Ala-Thr-Lys-Pro-Thr-Asp-Gly-Asn-Ala-Thr-Ala-Gly-Gly (2)

Abu-Abu-Abu-Lys-Pro-Thr-Asp-Gly-Asn-Abu-Abu-Abu-Gly-Gly (3)

Abu-Abu-Thr-Lys-Pro-Thr-Asp-Gly-Asp-Abu-Thr-Abu-Gly-Gly (4)

Abu-Lys-Pro-Thr-Asp-Gly-Asn-Abu-Gly-Gly (5)

Lys-Thr-Cys-Met-Thr-Thr-Ala-Gln-Gly-Thr-Ser-Met-Tyr-Pro-Ser-Cys-Gly (6)

Cys-Thr-Lys-Pro-Thr-Asp-Gly-Asn-Lys-Pro-Thr-Asp-Gly-Asn-Cys-Thr-Lys-Gly (7)

N-α-Boc protection was used through the synthesis with following side-chain protecting groups: Ser(Bzl), Thr(Bzl), Tyr(Bzl), Asp(Bzl), Lys(Z), Gln(Xan), and Cys(Acm).Dicyclohexylcarbodiimide was used to couple the Boc-amino acids: hydroxybenzotriazole was added tothe coupling reactions of glutamine and asparagine. All peptides were cleaved from the resin by treatment with liquid HF at $0^{o}C$ in the presence of anisole. In case of the cyclic peptides the use of acetamidomethyl residue as an S-protection for cysteine allowed to isolate and purify these analogues as linear peptides. The purified Acm-peptides were then deblocked using mercuric acetate (pH=4.5) and after addition of mercaptoethanol the reduced peptides were allowed to oxidize in air (4). Cyclic analogues were purified first by gel filtration on Sephadex G-25 and further by HPLC using semipreparative column. The linear analogues were purified by ionexchange chromatography on Sephadex SP-C25 (pH gradient).The ability of these analogues to induce humoral and cellular immunity to native HBsAg particles isolated from human blood,was tested in guinea pigs and rabbits. Guinea pigs in groups of five were each immunized with seven peptides conjugated with tetanus toxoid and adsorbed on aluminium gel. Three doses of 50 μg were given intramuscularly three times every two weeks.The humoral and cellular immunity to native HBsAg (isolated from a pool of sera of chronic carriers by the method of Pillot) (5) was checked two weeks after the last injection. Rabbits groups of 3 animals each received multiside intradermal injections of three noncoupled analogues only 1,6 and 7 emulsified with Freund's complete adjuvant. The doses of 100μg were given three times at two week intervals; simultaneously blood samples were collected for humoral immunity tests. Cellular immunity was evaluated two weeks after the last injection. Serum antibodies to HBsAg anti-HBs were tested by radio-immunoassay (Ausab, Abbot), cellular immunity was evaluated by in vitro macrophage (MIT) or leucocyte migration inhibition tests (LIT).

Results

The inhibition of macrophage migration in the presence of HBsAg occured in all immunized guinea pigs (Table 1), but not in the control of non immunized animals.

Table 1. Migration Inhibition of Guinea Pig Macrophages in the Presence of Native HBsAg.

peptides	mean % macrophage migration inhibition
1	56
2	49
3	41
4	51
5	52
6	69
7	24
control	no inhibition

Table 2. Anti-HBs Response of Rabbits and Leucocyte Migration Inhibition in Presence of Native HBsAg.

peptides	maximal level of anti-HBs in mIU/ml	mean % leucocyte migration inhibition
1	23	58
6	168	75
7	21	40
control		no inhibition

There was no response of immunized animals to normal human serum. In none of the guinea pigs, however, antibodies to HBsAg were detected after injection of peptides conjugated with tetanus toxoid. Antibodies to HBsAg were detected in all serum samples

from rabbits immunized with peptides 1,6 and 7 (Table 2). Cellular immune response to HBsAg as measured by LIT was detected in all rabbits(Table 2). The titer of anti-HBs was different in individual animals; the peak of the humoral response occured after the third injection. The highest level of anti-HBs was induced with peptide 6.
The immunization with all synthetic peptides tested, induced cellular immune response to the native HBsAg in both guinea pigs and rabbits, regardless of the immunization procedure. Antibodies to native HBsAg were induced by multiside intradermal injections with noncoupled peptides only. The highest humoral response was induced by a cyclic peptide 6.

Acknowledgement

This work was supported by the Polish Academy of Sciences (CPBR 3.13.4.3. and the Swedish National Research Council K-KU 3020-112).

References

1. Bull H.B,, K. Breese. 1975. Arch. Biochem. Biophys. 161, 665.
2. Dreesman G.R,, Y. Sanchez, I. Ionescu-Matiu, J.T. Sparrow, H.R. Six, D.L. Peterson, F.B. Hollinger, J.L. Melnick. 1982. Nature 295, 158.
3. Hopp T.P., K.R. Woods. 1981. Proc. Natl. Acad. Sci. USA 78, 3824.
4. Sandberg B.E., U. Ragnarsson. 1978. Int. J. Peptide Protein Res, 11, 238.
5. Pillot J., S. Goneffon, R.G. Keros, M.T. Rarmon. 1977. Acad. Sci. Paris Ser.D 285, 1151.

SYNTHESIS AND ANTIGENIC PROPERTIES OF PEPTIDES CORRESPONDING TO 200-210 SEQUENCE OF $p28^{sis}$ ONCOPROTEIN. COMPARATIVE STUDY OF N^{G}-PROTECTING GROUPS

I.I. Mikhaleva, V.V. Antonenko, I.A. Prudchenko
Shemyakin Institute of Bioorganic Chemistry USSR Academy of Sciences, Moscow, USSR

N.N. Mazurenko
Institute of Cancerogenesis National Oncology Research Center Academy of Medical Sciences, Moscow, USSR

Antibodies to oncogenic proteins which are produced in cells in very small amount provide a variety of possibilities for studying their localization,structure-functional organization, interaction with cellular targets, etc. Oncoprotein $p28^{sis}$, the expression product of simian sarcoma virus (SSV) sis oncogene (1,2) is of special interest due to its high homology with platelet-derived growth factor (3,4) (PDGF). This similarity is related to the fact that human sis proto-oncogene is the structural gene coding for one of the two related amino acid chains of this major connective tissue-derived cell mitogen (5,6). With the aim of obtaining $p28^{sis}$ specific antibodies we have synthesized peptides I and II corresponding to 200-210 region of the protein sequence:

```
200 201          206        210
   ArgArgProProLysGlyLysHisArgLys     (I)
ValArgArgProProLysGly                 (II)
```

The available empirical procedures were employed for selecting the exposed, presumably immunogenic segments of the protein, i.e. the Kyte-Doolittle (7) and Hopp-Woods (8) procedures for hydrophobicity (hydrophilicity) estimates and the Chou-Fasman method (9) for β-turn prediction. The relevant plots point to the segment 200-215 as a possible antigenic determinant.

The peptides I and II were synthesized by conventional methods in solution. In our initial plan we relied on acid labile side protection of the three Arg and the Lys residues present in the sequence.

Peptides 1986

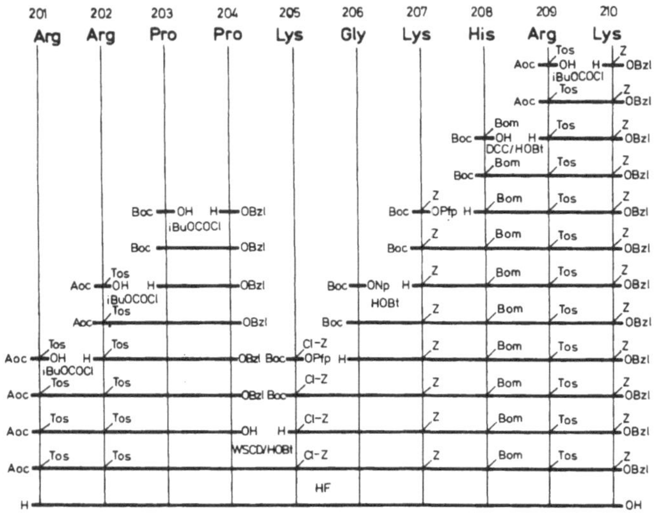

Fig. 1. Scheme for synthesis of peptide I. WSCD - water soluble carbodiimide, 1-cyclohexyl-3-(2-morpholinoethyl)-carbodiimide metho-p-toluenesulfonate; Bom - benzyloxymethyl

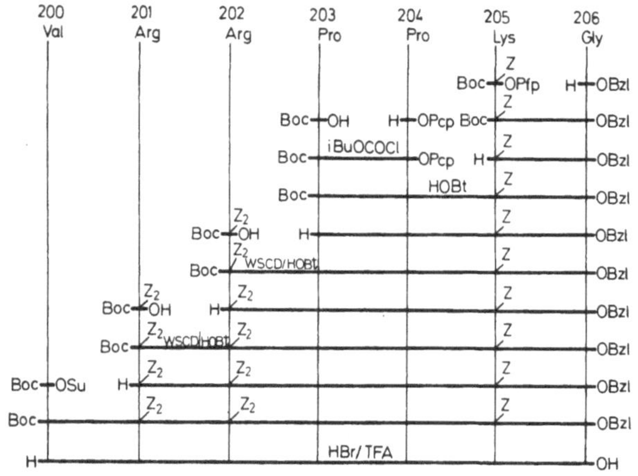

Fig. 2. Scheme for synthesis of peptide II

However the adamantyl derivative ZArg(Adc_2)OH and related peptides proved unsufficiently stable under usual conditions of the synthesis. For instance, coupling ZArg(Adc_2)OSu or ZArg(Z_2)Arg(Adc_2)OSu with HProProLys(Boc)GlyOH resulted in a complicated mixtures where the desired products could not be identified. Better results were obtained with traditional N^G-Tos and (to lesser extent) N^G-Z_2-derivatives. Peptide I was synthesized as shown in Scheme 1 (Fig. 1) by segment condensation of tetra- and hexapeptides prepared by stepwise chain elongation with N^{α}-Boc or Aoc, N^{im}-Bom, N^{ε}-Z or Cl-Z and α-COOH-OBzl protection. Coupling stages were performed by the mixed anhydride procedures,activated esters or DCC/HOBt method. The protected decapeptide was purified on Toyo-Pearl HW-40 column in DMF-AcOH (450:1). Final deprotection was carried out by HF/PhSMe (0^oC, 60 min). The peptide II was prepared according to Scheme 2 (Fig. 2). Partial decomposition of the protected hexapeptide 201-206 in methanol was clearly documented by TLC which we ascribe to instability of the Arg(Z_2) moiety in that solvent. Apparently neither Arg(Adc_2) nor Arg(Z_2) should be considered as entirely satisfactory derivatives. The unprotected peptides I and II were purified by reverse-phase HPLC on Ultrasphere ODS (5 μ) column in 0.01% TFA-CH_3CN (gradient 0-40%). Both had correct amino acid analysis and ^1H-NMR spectra.

The peptides I and II were conjugated via the C-terminal carboxyl group with keyhole limpet hemocianin (KLH) and bovine serum albumin (BSA) by WSCD method. Immunization of rabbits has been carried out by the I-KLH conjugate and complete Freund's adjuvant. Animals were boosted after 4 weeks, after 6 weeks antiserum was drawn and the antipeptide antibodies titers analysed by solid phase enzyme-linked immunosorbent assay (ELISA). The specificity of antiserum to sis oncogene expression product in cells transformed by SSV virus (marmoset cells HF-4/SSV and rat kidney cells NRK-SSV) was evaluated by means of radioimmunoprecipitation and immunoblotting. The antipeptide antiserum reacted with a p56 protein (which is a dimer of $p28^{sis}$) and a p46 protein (probably sis related (10)) in HF-4/SSV and NRK-SSV cell lysates in nonreducing conditions (Fig. 3, III, IV). In the presence of mercaptoethanol only $p28^{sis}$ was detected (Fig. 3, I).

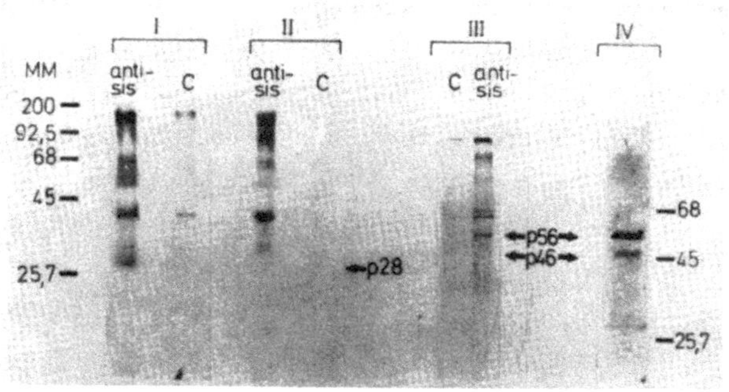

Fig. 3. Detection of $p28^{sis}$ protein in lysates of cells, transformed by SSV. I-III - immunoprecipitation of cell lysates (I - NRK-SSV, II - NRK, III - HF/SSV) by antipeptide antiserum and normal serum (C-control). Immune complexes were analysed by $NaDodSO_4$/polyacrylamide gel electrophoresis. Cells were labelled with $[^{35}S]$-methionine. IV - immunoblotting of NRK-SSV cell lysates with antipeptide antiserum. Experiments I and II were performed under reducing conditions, III and IV - in the absence of mercaptoethanol. MM - molecular mass, kD.

The antibodies obtained are currently being used for analysis of sis oncogene expression in human tumours.

References

1. Robbins, K.C., S.G. Devare, E.P. Reddy, S.A. Aaronson. 1982. Science 218, 1131.
2. Devare, S.G., E.P. Reddy, J.D. Law, K.C. Robbins, S.A. Aaronson. 1983. Proc. Natl. Acad. Sci. USA 80, 731.
3. Doolittle, R.F., M.W. Hunkapiller, L.E. Hood, S.G. Devare, K.C. Robbins, S.A. Aaronson, H.N. Antoanides. 1983. Science 221, 275.
4. Waterfield, M.D., G.T. Scrace, N. Whittle, A. Johnson, B. Westermark, C.N. Heldin, I.S. Huang, T.F. Deuel. 1983. Nature 304, 35.
5. Chiu, I.M., E.P. Reddy, D. Givol, K.C. Robbins, S.R. Tronick, S.A. Aaronson. 1984. Cell 37, 123.
6. Josephs, S.F., C. Guo, L. Ratner, Wong-Staal. 1984. Science 223, 487.
7. Kyte, J., R.F. Doolittle. 1982. J. Mol. Biol. 157, 105.
8. Hopp, T.P., K.R. Woods. 1981. Proc. Natl. Acad. Sci. USA, 78, 3824.
9. Chou, P.Y., G.D. Fasman. 1978. Adv. Enzymol., 45.
10. Robbins, K.C., H.N. Antoanides, S.G. Devare, M.W. Hunkapiller, S.A. Aaronson. 1983. Nature 305, 605.

SYNTHETIC SEQUENCES OF CHOLERA TOXIN B SUBUNIT ARE IMMUNOGENIC PER OS AND DO NOT INDUCE ORAL TOLERANCE

S. Pedoussaut, A. Guyon-Gruaz, A. Delmas, H. Halimi, G. Milhaud, P. Rivaille.

UA 163, C.H.U. St Antoine, 27 Rue Chaligny, 75012 Paris, France.

Introduction

Effective immunization against proteins with synthetic peptides, representing part of the protein sequence, usually requires the coupling of the peptides to carrier proteins. In the case of cholera toxin (CT), a protein composed of one B subunit, which binds to its receptor on intestinal cells (the ganglioside GM_1), and one A subunit, which activates adenylate cyclase, we have already shown that the two peptides representing the 30-50 and 50-75 sequences of CT B subunit were immunogenic without any carrier and adjuvant when administrated intaperitoneally to mice and that anti-50-75 antibodies were able to neutralize CT activity (1). (These two peptides were chosen as a function of their predicted surface localisation (2-3) and so as to contain tetrapeptides specific of CT and its related E. coli heat-labile enterotoxin, which should enhance their immunogenicity (4-5)).

```
H2N-*ALA-VAL-GLU-VAL-PRO-GLY-SER-GLN-HIS-ILE-ASP-SER-GLN-
      50                55                60
   LYS-LYS-ALA-ILE-GLU-ARG-MET-LYS-ASN-THR-LEU-ARG-ILE-
             65                70
   ALA-COOH
    75
H2N-*ALA-SER-LEU-ALA-GLY-LYS-ARG-GLU-MET-ALA-ILE-ILE-THR-
      30                35                40
   PHE-LYS-ASN-GLY-ALA-THR-PHE-GLU-VAL-COOH
             45                50
```

Fig.1 The two sequences 30-50 and 50-75 of cholera toxin B subunit

The two sequences were synthesized according to the solid phase method of Merrifield as already described by Delmas et al.(6), an extra 14 C labeled alanine was introduced at the N terminal position to facilitate the purification steps

As these two peptides are immunogenic in a free form and without adjuvant, and since

Peptides 1986

CT is known as a potent oral immunogen, we carried on this work with the study of oral immunization by these two peptides.

As we have already shown (table 1), both peptides are immunogenic when given orally (1).

PEPTIDE	PRIMING	BOOSTER 1	BOOSTER 2	BOOSTER 3
P 30-50	< 50	< 50	1 600	1 600
P 50-75	< 50	< 50	< 50	1 600

Table 1 :Antipeptide antisera titers of orally immunized mice.

Mice were immunized orally at day D, D + 10, D + 16, D + 22 with 5 µg of peptide dissolved in 200 µl water, sera were tested at day D + 6, D + 14, D + 20, D + 27.

Lack of oral tolerance to the peptides

Mucosal antigen exposure generally leads to local immunization, but induces poor seric antibody titers and a reduced seric response to systemic challenge with the same antigen. This induction of oral tolerance has often been demonstrated (7-8). But contrary to most protein antigens, CT stimulates not only mucosal but also systemic responses and does not induce oral tolerance (9).

Since our two peptides were immunogenic orally like CT, we intended to determine if they did not induce oral tolerance.

Immunization	1	2	3	4	5	6	7	8
50-75 IP				200 IP	800 IP	800 IP	1600 IP	800 IP
50-75 PO/IP	400 PO	400 PO	1600 PO	800 IP	3200 IP			
30-50 IP				50 IP	400 IP	400 IP	400 IP	1600 IP
30-50 PO/IP	<50 PO	100 PO	100 PO	100 IP	800 IP			

Immunizations wereperformed at day : D-79,D-49,D-19, D,D+30,D+60,D+79,D+109.

Mice were bled : 22 days after immunizations 1, 4 and 7; 15 days after immunizations 2,5 and 8;10 days after immunizations 3 and 6.

▒ no immunization

IP : Intraperitoneally
PO : Orally .

Table 2: Lack of oral tolerance to the peptides 30-50 and 50-75 : anti CT antibody titers.

Mice were immunized IP or PO with 20 µg peptide diluted in 100µl of water for PO injections,and saline for IP injections. PO immunized mice were given further IP boosters injections of the same peptide.

As shown in table 2, the titers obtained after 3 oral (PO) followed by 2 intraperitoneal (IP) immunizations with each peptide are higher than those obtained after 2 IP immunizations (and even higher than those obtained after 5 IP immunizations). Administering orally the peptides did not induce any tolerance to a subsequent intraperitoneal immunization and furthermore it enhanced the seric antibody titers when compared to their controls (mice unfed with peptide before IP immunizations). The systemic injection seems to act like a booster rather than a priming immunization.

It is to be noticed that anti-50-75 and anti-30-50 titers obtained with IP or PO administration were similar, though in this experiment the 30-50 peptide induced lower antisera titers than already described (1).

Discussion

The induction of antibodies by oral immunization with these peptides or lack of oral tolerance to them cannot be due to their binding to GM_1. Indeed in three different models (GM_1 coated on microtitration plates, covalently linked to cellulose, GM_1 of thymocytes membranes), we did not observe any binding of the peptide nor displacement, by the peptides, of the bound CT (data not shown) .

These results are in agreement with the observations of Elson et al. (10), who have already shown that the feeding of any antigen, binding to the intestinal mucosa, does not necessary result in a lack of oral tolerance to the antigen. These authors rather proposed that CT might accomplish these effects by shifting the regulatory environment in gut associated lymphoid tissues in favor of responsiveness rather than tolerance (10).

CT is a valuable probe for the study of oral tolerance. 30-50 and 50-75 sequences , which are only parts of this molecule, might be powerful tools for studiing the critical regions implicated in this mechanism.

Aknowledgements

We thank Dr. M.S Moukhtar for helpful advices,and R. Myrtill for technical assistance.
One of us (A.D) was supported by the Associatoin pour le developpement de la Recherche contre le Cancer (ARC).

References

1. Guyon-Gruaz, A., Delmas, A., Pedoussaut, S., Halimi, H., Milhaud, G., Raulais, D., Rivaille, P. Eur. J. Biochem. (in press).

2. Hopp, T. P., Woods, K. R. 1981. Proc.Natl. Acad. Sci. USA 78, 3824-3828.

3. Fraga, S. 1982. Can. J. Chem. 60, 2606-2610.

4. Delmas, A., Milhaud,G., Raulais, D., Rivaille, P. 1985. Mol. Immunol.22, 675-679.

5. Saroff, H. A. 1985. J. Theor. Biol. 115, 191-199.

6. Delmas, A., Guyon-Gruaz, A., Halimi, H., Raulais, D., Rivaille, P. 1985. Int. J. Peptide Protein Res. 25, 421-424.

7. Andre, C., Heremans, J. F.,Vaermans, J. P., Cambiaso, C. L. 1975. J. Exp. Med. 142, 1509-1519.

8. Challacombe, S. J., Tomasi T. B. Jr. 1980, J. Exp Med.152, 1459-1472.

9. Elson, C. O., Ealding, W. 1984. J. Immunol.132, 2736-2741.

10. Elson, C. O., Ealding, W. 1984. J. Immunol.133, 2892-2897.

SYNTHESIS OF LARGE NUMBERS OF COVALENTLY BOUND PEPTIDES IN ORDER TO DETERMINE THE SPECIFICITY OF PEPTIDE ANTISERA AT THE LEVEL OF SINGLE AMINO ACIDS

W.M.M. Schaaper, H. Lankhof, D.J.A. Meijer, W.C. Puijk, R.H. Meloen
Central Veterinary Institute, Edelhertweg 15, 8219 PH Lelystad, The Netherlands

INTRODUCTION

After raising antibodies against peptides it is of great importance to get detailed information about their specificity, especially in relation to the specificity of antibodies raised against the same sequences as part of the original structure (e.g. a virus capsid). Such information can be obtained by the synthesis of large numbers of peptide analogs ("replacement sets") (1,2).
With the aid of a neutralizing monoclonal antibody (MAb) we located (3) an important epitope, an octapeptide, in the C-terminal part of the surface protein VP1 of Foot-and-Mouth Disease Virus (FMDV). Although peptides of varying length from this area proved to be as active as the virus in competition experiments (4), only antibodies raised against a longer version showed neutralizing activity. The results obtained with these peptide antisera and the replacement set of peptides suggest that the anti-peptide response can be manipulated. This may be used to improve the anti-peptide response with the final aim to produce an antibody specificity similar to that of the monoclonal antibody.

PEPTIDES USED FOR TESTING ANTISERA

Large numbers of peptides were synthesized on polyethylene rods grafted with acrylic acid. 1,6-Diaminohexane was used as linking agent. Boc-amino acids were coupled via the DCC/HOBt method. The N-terminus was acetylated. The side-chain protection was removed with borontris(trifluoroacetate).
Replacement sets of peptides were synthesized by replacing the

Peptides 1986

first amino acid by 19 other amino acids, followed by the replacement of the second amino acid by 19 other amino acids and so on. The result for an octapeptide: 8 times the parent sequence plus 152 different analogs that can be tested directly in an Elisa. The Elisa was performed as discribed (1).

PEPTIDES USED FOR RAISING ANTISERA

Peptides used for raising antibodies were synthesized by conventional Solid Phase Synthesis procedures. Benzhydrylamine resin was used as solid phase. Boc-amino acids were coupled via their preformed symmetrical anhydrides. The peptides were cleaved by HF/anisole.

Conjugation of the peptides was performed in several ways. As carrier protein we used Keyhole Limpet Hemocyanin (KLH).

A. With glutaraldehyde (GDA, 2.7 eqv.) for the coupling of the peptide via several functional groups, especially -SH and $-NH_2$ groups.

B. With m-maleimidobenzoyl-N-hydroxysuccinimide ester (MBS, 1.2 eqv.) for the coupling of the peptide via its -SH groups.

C. No conjugation, but the introduction of N^{α},N^{ϵ}-dipalmitoyllysine as N-terminal amino acid during the peptide synthesis.

Rabbits were vaccinated once with peptide conjugate (1 mg of peptide), emulsified in Complete Freund's Adjuvant.

Table 1. Serological Results of Peptide Antisera

ANTISERUM against	COUPLING METHOD	ELISA Ac-YKQKIIAP□	ELISA VIRUS	MNT
VIRUS (MAb 18)	-	5.2	7.5	3.2
Ac-YKQKIIAP□	GDA	5.7	2.2	-
Ac-CYKQKIIAP□	MBS	5.1	2.3	-
$(Pa)_2$=KYKQKIIAP□	-	3.4	-	-
Ac-CRYKQKIIAPAKQ□	GDA	3.6	4.3	-
Ac-CRYKQKIIAPAKQ□	MBS	3.2	4.1	0.9

All titres are expressed as -log(final serum dilution), - for Elisa indicates < 2.0, - for MNT indicates < 0.3. Ac = Acetyl, Pa = Palmitoyl, MNT = Micro Neutralization Test, □ indicates the amide form.

RESULTS

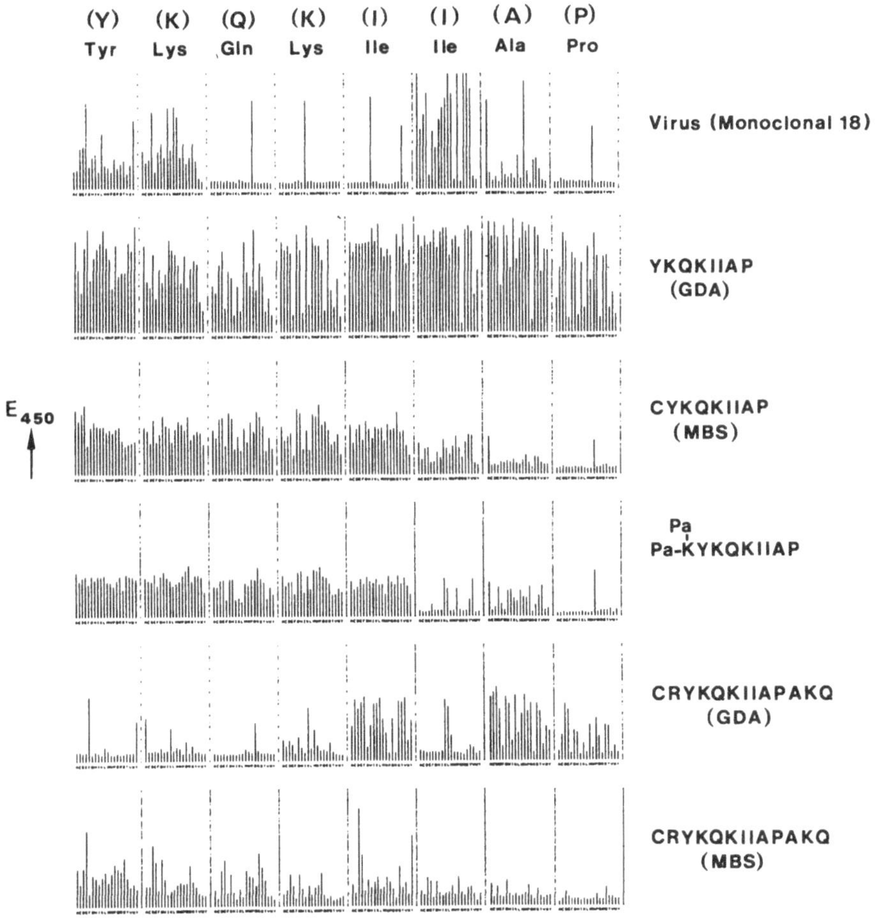

Fig. 1. Replacement set of peptides of the sequence Tyr-Lys-Gln-Lys -Ile-Ile-Ala-Pro testet against several antisera. The result for each peptide is shown as a vertical line proportional to the Elisa extinction. Each group of 20 lines corresponds to the complete replacement set for one of the 8 amino acid positions in the octapeptide. Within each group of 20 lines the left-hand line corresponds to the substitution of the original residue by Ala (A), and the successive lines are then in alphabethical order according to the single letter code for the amino acids (ACDEFGHIKLMNPQRSTVWY).

As shown in the figure above the monoclonal antibody interacts more or less specifically with six amino acid positions from the tested sequence. Antisera raised against the octapeptide YKQKIIAP, conjugated via GDA, hardly show any specificity. Conjugation of the peptide at the N-terminus either via Cys/MBS or via the introduction of a dipalmitoyllysin residue results in a specificity at the C-terminal end. Specificity to the other end is produced when an elongated version (CRYKQKIIAPAKQ) is coupled with the carrier via GDA. Conjugation of this longer peptide via MBS results in an antiserum with hardly any specificity in the tested sequence. However the result from the neutralization test indicates that in this latter serum a (small) proportion of the antibodies has neutralizing activity.

CONCLUDING REMARKS

1. Definite specificity patterns of polyclonal antisera raised against peptides can be detected with the aid of replacement sets of peptides.
2. The neutralizing monoclonal antibody raised against the whole virus shows a different specificity pattern than the antibodies raised against the peptides.
3. The specificity pattern of the peptide antisera may be directed by the way of coupling the peptide with a carrier protein.
4. Further attempts will be made to manipulate the immune response against peptides with the final aim to obtain a specificity pattern similar to that of the monoclonal antibody raised against the virus

REFERENCES

1. Geysen, H.M., R.H. Meloen, S.J. Barteling. 1984. Proc.Natl.Acad. Sci.USA 81, 3998.

2. Geysen, H.M., S.J. Barteling, R.H. Meloen. 1985. Proc.Natl.Acad. Sci.USA 82, 178.

3. Meloen, R.H., W.C. Puijk, D.J.A. Meijer, H. Lankhof, W.P.A. Posthumus, W.M.M. Schaaper. J. Gen. Vir., submitted.

4. Puijk, W.C., W.M.M. Schaaper, H. Lankhof, R.H. Meloen (in prep.)

SYNTHESIS AND IMMUNOLOGICAL ACTIVITY OF PEPTIDES OF THE THYMUS GLAND

W.Voelter, H.Echner, H.Kalbacher, T.Q.Dinh, A.Kapurniotu, M.Jahan, G.Grübler, P.Link, M.Mihelić, W.Stock
Abteilung für Physikalische Biochemie des Physiologisch-chemischen Instituts der Universität Tübingen, Hoppe-Seyler-Str.1, D-7400 Tübingen

Introduction

Several peptides (α thymosins, thymopoietins, ß thymosins), isolated from bovine thymus, amplify T cell immunity and induce terminal deoxynucleotidyl transferase in prothymocytes (1,2). Most recently the primary structure of rat prothymosin was determined and it has been shown that thymosin α_1 is produced in the thymus from a larger peptide. From the group of G.Goldstein three closely related peptides inducing selectively T cell differentiation were isolated from bovine thymus (thymopoietin I and II) and bovine spleen (thymopoietin III). It has been shown that thymopoietin 32-36 has almost identical biological activity in a number of assays as the parent peptide. The first biologically active and completely characterized polypeptide from the ß thymosin series, isolated from thymosin fraction 5, was thymosin $ß_4$. In the meantime, homologous peptides (thymosin $ß_8$, $ß_9$, $ß_{10}$, $ß_{11}$), were also structurally determined. For testing the thymus peptides in different bioassays, solution syntheses of thymopoietin II 32-36, the pagocytosis-stimulating peptide tuftsin, the selevtive B-cell-differentiating hormone bursin and fragments of thymosin $ß_9$ were developed (3-6).

Results

Several solution syntheses of thymopoietin 32-36 were performed in our group in the past (3-6). Herewith we want to inform about a rapid synthesis supplying the pentapeptide in high purity and

Peptides 1986

yields (Scheme 1). Fmoc-Tyr(Bu^t)-OH was attached to the p-benzyloxybenzyl alcohol resin (polystyrene - 1 % divinylbenzene (5 g)) with DCCI and 4-dimethylaminopyridine in CH_2Cl_2. The substitution content was 0.6 mmol/g resin as determined by amino acid analysis. The Fmoc group was cleaved from the tyrosine residue with 50 % piperidine/CH_2Cl_2 (1x5 min, 1x25 min). After washing with CH_2Cl_2 (2x2 min), DMF (2x2 min), H_2O/dioxane (1:2) (2x5 min), DMF (3x2 min), CH_2Cl_2 (3x2 min) Fmoc-Val-OH (excess of 3 equivalents) was coupled in DMF/CH_2Cl_2 (1:3) (45 min). After cleavage of the Fmoc group and washing, as described above, the following 3 Fmoc amino acids were coupled. The peptide resin was washed (3x) with CH_2Cl_2 and then subjected to CF_3COOH/CH_2Cl_2 (5,5:1) for 80 min. After filtration and washing (3x) with TFA and CH_2Cl_2 the combined filtrates were evaporated to a minimum content of TFA. The crude peptide was precipitated with ether. The dried precipitate was dissolved in 1N HAc and the solution treated with a Dowex 1x4 anion exchanger in the acetate form and finally lyophilized (yields: 2.15 g = 86 %). Amino acid analysis (6N HCl, 110°, 24h): Asp 1.04, Arg 1.02, Tyr 0.91, Val 0.97, Lys 1.06. For further purification the crude product is dissolved in H_2O, acidified with 1N H_2SO_4 and precipitated with isopropanol.

For our total synthesis of thymosin $ß_9$ the following fragments were synthesized (the side chains of Asp,Glu,Ser,Thr were protected by t.-butyl those of Lys by Boc residues): Ac-$\overset{1}{\text{Ala}}$-Asp-(OBu^t)-Lys(Boc)-Pro-Asp(OBu^t)-Leu-$\overset{7}{\text{Gly}}$-OH, Z-$\overset{8}{\text{Glu}}$(OBu^t)-Ile-Asn-Ser(Bu^t)-Phe-Asp(OBu^t)-$\overset{14}{\text{Lys}}$(Boc)-OH, Z-$\overset{15}{\text{Ala}}$-Lys(Boc)-$\overset{17}{\text{Leu}}$-OH, Z-$\overset{18}{\text{Lys}}$(Boc)-Lys(Boc)-Thr(Bu^t)-$\overset{21}{\text{Glu}}$(OBu^t)-OH, Z-$\overset{22}{\text{Thr}}$(Bu^t)-Gln-$\overset{24}{\text{Glu}}$-(OBu^t)-OH, Z-$\overset{25}{\text{Lys}}$(Boc)-Asn-Thr(Bu^t)-Leu-$\overset{29}{\text{Pro}}$-OH, Z-$\overset{30}{\text{Thr}}$(Bu^t)-Lys-(Boc)-$\overset{32}{\text{Glu}}$(OBu^t)-OH, Z-$\overset{33}{\text{Thr}}$(Bu^t)-Ile-Glu(OBu^t)-Gln-$\overset{37}{\text{Glu}}$(OBu^t)-OH and H-$\overset{38}{\text{Lys}}$(Boc)-Gln-Ala-$\overset{41}{\text{Lys}}$(Boc)-OBu^t. Detailed information about the applied coupling and cleavage methods for the syntheses of the thymosin $ß_9$ fragments is seen from Scheme 2 which, as an example, demonstrates the strategy for the synthesis of thymosin $ß_9$ fragment 30-41 (4,6,7). Several of the synthetic peptides and thymosin $ß_4$ and $ß_9$ (2), isolated from bovine thymus, were tested in the rosette assay (6). The concentration dependency of the activity is seen from Figure 1.

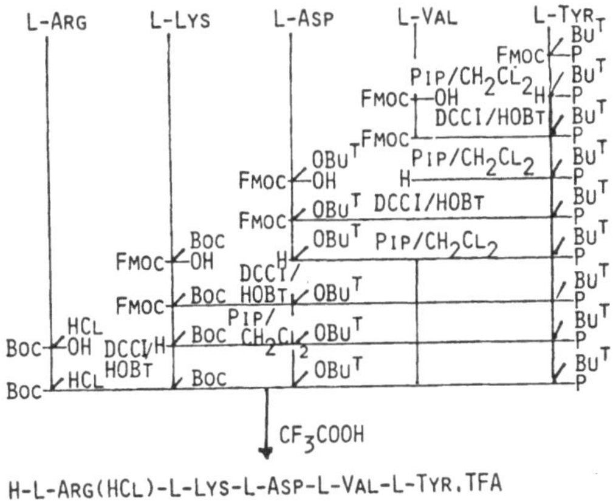

Scheme 1. Solid phase synthesis of thymopoietin pentapeptide 32-36.

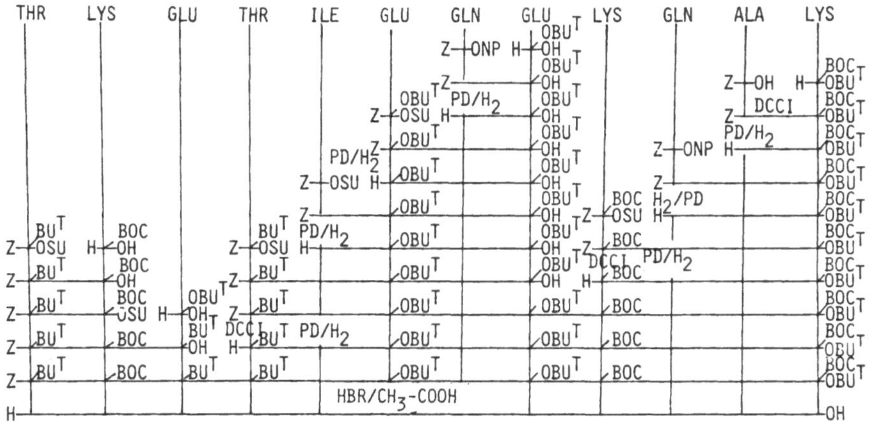

Scheme 2. Solution synthesis of thymosin β_9 fragment 30-41.

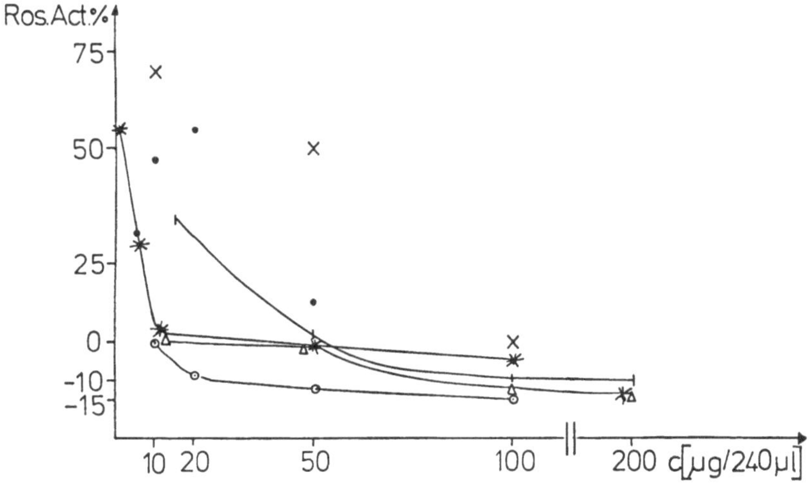

Fig.1. Concentration dependency of the induction of E-rosette formation (%) of human T lymphocytes ($5x10^6$ lymphocytes/ml and sheep red blood cells (10^9 red blood cells/ml) of different thymus peptides after inhibition with α-amanitin (2.0 µg/ $5x10^5$ lymphocytes) (x thymopoietin pentapeptide 32-36; ⊙ thymosin β_4; · thymosin β_9; ⁎ thymosin β_9 30-41; ı thymosin β_9 33-41; Δ thymosin β_9 33-37).

References

1. Heinzel, W., W.Voelter. 1981. Chem.Ztg.105, 291.
2. Hannappel,E. 1983. Habilitationsschrift, Erlangen-Nürnberg.
3. Friedl,C., W.Voelter. 1984. Chem.Ztg.108, 323.
4. Voelter,W., H.Echner, H.Kalbacher. 1986. In: Chemistry of Peptides and Proteins, Vol.3 (W.Voelter, E.Bayer, Y.Ovchinnikov, V.Ivanov, eds) Walter de Gruyter, Berlin-New York) p.55.
5. Voelter,W., H.Kalbacher, H.Echner, H.Wagner, Liebigs Ann. Chem. (in press).
6. Voelter,W., H.Echner,H.Kalbacher, T.Q.Dinh, A.Kapurniotu, M.Jahan, P.Link, M.Mihelič. 1986. In: Natural Product Chemistry (A.U.Rahman, ed., Elsevier, Amsterdam - New York-Oxford) (in press).
7. Voelter,W., H.Kalbacher, M.Mihelić, E.Hannappel. 1986. In: Proceedings of the Shanghai-Munich-Symposium on Peptide and Protein Chemistry (E.Wünsch, L.Moroder, eds) (in press).

STRUCTURE-ACTIVITY RELATIONSHIPS STUDIES OF DIPHTHERIA TOXIN USING SYNTHETIC PEPTIDES THAT ELICIT TOXIN-REACTIVE ANTIBODIES

A. Michel, D. Muyshont

Laboratory of biological chemistry, Faculty of Sciences, University of Mons - B - 7000-Mons (Belgium)

Introduction

Diphtheria toxin is a highly toxic protein for eukaryotic cells which acts by blocking protein synthesis. From the site of its synthesis in corynebacterium diphtheriae to its target in the cytoplasm, diphtheria toxin has to cross two cytoplasmic membranes. It first traverses the corynebacterium membrane as a protoxin with a transient n-terminal signal peptide. Extracellular toxin is a single-chain protein (intact toxin, MW 58390) that is activated by limited proteolysis which occurs in a highly specific region of the peptide chain, two functionally distinct fragments : A (MW 21150, the amino-terminal part) and B (MW 37240, the C-terminal part), still attached by a disulfide bridge are therefore generated (nicked toxin). Fragment B (FB) binds then to specific surface receptors on sensitive cells and mediates the export of fragment a from acidic endosomes into the cell cytoplasm where A inhibits the protein synthesis by catalytic ADP-ribosylation of elongation factor 2. The receptor-recognition steps of diphtheria toxin involves the MW 15000 C-terminus of fragment B while the interaction with the cell membrane bilayer requires the MW 23000 N-terminal portion of FB. Recently, like diphtheria toxin and the latter MW 23000 region, CB1, the cyanogen bromide peptide (MW 13000 residues 147-266) located in the middle of fragment B has been shown to induce pore formation in phospholipid bilayers at low pH, pores through which, in the whole toxin, fragment A might find a passage across the cytoplasmic or vesicular membrane (1). The N-terminal part of CB1 possesses hydrophobic 26 amino acids residues domain (2) (residues 153-178) that strongly ressembles the transverse lipid-associating domain of intrinsic membrane proteins. This region of CB1 is probably responsible for the

Peptides 1986

property of diphtheria toxin, FB and CB1 itself to form transmembrane channels at low pH. Lowering of the pH causes conformational modification in the polypeptide chain of the fragment B portion which allows accessibility of the hydrophobic domain. On the other hand, in the C-terminal region of CB1 a cluster of 4 prolyl residues embedded in a strongly hydrophobic 14 residues segment (residues 233-246) is found. Deleers et al. have proposed that low pH could induce conformational modification in this region through a cis-trans proline isomerization process leading to the anchoring of toxin and the formation of transmembrane channels with the hydrophobic domain (1). In order to precise the role of this proline-rich CB1 region in the mechanism of action of toxin we have synthesized three peptides containing this amino acids sequence (residues 226 to 255), some of their immunogenic and conformational properties are presented in the following.

Results and discussion

Synthetic peptides. The three peptides synthesized - named R 323, R 328 and R 329 - contain 15, 30 and 31 amino acids respectively (Figure 1) and correspond to the sequence 226-255 of diphtheria toxin fragment B, to which a cysteinyl residue was added at the C-terminal end.

← R 329 —
← R 328 —

NH2-LYS-ILE-THR-ALA-GLU-ASN-THR-PRO-LEU-PRO-ILE-ALA-GLY-VAL-LEU-LEU-

— R 329 →
— R 328 →
← R 323 →

PRO-THR-ILE-PRO-GLY-LYS-LEU-ASP-VAL-ASN-LYS-SER-LYS-THR-CYS-COOH

Figure 1
Amino acid sequence of 226-255 peptide of diphtheria toxin fragment B and relevant synthetic peptides

Synthesis and purification

All the peptides were synthesized by the classical solid-phase method using a beckman model 990 B synthesizer. N-α-BOC protection was used for all amino acids and DCC was chosen as coupling agent. BOC-Asn was introduced via the DCC/HOBT procedure. Peptides were cleaved from the resin by anhydrous HF according to Heath et al.(3) (low HF procedure). After extraction from the resin and lyophilization, the peptides were purified by gel filtration on Sephadex G-25 F, followed by ion-exchange chromatography on CM-cellulose CMC-52. All peptides were found homogeneous by TLC and amino acid analyses of acidic and enzymatic hydrolyses.

Coupling of peptides to a protein carrier

Purified peptides were covalently linked to bovine serum albumin (BSA) using two different methods : a) They were coupled to BSA via their amino groups using glutaraldehyde. In this case the cysteine of the peptides was first S-carboxamidomethylated using ^{14}C-iodoacetamide. The radioactivity measurement allowed us to calculate the peptide/carrier ratio. b) Peptides were coupled to BSA through the C-terminal cysteine using m-maleimido benzoyl-N-hydro xysuccinimide ester (MBS) as coupling agent. Substitution of BSA was estimated by amino acid analysis. The degree of incorporation of peptides into conjugates is ranging from 3 to 12 moles peptide/mole BSA in the glutaraldehyde method (according to the reaction time and the reagent excess) and is about 2,5-3 moles peptide/mole BSA for the MBS method.
Assay for measuring protein synthesis inhibition cells was carried out on Vero cells (African green monkey kidney cells) as described previously (4).

Rabbits immunization

Peptide-carrier conjugates and free R 328 and 329 were injected subcutaneously into the rabbits. On day 1 rabbits received 50-100 μg of peptide as conjugate or non-coupled in complete freund's adjuvant water-in-oil emulsion. On day 14 and then from week to week the same amount of antigen was administered in incomplete freund's adjuvant. Sera were collected on day 42. The enzyme-linked immunosorbent assay (ELISA) was used to test all antisera for their ability to react with the peptide used for immunization. Antibody was bound to antigen immobilized on polystyrene plates and the antigen-antibody complex was detected using swine anti-rabbit serum coupled to horse-radish peroxidase. O-phenylenediamine was used as enzyme substrate. Figure 2 shows that BSA-coupled peptides as well as R 328 and R 329 non coupled are able to produce in all cases tested a anti-peptide immune response. Moreover we found that anti-peptide antibodies are able to react with the whole diphtheria toxin and its fragment B, particularly when they are directed against R 328 and R 329 (coupled or not). On the contrary however antibodies raised against diphtheria toxoid react only weakly with the synthetic peptides.

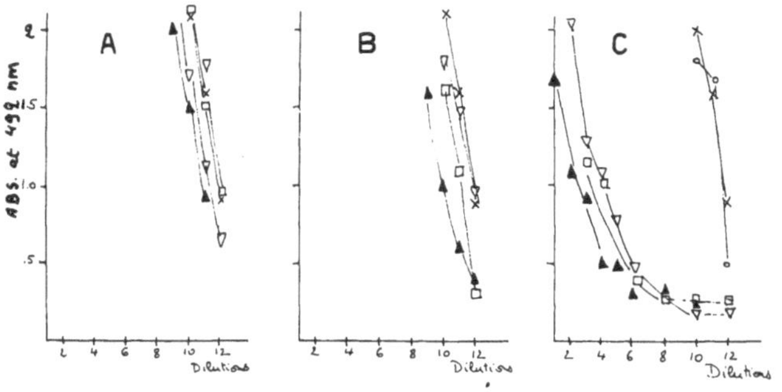

Figure 2 : Antigen-linked immunosorbent assay of rabbit antisera (ELISA). Twofold serial dilutions of each antiserum were prepared. Dilution numbers 1-12 correspond to dilutions 1/16-1/32000 for anti-peptides antisera. Dilution numbers 1-12 correspond to dilutions 1/256-1/520000 for anti-diphtheria toxin antiserum.
A- Antisera tested for their ability to react with the corresponding peptide used for immunization (immobilized on the plates). (X—X) anti-diphtheria toxin - (▲—▲) anti-R323 - (□—□) anti-R328 - (▽—▽) anti-R329. B- Antisera tested for their ability to react with diphtheria toxin (used as antigen), similar results were obtained with fragment B as antigen. (X—X) anti-diphtheria toxin - (▲—▲) anti-R323 - (□—□) anti-R328 - (▽—▽) anti-R329. C- Anti-diphtheria toxin antiserum tested for its ability to recognize synthetic peptides (used as antigen immobilized on the plates). The antigen is (X—X) toxin - (○—○) fragment B - (▲—▲) R 323 - (□—□) R 328 - (▽—▽) R 329.

Our results show that purified rabbit antibodies raised against synthetic peptides corresponding to the sequence 226-255 of diphtheria toxin fragment B are able to recognize the whole toxin molecule or fragment B. This support the concept that that amino acids sequence is likely an exposed region at the surface of the toxin molecule. These results are in agreement with the hypothesis of Deleers and claim for a possible role of this particular region in the mechanism of action of diphtheria toxin into cell. The fact that antibodies directed against diphtheria toxoid react only weakly with the three peptides, seems indicate that the region 226-255 does not constitue an immunodominant antigenic site. On the other hand, experiments on the intoxication of Vero cells cultures indicate that these anti-peptides antibodies have no neutralizing effect on the lethal action of diphtheria toxin.

Fragment B region 256-282 : This region corresponds to the sequence : His-Ile-Ser-Val-Asn-Gly-Arg-Lys-Ile-Arg-Met-Arg-Cys-Arg-Ala-Ile-Asp-Gly-Asp-Val-Thr-Phe-Cys-Arg-Pro-Lys-Ser. Up to seven peptides covering this protein region were also synthesized. Indeed, due to the accumulation of basic residues, this region seems to be a good candidate for the oligonucleotides binding site P described by Lory et al. (5), site which could be included in the cell binding site. HF-cleaved peptides were cyclized ($K_3Fe(CN)_6$ oxydation procedure) and purified by semi-preparative HPLC (Figure 3). Amino acids analyses after enzymatic hydrolyses reveal the presence of a $\alpha \rightarrow \beta$ rearrangement in the sequence Asp-Gly.

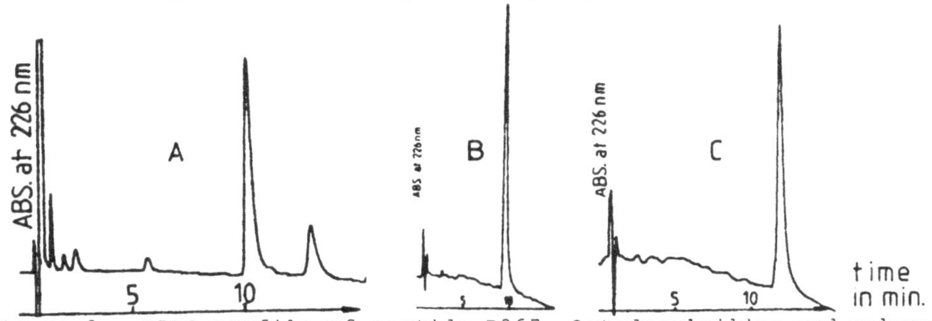

Figure 3 : HPLC profile of peptide R367. Octadecylsilica gel column (4x250mm) Macherey-Nagel 120 Å-5μ. Solvent : A :HCOOH 0,1 M.Triethylamine pH 4.0-B: CH_3CN.60% in A. Gradient 20-35% B in 15 min.
A:crude product-mixture of α,β isomers of R367-B: purified isomer α
C: purified isomer β.

Assays on cells cultures done with peptide R-367 (sequence 265-282, thus containing most of the basic residues) show that that amino acids sequence does not compete with the whole diphtheria toxin molecule (Figure 4) for its ability to inhibit protein synthesis. This indicates that region 256-282 of fragment B should not be implied in the cell binding site.

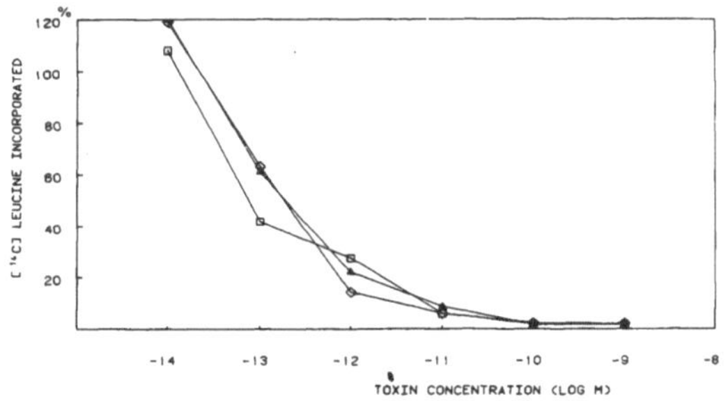

Figure 4 : Protein synthesis inhibition on Vero cells by diphtheria toxin alone (◇—◇) or in presence of peptide R 367 10^{-7}M (□—□) and 10^{-6}M (△—△).

Studies using antibodies raised against these peptides (coupled to protein carriers through a C-terminal added tyrosyle residue) are now in progress.

References

1. M. Deleers, N. Beugnier, P. Falmagne, V. Cabiaux and J.M. Ruysschaert. 1983. Febs Letters 160, 82.

2. L. Greenfield, M.J. Bjorn, G. Horn, D. Fong, G.A. Buck, R.J. Collier and D.A. Kaplan. 1983. Proc. Natl. Acad. Sci. USA 80, 6853.

3. W.F. Heath, J.P. Tam and R.B. Merrifield. 1982. J. Chem. Soc. Chem. Commun., 896.

4. K. Sandvig and S. Olsnes. 1981. J.Biol.Chem. 256, 9068.

5. S. Lory, S.F. Carroll and R.J. Collier. 1980. J.Biol.Chem. 255, 12016.

PREPARATION OF SUBSTRATE INHIBITORS OF THE AMIDATING ENZYME IN PORCINE PITUITARY

A.F. Bradbury, D.G. Smyth

National Institute for Medical Research, The Ridgeway, Mill Hill, London NW7 1AA

Introduction

The existence of an amidating enzyme was first demonstrated by the use of synthetic peptide substrates stable to proteolytic degradation (1). The mechanism of action of the enzyme was shown to be oxidative and it exhibited a mandatory substrate requirement for C-terminal glycine (2). In the present report we describe the synthesis of a number of potential inhibitors of the enzyme which have structural similarity to the C-terminal glycine. The compounds have been tested for their ability to inhibit the conversion of D-Tyr Val Gly to D-Tyr Val amide. It is of interest that the most potent inhibitor resembles a possible intermediate formed during the amidation reaction.

Experimental

A series of hydrazones of glyoxylic acid was prepared by the reaction of N-substituted hydrazines with glyoxylic acid. The reactions were carried out for 2 hours with 0.1M reactants at neutral pH in aqueous or aqueous ethanol solution and the products were isolated by precipitation at pH 3 followed by recrystallization from ethanol. The melting points of the phenylhydrazones of glyoxylic acid and pyruvic acid were in agreement with literature values and elemental analyses were obtained for those hydrazones that had not been reported previously. Similar reaction conditions were used for the preparation of semicarbazones and thio-semicarbazones of glyoxylic acid. The reduced form of glyoxylate phenylhydrazone was prepared by hydrogenation in methanol using 10% palladium on charcoal as a catalyst. Peptide substrates were

Peptides 1986

prepared by solid phase procedures involving fluorenylmethyloxycarbonylamino acids and polyamide resin (3). D-Tyrosylaminoethoxyethanoate was prepared by coupling t-BOC D-tyrosine to aminoethoxyethanoate methyl ester with DCC followed by deprotection with TFA and saponification with NaOH. 3-(Acetylphenylalanyl) propanoate was prepared by Dr. D.F. Dyckes*. Other inhibitors which possessed the CH_2O group in place of the CONH peptide bond of glycine were purchased from Aldrich Ltd.

The inhibitory properties of these compounds were determined by testing their ability to reduce the conversion of |^{125}I|-D-Tyr Val Gly to |^{125}I|-D-Tyr Val amide. The assays were carried out at pH 6.8 in the presence of 10μM Cu^{++}, 1mM dopamine, bovine serum albumen (1mg/ml) and catalase (20μg/ml) and the rates of amidation were determined by chromatography on columns of SP Sephadex C25 (4). Values for K_i were calculated from Lineweaver-Burke plots obtained from the results of experiments employing a series of concentrations of D-Tyr Val Gly.

Results

Table 1 shows the degree of inhibition of amidation of |^{125}I|-D-Tyr Val Gly produced by a number of substrate analogues. The most potent inhibitor was the phenylhydrazone of glyoxylic acid which exhibited a K_i value of 15μmolar, indicating that this compound has a higher affinity for the amidating enzyme than the peptide substrate D-Tyr Val Gly. It may be noted that the structure of the phenylhydrazone reflects the oxidation state of a postulated intermediate in the amidation reaction, suggesting that amidation may be product inhibited.

possible intermediates	- CO.NH - CH - COOH OH
generated during amidation	- CO.N = CH - COOH
glyoxylic acid phenylhydrazone	- Ph NH.N = CH - COOH
pyruvic acid phenylhydrazone	- pH NH.N = C - COOH CH_3

*University of Texas, U.S.A.

Table 1. Comparison of inhibitors of enzymic amidation

Compound	Concentration m.molar	% inhibition
glyoxylate phenylhydrazone	0.1	96
" " (reduced)	0.1	17
" benzylhydrazone	0.1	52
" 2-pyridylhydrazone	0.1	45
" benzoylhydrazone	0.1	8
" semicarbazone	0.1	33
" phenylsemicarbazone	0.1	50
" thiosemicarbazone	0.1	83
" phenylthiosemicarbazone	0.1	93
D-Tyr Val Gly	0.1	45
pyruvate phenylhydrazone	0.1	0
D-Tyr Gly Gly	2	82
D-Tyr aminoethoxyethanoate	2	38
acetylaminoethoxyethanoate	2	30
dodecyloxyethoxyethanoate	2	85
3(acetylphenylalanyl) propanoate	2	16

The phenylhydrazone of pyruvic acid gave no evidence of inhibition at 1mM concentration. This is in line with a previous finding that peptides containing C-terminal alanine do not serve as substrates for amidation (2) and is in agreement with the present data which shows that the phenylhydrazone of glyoxylic acid is a competitive inhibitor of the enzyme. The high affinity of this phenylhydrazone suggests that it may prove useful for facile purification of the amidating enzyme by affinity chromatography.

References

1. Bradbury, A.F., Finnie, M.D.A. and Smyth, D.G. 1982. Nature (London) 298, 686.

2. Bradbury, A.F. and Smyth, D.G. 1983. Biochem. Biophys. Res. Comm. 112, 372.

3. Atherton, E., Fox, H., Harkiss, D., Logan, C.J., Sheppard, R.C. and Williams, B.J. 1978. J.C.S. Chem. Comm. 537.

4. Landymore-Lim, A.E.N., Bradbury, A.F. and Smyth, D.G. 1983. Biochem. Biophys. Res. Comm. 117, 289.

RE-APPRAISAL OF THE SERINE PROTEASE CATALYTIC MECHANISM: POSSIBLE INVOLVEMENT OF AN ACYL-HISTIDINE INTERMEDIATE

Graham J. Moore

Department of Medical Biochemistry, University of Calgary
Calgary, Alberta, Canada T2N 4N1

Introduction

Several mechanisms have been enumerated to explain the catalytic hydrolysis of peptide substrates by serine proteases(1-6). In all cases, a carbonyl group in the substrate was predicted to derivatize the active-site Ser hydroxyl, invoking an acyl-serine intermediate in the mechanism for catalytic hydrolysis. In the present report a mechanism is suggested in which the active-site His residue is acylated by the substrate, and the formation of an acyl-serine intermediate is not normally involved in the catalytic hydrolysis of peptide substrates.

Investigations on the octapeptide angiotensin II (ANG II), Asp-Arg-Val-Tyr-Ile-His-Pro-Phe, have suggested that ANG II contains a phenoxyl--imidazole--carboxylate interaction which is analogous to the proposed Ser--His--Asp charge relay system (2) at the active site of serine proteases, thoughbeit the latter involves a Ser hydroxyl rather than a Tyr hydroxyl group (7,8). Studies on the kinetics of acetylation and deacetylation of ANG II at physiological pH (7) have illustrated that although both the imidazole and phenoxyl groups in ANG II are superreactive, the imidazole in ANG II appears to acetylate and deacetylate faster than the phenoxyl group, suggesting that the order of reactivity for the free amino acids, i.e. His > Tyr, is preserved in the network interaction of ANG II. Since the order of nucleophilicity for free amino acids is His > Tyr > Ser at neutral pH, it is suggested, based on the findings with ANG II, that the active site His residue in serine proteases should be considerably more nucleophilic than the active site Ser residue. Assuming that accessibility and steric factors are not rate limiting, nucleophilic attack by serine proteases on a substrate should preferentially involve the active-site His residue,

Peptides 1986

and the resulting acyl-intermediate complex should connect the His N_3 atom with a carbonyl-carbon atom in the substrate. Others have suggested that phenyl benzoates preferentially acylate the active-site imidazole of chymotrypsin (6).

Results

The following reaction scheme illustrates a catalytic mechanism involving acylation of the active-site His residue in serine proteases:

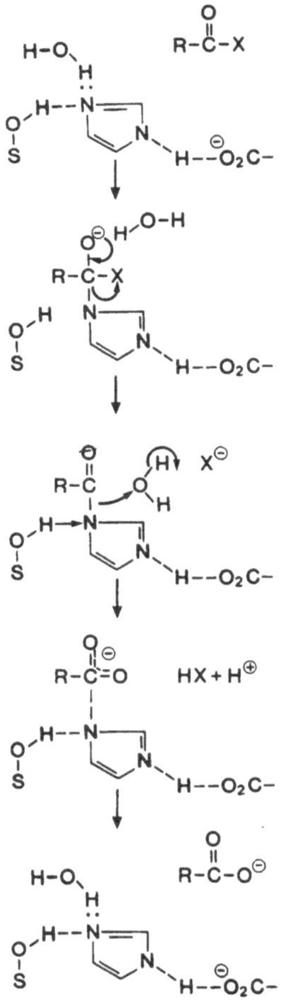

(I)

(II)

(III)

(IV)

(V)

(I) The network interaction originally proposed by Blow et al (2) involving sequestration of one molecule of water, as found at the active site by crystallography. Abstraction of the His N_1 proton by CO_2^- imparts marked basicity and nucleophilicity to the His N_3, essentially producing a superamine/superimine. The active-site Ser residue (SOH) and the water molecule interact with the His N_3; this interaction will be dependent on the proximity of SOH and should favour the stronger proton donor, particularly if the SOH group moves close to the His side-chain upon binding of the substrate. It is reasonable, particularly for serine proteases in which SOH is not very close to the His N_3 in the native enzyme structure (5), that binding of substrate could cause an "induced fit" resulting in optimum positioning of SOH relative to His N_3.

(II) Since His N_3 is predicted to be a stronger nucleophile than SOH, acylation of His N_3 by the substrate should occur preferentially and displace SOH, resulting in the formation of a tetrahedral acyl intermediate (imidate).

(III) As is the case for free His (7), deacylation is predicted to be slower than acylation. SOH should expedite the rate-limiting deacylation step by acting as a proton donor or acid catalyst; without SOH, the rate of deacylation of His would be nonfacilitated and considerably slower (7).

(IV) A quaternary imino intermediate is formed; HX appears before $R\text{-}CO_2^-$. Catalysis results in liberation of 1 hydrogen ion per molecule of substrate hydrolysed.

(V) $R\text{-}CO_2^-$ is released and SOH moves to its starting position. Another water molecule takes up its position close to the His N_3.

The basic mechanism of peptide hydrolysis outlined above is an aminolysis mechanism (1) or, more specifically, <u>acid-catalyzed superaminolysis</u> - in which the acid catalyst or proton donor is the Ser hydroxyl (SOH), and the superamine is the imidazole N_3 (which has been activated by carboxylate interaction with the imidazole N_1). Recent energy calculations (Rauk and Moore, unpublished work) suggest that the Ser hydroxyl also plays an important role in activating the imidazole, by causing transfer of the negative charge on the carboxylate to the imidazole ring. Thus superamine attack of the His N_3 anion on the carbonyl carbon of the substrate produces a tetrahedral intermediate (II), the breakdown of which (II-IV) is expedited by the ideally poised Ser hydroxyl proton. Expulsion of X (step II III) is known to be rate limiting, and subject to general acid catalysis, in the aminolysis of aliphatic esters (1) - presumably such considerations can be extended to amide (peptide) hydrolysis. It is known that the rate of deacylation of His increases as the pH is increased from 6.6 to 9.8 (7), suggesting that for serine proteases such as trypsin, chymotrypsin and elastase, which have pH optima of 8-9, factors such as denaturation-induced misalignment of binding/catalytic groups may cause the observed upper limit of pH dependence.

Discussion

Previously postulated acylation mechanisms have been largely based on the observation that non-substrate-like reagents such as DFP and PMSF inactive serine proteases by irreversibly acylating only the active-site Ser residue. This may be because these reagents are, by virtue of stability considerations, preferentially hydroxyl directed; the presumption that DFP forms a stable transition-state analogue may be incorrect. Acyl-histidine derivatives are largely unstable, particularly by comparison to acyl-serine derivatives; acylating agents such as DFP and PMSF might form both an unstable acyl-His derivative and, more slowly, a stable acyl-Ser derivative (the involvement of a transfer mechanism (6) analogous to base-catalyzed transesterification for these small reagents cannot be ruled out). Peptidic alkylating agents such as TPCK, on the other hand, should be able to mimick a true substrate, induce an appropriate conformation, and form a stable derivative with His. Accordingly, TPCK-His, rather than DFP-Ser, would represent a stable transition state analogue.

One other factor which argues against a catalytic mechanism involving an acyl-serine intermediate comes from the knowledge that deacetylation of the Tyr residue in the charge transfer system of ANG II is practically negligible at physiological pH (7). The chemistry of aliphatic and aromatic esters dictates that acyl-serine derivatives are much more stable than acyl-tyrosine derivatives. Consequently an acyl-serine intermediate ("distorted" transition states notwithstanding) would not represent a very practical situation in an enzyme catalytic mechanism designed for speed and efficiency. In contrast, deacetylation of the His residue in the CDN of ANG II at neutral pH appears to be extremely rapid(7)

Given that the order of reactivity of the groups in the catalytic triad of serine proteases is as follows: His acylation > His deacylation > Ser acylation > Ser deacylation, then Ser acylation would be practically non-existent for acylating agents which form very unstable derivatives with His (e.g. peptide substrates). However, if an acylating agent forms a more stable bond with His (e.g. DFP), such that the rate of Ser acylation approaches the rate of His deacylation, then a [stable] acyl-Ser derivative will be observed.

Acknowledgement

This work was supported by the Alberta Heart Foundation, the Alberta Heritage Foundation for Medical Research and the Medical Research Council of Canada. We thank Joanne Ward for typing the manuscript.

References

(1). Satterthwait, A.C. and Jencks, W.P.. 1974. J. Am. Chem. Soc. 96, 7018-7031.

(2) Blow, D.N., Birktoft, J.J. and Hartley, B.S.. 1969. Nature 221, 337-340.

(3) Wang, J.H.. 1970. Proc. Natl. Acad. Sci. U.S.A. 66, 874-881.

(4) Polgar, L. & Bender, M.L.. 1969 Proc. Natl. Acad. Sci. U.S.A. 64, 1336-1340.

(5) Matthews, D.A., Alden, R.A., Birktoft, J.J., Freer, S.T. and Kraut, J. 1977. J. Biol. Chem. 24, 8875-8883.

(6) Hubbard, C.D. and Shoupe, T.S.. 1977. J. Biol. Chem. 252 1633 -1638.

(7) Moore, G.J.. 1985. Int. J. Pept. Prot. Res. 26, 469-481.

(8) Moore, G.J. & Matsoukas, J.M.. 1985. Bioscience Reports 5, 407-416.

PREPARATION AND PROPERTIES OF SELECTIVELY ε-NH_2 ACYLATED PANCREATIC PHOSPHOLIPASES A_2

Arend J. Slotboom, Frans van der Wiele, Wim Atsma, Jan van Binsbergen, Margreet van Linde, Ben Roelofsen and Gerard H. de Haas

Laboratory of Biochemistry, State University of Utrecht, Padualaan 8, NL-3584 CH Utrecht, The Netherlands

Introduction

Phospholipase A_2 (EC 3.1.1.4) (PLA) is a low molecular weight enzym (14 kD) which hydrolyzes the 2-acyl ester linkage in 3-*sn*-phosphoglycerides [for a review see (1)]. The two richest sources are the mammalian pancreas and the venom of snakes. Notwithstanding the high sequence homology between snake venom and pancreatic PLAs and their similar 3-D structures some interesting differences have been noted. One of the most intriguing differences is the preferential hydrolysis of neutral substrates by the snake venom PLAs, whereas the pancreatic PLAs clearly prefer negatively charged substrates and they hardly attack long-chain zwitterionic lecithin bilayers. These differences might be related to the molecular packing characteristics of the various phospholipids. Natural lecithins form closely packed bilayer structures while the anionic phospholipids in the presence of Ca^{2+} often organize themselves in the hexagonal phase. Compared to most venom enzymes, the pancreatic phospholipases have been shown to possess very weak penetrating properties on monomolecular surface films, and possibly they are able to enter into more loosely packed hexagonal structures, but not in the bilayer structure. The weak penetrating properties of bovine and porcine pancreatic PLAs have been improved

Peptides 1986

considerably by covalent introduction of different long-chain fatty acids to the ε-NH_2 group of a particular Lys residue located in the lipid-binding domain of these enzymes.

Results and discussion

Reaction of methylacetimidate with porcine pancreatic prophospholipase A_2 (Prec) results in amidination of all 9 ε-NH_2 lysyl groups (Amprec-9). When the same reaction is done in the presence of L-dioctanoylglycerosulfate, one ε-NH_2 lysyl group is protected against amidination. In the resulting Amprec-8 the ε-NH_2 group of Lys-116 was found not to be amidinated. Subsequent reaction of Amprec-8 with the *N*-hydroxysuccinimide esters of *pal*mitic or *lau*ric acid yielded the corresponding Lys-116-ε-NH_2 acyl Amprecs-8. These acylated zymogens were converted by limited proteolysis into the corresponding enzymes *viz. Pal*-Ampa-8 (porcine) and *Lau*-Ampa-8 (porcine), respectively.

Lys-10-ε-NH_2 palmitoyl bovine Ampa was obtained by semisynthesis using the following strategy. After complete amidination of all 11 ε-NH_2 lysyl groups in bovine prophospholipase A_2 with methylacetimidate the resulting Amprec-11 was subjected to CNBr cleavage at the unique Met-8 residue. Ile-9 and ε-amidino-Lys-10 were then removed by Edman degradation yielding des (Ala-1-Lys-10) Ampa-10. To this fragment α-*N*-*t*-Boc-ε-NH_2-palmitoyl Lys-10 was coupled using the mixed-anhydride procedure. After removal of the α-*t*-Boc group by TFA, Ile-9 and Met-8 were then restitched as their *N*-hydroxysuccinimide esters using *t*-Boc protection for the α-NH_2 functions. Removal of the α-*t*-Boc groups was effected by treatment with TFA. Finally the N-terminal heptapeptide α-*t*-Boc Ala^1.Leu.Trp(For).Gln.Phe.Arg.Gly^7, prepared by solid-phase peptide synthesis, was reintroduced by the mixed-anhydride procedure. The resulting protected enzyme was purified by ion-exchange chromatography. Deformylation and removal of the α-*t*-Boc group was affected by treatment with NH_4HCO_3 and TFA, respectively. The resulting Lys-10-ε-NH_2 *pal*mitoyl Ampa-10 [Pal-Ampa-10 (bovine)] was purified to homogeneity by ion-exchange chromatography.

Table 1, Kinetic Properties of Acylated Phospholipases A_2

	V_{ma} a) (μequiv.$min^{-1}mg^{-1}$)	Sp.Act. b) (μequiv.$min^{-1}mg^{-1}$)
porcine		
Pal-Ampa-8	2250	360
Lau-Ampa-8	2200	270
Ampa-9	2200	250
Ampa-8	2500	650
bovine		
Pal-Ampa-10	5600	40
Ampa-11	5900	120

a) Determined at pH 6.0 using L-dioctanoyllecithin micelles
b) Determined at pH 8.0 using egg-yolk lecithin and Na-deoxycholate

As can be seen in Table 1 all three acylated Ampas possess good enzymatic activities as compared to the corresponding non-acylated Ampas. The use of monomolecular surface films of lecithins in a convenient way to determine the penetrating power of phospholipase A_2. The non-acylated fully ε-amidinated phospholipases A_2 penetrate monolayers of L-didecanoyllecithin upto surface pressures of about 10 dynes/cm (Figure 1). At slightly higher packing densities, however, the lecithin

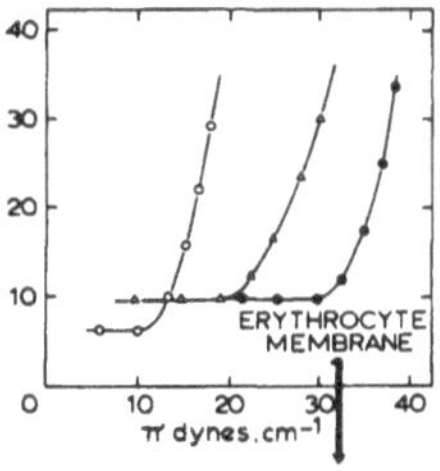

Figure 1. Influence of surfacepressure (π) of a L-didecanoyllecithin monolayer on the induc-tion time (τ) of the kinetics of hydrolysis by Ampa-9 (porcine) and Ampa-11 (bovine), (-O-); Lau-Ampa-8 (porcine) and Pal-Ampa-10 (bovine), (-Δ-); Pal-Ampa-8 (porcine), (-•-).

film behaves as a "wall": the enzyme encounters increasing difficulties to penetrate and long lag times are observed. Lau-Ampa-8 (porcine) and Pal-Ampa-10 (bovine) can easily penetrate much more densily packed monolayers: upto surface pressures of 22 dynes/cm no significant increase in lag time is observed. So far Pal-Ampa-8 (porcine) turned out to possess the most powerful penetrating capacity. This enzyme easily anchors the lecithin film upto surface pressures of 32 dynes/cm. This behaviour raised the question: is Pal-Ampa-8 (porcine) able to attack a biological membrane where the lateral surface pressure has been estimated to be 31-35 dynes/cm as in the case of the human erythrocyte? Short-time incubation of the human erythrocyte cells with Pal-Ampa-8 (porcine) results in a dramatic morphological change of the erythrocytes. The transformation into so-called "echinocytes" is due to the phospholipase A_2 catalyzed conversion of 70% of the outer monolayer lecithin into lysolecithin and fatty acid, which causes an increase in lateral surface pressure in that layer. The non-acylated ε-amidinated phospholipases A_2 and even the more strongly penetrating Lau-Ampa-8 (porcine) and Pal-Ampa-10 (bovine) do not show these effects.

Acknowledgement

These investigations were carried out under the auspices of The Netherlands Foundation for Chemical Research (S.O.N.) and with financial aid from The Netherlands Organization for the Advancement of Pure Research (Z.W.O.).

References

1. Slotboom, A.J., H.M. Verheij and G.H. de Haas. 1982. In: Phospholipids, New Biochemistry, Vol. 4 (J.N. Hawthorne and G.B. Ansell, eds.). Elsevier, Amsterdam, pp. 359-434.

INTRAMOLECULARLY QUENCHED FLUOROGENIC SUBSTRATES FOR AMINOPEPTIDASE P IN HUMAN SERUM

E.J. HOLTZMAN

Department of Medicine, The Chaim Sheba Medical Center, Tel-Hashomer, Israel

A. YARON

Department of Biophysics, The Weizmann Institute of Science, Rehovot, Israel

Introduction

Synthetic chromogenic and fluorogenic substrates for proteolytic enzymes are finding increasing application in basic and clinical research. Intramolecularly quenched substrates (1) represent a class of substrates in which the fluorescence of a group situated in one part of a peptide is quenched by a chromophore placed in another part of the molecule. Cleavage of the peptide chain at any bond between the two groups interrupts the intramolecular quenching interaction and the resulting emission measures the extent of the substrate cleavage. Only the first cleavage is observed, since subsequent degradation of the molecule has no influence on the emission released. Application of this type of substrate was particularly useful for developing a specific and sensitive assay for aminopeptidase P (aminoacylprolyl-peptide hydrolase, EC 3.4.11.9) (2-4). This exopeptidase catalyses the hydrolysis of peptide bonds linking any N-terminal amino acid residue with a prolyl residue in peptides of low and high molecular weight. Due to these specificity requirements the common design of fluorogenic substrates, in which a chromophore is engaged in the cleavable peptide bond, cannot be applied with aminopeptidase P. We therefore synthesized intramolecularly quenched substrates in which the peptide nature of the Xaa-Pro bond is preserved.

Peptides 1986

Results and Discussion

In previous investigations we used successfully the 2-aminobenzoyl group (ABz) as the fluorophore and the 4-nitro-L-phenylalanyl residue [Phe(p-NO_2)] as the fluorescence-quenching group in the synthesis of intramolecularly quenched fluorogenic substrates (5). The applicability of the Phe(p-NO_2) residue as the N-terminal amino acid in aminopeptidase P substrates was first demonstrated by its rapid release from Phe(p-NO_2)-Pro-Ala by the enzyme purified from Escherichia coli. Using ethylene diamine (-NH-CH_2-CH_2-NH-,Eda) as the spacer, compounds I, II and III were prepared:

Phe(p-NO_2)-Pro-Eda-ABz (I)
Phe(p-NO_2)-Pro-Pro-Eda-ABz (II)
Lys(Dnp)-Pro-Pro-Eda-ABz (III)

in which Dnp is 2,4-dinitrophenyl.

There is no overlap between the ABz emission band centered at 412 nm (λ_{ex}=310 nm) and the absorption of the p-nitrophenylalanine chromophore (λ_{max}=280 nm). Therefore, the intramolecular quenching interaction in I and II is not mediated by the Förster radiationless energy-transfer. The collisional quenching results in a 7.5- and 3.4-fold decrease of the ABz emission, respectively. A more efficient quenching (160 X) was achieved with III in which the Dnp absorption band (λ_{max}=360 nm) overlaps the ABz emission, thereby permitting the radiationless energy transfer.

The compounds I and II were shown (6) to be hydrolyzed by the purified bacterial aminopeptidase P, when incubated in presence of Co^{++} ions at pH 8.2, by electrophoretic analysis of the hydrolysates which revealed the expected products. Hydrolysis of the substrates was accompanied by an increase of fluorescence, reaching a maximum which was 7.5-times and 3.4-times higher than the initial fluorescence of I and II, respectively. This permitted initial rate measurements and determination of the kinetic constants of I (K_m=3.0 mM, k_{cat}=135 s^{-1}) and II (K_m=0.22 mM, k_{cat}=20 s^{-1}). The specificity was shown to be high for II by incubation with human serum and with calf-lung extracts. The amounts of p-nitro-L-phenylalanine released were determined by quantitative amino acid analysis at various time intervals. As expected for an aminopeptidase P

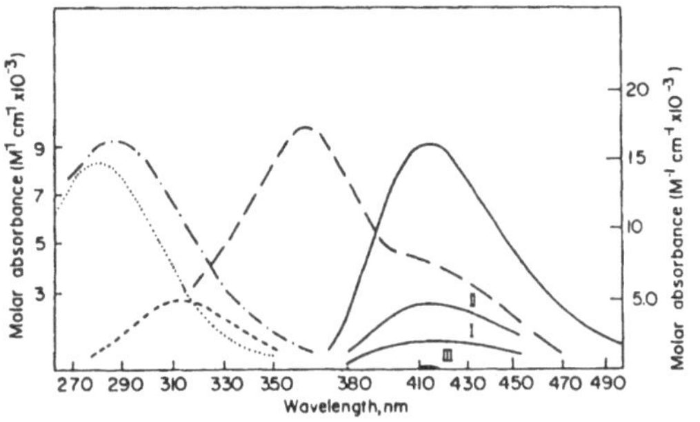

Fig. 1. Spectral properties of substrates I, II and III. Absorption spectra of Phe(p-NO_2)-Pro (·······), H_2N-CH_2-CH_2-NH-ABz (- - - -), substrate I and II (—·—·—) and substrate III (— — —) in 50 mM Tris buffer, pH 8.0. The scale for absorbance of substrate III is given on the right ordinate, the other absorbancies on the left. Emission of ABz-Gly, substrates I, II and III (——) are drawn on a scale of arbitrary units.

activity, the number of Phe(p-NO_2) residues released was equal to the number of initial cleavages of substrate II molecules as determined by fluorimetry at each time interval. In contrast to II, the release of Phe(p-NO_2) from substrate I was slower than the increase of unquenched 2-aminobenzoyl groups, indicating that the initial cleavage of a part of substrate I molecules occurred at bonds other than the Phe(p-NO_2)-Pro bond, due to additional enzymatic activities present in the mammalian enzyme preparations.

Of the two substrates, compound I was more efficiently quenched and had a higher k_{cat} value. In crude extracts, however, compound II is the substrate of choice despite its lower sensitivity since the presence of a sequence of two proline residues rendered the peptide hydrolysable by aminopeptidase P only. The requirement of a free amino group was demonstrated by the resistance of the acetylated substrate II to the crude enzyme preparations. This also showed absence in the tissues of activities - other than aminopeptidase P - capable to degrade the structure of substrate II.

Preservation of the basic structure of compound II, that endows the peptide with high specificity, with an increase in sensitivity, was achieved with substrate III. In the assay developed, this substrate (0.5 mM) was incubated with human serum (10µ) in 0.2 M Tris-HCl buffer, pH 8.0, containing $MnSO_4$ (4.6 mM) and trisodium citrate

(9.3 mM). The reaction was terminated by adding a mixture of dithiothreitol (1 mM) and EDTA (50 mM) in the above buffer (4.3 ml). The extent of hydrolysis was calculated from the fluorescence increment (λ_{ex}=320, λ_{em}=412 nm) as compared to an incubation mixture handled identically, except for adding the serum after the incubation. A solution of 2-aminobenzoylglycine was used as the standard.

Kinetic measurements with III showed a linear dependence of the hydrolysis rate on time up to 5% substrate-conversion and on enzyme concentration up to at least 20 µl of human serum. The K_m value of 100 ±3.0 µM was calculated from initial rate measurements for the human serum enzyme. Aminopeptidase P activity in the sera of 53 healthy adults was found to be 37.4 ± 2.7 (S.E. of mean) with a standard deviation of 19 units per ml serum.

The enzyme acts on biologically active peptides such as bradykinin, substance P and tuftsin (7). There are many potential substrates known that contain the N-terminal sequence Xaa-Pro which may be expected to be hydrolyzed by this enzyme. The assays used so far depended on the quantitative measurements of an amino acid released. The presence of amino acids in crude biological liquids and tissues limited the assay sensitivity. The new method may be expected to promote the study of the biological role of aminopeptidase P.

References

1. Yaron, A., A. Carmel, E. Katchalski-Katzir. 1979. Intramolecularly quenched fluorogenic substrates for hydrolytic enzymes. Anal. Biochem. 95, 228-235.

2. Yaron, A., D. Mlynar. 1968. Aminopeptidase P. Biochem. Biophys. Res. Commun. 32, 658-663.

3. Dehm, P., A. Nordwig. 1970. The cleavage of prolyl peptides by kidney peptidases. Partial purification of a "X-prolyl-aminopeptidase" from swine kidney microsomes. Eur. J. Biochem. 17, 364-371.

4. Sidorowicz, W., J. Szechinski, P.C. Canizaro, F.J. Behal. 1984. Cleavage of the Arg^1-Pro^2 bond of bradykinin by a human lung peptidase: isolation, characterization, and inhibition by several β-lactam antibiotics[1] (41828). Proc. Soc. Exptl. Biol. Med. 175, 503-509.

5. Carmel, A., A. Yaron. 1978. Intramolecularly quenched fluorescent tripeptide as a fluorogenic substrate of angiotensin-I-converting enzyme and of bacterial dipeptidyl carboxypeptidase. Eur. J. Biochem. 87, 265-273.

6. Fleminger, G., A. Carmel, D. Goldenberg, A. Yaron. 1982. Fluorogenic substrates for bacterial aminopeptidase P and its analogs detected in human serum and calf lung. Eur. J. Biochem. 125, 609-615.

7. Fleminger, G. and A. Yaron. 1983. Sequential hydrolysis of proline-containing peptides with immobilized aminopeptidases. Biochim. Biophys. Acta 743, 437-446.

SENSITIVE CHYMOTRYPSIN DETERMINATION USING FLUOROGENIC SUBSTRATES CONTAINING 2-QUINOLINONE DERIVATIVES

G. Kokotos, C. Tzougraki

Department of Chemistry, University of Athens, 13A Navarinou st., Athens 106 80, Greece

R. Geiger

Abteilung für Klinische Chemie und Klinische Biochemie, Universität München, Nussbaumstrasse 20, 8000 München 2, Deutschland

Introduction

We have recently reported a convenient and sensitive assay for chymotrypsin using the substrate Glt-Phe-AMQ (1) (1) based on the enzymatic release of the highly fluorescent compound 7-amino-4-methyl-2-quinolinone (AMQ, 2) (Scheme 1). In continuing our search we synthesized and examined the fluorescence properties of 7-amino-4-trifluoromethyl-2-quinolinone (AFQ, 3) (Scheme 1) and the corresponding substrate Glt - Phe - AFQ (4) expecting that by the replacement of the methyl group in AMQ with the electron withdrawing trifluoromethyl group, the excitation and emission maxima would shift to longer wavelengths. In the present study we also tried to increase the sensitivity of the assay by enhancing the substrate-enzyme binding affinity through the extension of the peptide backbone of 1 and 4 with the insertion of the hydrophobic residue leucine. We now present the fluorescence properties, the kinetic parameters and the sensitivity of the substrates Glt-Phe-AFQ (4), Glt-Leu-Phe-AFQ (5) and Glt-Leu-Phe-AMQ (6).

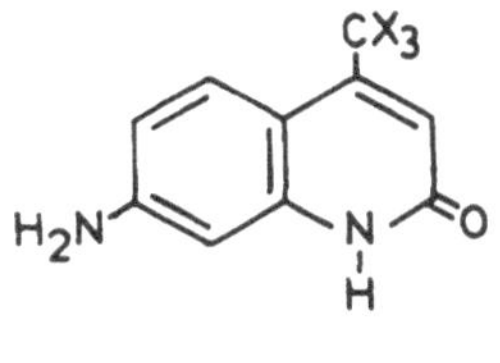

2 X= H

3 X= F

Scheme 1

Peptides 1986

Table 1. Fluorescence Properties of 2-Quinolinone Derivatives and the Corresponding Substrates[a].

No.	Compound	Maxima (nm) λex	λem	Relative fluorescence[b]
2	AMQ	344	418	1.00
3	AFQ	360	470	0.31
1	Glt-Phe-AMQ (1)	333	370	0.14
6	Glt-Leu-Phe-AMQ	333	370	0.14
4	Glt-Phe-AFQ	340	390	0.07
5	Glt-Leu-Phe-AFQ	340	390	0.07

a 1.0 µM solution in 0.05 M TES, pH 8, containing 1% DMSO.
b The fluorescence intensities are relative to the intensity of AMQ, on an equimolar basis.

Results

The compound AFQ was prepared by condensation of 1,3-phenylenediamine with 4,4,4-trifluoroacetoacetate and the substrates by conventional methods of peptide synthesis in good yield and purity. Details of their syntheses will be reported elsewhere.

The excitation and emission maxima and the fluorescence intensities of the fluorophores AMQ, AFQ and the corresponding substrates 6, 4 and 5 are given in Table 1. AFQ had a fluorescence maximum at 470 nm when excited at 360 nm. As expected, a red shift of 52 nm and a decrease in fluorescence intensity was observed when the 4-methyl group in AMQ was replaced with the trifluoromethyl group in AFQ. The excitation and emission maxima and the fluorescence intensity of AFQ were distinctly different from those of its N-acyl derivatives 4 and 5 (Fig. 1a). When λex and λem were set at 380 nm and 480 nm, respectively, AFQ retained 41% of its maximum fluorescence, possessing a relative fluorescence 850-fold greater than that of an equimolar solution of the substrates. The enzymatic hydrolyses of substrates 4 and 5 were also monitored at these wavelengths, where the presence of the substrates did not interfere (Fig. 1b), while λex and λem settings of 360 nm and 435 nm, respectively, were used for the hydrolysis of 6 (1). The enzymatic hydrolyses of 0.5 mM solutions of the substrates were conducted at 25°C in a buffer solution 0.1 M Tris-HCl, pH 7.8, containing 0.01 M $CaCl_2$, 0.05% Triton X-100 and 10% DMSO.

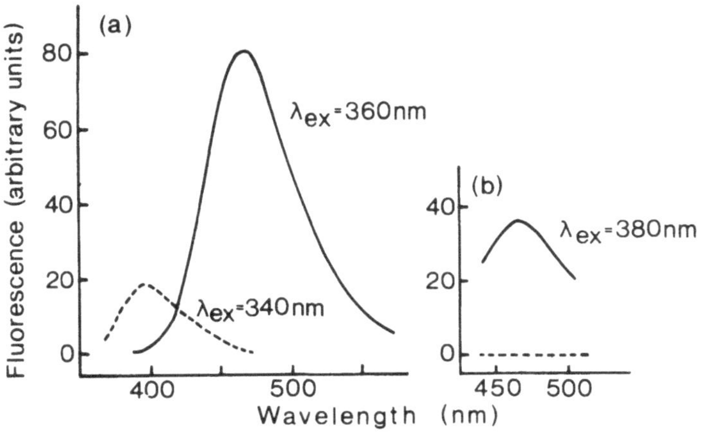

Figure 1. Fluorescence spectra of AFQ —— and Glt-Leu-Phe-AFQ --- 1.0 μM solutions in 0.05 M TES, pH 8, containing 1% DMSO.

The kinetic parameters of the new substrates, as well as the detection limits of chymotrypsin obtained in continuous assays with short incubation times (5-10 min) are listed in Table 2. Substrates 5 and 6 have 27-fold and 150-fold higher kcat/Km values than the substrates 4 and 1 respectively. This finding can be explained by specific enzyme-substrate "secondary interactions" (2), due to the presence of the hydrophobic leucine in the P_2 position of substrates 5 and 6.

Table 2. Kinetic Parameters and Maximum Sensitivity of Some Fluorogenic Substrates for Chymotrypsin.

No.	Substrate	Km (mM)	kcat/Km (mM^{-1} s^{-1})	Detection limit (ng/ml)	Ref.
6	Glt-Leu-Phe-AMQ	0.62	7.100	0.7	
1	Glt-Phe-AMQ	0.5	0.047	10	(1)
5	Glt-Leu-Phe-AFQ	0.33	2.670	10	
4	Glt-Phe-AFQ	0.28	0.099	35	
7	Glt-Phe-AMC	0.67	0.078	500	(3)
8	TFA·Ala-Ala-Phe-AMC	0.5	1.660	25	(3)
9	Glt-Phe-AQ	1.77	0.079	1,000	(4)
10	Z-Phe-OMC	0.013	446.000	2	(5)

The sensitivity of the assay using substrates 5 and 6 was 3.5 to 14 times higher than that using 4 and 1. This is in correlation with the higher kcat/Km values of the first pair. Comparison with other fluorogenic substrates listed in Table 2 shows that the substrates containing the 2-quinolinone derivatives AMQ and AFQ are much more sensitive than those containing the coumarin derivative 7-amino-4-methylcoumarin (AMC), which is, to date, the most commonly used fluorescent marker. The use of ester substrate 10, in a non-continuous assay, has yielded the lowest detection limit thus far. However, some limitations have been reported concerning the specificity and stability of 10 (5).

The new substrates were specific for chymotrypsin showing no detectable hydrolysis with porcine leukocyte elastase (10 µg) and were also stable under the assay conditions. The rates of hydrolysis of 4 and 6 were proportional to chymotrypsin concentration - up to at least 500 ng, while that of 5 was proportional up to 200 ng.

In conclusion all three new substrates 4, 5 and 6, are suitable for fluorometric chymotrypsin determination. In particular, 6 is the best substrate found to date; it is ideally suited since it is extremely sensitive, convenient and useful for the continuous assay of chymotrypsin. Both 2-quinolinone derivatives, AMQ and AFQ will be exceptionally useful leaving groups in fluorogenic substrates for other proteases.

REFERENCES

1. Kokotos, G. and C. Tzougraki : Int. J. Pept. Prot. Res. (in press).

2. Jakubke, H.-D., H. Däumer, A. Könnecke, P. Kuhl and J. Fischer. 1980. Experientia 36, 1039 - 1040.

3. Zimmerman, M., B. Ashe, E. C. Yurewicz and G. Patel. 1977. Anal. Biochem. 78, 47 - 51.

4. Brynes, P. J., P. Bevilacqua and A. Green. 1981. Anal. Biochem. 116, 408 - 413.

5. Kuromizu, K., Y. Shimokawa, O. Abe and N. Izumiya. 1985. Anal. Biochem. 151, 534 - 539.

SYNTHESIS OF GLYCOPEPTIDES USING THE ALLYL ESTER AND THE ALLYLOXYCARBONYL GROUP AS PROTECTING FUNCTIONS

Horst Kunz, Herbert Waldmann and Carlo Unverzagt

Institut für Organische Chemie der Universität Mainz,
Joh.-Joachim-Becher-Weg 18-20, D-6500 Mainz, F.R.Germany

Most of the proteins produced in eukaryotic cells are glycosylated. The oligosaccharide side chains glycosidically linked to certain trifunctional amino acids of the peptide part can exhibit different functions in biological processes. They influence the solubility and the conformation of the glycoproteins, and they can protect the proteins against the attack of proteolytic enzyms. Furthermore, the glycoproteins and their carbohydrate parts obviously play central roles as the recognition signals and antigens in important processes of biological selectivity, in particular, in the recognition of serum components on membranes and in cell-cell recognition. These recognition phenomena include, for instance, the regulation of the cell growth and the cell differentiation and are also a vital part of the immune response[1].

In contrast to pure peptides and proteins, glycoproteins and glycopeptides are not easily available by gene technological methods. The chemical synthesis of glycopeptides is a complex preparative task since these compounds contain numerous functional groups and the chemically labile glycosidic bonds. These acetalic bonds can be cleaved or anomerized by strong acids and, furthermore, the O-glycosyl serine and threonine peptides are also destroyed under moderately basic conditions by the characteristic ß-elimination.

Therefore, glycopeptide synthesis requires protecting groups which are selectively removable under almost neutral conditions.

Peptides 1986

To meet these demands we have developed the allylic protecting principle in peptide and glycopeptide synthesis.
The allyl esters of amino acids are easily available by azeotropic esterfication[2] or by allylation of the salts[2,3]. This carboxy protection is stable to moderate bases and acids. Thus, the tert-butyloxycarbonyl group can be cleaved selectively in the presence of the allyl ester either with trifloroacetic acid or with hydrogen chloride in dichloromethane.
On the other hand, the allyl ester protection can be removed selectively by rhodium-I-catalysis under neutral conditions[2], or, even more mildly, by palladium-O-catalysed allyl transfer to an accepting nucleophile[4]. In this way, for example, by treatment of the very sensitive galactosaminyl threonine derivative 1 with catalytic amounts of the palladium-O-catalyst in tetrahydrofuran in the presence of morpholine as the allyl acceptor at room temperature, one can establish a completely selective and quantitative cleavage of the allyl ester[4] within 30 minutes.

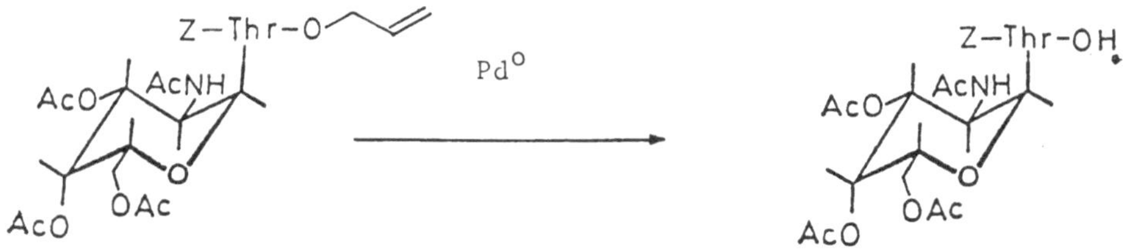

1 2

96%

Pd^{0} : $[(C_6H_5)_3P]_4Pd^{0}$, morpholine/tetrahydrofuran
room temperature, 30 min

Under these conditions the other protecting groups and the glycosidic bond (which is sensitive to both acids and bases) remain completely unchanged. The small allyl ester has the further advantage that it doesn't hinder the condensation reactions either by steric hinderance or by decreasing the solubility.
We could show that the allyloxycarbonyl(Aloc)group[5] as the amino protecting group has all the advantages of the allyl ester[6,7]. It is stable to trifluoroacetic acid so that the tert-butyl ester can be cleaved selectively.

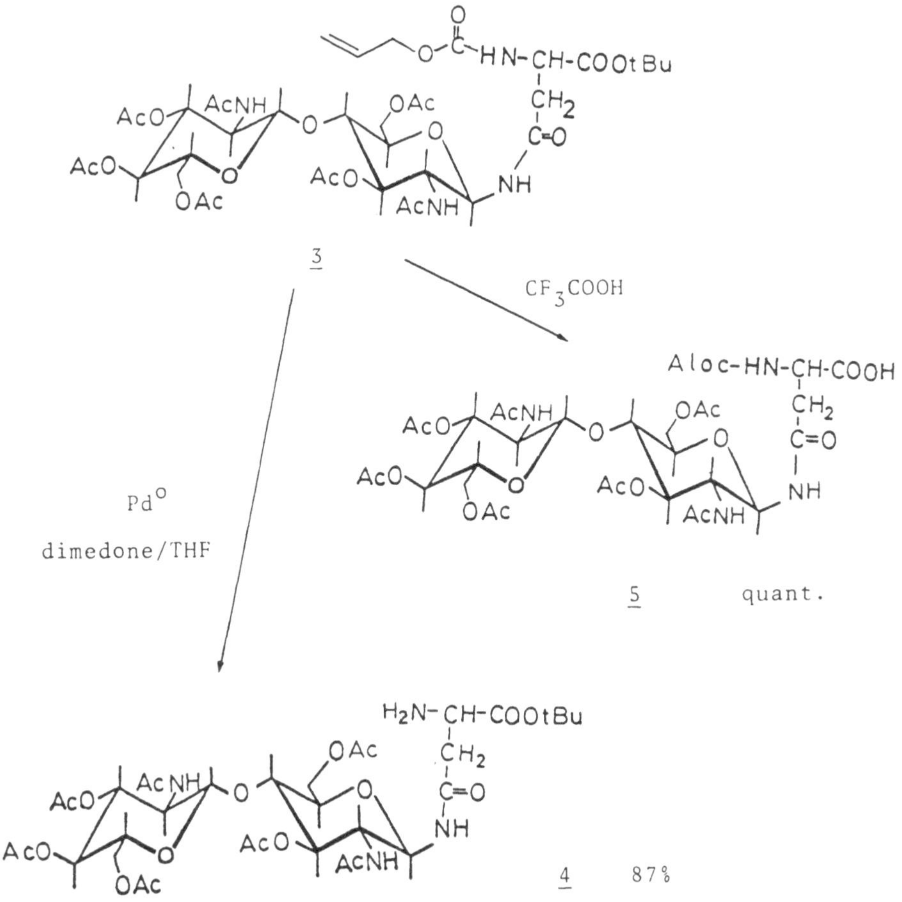

3

CF_3COOH

Pd^o

dimedone/THF

5 quant.

4 87%

On the other hand, the Aloc-group which was earlier considered to be not completely removable[5], can selectively and quantitatively be cleaved even from the complex chitobiose asparagine derivative 3.
In the case of the amino deprotection it is favorable to use a non basic nucleophile, for instance, dimedone, which can easily by separated from the liberated amino compound 4.

References

1. G.Ashwell. 1977. Trends Biochem.Sci. 2, N-186 .

2. H.Waldmann, H.Kunz. 1983. Liebigs Ann.Chem., 1712.

3. H.Kunz, H.Waldmann. 1985. Chem.Commun., 618.

4. H.Kunz,H.Waldmann. 1984. Angew.Chem. 96, 49; Angew.Chem.Int.Ed.Engl. 23 , 71.

5. C.M.Stevens,R.Watanabe. 1950. J.Am.Chem.Soc.72, 725.

6. H.Kunz, C.Unverzagt. 1984. Angew.Chem. 96, 426; Angew.Chem.Int.Ed.Engl. 23, 436.

7. H.Kunz,H.Waldmann,C.Unverzagt. 1985. Int.J.Peptide Protein Res. 25, 493.

EVALUATION OF RECEPTOR SITES FOR PLASMODIUM FALCIPARUM (MALARIA) ON HUMAN ERYTHROCYTES. CHEMICAL SYNTHESIS OF PEPTIDES RELEVANT TO GLYCOPHORIN A.

A. Tounkara, F. Piller, D. Blanchard, J.P. Cartron
Unité INSERM U76, CNTS-Institut, 6 rue Alexandre Cabanel 75015 PARIS

B. Ferrari, J.M. Lacombe, A. Pavia
Laboratoire de Chimie Bioorganique, Université d'Avignon, 33 rue Louis Pasteur
84000 Avignon, France

Introduction

A number of investigations have suggested that the red cell membrane sialoglycoproteins (glycophorins A, B and C) are receptors for the entry of Plasmodium falciparum merozoites into human red cells. This is based on in vitro invasion assays using erythrocytes with variant glycophorins and enzyme treated cells, as well as on inhibition of invasion by purified glycophorins, glycophorin derived peptides and by anti-glycophorin antibodies (for review see 1). The relative contribution of the carbohydrate and peptide domains of glycophorin A have been explored (1-4). The carbohydrate domains apparently interact only weakly with the parasite, but a tryptic fragment, T_6 (position 62-97), containing the hydrophobic intramembraneous region of glycophorin A was found to be a good inhibitor of invasion (2,3). Since non-specific hydrophobic interaction may influence the inhibition phenomena (1,4) we have examined the role of synthetic peptides derived from the sequence 55 to 76 of glycophorin A in the process of invasion. In addition, the role of glycophorins from native and variant red cells (Tn, Cad), have been studied.

Methods

Synchronized cultures of ring forms of P. falciparum (6) were diluted with uninfected human red cells at a 2.5% final suspension and 0.2% parasitemia and incubated in microtiter plates (100 µl final volume) according to Traeger and Jensen (7). The medium used for cultures was RPMI 1640 supplemented by 10% normal human serum. Inhibitors (glycoproteins or peptides) and appropriate controls in neutral solution were prepared in RPMI 1640 with 10% human serum. At the beginning of the schizonte stage (24h) the culture medium was carefully removed and replaced by the solution of inhibitors. During the next 24h incubation at 37°C, the parasites develop into schizontes and merozoites which invade uninfested erythrocytes.

Peptides 1986

After removal of the inhibitors the amount of newly infected cells was measured by the incorporation of ^{3}H-hypoxanthine over a periode of 24h (8). All assays were performed in triplicate and parasite counts were controlled in separate wells at all stages of the microculture on Giemsa-stained thin blood smears.

The following peptides used as inhibitors of invasion were obtained by an active ester stepwise coupling strategy beginning at the C-terminus and proceeding towards the N-terminus.

1 ^{55}Glu-Glu-Glu-Thr-^{59}Gly
2 ^{60}Glu-Arg-Val-Gln-^{64}Leu
3 ^{65}Ala-His-His-Phe-^{69}Ser
4 ^{70}GLu-Pro-Glu-Ile-^{74}Thr
5 ^{58}Thr-Gly-Glu-Arg-^{62}Val
6 ^{63}Gln-Leu-Ala-His-^{67}His
7 ^{67}His-Phe-Ser-Glu-^{71}Pro
8 ^{72}Glu-Ile-Thr-Ile-^{76}Ile

9 ^{62}Val-Gln-Leu-Ala-His-His-Phe-Ser-^{70}Glu

Peptide coupling was achieved by using N-hydroxysuccinimidyl (OSu) active ester, whereas $-NH_2$ was temporary protected by a benzyloxycarbonyl (Z) group. Deprotection at every step was performed by treatment of the peptides with hydrogen in the presence of 10% palladium on charcoal. N- and C-terminal residues of above peptides were obtained as ter-butyloxycarbonyl (BOC) and ter-butyl (Bu^t) derivatives respectively. In addition Bu^t derivatives were chosen to protect the side chain carboxyl and hydroxyl functions of glutamic acid and of serine/threonine residues. The guanidino group of Arg-61 as well as the imidazole ring of His-66 and 67 were protected by an adamantyloxycarbonyl derivative (Adoc). Protecting groups were selected in order to allow deprotection of the final peptides in one step by treatment with trifluoroacetic acid. Fragment 63-67 could not be obtained according to this strategy ; removal of Z protecting group from Z-His (Adoc)-His-(Adoc)-OBu^t was unsuccessful.

Intermediate peptides were purified by silica-gel column chromatography or gel filtration and the structure assigned by natural abundance ^{13}C-n.m.r spectroscopy. H.P.L.C. monitoring of fully deprotected peptides showed all of the above compounds to possess a high degree of purity (> 88%)..Others peptides related to either glycophorin A (n° **10** = ^{26}Asn-Asp-Thr-Lys-His-^{31}Lys ; n° **11** = ^{1}Leu-Ser-Thr-Asn-Gly-^{6}Val) or glycophorin B (n° **12** = ^{26}Asn-Gly-Glu-Thr-Gly-Gln-Leu-^{33}Val) were also synthetized and tested as controls.

Results and Discussion

Earlier studies demonstrated that invasion of human red cells by merozoites was inhibited to 50% by 20 to 50 ug/ml of purified native or desialized glycophorins A or B isolated from human erythrocytes (not shown and also found by others (4)). Since blood group Tn and Cad red cells are refractory to invasion by the parasites (9,10) and carry glycophorins with an abnormal carbohydrate composition both terminating with a GalNAc residue (11,12), we have examined the behavior of the glycophorin A molecules purified from these cells in the invasion process. Surprisingly, glycophorin A isolated from Tn cells (patient Ba.) was as good an inhibitor as normal glycophorin, whereas glycophorin A from Cad erythrocytes (donor Cad.) was not inhibitory. These results indicate that the invasion of red cells by merozoites can be mediated by factors others that those related to the carbohydrate structure carried by the red cell membrane glycophorins.

It is known that the carbohydrate free tryptic fragment T_6 of glycophorin A (residues 62 to 97) inhibits invasion (2,3) and carries the hydrophobic intramembraneous domain (residues 72 to 92) of the glycoprotein. Since the receptor for the parasite is very likely located on the portion protruding from the membrane, we have probed the region 55 to 76 of glycophorin A by using synthetic peptides as inhibitors of the invasion process.

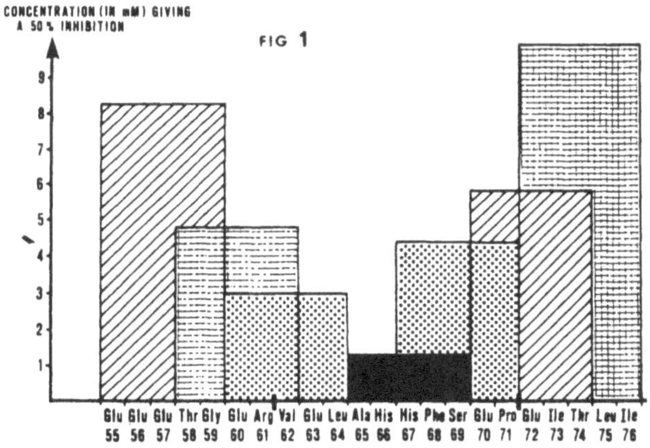

IN VITRO INHIBITION OF THE INVASION OF HUMAN RED CELLS BY PLASMODIUM FALCIPARUM WITH PEPTIDES RELATED TO GLYCOPHORIN A

As shown in Fig. 1, 1mM of peptide **3** inhibited invasion by 50%. Peptides **2** and **7** were also good inhibitors but others located either upstream (peptide **1**) or downstream (peptide **8**) showed less inhibition. Peptide **9** which covers the region 62-70 was inhibitory at 1mM concentration. It was observed in addition that peptides **2** to **9** inhibited almost completely the invasion at 20mM concentration. Control peptides related to glycophorin B or other regions of glycophorin A (peptides **10**, **11**, **12**) inhibited only weakly (20 to 30%) at 20mM concentration (not shown). This is indirect evidence that the peptides used here were not toxic for parasite growth, which is consistent with the controls by microscopic examination of blood smears after Giemsa staining.

Our results suggest that the peptide sequence within residues 65-69 of glycophorin A is involved in the interaction between the merozoites and the red blood cells. This is probably not due to a charge effect since peptide n° **10** which contained a sequence Lys-His-Lys is not inhibitory. It is interesting to note that glycophorin B, which is also a putative receptor for P. falciparum, carries a sequence ^{32}Leu-Val-His-Arg-Phe-^{37}Thr-, which is closely related to region 65-69 of glycophorin A. There is, however, no obvious analogous sequence on glycophorin C (13). Recently, Dahr et al. (14) reported that the blood group Wr^b antigen

is located between amino-acids 62-72 of glycophorin A, a region which includes the sequence 65-69 represented in peptide **3**. However, comparative amino-acid analyses of glycophorin A isolated from Wr(a+b-) and Wr(a-b+) red cells did not indicate any significant alteration within the peptide sequence 64-68, suggesting that the Wr^b antigen is located elsewhere on this glycoprotein (14). The findings indicating that the Wr^b antigen is not located in this region is consistent with the recent observation that Wr (a+b-) cells are normally infested by P. falciparum merozoites (15-17) as well as with the results presented here.
In conclusion we suggest that non carbohydrate containing peptide sequences on the red cells are also involved in the entry of merozoites from P. falciparum into human erythrocytes. This is in good agreement with other evidence for receptor heterogeneity which has been recently reported (18).

References

1. Hermentin P., Enders B. 1984. Behring Inst. Mitt. **76**, 121.
2. Breuer W.V., Baruch K.I., Ginsburg H., Cabantchik Z.I. 1983. Infect. Immun. **42**, 133.
3. Perkins M.E. 1984. Mol. Biochem. Parasitol. **10**, 67.
4. Vanderberg J.P., Gupta S.K., Schulman S., Oppenheim J.D., Furthmayr H. 1985. Infect. Immun. **47**, 201.
5. Breuer W.V., Ginsburg H., Cabantchik Z.I. 1984 Mol. Biochem. Parasitol. **12** : 125.
6. Lambros C., Vanderberg J.P. 1979. J. Parasitol. **65**, 418.
7. Traeger W., Jensen J.B. 1976. Science **193**, 634.
8. Chulay J.D., Haynes J.D., Diggs C.L. 1983. Exp. Parasitol. **55**, 138.
9. Pasvol G., Jungery M., Weatherall D.J., Parsons S.F., Anstee D.J., Tanner M.J.A. 1982. Lancet **ii**, 947.
10. Cartron J.P., Prou O., Luilier M., Soulier J.P. 1983. Br. J. Haematol. **55**, 639.
11. Dahr W., Uhlenbruck G., Gunson H.H., Van der Hart M. **24**, 36.
12. Blanchard D., Cartron J.P., Fournet B., Montreuil J., Van Halbeek H., Vliegenthart 1983. J. Biol. Chem. **258**, 7691.
13. Colin Y., Rahuel C., London J., Romeo P.H., d'Auriol L., Galibert F., Cartron J.P. 1986. J. Biol. Chem. **261**, 229.
14. Dahr W., Wilkinson S., Issitt P.D., Morel P. 1983. 10th Int. Congr.Soc. Forensic Haematogenetics, München, FRG, Oct 11-15, Schmidt und Meyer, Würzburg, pp 135-140.
15. Cartron J.P., Tounkara A., Prou O., Luilier M., Soulier J.P. 1984. Lancet **ii**, 466.
16. Hermentin P., Enders B., Neunziger G., Dahr W., Wilkinson S., Miller L.H., Pasvol G. 1984. Lancet **ii**, 466.
17. Facer C.A., Mitchell G.H. 1984. Lancet **ii**, 758.
18. Mitchell G.H., Hadley T.J., Mc Ginnis M.H., Klotz F.N., Miller L.H. 1986. Blood **67**, 1519.

STRUCTURAL REQUIREMENTS FOR DESENSITIZATION OF SMOOTH MUSCLE TO ANGIOTENSIN II

M.E.M. Oshiro, T.B. Paiva, C.R. Nakaie , A.C.M. Paiva
Department of Biophysics, Escola Paulista de Medicina
04034 São Paulo, SP, Brazil

Introduction

The spasmogenic effect of angiotensin II (Asp-Arg-Val-Tyr-Ile-His-Pro-Phe) (ANG II) upon smooth muscles is characterized by an acute agonist-induced desensitization, also known as tachyphylaxis. This desensitization is specific and probably involves changes at the receptor level, whose study may contribute to the understanding of the mechanism of action of the hormone. Previous studies of the structure-desensitization relationships have established that the protonated N-terminal amino group, the guanidinium group of the Arg side-chain (1, 2) and the phenolic hydroxyl of the Tyr side-chain (3) are essential for the ability of the hormone to induce desensitization. On the other hand, we have obtained evidence indicating that changes in sodium transport may participate in the mechanism of the desensitization, and found that the behaviour of tissues desensitized by ANG II is similar to that of those treated with amiloride, an inhibitor of sodium transport (unpublished results). The fact that a protonated guanidino moiety in amiloride is essential for its sodium-blocking activity (4) has suggested to us that the guanidino group of ANG II may interact with sites on the receptor that are important for sodium translocation in the smooth muscle cell membrane. With this hypothesis in mind, we have synthesized ANG II analogues with modified guanidino groups and studied their biological activities and their desensitizing potencies, using the longitudinal smooth muscle of the guinea-pig ileum for the biological assays. We have also synthesized ANG II analogues with modifications in position 4 and investigated their desensitizing properties.

Peptides 1986

Methods

ANG II and its analogues were sўnthesized by the solid phase method (5), except for |Asn^1,Phe^4|ANG II which was a gift from Dr. R. Schwyzer, Ciba Ltd., Basel, and for |4-(3-benzyltyrosine)|-ANG II (|Tyr(Bzl)^4|ANG II) whose obtention has been described earlier (6).

Agonistic activity of the peptides was determined in the isolated guinea-pig ileum preparation (1) by matching the contractile responses to single and double doses of analogue and ANG II, and the analogue´s potency was expressed as a percentage of the ANG II activity. Desensitization potency was determined by a standard treatment with four successive administrations (90 s contacts at 5 min intervals) of supramaximally effective doses (10 x ED_{50}) and was expressed as the percent decrease in the amplitude of the fourth response relative to the first.

Results and Discussion

Position-2 analogues

Five analogues with modifications in the arginine side-chain were synthesized and tested for their agonistic activity and desensitizing potency (Table 1, compounds 2-6). Replacement of the guanidino group by an amino (compound 2) or an aminomethyl (compound 3) group greatly reduced the agonistic activity and the resulting analogues were unable to induce desensitization in the guinea-pig ileum, even after further replacement of the N-terminal aspartyl by a sarcosyl residue (compound 4), a change that greatly increases the desensitizing potency of ANG II (7). Thus, the presence of a guanidino group at the position 2 side-chain is a requisite for the desensitizing property. However, the two modifications of the guanidino group in compounds 5 and 6 (Table 1), besides having a smaller effect on ANG II´s agonistic activity, had different effects on the desensitizing potency: the nitro derivative (compound 5) did not induce desensitization whereas the tosyl derivative (compound 6) was the most potent desensitizer among the analogues studied. This parallels the finding that aromatic substitutions on the guanidino group of amiloride increase its sodium-blocking activity (4), and is suggestive of a role of ANG II´s Arg side-chain in its interaction with sodium transport processes.

Position-4 analogues

Study of the five analogues modified at position 4 (Table 1, compounds 7-11) showed that removal of the phenolic hydroxyl group (compounds 7, 8), or its replacement by a nitro group (compound 9), or its blocking by a methyl group (compound 11), or introduction of a bulky substitution at the aromatic ring of the

Table 1. Agonistic activity and desensitizing potency of ANG II analogues in the isolated guinea-pig ileum

Compound	Agonistic activity[a]	Desensitizing potency[b]
1. ANG II	100	31 ± 5 (20)
2. $\|Orn^2\|$ANG II	3 ± 1 (3)	-5 ± 3 (5)
3. $\|Lys^2\|$ANG II	6 ± 2 (4)	-4 ± 4 (5)
4. $\|Sar^1,Lys^2\|$ANG II	29 ± 4 (6)	0 ± 3 (4)
5. $\|Arg(NO_2)^2\|$ANG II	30 ± 2 (5)	-10 ± 6 (8)
6. $\|Arg(Tosyl)^2\|$ANG II	16 ± 2 (5)	40 ± 5 (7)
7. $\|Phe^4\|$ANG II	3.9 ± 0.3 (5)	-10 ± 5 (5)
8. $\|Asn^1,Phe^4\|$ANG II	0.11 ± 0.02 (4)	-7 ± 4 (5)
9. $\|Phe(p\text{-}NO_2)^4\|$ANG II	0.21 ± 0.05 (5)	5 ± 3 (5)
10. $\|Tyr(Bzl)^4\|$ANG II	0.6 ± 0.1 (5)	25 ± 3 (5)
11. $\|Tyr(O\text{-}Me)^4\|$ANG II	1.4 ± 0.1 (5)	17 ± 4 (5)

[a]Expressed as percentage of ANG II activity; [b]Percentage decrease in response after 4 successive standard treatments. Negative values indicate increased responses

tyrosine side-chain, all reduced biological activity by over 95%. |Phe^4|ANG II, |Asn^1,Phe^4|ANG II and |Phe(p-NO_2)^4|ANG II did not induce significant desensitization, whereas |Tyr(Bzl)^4|ANG II and |Tyr(O-Me)^4| ANG II were moderate desensitizers. This indicates that the presence of the phenolic hydroxyl oxygen may be important for the manifestation of the desensitization, and that the receptor site responsible for this phenomenon may accommodate bulky substituents in the position-4 side-chain of the hormone.

The finding that angiotensin desensitization of smooth muscle is dependent on the structure of the hormone, and is not correlated with the agonistic potencies of the analogues, indicate that the desensitization is due to changes that occur at the level of the hormone-receptor interaction and not at later steps in the stimulus-response chain of events.

References

1. Miasiro, N., M.E.M. Oshiro, T.B. Paiva, A.C.M. Paiva. 1983. Eur. J. Pharmacol. 87 , 397-406.

2. Oshiro, M.E.M., N. Miasiro, T.B. Paiva, A.C.M. Paiva. 1984. Blood Vessels 21 , 72-79.

3. Moore, G.J., K.J. Franklin, D.M. Nystrom, M.H. Gogari. 1985. Can. J. Physiol. Pharmacol. 63 , 966-971.

4. Cuthbert, A.W., G.M. Fanelli. 1978. Br. J. Pharmacol. 63 , 139-149.

5. Barany, G., M.R.B. Merrifield. 1980. In: The Peptides, vol. 2 (E. Gross and J. Meienhoffer, eds.). Academic Press, New York, pp. 3-254.

6. Nakaie, C.R., M.E.M. Oshiro, G. Goissis, A.C.M. Paiva. 1977. Biochim. Biophys. Acta. 495 , 151-158.

7. Paiva, T.B., M.E. Miyamoto, L. Juliano, A.C.M. Paiva. 1977. J. Pharmacol. Exper. Therap. 202 , 294-300.

STRUCTURE AKTIVITY RELATIONSHIP ON NEW PEPTIDE INHIBITORS OF THE ANGIOTENSIN CONVERTING ENZYME

S. Reißmann, C. Schwuchow, H. Arold

Friedrich-Schiller-University, Department of Biology, DDR - 6900 Jena, Goetheallee 26

M.P. Filatova, N.A. Krit

Institute of Biological and Medicinal Chemistry, AMN, USSR - Moscow, Pogodinskaja Ul. 10

W.-E. Siems, G. Heder

Institute of Drug Research, Academy of Sciences, DDR - 1136 Berlin-Friedrichsfelde, A.-Kowalke-Str. 4

I. Paegelow

Wilhelm-Pieck-University, Institute of Pharmacology and Toxikology, DDR - 2500 Rostock, Leninallee 70

Introduction

For further investigations in the renin-angiotensin system we have studied the influence of modifications and substitutions in different angiotensin-converting enzyme inhibitors. For this purpose are synthesized analogues of the bradykinin potentiating peptides $BPP_{9\alpha}$ and $BPP_{5\alpha}$, their partial sequences and acylated tripeptides with nonproteinogenic amino acids.

Results

The syntheses of the nona-, penta- and tripeptide analogues were carried out by conventional methods. Studies on the structure-activity relationship show that variations in the side chains of the nonapeptide $BPP_{9\alpha}$ Pyr-Trp-Pro-Arg-Pro-Gln-Ile-Pro-Pro have only a slight influence on the biological activity (1).

The most striking finding is the influence of position 1 (table 1).

Peptides 1986

Table 1 Role of the N-Terminus

No	Compound	Inhibition of the ACE	Potentiation of the bradykinin action on GPI (perc.)
1	$BPP_{9\alpha}$	$3 \cdot 10^{-9}$ M	100
2	[1-Pro]-$BPP_{9\alpha}$	$5 \cdot 10^{-10}$ M	115
3	HPP-[1-Pro]-$BPP_{9\alpha}$	n.d.	150 - 300

The analogues with Pro in position 1 is 6 times as active than BPP . Replacement of Pyr by Gln or L- and D-hydroproline reduces the potency. Acylations of the N-terminus decrease the activity in the most cases. Only the p-hydroxy-phenyl-propionyl-residue (HPP; compound 3) increases the potentiating activity. The importance of the N-terminal sequence was supported by the high inhibitory activity of the [2-desPro]-bradykinin (3). In this analogues the modification in the N-terminal sequence enhances the activity from $2{,}7 \cdot 10^{-6}$ to $4{,}5 \cdot 10^{-12}$M. The finding about the important role of the N-terminal sequence contrasts with the assumed model for the active centre of the ACE. The computed (2) and estimated (1) folded structure of the BPP seems to contrast, too, but allowed to explain the role of the N-terminal sequence for the biological activity.

The inhibition of the ACE by partial sequences of the BPP has us encouraged to discover the potency of free and acylated tripeptides. One of the key ideas in this field was the use of nonproteinogenic amino acids for stabilizing the peptides against enzymatic degradation. We have used D-, N-methyl-, β-substituted and hydrophobic amino acids (figure 1).

Combination with acyl residues enhances or reduces the inhibitory activity of the tripeptides. We used bulky and hydrophobic carbonic acids or dicarbonic acids e.g. adamantane-1-carbonic-, benzoic-, p-hydroxyphenyl-propionic-, phthalic- and norbornene-dicarbonic acid.

D-Amino acids: D-Phe, D-Ala, D-Trp, D-Val

N-Methyl amino acids: NMePhe, NMeAla

β-substituted amino acids: DL-βPhabu DL-εPhser DL-ε-βHyNleu Ahc

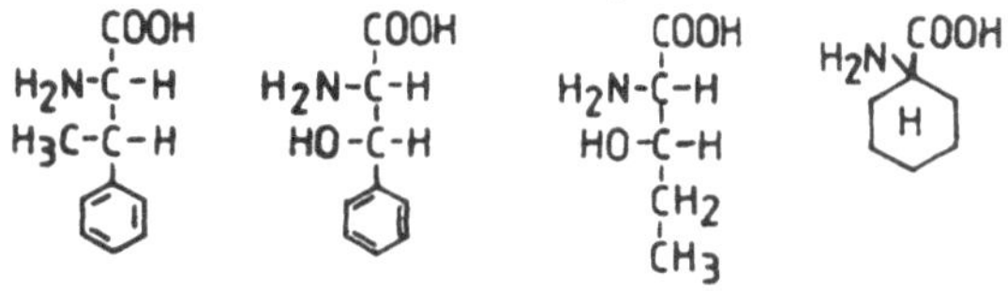

Modified aromatic amino acids: DMePhe

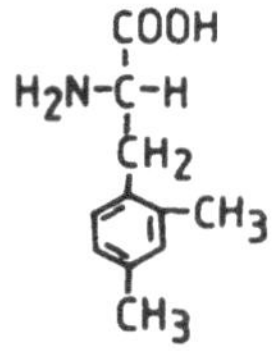

Substitutions in the penultimate position at the C-terminus lowers as well in the nona- and pentapeptide as in tripeptides the inhibitory activity (table 2).

Table 2 Substitutions in the Position (X)- Pro

No.	Compound		IC_{50} (μM)
1	Pyr-Trp-Pro-Arg-Pro-Gln-Ile-Pro-Pro	BPF$_{9\alpha}$	0,003
2	Pyr ——— Ala ——		0,55
3	Pyr ——— NMeAla ——		0,6
4	Pyr-Lys-Trp-Ala-Pro	BPF$_{5\alpha}$	0,5
5	Pyr ——— DAla ——		20
6	BZ-Phe-Ala-Pro		2,8
7	BZ — NMeAla ——		inactive
8	BZ —— Arg ——		20
9	BZ-DL-Cys ——		inactive
10	BZ-DL-Cys ——		

In the prepenultimate position amino acid exchanges are allowed (table 3).

Table 3 Substitutions in the Position (x) -AS-Pro

No.	Compound		IC_{50} (µM)
1	Pyr-Trp-Pro-Arg-Pro-Gln-Ile-Pro-Pro	$BPP_{9\alpha}$	0,003
2	Pyr ——— Ile ———		0,010
3	BZ-Phe-Ala-Pro		2,8
4	BZ-NMePhe ———		50
5	BZ-DPhe ———		50
6	BZ-βPhabu ———		12
7	BZ-βPhser ———		25
8	BZ-Cha ———		20
9	BZ-Trp ———		5,5
10	BZ-HyNleu ———		7

The best results we could obtain by the combination of tripeptides, containing hydrophilic side chains, with aromatic acyl residues (4). The combination of hydrophobic peptides with hydrophobic acyl residues leads to inactive compounds. The potency of the acylated tripeptides results from the binding of different structural elements to the enzyme. There exist no general influence of an acyl residue on different tripeptides.

Some of the synthesized tripeptides are tested on normotensive rats by intravenois or oral application. The compounds 5, 6, 7 and 10 reduce the angiotensin-I- induced blood pressure enhancement. We found, that only acylated tripeptides are active by oral application. The duration of the ACE-inhibition differs from 20 min to some hours. There is no good correlation between the inhibition of the isolated enzyme and the oral activity (4). It seems to be commonly assumed, that the potentiation of the bradykinin action results from the inhibition of the bradykinin inactivation. But there exist some findings from our own studies and from the literature, which disagree with this explanation. Thus the bradykinin potentiating activities and the inhibitory potencies differ in some cases more than in 1 or 2 orders of magnitude. These results indicate different structural requirements for these two kinds of biological activity. From our preliminary binding studies we might draw the conclusion, that the potentiation of the bradykinin action on the smooth muscle is to some extent based on a increased receptor affinity to bradykinin.

Abbrevations

HPP p-hydroxyphenyl-propionyl-, BZ benzoyl-, NMePhe L-N-methyl-phenylalanine, Cha cyclohexyl-alanine, βPhser DL-erythro-β-phenylserine, βPhabu DL-erythro-α-amino-β-phenylbutyric acid, HyNle DL-erythro-β-hydroxy-norleucine

References

1. Reißmann, S., M.P. Filatova, N.A. Krit, T.A. Aleksandrova, E. Birckner, M. Friedrich, I. Fric, I. Paegelow, W.-E. Siems, G. Heder, H. Arold: 1985. In: Proceedings of the 16th FEBS Congress, Part B. VNU Science Press. pp. 449-455.

2. Sevastjanova, N.N., G.M. Lipkind, S.F. Arkhipova, E.M. Popov. 1977. Bioorg. Khim. 3, 473.

3. Okuda, M., K. Arakawa. 1985. J. Biochem. 98, 621.

4. Reißmann, S.: Pharmazie (in preparation)

FURTHER SYNTHETIC STUDIES ON POSITION 1 OF ANGIOTENSIN II

P.Cordopatis and D.Theodoropoulos
Laboratory of Organic Chemistry, Department of Chemistry, University of Patras, 26200 Greece

Introduction

That the amino and carboxyl groups of the aspartyl residue in angiotensin II, Asp-Arg-Val-Tyr-Ile-His-Pro-Phe, are not essential for pressor activity although they do possess features which are necessary for maximal activity has been supported from our earlier work from this laboratory (1,2). Thus,the replacement of the 1-Asp with α-aminoisobutyric acid (Aib) gave analogue with 3.2±1.3 fold higher potency than Ciba-Hypertensin but ~60% relative to [Asp[1]]-AII (3). On the other hand [β-Malyl[1]]-AII, in which the N-terminal amino group was replaced by a hydroxyl group, retain ~45% of the pressor activity of the parent compound (4). An explanation for the loss of activity would be a direct steric interference of the modification with the receptor which might impaire ideal receptor binding of the analogue and thus reduce its potency. Also, these findings further confirm the results of previous investigations (5,6) on the importance of the N-terminal α-nitrogen and its basicity for maximum agonistic properties of angiotensin II and analogues.

In an extention of our previous work on position 1 of angiotensin II the analogue

I. [N-Methyl-Aib[1]]-angiotensin II

was synthesized by the solid-phase method, in order to elucidate further the influence of the α-carbon substitution with simultaneously increase of basicity of the α-amino group on position 1. Furthermore, we have prepared the analogue

II. [L-DOPA[1]]-angiotensin II

in which the aspartic acid residue at position 1 has been replaced by L-3,4-dihydroxyphenylalanine. The effects of these changes on

Peptides 1986

characteristic biological properties of angiotensin II are present in Table 1.

Experimental Section

The peptides were prepared by the solid-phase method (7). The protocol used for the synthesis was the same as previously described (3,4). N-Me-Aib, DCCI and HOBt (4 molar excess, 1:1:1) in CH_2Cl_2: DMF (4:1) were shaken with the resin overnight until a negative ninhydrin test was obtained. For the incorporation of L-DOPA coupling was performed with 3 equiv. of Boc-L-DOPA, DCCI and HOBt for 5h. The following side-chain protected amino acid derivatives were used: Boc-His(Tos), Boc-Tyr(2-Br-Z), Boc-Arg(NO_2). Boc-L-DOPA was prepared according to Banerjee and Ressler (8) from L-DOPA-OCH_3·HCl which was converted to Boc-L-DOPA-OCH_3 by the method of Nagasawa et al. (9).Cleavage of the peptides from the resin and removal of the protecting groups was accomplished with HF containing 10% by volume of anisole for 1h at 0^oC. After removal of HF and drying under vacuum the resin was washed several times with ether and then was extracted with acetic acid (2M). Lyophilization of the latter extract yielded the crude peptide. Both analogues were purified by gel chromatography on sephadex G-15 using 0.2N acetic acid as the eluent, followed by preparative HPLC on C-18 with gradient 5-35% CH_3CN in 0.05% TFA-water.

The compounds were tested on rabbit aorta strips, liver membranes and in the rat blood-pressure test (in vivo) in nephrectomized, male rats anesthetized with urethane. The results of the three biological assays are present in Table 1.

Results-Conclusions

The replacement of aspartic acid in position 1 of angiotensin II with N-methyl-Aib or L-DOPA significantly reduces the agonistic potency of the analogues in vivo and rather drastically in vitro (Table 1). More importantly, substitution of Aib with N-methyl-Aib in position 1 of [Aib^1]-AII gives analogue with 4 times lower in vivo (rat blood pressure) activity.

Thus it appears, at present, that:

Table 1: Comparison of the Biological Activities of Angiotensin II and Its 1-Position Analogues on Three Pharmacological Preparations[α].

Peptide	Rabbit Aorta				Liver Membranes			Blood Pressure	
	PD_2	α_E	R.A. (%)	n	EC_{50}	R.A. (%)	n	RBP (%)	n
Angiotensin II	7.976±0.149	1.0	100	5	$2.1 \cdot 10^{-9}$	100	3	100	3
[DOPA[1]]-AII	6.7905±0.1845	1.0	6.5	5	$5.9 \cdot 10^{-8}$	35.6	3	20	3
[N-Me-Aib[1]]-AII	6.4305±0.1755	1.0	2.8	5	$4.3 \cdot 10^{-8}$	4.9	3	13	3

[α] Key: PD_2 is the negative log. of the dose that produces half maximal contraction; α_E is the intrinsic activity; R.A. is the relative affinity. EC_{50} is the dose that displaces half of the radioactive label [Sar[1], ^{125}I Tyr[4]]-AII; RBP is the relative blood pressure potency and n is the number of experiences.

a. An increase in basicity at the N-terminal nitrogen atom of angiotensin II is not the only reason for increased potency (6).

b. Bulky groups in position 1 may affect the spatial orientation of essential residues like 3-Val and 4-Tyr. This in turn could produce changes in the preferred backbone conformation of the hormone and explain differences in biological activities observed for the two analogues.

Acknowledgement

We wish to express our appreciation to Prof. E.Escher, University of Sherbrooke, Canada, for the biological assays.

References

1. Cordopatis, P., J.Brison, J.Dirkx, D.Theodoropoulos. 1979. Experientia 35, 849.

2. Cordopatis, P., D.Theodoropoulos, G.Guillemette, E.Escher. 1981. J.Med.Chem. 24, 209.

3. Cordopatis, P., D.Theodoropoulos. 1983. Experientia 39, 106.

4. Cordopatis, P., J.Matsoukas, A.Michel, J.Janen, D.Theodoropoulos. 1983. Experientia 39, 353.

5. Regoli, D., W.K.Park, F.Rioux. 1974. Pharmac.Rev. 26, 69.

6. Khosla, M.C., M.M.Hall, R.R.Smeby, F.M.Bumpus. 1974. J.Med.Chem. 17, 1156.

7. Merrifield, R.B. 1969. J.Am.Chem.Soc. 85, 2149.

8. Banerjee, S.N., C.Ressler. 1976. J.Org.Chem. 41, 3056.

9. Nagasawa, T., K.Kuroiwa, K.Narita, Y.Isowa. 1973. Bull.Chem.Soc. Jap. 46, 1269.

CONFIRMATION OF DIMERIC STRUCTURE OF HUMAN BETA-ATRIAL NATRIURETIC POLYPEPTIDE (β-hANP) BY SYNTHESIS

Tamayo Hatoh, Kenji Kangawa and Hisayuki Matsuo

Department of Biochemistry, Miyazaki Medical College, Kiyotake, Miyazaki 889-16, JAPAN

Atrial natriuretic polypeptide (ANP) is a hormone synthesized in and secreted from mammalian atrium, eliciting a key role in the regulation of extracellular fluid volume and electrolyte balance. In human atrial tissue, we have identified three distinct but related peptides, named α-, β- and γ-hANP, all of which exhibit natriuretic-diuretic, hypotensive and smooth-muscle relaxant activities: α-hANP is a 28-residue peptide containing an intramolecular disulphide linkage, while γ-hANP of 126 residues is an N-terminally extended peptide of α-hANP, which is directly processed as the largest hANP from a 151-residue precursor(1,2). Along with these ANPs, β-hANP comprising 56 residues has been identified as an anti-parallel dimeric form of α-hANP(3). Interestingly, such a dimeric ANP has not so far been found in the mammals other than human. Identification of β-hANP in human atria raised several questions as to how the peptide is synthesized in vivo, what its physiological significance is, and whether β-hANP is circulating in the blood or not. Synthesis of β-hANP should be very important for providing a tool to answer the above questions. In the present paper, we will report syntheses of parallel and anti-parallel dimers of α-hANP. By comparing the synthesized dimers, identity of the anti-parallel dimer with natural β-hANP was definitely verified.

Results and discussion

<u>Syntheses of anti-parallel and parallel dimers of α-hANP</u>: Many attempts for synthesizing β-hANP by direct dimerization of reduced

Peptides 1986

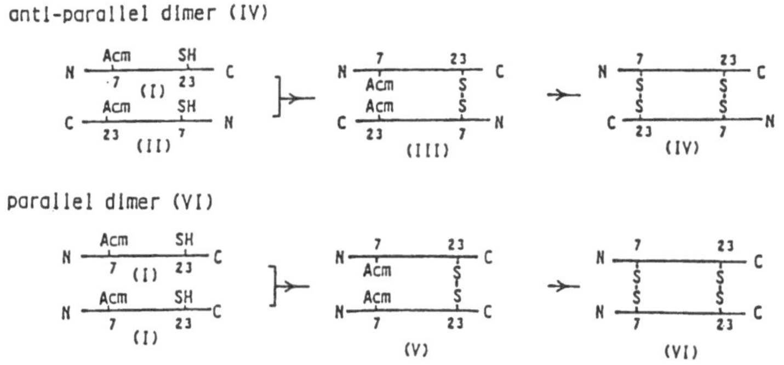

Fig. 1. Synthetic routes

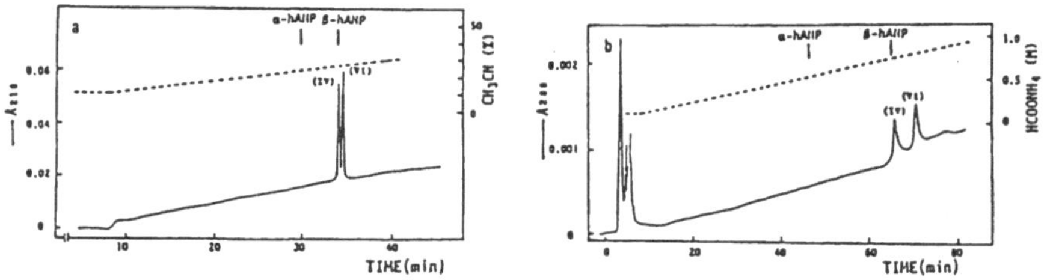

Fig. 2.

a) RP-HPLC of purified dimers (IV and VI)
Sample size: 1 µg each. Column: TSK LS410 ODS (4x150 mm: Toyosoda) Flow rate: 1.5 ml/min. Solvent system: linear gradient from A to B for 80 min. H_2O: CH_3CN: 10% TFA = A) 90:10: 1; B) 40:60: 1

b) Cation-exchange HPLC of purofoed dimers (IV and VI)
Sample size: 3 µg each. Column: TSK CM-2SW(4.6x 250 mm: Toyosoda) Flow rate: 1ml/min. Solvent system: linear gradient from A to B for 80 min. A) 10 mM $HCOONH_4$(pH 6.6) : CH_3CN = 90:10, B) 1M $HCOONH_4$(pH 6.6): CH_3CN = 90:10

α-hANP were made under various conditions but in vain, since reduced α-hANP was surprisingly easy to regenerate α-hANP under the conditions for disulphide formation. This fact suggests that two Cys residues in a reduced form of α-hANP may possess a favorable conformation for the formation of an intramolecular disulphide linkage. After all, syntheses of anti-parallel and parallel dimers of α-hANP have been accomplished by stepwise formation of two sets of disulphide crosslinking, starting from peptide I, [Cys(Acm)-7, Cys(SH)-23]-α-hANP and peptide II, [Cys(SH)-7', Cys(Acm)-23']-α-hANP, as shown in Fig. 1. By air-oxidation of an equimolar mixture of peptide I and II, a heterodimer (peptide III) leading to the anti-parallel dimer(β-hANP), which possesses a disulphide linkage between peptide I and II was produced as a major product in a yield of 60%, predominant over homodimers derived solely from either peptide I or peptide II. A disulphide linkage newly formed in peptide III was identified by trypsinization yielding a tryptic fragment peptide with a Cys-7-Cys-23'disulphide bond. Treatment of the heterodimer (III) thus obtained with iodine smoothly resulted into the formation of the second set of disulphide linkage between Cys-23 and Cys-7' to give an anti-parallel dimer of α-hANP (peptide IV) corresponding to the structure proposed for natural β-hANP.

On the other hand, an alternative parallel dimer of α-hANP was also synthesized similarly by stepwise disulphide formation, starting solely from peptide I, via a homodimer V. Note that conversion of homodimer V to a final parallel dimer VI by the action of iodine was accomplished only in the high dilution conditions, differently from the case of heterodimer III.

Structural confirmation: The anti-parallel and parallel dimers of α-hANP thus synthesized, are quite distinguishable from each other in their chromatographic behaviours (Fig. 2). Conclusive evidence for the structural difference between both dimers was obtained by comparing their tryptic peptide maps prepared from each peptide. An anti-parallel dimer by trypsinization produces a single peptide having a Cys-7-Cys-23' disulphide bond, along with four other peptides. On the other hand, trypsinization of a parallel dimer produced two distinct disulphide containing peptides, having a Cys-7-Cys-7' bond and a Cys-23-Cys-23' linkage, respectively, along with four other peptides, all of which were identical with those obtained from an anti-parallel dimer. Thus, it has been unambiguously verified that peptide IV is an anti-parallel dimer distinct from a parallel dimer VI.

Identity of anti-parallel dimer with natural β-hANP: Synthetic anti-parallel peptide IV is chromatographically identical with natural β-hANP, while parallel peptide VI is not (Fig. 2). And, a peptide map prepared from synthetic peptide IV by trypsin digest, was proved to be quite superimposable with that from natural β-hANP. Thus, it has been proven that an anti-parallel dimer(IV) of α-hANP is exactly identical with natural β-hANP.

Biological Activities: Natriuretic-diuretic, hypotensive and chick rectum relaxant activities of the synthetic anti-parallel dimer is in a good agreement with those of natural β-hANP. Interestingly, bioactivities of a parallel dimer VI was found to be not so different from those of an anti-parallel dimer, even though they are about one fourth as potent as α-hANP. This fact suggests that both dimers retain a structural conformation resembling the ring-structure of α-hANP which is thought to be essential for activity.

References

1 Kangawa, K., H. Matsuo. 1984. Biochem. Biophys. Res. Commun. 118, 131.

2 Kangawa, K., A. Fukuda, H. Matsuo. 1985. Nature 313, 397.

3 Oikawa, S., M. Imai, A. Ueno, S. Tanaka, T. Noguchi, H. Nakazato, K. Kangawa, A. Fukuda, H. Matsuo. 1984. Nature 309, 724.

SMALL PEPTIDES MAY BE TISSUE SPECIFIC MITOSIS INHIBITORS INDUCING TERMINAL DIFFERENTIATION.

K.L. Reichelt, K. Elgjo, J.E. Paulsen and Ø. Skraastad.

Pediatric Research Institute and Dept. of Pathology, Rikshospitalet, 0027 Oslo 1, Norway.

Introduction

It is remarkable that tissues stop growing when of correct genetic and functional size. This effect is apparantly caused by paracrine action for each specific tissue (1,2). We have isolated several small MW peptides from various tissues that cause a) Mitosis inhibition, b) Decreased Thymidine incorporation and c) Terminal differentiation (3,4).

Methods

Tissues were extracted by cold milling as described (3) and freeze dried. The resultant powder if active, was treated to 1M acetic acid, centrifuged at 5000 g 30 min at 0°C and filtered and purified sequentially as follows:

a) Sephadex G-25 superfine gel 1.6 x 09 cm in 0.5 M acetic acid. application volume 10 ml and rate of eluation 3 ml/10 min. This step seperates macromolecules (carriers) from oligo peptides and amino acids.

b) Active material from the previous run as dissolved in 1M acetic acid 40 mm HCl and run and Fractogel MG 2000 in the same buffer. This column has dimensions 1.6 x 80 and has very powerful peptide separating ability. Application volume 6 ml and rate of eluation 3 ml/10 min. UV 254 and 280 were continuously monitored by Isco UA 5 machine and offline ninhydrin (see later).

c) CG 50 in the H^+ form of dimensions 1 x 20 cm or also the powerful cation exchanger Dow 50 in the H^+ form was used. Eluted

Peptides 1986

stepwise in 40 ml H_2O and 40 ml 4 M formic acid. The active compound eluted from 5 - 15 ml H_2O on both cation exchangers and must be cyclic or N substituted. Lack of ninhydrin colour without hydrolysis likewise.

d) HPLC was run on an LCD apparatus using Partisil M9 10/25 ODS in system 1: 20 mM TFA pH 2 with 3% by volume n-propanol was run isocratically for 16 minutes and a 15 min gradient from 3 to 48% n propanol. UV at 205 nm was read and the rate of eluation was 1 ml/min. In system 2:20 mM TFA pH 2 with 9% propanol was run as gradient against 42% n propanol after a 10 min. isocratic run. UV205 nm was used to register the peptide peaks.

The amounts of peptides were anlyzed by drying 10% aliquots of each fraction form each column overnight, hydrolyzing in 2 M KOH for 2 h, and neutralizing with HCL. The ninhydrin colour was then developed in a cyanide/acetate buffer which assures close to equimolecular absorption coefficient (5).

Structural studies: The N terminal pyroglu amino acid was split off by pyroglu aminopeptidase and also hydrolysis in 4M HCL in waterfree methanol (3). Total amino acid composition was analyzed by Chromaspeck amino acid analysis. The 2nd and 3rd amino acids determined by Edman/dansylation technique.

Results

Fig. 1 illustrates the difference for three tissues on Primary G 25 fractionation. It is clear that the activity is low MW. If run at neutral pH the activity is found in association with several protein components (3). Fig. 2 shows the fractogel pattern after cation and late elution HPLC. The peptide was purified until HPLC wise pure and with an constant amino acid ratio. Only 1 N-terminal and 1 C-terminal amino acid. Custom synthesis was kindly performed at Peninsula lab. and complete identity was assured functionally and chemically of a pentapeptide Pyroglu-glu-Asp-ser-glyoH (6). The liver chalone is a different pentapeptide and the colon peptide is a tripeptide (data to be published).

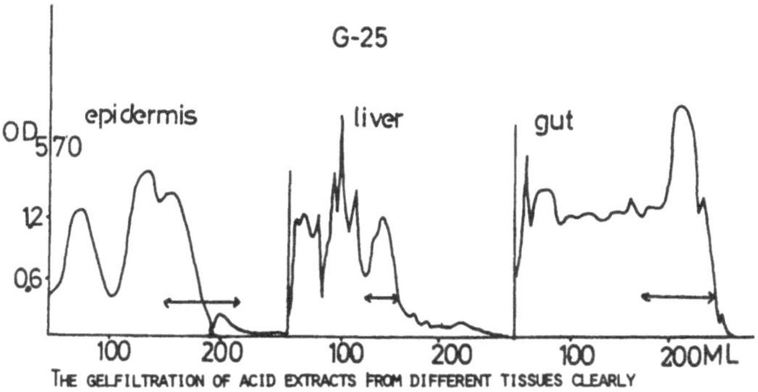

THE GELFILTRATION OF ACID EXTRACTS FROM DIFFERENT TISSUES CLEARLY INDICATES THE DIFFERENT TISSUE PROFILES .

FIG 1

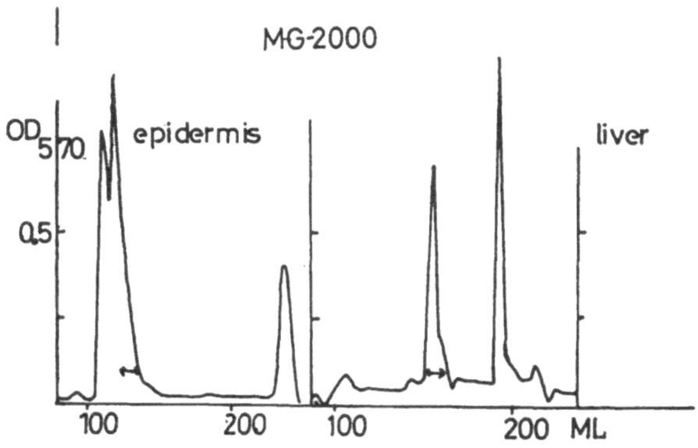

FIG 2 .FRACTOGEL PATTERN AFTER G -25 , DOW -50 CATION , AND 3- 42% N PROAPNOL C 18 REVERSE PHASE HPLC IN 20MM TFA PH 2. THE ACID CONDITIONS HERE USED 8 I M ACETIC ACID , 40 MM HCL) ENSURES PEPTIDE DISSOCIATION.

FIG 2

Table 1. Effect on primary cultures of repeated treatments with peptide

Peptide concentration (M)	Mitotic rate per 1,000 cells/3 h	Effect
10^{-10}	10.7±2.9 (3)	-33%*
10^{-12}	8.5±1.9 (6)	-45%*
Controls	15.5±1.2 (6)	

Cells were grown in high calcium (1.2 mM Ca^{2+}) medium for 3 days. Peptide at the indicated doses was added at t=0, t=1, and t=2 h. VLB was added together with the first dose of peptide. The cells were fixed at 3 h. Number of dishes given in parenthesis. * = $p < 0.05$.

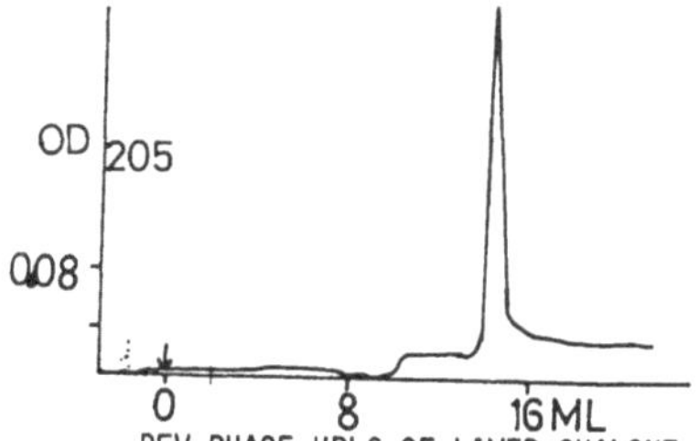

REV. PHASE HPLC OF LIVER CHALONE. 20mM TFA 12% N PROPANOL.

Table 2. Effects of the isolated hepatic peptide on DNA synthesis and mitosis after partial hepatectomy of hairless mice.

Peptide	% of controls					
pmol	SA[a]	LI[b]	Mit[c]	weight[d]	SA/LI	AS[e]
1	44±10	33±10	6± 6	86± 3	147±35	82±10
10	36± 7	23± 6	18± 7	93± 5	118±19	107±13
100	57±18	37± 7	25±10	78± 6	137±24	102±16
1 000	57±16	56±10	42±18	89± 4	88±13	101±16

The peptide was injected at 28 h and the animals were killed at 48 h after partial hepatectomy. The control levels of SA, LI, Mit, weight, SA/LI and AS were 122 ± 6 cpm/ug DNA, -56 ± 32 per 1 000 cells, 55 ± 56 per 20 vision fields, 537 ± 32 mg wet weight, 2.41 - 0.63 and 1226 ± 348 cpm/ug DNA respectively. N = 7 for the controls and N = 4 for the experimental groups.

a) Specific activity of DNA, ^3H-thymidine incorporated (acid-insoluble)
b) Labilling index
c) Colcemid arrested mitoses which lasts four hours
d) Liver weight at sacrifice
e) Acid-soluble activity of ^3H-thymidine

The peptide inhibits mitosis of epithelial cells and causes terminal differentation measured by cornified envelopes (4). In table 1 the action is specific for cornified epithelial cells. The peptide also inhibits the thymidine incorportaion in cells in vivo also when induced to growth (6). The thymidine incorporation is also decreased exemplified for liver (table 2) and this in spite of a post-hepatectomy surge (table 2) (7). The epidermal peptide is also active against human cells in culture (8).

References

1. Bullough, W.S. 1962. The control of mitotic activity in adult mammalian tissues. Biol. Rev. 37, 307-342.

2. Iversen, O.H. 1981. The chalones handbook of experimental pharmacology 57, 492-550.

3. Elgjo, K. and Reichelt, K.L., 1984. Purification and characterization of a mitosis inhibiting epidermal peptide. Cell. Biol. Int. Rep. 8, 379-382.

4. Elgjo, K., Reichelt, K.L., Hennings, H., Michael, D. and Yuspa, S.H. 1986. Purified epidermal oligopeptide inhibits proliferation and enchanges terminal differentiation in cultured mouse epidermal cells. J. Inest. Dermat. in press.

5. Reichelt, K.L. and Kvamme, E. 1973. Histoamine dependent formation of N-acetyl-aspartyl peptides in mouse brain. J. Nevrochem. 21, 849-859.

6. Elgjo, K., Edminson, P.D. and Reichelt, K.L. 1986. Endogenous peptide sin epidermal growth control in biological regulation of cell proliferation. Plenum Press, N.Y., in press.

7. Paulsen, J.E. and Reichelt, K.L. 1986. Purification and characterization of a growth inhibitory hepatic. Peptide, submitted.

8. Jensen, P.K.A. and Bolund, L. 1986. Changes in proliferating cell subpopulations and mitotic activity in human epidermal cultures treated with epithelal growth inhibitors. J. Invest. Derm. 86, 46-50.

SYNTHESIS AND BIOLOGICAL PROPERTIES OF ANTIPARALLEL AND PARALLEL DIMERS OF α-HUMAN ATRIAL NATRIURETIC PEPTIDE

Naoyoshi Chino, Kumiko Yoshizawa-Kumagaye, Yuko Noda, Takushi X. Watanabe, Terutoshi Kimura, and Shumpei Sakakibara

Peptide Institute Inc., Protein Research Foundation, Minoh-shi Osaka 562, Japan

Introduction

Three types of natriuretic peptides (α-, β- and γ-hANPs) were isolated from human atrial tissue by Kangawa et al. in 1984, and α-hANP was found to be a 28-peptide with an intramolecular disulfide bond (1). Later, they also established that β- and γ-hANPs were 56- and 126-peptides, respectively(2), with γ-hANP being a precursor peptide having the sequence of α-hANP at the C-terminus. The structure of β-hANP was then predicted to be an antiparallel dimer of α-hANP. Kangawa et al. also hypothesized that β-hANP must be an endogenous factor because it is always detectable by HPLC together with α-hANP in the initial extract of human atrial tissues, and no evidence of β-hANP being formed from α-hANP was found under the conditions of its isolation or purification. β-hANP was reported to show similar biological properties to α-hANP, but its onset was a little slower and its duration was longer than α-hANP(2). In the present study, antiparallel and parallel dimers of α-hANP were synthesized by the classical solution procedure to confirm the findings of Kangawa et al. for β-hANP with synthetic materials.

Synthesis of peptides

Antiparallel and parallel dimers of α-hANP were synthesized from two fully protected peptides I and II having the following structure:

Peptides 1986

```
    Bzl      Tos Tos Bzl Bzl R1                 Tos      cHx Tos
Boc-Ser-Leu-Arg-Arg-Ser-Ser-Cys-Phe-Gly-Gly-Arg-Met-Asp-Arg-Ile-

                Bzl             R2      Bzl     Tos BrZ
Gly-Ala-Gln-Ser-Gly-Leu-Gly-Cys-Asn-Ser-Phe-Arg-Tyr-OBzl
```

where R1 and R2 are acetamidomethyl(Acm) and 3-nitro-2-pyridine-sulfenyl(Npys) (3) for peptide I, and 4-methylbenzyl(MeBzl) and Acm for peptide II. Each fully protected peptide was synthesized as reported for the synthesis of α-hANP(4). Deprotection and selective disulfide bond formation were carried out as shown below:

Peptide I and II (1:1)

1. TFA
2. HF/p-cresol
3. NH_4OAc(pH 6.0)

```
SLRRSSCFGGRMDRIGAQSGLGCNSFRY
      |                Acm
      |                Acm
 YRFSNCGLGSQAGIRDMRGGFCSSRRLS
```

1. CMC-chromatography
2. I_2

```
SLRRSSCFGGRMDRIGAQSGLGCNSFRY
      |                |
 YRFSNCGLGSQAGIRDMRGGFCSSRRLS
```

Antiparallel Dimer

Peptide II (2 eq)

1. TFA
2. HF/p-cresol
3. $K_3Fe(CN)_6$

```
SLRRSSCFGGRMDRIGAQSGLGCNSFRY
      |                Acm
      |                Acm
SLRRSSCFGGRMDRIGAQSGLGCNSFRY
```

1. CMC-chromatography
2. I_2

```
SLRRSSCFGGRMDRIGAQSGLGCNSFRY
      |                |
SLRRSSCFGGRMDRIGAQSGLGCNSFRY
```

Parallel Dimer

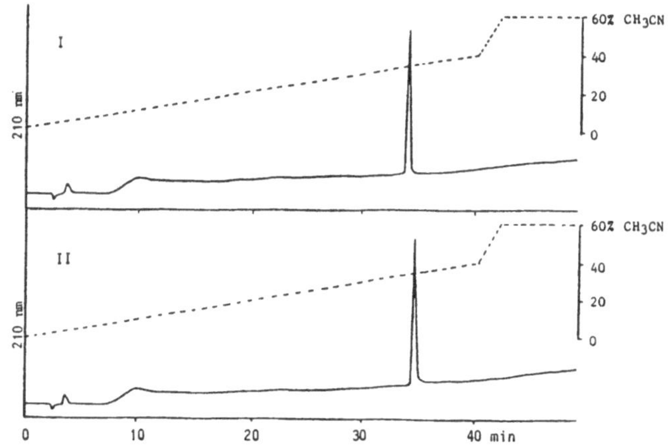

Fig. 1. HPLC profile of synthetic antiparallel and parallel dimers. I, antiparallel dimer; II, parallel dimer. Column: Nucleosil 5C18 (4.6 x 150 mm). Eluant: 0.1% TFA in MeCN/water(see text for gradient). Flow rate: 1 ml/min.

The Npys group was first introduced by Matsueda et al.(3) as a useful protective group for the thiol function of Cys; it is stable in TFA or HF but reacts smoothly with a free SH group to form their mixed disulfide bond. After the dimerization reaction, the major product was isolated to homogeneity by preparative HPLC (Fig. 1). The tryptic peptide mapping of the antiparallel dimer was the same as that of α-hANP, but that of the parallel dimer was quite different. These observations confirmed that our synthetic peptides have the expected structures of antiparallel and parallel dimers of α-hANP. The retention time of the synthetic antiparallel dimer was confirmed to be identical with that of natural β-hANP on HPLC by the group of Matsuo.

Biological Properties

Spasmolytic activities were determined using helical strips of rat thoracic aorta or chick rectum by a modified method of Currie et al.(5) Natriuretic activity was measured in anesthetized male Sprague-Dawley strain rats following the method of de Bold et al. (6). The relative potency of each peptide was determined using α-hANP as a standard. In each spasmolytic measurement, the potency ratio was calculated using the ED_{50} values measured by a test sample and a standard in the same assay preparation. In the natriuretic activity measurement, the potency ratio was calculated using the response of a test sample (1 or 10 nmol/kg, i.v.) and a standard in the same assay preparation. The results are summarized in Table 1.

Conversion Test

The antiparallel dimer(β-hANP) is present in the atrial tissue as an endogenous entity according to Kangawa et al. To confirm that it is not an artifact, the stability of synthetic dimers and α-hANP was examined under various conditions including those for isolating natural β-hANP. We found that at most about 2-3% α-hANP formation from both dimers was detectable, but not even a trace of either dimer was detectable from α-hANP. From dihydro-α-hANP,

Table. 1. Biological Activities of Synthetic Antiparallel and Parallel Dimers of α-hANP.

	Spasmolytic Activity		Natriuretic Activity	
	Chick Rectum	Rat Aorta	1 nmol/Kg	10 nmol/Kg
α-hANP	100[a]	100[b]	100[c]	100[d]
Antiparallel Dimer	34.4 ± 4.9 (6)	16.3 ± 2.6 (6)	18.7 ± 7.0	97.5 ± 18.4 (6)
Parallel Dimer	29.1 ± 3.1 (6)	30.8 ± 2.9 (6)	32.0 ± 11.7	94.6 ± 21.0 (5)

Values are mean ± S.E. Numbers of experiments are in parentheses.
a, $ED_{50} = 3.1 \pm 0.4 \times 10^{-9}M$ (n = 12); b, $ED_{50} = 1.0 \pm 0.1 \times 10^{-9}M$ (n = 12);
c, $\Delta U_{Na}V = 1.4 \pm 0.4$ μEq/min (n = 11); d, $\Delta U_{Na}V = 4.6 \pm 0.9$ μEq/min (n = 11).

however, small quantities of both parallel and antiparallel dimers were detected. From these observations, we concluded that a trace of β-hANP may be formed during the biosynthesis of α-hANP in human cells, but specific formation of a relatively large amount of β-hANP must be due to some enzymatic reactions. The significance of β-hANP in human atrial tissue is yet to be elucidated.

References

1. K. Kangawa and H. Matsuo. Biochem. Biophys. Res. Commun. 1984, 118, 131.
2. K. Kangawa, A. Fukuda, and H. Matsuo. Nature 1985, 313, 397.
3. R. Matsueda, T. Kimura, E.T. Kaiser and G.R. Matsueda. Chem. Lett. 1981, 737.
4. N. Chino, Y. Nishiuchi, Y. Masui, Y. Noda, T.X. Watanabe, T. Kimura, and S. Sakakibara. 1985. In: Peptide Chemistry 1984 (N. Izumiya, ed.). Protein Research Foundation, Mihoh Osaka, pp. 241.
5. H.G. Currie, D.M. Geller, R.B. Cole, J.G. Boylan, W. Yu Sheng, S.W. Holmberg, and P. Needleman. Science 1983, 221, 71.
6. A.J. de Bold, H.B. Borenstein, A.T. Veress, H. Sonnenberg. Life Sci. 1981, 28, 89.

STRUCTURE-ACTIVITY RELATIONSHIP OF LEPIDOPTERAN, SELF-DEFENCE PEPTIDE PRODUCED BY SILKWORM.

T. Shiba, T. Nakai, M. Kitazawa, and T. Teshima
Department of Chemistry, Faculty of Science, Osaka University, Toyonaka, Osaka 560, Japan

Introduction

Silkworm, Bombyx mori, produces an immune responce substance called lepidopteran in its hemolymph when it was vaccinated with killed E. coli (1). Structures of lepidopteran A, B, and C were determined as shown in Fig. 1 (2,3). Lepidopterans possess wide antibacterial spectra against Gram-negative as well as Gram-positive bacteria. Total synthesis of lepidopteran A and its structure-activity relationship will be presented.

	1–19
Lepidopteran A	H-Arg-Trp-Lys-Ile-Phe-Lys-Lys-Ile-Glu-Lys-Met-Gly-Arg-Asn-Ile-Arg-Asp-Gly-Ile-
Lepidopteran B	H-Arg-Trp-Lys-Ile-Phe-Lys-Lys-Ile-Glu-Lys-Met-Gly-Arg-Asn-Ile-Arg-Asp-Gly-Ile-
Lepidopteran C	H-Arg-Trp-Lys-Leu-Phe-Lys-Lys-Ile-Glu-Lys-Val-Gly-Arg-Asn-Val-Arg-Asp-Gly-Leu-

	20–35
A	Val-Lys-Ala-Gly-Pro-Ala-Ile-Glu-Val-Leu-Gly-Ser-Ala-Lys-Ala-Ile-NH_2
B	Val-Hyl-Ala-Gly-Pro-Ala-Ile-Glu-Val-Leu-Gly-Ser-Ala-Lys-Ala-Ile-NH_2
C	Ile-Hyl-Ala-Gly-Pro-Ala-Ile-Ala-Val-Ile-Gly-Gln-Ala-Lys-Ser-Leu-NH_2

Hyl : δ-hydroxylysine

Fig. 1

Total Synthesis of Lepidopteran A

Lepidopteran A was synthesized in order to confirm the proposed

Peptides 1986

structure by the solution method applying maximum protection procedure. Protecting groups of functional groups in side chain were as follows: p-toluenesulfonyl for Arg, formyl for Trp, 2-chlorobenzyloxycarbonyl for Lys, benzyl for Glu and Ser, and cyclohexyl for Asp. Fragment condensations were carried out as shown in Fig. 2 using water-soluble carbodiimide — 1-hydroxybenzotriazole as a condensation reagent. All C-terminal amino acids in fragments were Gly except in fragment (1-6) which has Lys^6 at C-terminus. The content of D-Lys^6 in the condensation product (1-12) was determined to be 1.6 % by HPLC analysis after removal of all protecting groups, which indicated negligible racemization in this reaction. The protected lepidopteran A was deprotected with trifluoroacetic acid followed by two step HF treatments to give crude lepidopteran A which was purified by HPLC. Synthetic lepidopteran A was identical with natural one in all respects (retention time of HPLC, HPLC pattern of trypsin digests, UV spectrum, and antibacterial activity).

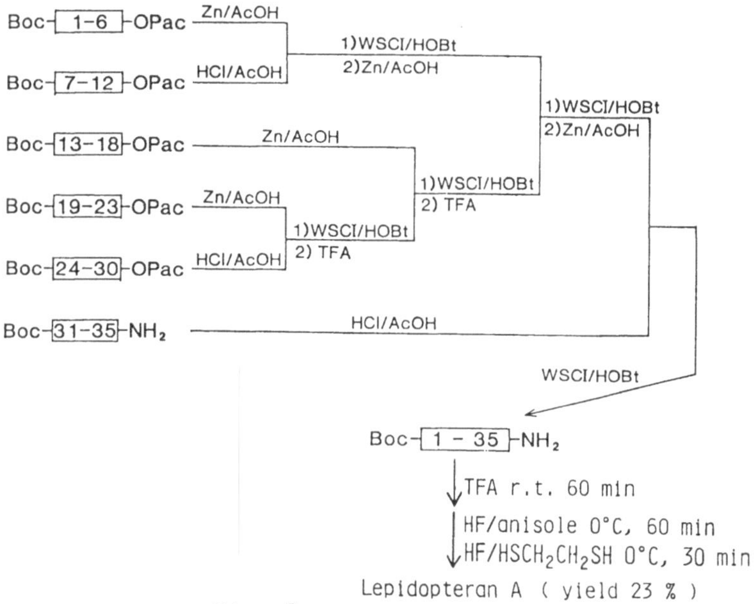

Fig. 2

Structure-Activity Relationship

We synthesized several related peptides listed in Table 1 which shows antibacterial activity of each compound. Although fragments (1-23) and (1-30) exhibited comparable activities to that of lepidopteran A, fragments (1-12) and (1-18) showed weak activities. The peptides lacking N-terminal region were almost inactive even if they still possess considerably long peptide chain such as (7-23) and (13-30). Moreover, fragments (2-23) and (3-23) were much less active than (1-23), and finally a fragment (4-23) became inactive. From these results, N-terminal region including especially Arg^1 and/or Trp^2 was found to be significantly essential for exhibition of antibacterial activity. The CD spectra showed that the herical contents of the active peptides, (1-12), (1-18), (1-23), and (1-30), measured from the absolute value of $[\theta]_{222}$, increased in an order of the antibacterial activity. On the other hand, the non N-terminal peptide showed no activity in spite of high helical contents. The conclusion is that (i) basic N-terminal region of Arg^1 and/or Trp^2 is required necessarily for antibacterial

Table 1. Antibacterial Activities of Related Peptides of Lepidopteran A (MIC µg/ml)

Test organism	1-6	1-12	1-18	1-23	1-30	1-35	1-23	2-23	3-23	4-23
Proteus mirabilis 129	-	250	-	-	250	16	-	-	-	-
Proteus mirabilis 1287	-	-	-	-	-	32	-	-	-	-
Escherichia coli NIHJ	-	250	125	8	4	4	8	64	250	-
Escherichia coli NIHJ JC-2	-	250	125	8	4	2	16	64	-	-
Salmonella enteritidis	-	250	250	16	8	4	64	250	-	-
Shigella sonnei EW-33	-	250	125	8	4	4	N.T	N.T	N.T	N.T
Pseudomonas aeruginosa	-	-	250	62	16	8	125	-	-	-
Pseudomonas aeruginosa 99	-	-	-	62	32	8	125	-	-	-
Enterobacter aerogenes	-	250	250	250	16	8	250	-	-	-
Staphylococcus aureus	-	-	-	-	250	250	-	-	-	-
Bacillus subtilis PCI 219	-	-	-	125	62	16	125	-	-	-
Streptococcus hemolyticus	-	250	-	-	-	-	-	-	-	-

- : >250, N.T : not tested.

7-12, 13-18, 19-23, 24-30, 31-35
7-18, 13-23, 19-30, 24-35, 28-35
7-23, 13-30, 19-35
} >250 against all test organisms

activity; (ii) hydrohobic C-terminal region played an auxiliary role to enhance the activity; (iii) the formation of α-helix seemed to be not directly related to exhibition of antibacterial activity.

According to this conclusion, we then attempted to synthesize a novel kind of derivatives which possess long aliphatic amide chain at C-terminus of fragment peptide (1-18) corresponding to the least active structure, in order to strengthen the activity. As shown in Table 2, although C_{18} derivative was inactive, activities of C_{16}, C_{14}, and C_{12} amide derivatives increased in this order. The interaction of aliphatic chain C_{12} to cullular membrane seemed to be very characteristic for enhancement of the activity.

Table 2. Antibacterial Activities of the Derivatives with Long Aliphatic Amide Chain (MIC μg/ml)

Test organism	H-(1-18)-NH-R R: C_{18}	C_{16}	C_{14}	C_{12}	C_2	Lepidopteran A
Proteus mirabilis 129	-	-	-	-	-	32
Proteus mirabilis 1287	-	-	-	-	-	32
Escherichia coli NIHJ	-	125	32	16	16	4
Escherichia coli NIHJ JC-2	-	64	32	8	16	4
Salmonella enteritidis IFO 3313	-	125	32	8	8	4
Pseudomonas aeruginosa	-	250	64	16	250	8
Pseudomonas aeruginosa 99	-	64	32	8	64	8
Enterobacter aerogenes IFO 13534	-	-	125	16	125	8
Staphylococcus aureus FDA 209P	-	64	32	8	16	125
Bacillus subtilis PCI 219	-	32	8	2	32	32
Streptococcus hemolyticus	-	-	-	-	-	-

-: >250, C_{18}:$(CH_2)_{17}CH_3$, C_{16}:$(CH_2)_{15}CH_3$, C_{14}:$(CH_2)_{13}CH_3$, C_{12}:$(CH_2)_{11}CH_3$, C_2: CH_2CH_3

References

1. Shiba, T., Y. Ueki, I. Kubota, T. Teshima, Y. Sugiyama, Y. Oba, M. Kikuchi. 1984. Peptide Chemistry 1983, 209.
2. Teshima, T., Y. Ueki, T. Nakai, T. Shiba, M. Kikuchi. 1986. Tetrahedron 42, 829.
3. The structure of lepidopteran C was presented at the 52th National Meeting of the Chemical Society of Japan, Kyoto, 1986, 2N43.

BRADYKININ ANTAGONISTS CONTAINING HYDROXYPROLINE

Raymond J. Vavrek and John M. Stewart
Biochemistry Department, University of Colorado Medical School
Denver, CO 80262 USA

Introduction

Our description of the first sequence-related competetive antagonists of the classic pharmacological activities of bradykinin (BK, Arg-Pro-Pro-Gly-Phe-Ser-Pro-Phe-Arg) by BK analogs containing a DPhe residue replacing the Pro residue at position seven (1) included one additional antagonist modification, the replacement of the Phe residues at positions 5 and 8 with beta-2-thienylalanine (Thi). Subsequently, we showed that in certain instances basic residues could be added at the N-terminus of DPhe-containing antagonists (2), that the D-isomer of beta-2-pyridylalanine (Pal) or Thi at position seven, or substitution of the Ser-Pro sequence at positions 6 and 7 with adjacent D-aromatic amino acids, could produce antagonist analogs (3), and that 4-hydroxyproline (Hyp) could replace Pro at positions 2 and 3 in antagonist sequences with favorable results (3,4). We now expand the inclusion of Hyp into several new classes of BK antagonist analogs containing the D-isomer of halogenated Phe residues at position 7, and present complete data on the assay of these analogs on smooth muscle and rat blood pressure (RBP) by published methods (1). Peptides were synthesized by standard solid phase methods; they were purified by countercurrent distribution and characterized by tlc and amino acid analysis.

Results and Discussion

Table 1 lists the analogs with their biological activities. Generally, we have found that it is much easier to develop antagonists of BK smooth muscle activity on the GPI than on the

Peptides 1986

Table 1 - Bradykinin Antagonists Containing Hydroxyproline[a]

ANALOG	RUT	GPI	RBP-IA	RBP-IV	%D
BK	100%	100%	100%	100%	99%
[DPhe^{7}]-BK[b,d]	4%	5.9(5)	2	4	36%
[Hyp^{2}-DPhe^{7}]-BK[d]	0.8%	5.8(5)	I(B)	I(B)	-
[Hyp^{3}-DPhe^{7}]-BK[d]	0.4%	5.6(5)	I(B)	I(B)	-
[Hyp^{2,3}-DPhe^{7}]-BK[d]	0.4%	5.3(5)	I(B)	I(B)	-
Lys-Lys-[DPhe^{7}]-BK[c,d]	0.3%	5.6(5)	I(B)	I(B)	-
Lys-Lys-[Hyp^{2}-DPhe^{7}]-BK[d]	0.1%	5.7(5)	I(B)	I(B)	-
Lys-Lys-[Hyp^{3}-DPhe^{7}]-BK[d]	0.4%	5.6(5)	I(B)	I(B)	-
Lys-Lys-[Hyp^{2,3}-DPhe^{7}]-BK[d]	0.2%	5.4(5)	I(B)	I(B)	-
DArg-[DPhe^{7}]-BK[c,d]	0.09%	6.3(5)	I(B)	I(B)	-
DArg-[Hyp^{2}-DPhe^{7}]-BK[d]	0.5%	6.3(6)	I(B)	I(B)	-
DArg-[Hyp^{3}-DPhe^{7}]-BK[d]	6.5(5)	5.9(7)	I(B)	I(B)	-
DArg-[Hyp^{2,3}-DPhe^{7}]-BK[d]	6.5(5)	6.3(6)	I(B)	I(B)	-
[FDF^{7}]-BK	3%	5.4(6)	0.3%	7%	38%
[Hyp^{2}-FDF^{7}]-BK	0.5%	5.9(5)	0.1%	13%	0
[Hyp^{3}-FDF^{7}]-BK	4%	5.2(4)	0.1%	3%	50%
[Hyp^{2,3}-FDF^{7}]-BK	2%	5.4(4)	0.02%	0.7%	33%
DArg-[FDF^{7}]-BK	2%	6.1(6)	I(B)	I(B)	-
DArg-[Hyp^{2}-FDF^{7}]-BK	1%	5.6(6)	I(B)	I(B)	-
DArg-[Hyp^{3}-FDF^{7}]-BK	2%	6.0(6)	I(B)	I(B)	-
DArg-[Hyp^{2,3}-FDF^{7}]-BK	0.5%	5.7(6)	I(B)	I(B)	-
[Thi^{5,8}-CDF^{7}]-BK	2%	5.3(4)	0.5%	13%	47%
[Hyp^{3}-Thi^{5,8}-CDF^{7}]-BK	3%	5.1(6)	0.2%	11%	0
[Hyp^{2,3}-Thi^{5,8}-CDF^{7}]-BK	0.4%	5.3(5)	0.02%	0.6%	35%
DArg-[Thi^{5,8}-CDF^{7}]-BK	0.6%	6.1(7)	0.2%	5%	48%
DArg-[Hyp^{3}-Thi^{5,8}-CDF^{7}]-BK	5%	5.8(4)	I(B)	I(B)	-
DArg-[Hyp^{2,3}-Thi^{5,8}-CDF^{7}]-BK	0.2%	5.7(6)	I(B)	I(B)	-

a. Agonist activities are given relative to BK=100%; RUT (rat uterus) and GPI (guinea pig ileum) antagonist activities are given as pA_2 values (5) and are underlined, with the number of tissues in parentheses. The pD_2 of BK on the RUT is 8.85, and on the GPI is 7.22. In the RBP (rat blood pressure) assay, I(B) indicates antagonism of BK when the analog is administered as a bolus mixed with a 25mm depressor dose of BK. The percent pulmonary destruction (%D) for RBP agonists is determined from the IA and IV data (1,6). FDF=p-fluoro-DPhe; CDF=p-chloro-DPhe; Thi=ß-(2-thienyl)-LAla. b. Ref. (1). c. Ref. (2). d. Ref. (3).

RUT, leading to two obvious conclusions. First, one is apparently dealing with different receptor types in these two smooth muscle preparations. Second, no single tissue can give a universal answer to the question of BK antagonism since certain analogs have shown antagonist activity on only the RUT or the GPI. While the uterus is still the organ of choice for screening BK agonist analogs, primarily due to the lack of proteolytic enzymes which degrade BK and BK-related analogs, the ileum is a better tissue for screening potential antagonists. We find good qualitative correlation between BK antagonist activity on the ileum and in the rat blood pressure assay, although precise measurement of antagonist potency _in vivo_ is difficult. The key modification of the BK sequence to produce antagonists is the inclusion of a D-aromatic residue at position seven. Presence of a D-residue in position seven of BK agonist analogs can completely block pulmonary degradation; this modification confers resistance to the action of angiotensin converting enzyme (ACE), which is the primary cause of BK breakdown in the pulmonary circulation. Addition of basic residues to the N-terminus of BK can also inhibit action of ACE, and this modification has enhanced the potency of some BK antagonist analogs.

The structure-activity relationships of Hyp containing antagonists indicate that a DArg residue added at the N-terminus of these analogs generally produces more consistent BK antagonism in the RBP assay. DPhe and _p_-fluoro-DPhe (FDF) at position seven have qualitatively and quantitatively similar effects on producing BK agonist or antagonist analogs, except that in analogs without a DArg residue at the N-terminus the FDF analogs with Hyp are quite selective for antagonism on the GPI. Hyp substitution in _p_-chloro-DPhe (CDF)-containing analogs which also contain a DArg residue at the N-terminus and Thi at positions 5 and 8 yields analogs which are more consistently antagonists in the RBP assay, but CDF^7-antagonists are generally less potent than DPhe or FDF analogs. Finally, the only analogs in Table 1 which show BK antagonism on the uterus contain both the DArg addition at the N-terminus and Hyp substitution at position 3. This is seen only with $DPhe^7$-containing antagonists, and parallel antagonism is not seen in analogs with FDF or CDF at position 7. With pA_2 values of 6.5,

DArg-[Hyp^3-DPhe^7]-BK and DArg-[Hyp^{2,3}-DPhe^7]-BK represent the most potent antagonists of BK activity on the rat uterus. Certain of these Hyp-containing antagonists have already been applied to other biological systems (7).

Acknowledgment

The authors thank Virginia Sweeney for the amino acid analyses, and Frances Shepperdson and Garima Vashistha for the bioassays. This research was supported by grants from NHLBI-NIH (#HL-26284), and from Nova Pharmaceutical Corporation.

References

1. Vavrek, R.J. and J.M. Stewart. 1985. Competetive antagonists of bradykinin. Peptides 6, 161-164.
2. Stewart, J.M. and R.J. Vavrek. 1985. In: Kinins - IV (L.M. Greenbaum, ed). Plenum (in press).
3. Stewart J.M. and R.J. Vavrek. 1986. In: Proceedings of the International Symposium on the Kallikrein-Kinin System and Inflammation (G. Kunkel, ed). Allergology (in press).
4. Vavrek, R.J. and J.M. Stewart. 1986. In: Peptides, Proceedings of the 9th American Peptide Symposium (C.M. Deber, V.J. Hruby, K.D. Kopple, eds). Pierce Chemical Company, pp. 655-658.
5. Schild, H.O.. 1947. pA, a new scale for the measurement of drug antagonism. Brit. J. Pharmacol. 2, 189-206.
6. Ryan, J.W., J. Roblero and J.M. Stewart. 1969. Inactivation of bradykinin in the pulmonary circulation. Biochem. J. 110, 795-797.
7. Longridge, D.J., M. Schachter, J.M. Stewart, Y. Uchida and R.J. Vavrek. 1986. Synthetic competitive antagonists of bradykinin on vascular permeability in the rabbit. J. Physiol. 372, 73P.

SYNTHESIS OF CHIRAL TETRAMIC ACID, THE INTERMEDIATE IN THE STEREO-CONTROLLED PREPARATION OF STATINE ANALOGUES, KEY COMPONENTS IN ASPARTIC PROTEINASES INHIBITORS.

P. Jouin, B. Castro, J. Poncet.

CCIPE, rue de la Cardonille, 34094 Montpellier Cedex, France.

D. Nisato.

SANOFI Recherche, rue du Pr. J. Blayac, 34082 Montpellier Cedex, France.

Introduction

Statine, (3S,4S)-3-Hydroxy-4-amino-6-methylheptanoïc acid (1) is present in the general aspartyl protease inhibitor pepstatin (2). However pepstatin was found to be less effective against renin ($ID_{50} = 10^{-6}$ M) than against pepsin ($ID_{50} = 10^{-8}$ M) and cathepsin D ($ID_{50} = 10^{-10}$ M).[1] Others (3S,4S)-3-Hydroxy-4-amino-carboxylic acids have been proved to be more potent than statine itself in the elaboration of inhibitors of renin ; this included the α-amino-β-hydroxy acid dorived from phenylalanine (AHPPA 3) and the reduced analogue derived from cyclohexylalanine (ACHPA 4).[2] In all cases only the S,S isomer was able to act as transition state analogue in the proteolytic step.[3]

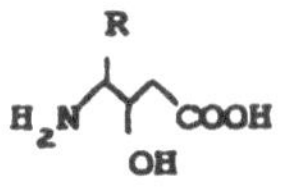

(1) : R = $CH_2-CH-(CH_3)_2$

(3) : R = $CH_2-C_6H_5$

(4) : R = $CH_2-C_6H_{11}$

Iva Val Val Sta Ala Sta

pepstatin (2)

Previously described methods for the preparation of statine are largely based on the aldol condensation of an N-protected-amino aldehyde or on the reduction of the corresponding β-ketoester.[4] These methods lead to enantiomeric mixture and their extrapolation to a large scale preparation was very critical. We demonstrated that a stereocontrolled synthesis of statine occured by specific *cis*-reduction of the corresponding chiral N-protected tetramic acid (7).[5] In that paper we compared the two different pathways to the so-called chiral tetramic acid intermodiate: the cyclisation procedure that we elaborated (*scheme 1*) and the previously described Dieckmann cyclisation (*scheme 2*),[6] in order to determine the more convenient access to enantiomerically pure statine analogues.

Results

We successfully realized the stereospecific synthesis of statine (1), starting from (L)-N-Boc-leucine (5) (*scheme 1*). The N-protected aminoacid activated with isopropenyl chloroformate (IPCF) was coupled with Meldrum's acid and the adduct was thermally cyclized under neutral conditions into the corresponding chiral

Peptides 1986

tetramic acid (7). Since the production of pure chiral tetramic acids has not been proved so far, the reference data were not reliable, and the control of the chirality was ascertained according to the following correlation. The reduction of the compound (7) was proved, by spectroscopic methods, to be specifically *cis* and led to the N-Boc *threo*-4-hydroxy-pyrrolidin-2-one derivative (8), which was easily hydrolyzed into the enantiomerically pure N-Boc statine (1) in 63 % overall yield. Likewise, this pathway has been followed for the preparation of statine analogues including AHPPA (3), ACHPA (4) and derivatives with fonctionalized γ-chain when starting from aminoacids with protected chain.[5]

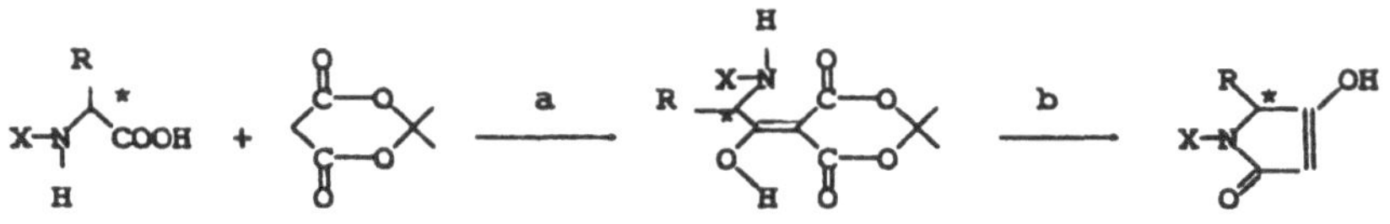

(5) : X = Boc , R = isopropyl (6) (7)

(9) : X = Boc , R = benzyl (10)

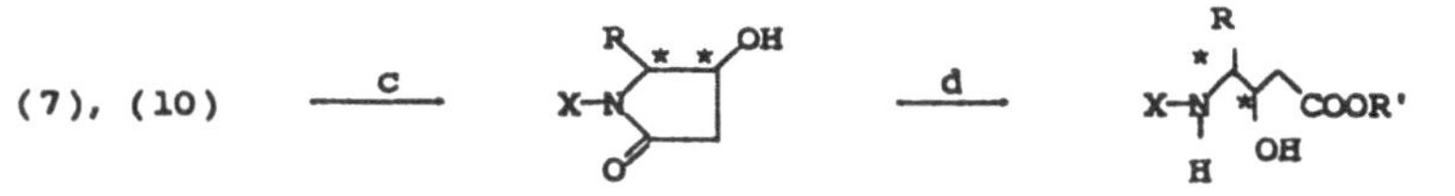

(8) : X = Boc , R = isopropyl (1)

e (11) : X = Boc , R = benzyl (3)

(12) : X = H , R = benzyl

a: i- CH_2Cl_2, DMAP ; ii- IPCF. b: Δ CH_3OH. c: $NaBH_4$, AcOH, CH_2Cl_2;
d: (X =Boc) NaOR'; (X = Z) HCl, HOR'. e: TFA, CH_2Cl_2.

Scheme 1 : Stereocontrolled synthesis of statine analogues.

The preparation of tetramic acids by the Dieckmann cyclisation strategy could be reexaminated with in hands enantiomerically pure references. To evaluate the extent of racemization that occured during the synthesis of (16) by the Dieckmann process, we choosed to correlate this tetramic acid with the reduced derivative (12) ($[\alpha]_D^{20}$= -44°). This reference compound was obtained by acidic deprotection of the amide fonction of the N-Boc *threo*-4-hydroxy-pyrrolidin-2-one derivative (11).

Methyl-(L)-phenylalaninate (13) was easily transformed into the N-acyl compounds (**14a,b,c**) by known procedures. Compounds (14) were cyclized under basic conditions into C-3 acyl tetramic derivatives (15) which were isolated as sodium or potassium salts (*Scheme* 2).[7] The compound (15c) could not be isolated in pure form.

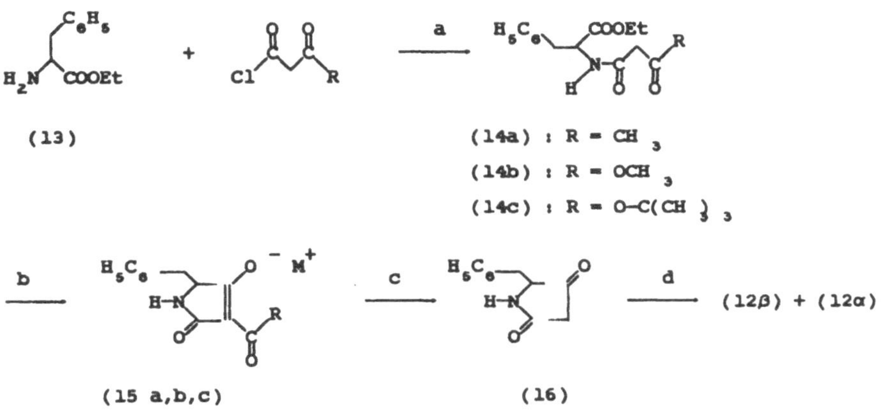

a: K_2CO_3; b: NaOEt; c: hydrolysis ; d: reduction.

Scheme 2 : Synthesis of tetramic acids via Dieckmann cyclisation.

During the cyclisation step, it appeared that unreproducible optical rotations were obtained for compounds (15) in each case, if no special care was taken. This drawback could be avoided by controlling rigorously the stoichiometry and the reaction time. In these conditions, optimal rotations could be measured (*table 1*).

initial compound	base	time	temp.	final compound	yield	optical rotation [1]
(14a)	NaOMe	3 h	20 °C	(15a)	71%	-126 °
(14b)	NaOMe	1 h	reflux	(15b)	77%	-192 °
(14c)	KOtBu	10 mn	20 °C	(15c)	—	—

1) measured at 20 °C, at 589 µm, in methanol (c = 1)

Table 1 : Dieckmann cyclisation.

The conversion of (**15a**), (**15b**) and (**15c**) into the 4-hydroxy-5-benzyl-pyrrolidin-2-one (**12**) proceeded by their deacylation at the 3 position and reduction of the corresponding tetramic acid (**16**). Decarboxylation of (**15a**) was best accomplished by a short reflux (15 mn) in 25 mM sulfuric acid: deacylation of (**15b**) occured in similar conditions but under longer reflux (4 h). Finally, the tetramic acid (**16**) was prepared from the salt (**15c**) after trifluoroacetic treatment. The raw products respectively obtained from each compound (**15a**), (**15b**) and (**15c**) were reduced into the alcohol (**12**).

We observed an unexpected non-stereospecific reduction of tetramic acid (**16**) although the *cis*-specificity of the reduction of N-protected tetramic acid (**10**) by catalytic hydrogenation as well as sodium hydride reduction, was demonstrated. By the way, the two alcohols (**12β**) ($[\alpha]_D$=-44°), and (**12α**) ($[\alpha]_D$=+16°) were obtained by reduction of the unprotected tetramic acid prepared from (**10**). By another way, the optical rotations measured for componds (**12α**) and (**12β**) prepared from (**15a**), (**15b**) and (**15c**) are listed in *table* 2.

initial compound	optical rotation [1] (12α)	(12β)	enantiomeric excess (ee)
(15a)	−31 °	+11 °	70
(15b)	0	0	0
(15c)	−27 °	+10 °	61

[1]) measured at 20 °C, at 589 μm, in methanol (c = 1)

*Table 2. Enantiomeric purities of compounds (**16**).*

The enantiomeric purities (ee) of these alcohols were calculated by comparison with the specific optical rotations of the enantiomerically pure (**12α**) (+16°) and (**12β**) (−44°). The ee measured in each case, indicated the degree of racemization during the formation of compound (**16**), including the cyclisation, and the deacylation or decarboxylation steps. From these data, it was obvious that unacceptable racemization was observed in every methods that have been tested, following the Dieckmann's route.

Conclusion

The Dieckmann cyclisation strategy did not appeared to be suitable for the synthesis of chiral tetramic acids. Moreover, only the N-protected tetramic acids are stereospecifically reduced. Our method[5] seemed to be, at this time, the sole enantiospecific access to these latter compounds, from which pure enantiomeric statine analogues could be easily obtained.

References

[1] R.J. Workman, D.W. Burkitt, *Archs. Biochem. Biophys.*, 1979, **194**, 157.

[2] J. Boger, L.S. Payne, D.S. Perlow, N.S. Lohr, M. Poe, E.H. Blaine, E.H. Ulm, T.W. Schorn, B.I. Lamont, T.-Y. Lin, M. Kawai, D.H. Rich, *J. Med. Chem.*, 1985, **28**, 1779.

[3] D.H. Rich, *J. Med. Chem.*, 1985, **28**, 263.

[4] K.E. Rittle, C.F. Homnick, G.S. Ponticello, B.E. Evans, *J. Org. Chem.*, 1982, **47**, 3016; R. Steulmann, H. Klostermeyer, *Liebigs Ann. Chem.*, 1975, 2245.

[5] P. Jouin, B. Castro, D. Nisato, 1985, European patent (1985); P. Jouin, B. Castro, D. Nisato, *J. Chem. Soc., Perkin Trans. 1*, in press.

[6] T. Katsuki, M. Yamaguchi, *Bull. Chem. Soc. Jpn.*, 1976, **49**, 3287.

[7] P.G. Williard, S.E. de Laszlo, *J. Org. Chem.*, 1984, **49**, 3489.

AMINO ACIDS ON THE BASIS OF CYCLIC β-DIKETONES - SYNTHESIS AND PHOTOPHYSICAL CHARACTERISTICS

St. Minchev, N. Sofroniev
Department of Chemistry, Higher Pedagogical Institute, Shoumen, Bulgaria

I. Timtcheva, P. Nikolov
Institute of Organic Chemistry with Centre of Phytochemistry, Bulgarian Academy of Sciences, 1113 - Sofia, Bulgaria

Introduction

The 2-p-phenylsubstituted, as well as the 2-phenyl-2-substituted-1,3-indandiones do not fluoresce in solution (1,2). They photoisomerize to the corresponding 3-arylmethylene-1(3H)-isobenzofuranones (benzylydene phthalides) upon steady-state and pulse irradiation with UV light (1,2,3). The diketo tautomeric form of the indandiones is responsible for their photoisomerization to benzylidenephthalides (4). In the present communication the synthesis and results of the photophysical investigations of some amino acids on the base of cyclic β-diketones are described.

Results

Table 1 shows all compounds investigated. The model compounds (1) and (7) are described in the literature, while (2-6) and (8-11) are synthesized according to (5). The D,L-α-amino acids (12) and (13) are obtained from the corresponding benzaldehyde (A or B) and phthalic anhydride:

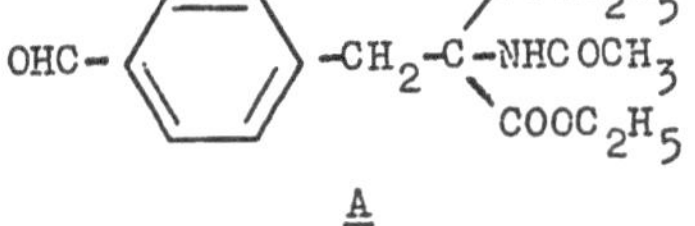

A

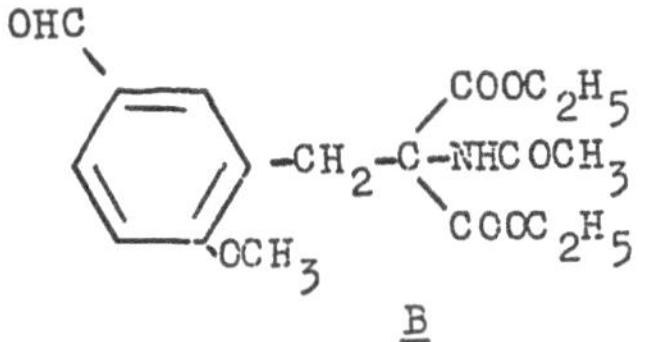

B

The reaction of 6-nitrophthalide with benzaldehyde (B), followed by isomerization and reduction of the nitro group, gives D,L-compound (14) -

Peptides 1986

Table 2. In contrast to the unsubstituted in the phthaloyl fragment indandiones all 4- and 5-aminoindandiones are photostable upon steady-state and pulse irradiation with UV light and fluoresce intensively in the region 20000 - 23000 cm^{-1}. Table 3 shows the absorbtion and fluorescence characteristics of some compounds in ethanol and 1,4-dioxane. The analysis of the spectral characteristics of the investigated 4- and 5-amino-2-aryl-1,3-indandiones shows, that their diketo-form is responsible for the fluorescence observed. This conclusion is based on the following experimental results:

1. The 4- and 5-amino-2-aryl-2-carboxymethyl-1,3-indandiones, which have a fixed diketo structure fluoresce; their excitation spectra coincide with the absorption ones.

2. Two facts for the cases, when a keto-enol tautomerism is possible should be pointed out:

a) the excitation in the longest wavelenght absorption maximum (21000 cm^{-1}), which results from absorption of the enol-form, does not lead to fluorescence;

b) the excitation spectra of these compounds differ from the absorption spectra (which are, a superposition of the bands of the keto- and enol-forms) and coincide with the bands, which are due to absorption of the diketo-form only.

3. There is a mirror symmetry between the fluorescence spectrum and the longest-wavelenght absorption band of the diketo-form.

4. The quantum yield of the compounds with a fixed diketo-form is greater than that of the compounds, for which a keto-enol tautomerism is possible.

Upon freezing of the ethanolic solutions of the investigated aminoindandiones the fluorescence maxima move to the blue with about 10 nm the fluorescence intensity slightly increases (2-5 times). Phosphorescence is observed only from the frozen solutions of compounds (5, 6 and 11); the phosphorescence maximum is in the region 19500 - 20000 cm^{-1}.

The replacement of the amino group in the phthaloyl fragment with a hydroxy- or a nitro one leads practically to quenching of the fluorescence ($Q_f <$ 0.001), while the unsubstituted in the phthaloyl fragment 1,3-indandiones do not fluoresce at all (1,2). Consequently it could be assumed that, besides the diketo-form, the intense fluorescence of the investigated indandiones is unambigiously connected with the amino group in the phthaloyl fragment. This assumption is supported by the fact, that the acylation of the amino group leads to a significant diminishment of the fluorescence quantum yield (Q_f less than 0.001, like the unsubstituted in the phthaloyl fragment compounds). The indandiones which have an amino group at position

2 or at p-position in the phenyl ring do not fluorescence either.

Table 1

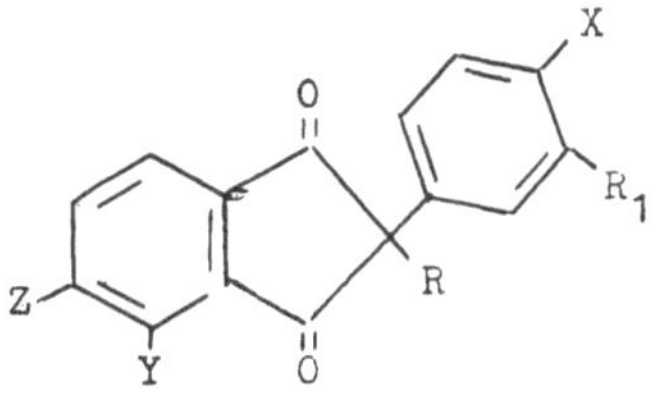

Nº	X	Y	Z	R	R_1
1	H	NH_2	H	H	H
2	CH_3	NH_2	H	H	H
3	OCH_3	NH_2	H	H	H
4	H	NH_2	H	CH_2COOH	H
5	CH_3	NH_2	H	CH_2COOH	H
6	OCH_3	NH_2	H	CH_2COOH	H
7	H	H	NH_2	H	H
8	H	H	NH_2	CH_2COOH	H
9	CH_3	H	NH_2	CH_2COOH	H
10	OCH_3	H	NH_2	CH_2COOH	H
11	Cl	H	NH_2	CH_2COOH	H
12	OCH_3	H	H	H	$CH_2-CH(NH_2)-COOH$
13	$CH_2-CH(NH_2)-COOH$	H	H	H	H
14	OCH_3	H	NH_2	H	$CH_2-C(COOC_2H_5)_2-NHCOCH_3$

Table 2

Nº	M.p., °C	Yield, %	Analyses N,% calc.	found	R_f,TLC
12	279–81	45.1	4.13	4.31	0.52^A
13	264–66	44.9	4.53	4.62	0.50^A
14	>350	76.0	5.63	5.71	0.72^B

System A - n-butanol:acetic acid:pyridine:water (15:3:10:12)
System B - n-butanol:acetic acid:pyridine:water (4:1:1:2)

Table 3
Experimental spectral characteristics of the compounds investigated at 293°K in solvents of different polarity. ν^A, ν^F - energy of the Frank - Condon absorption and fluorescence transition $[\nu] = cm^{-1}$; ε-molar absorptivity, $[\varepsilon] = l.mol^{-1}.cm^{-1}$; Q_f - fluorescence quantum yield.

N°	ethanol				1,4-dioxane	
	ν^A	ε	ν^F	Q_f	ν^A	ν^F
1	25800	5480	21320	0.14	26316	22730
2	25900	3590	21410	0.02	25970	22220
3	25910	6550	21230	0.03	25970	21980
4	25880	7560	21250	0.40	26316	22470
5	26040	7850	21500	0.35	26450	22470
6	26040	6060	21280	0.14	26380	22470
7	31120	20320	20320	0.10	32050	-
8	33100	18470	20140	0.10	31250	22470
9	31120	17920	20240	0.17	31746	22720
10	31150	22070	22120	0.06	31746	22730
11	30980	15460	20200	0.20	31827	22830

References

1. Zechner, J., N. Getoff, I. Timtcheva, F. Fratev, St. Minchev. 1983. Z. Naturforsch. 38a, 1337.
2. Timtcheva, I., P. Nikolov, J. Zechner, N. Getoff, St. Minchev. J. Photochem. (in press)
3. Zechner, J., G. Grabner, G. Kohler, N. Getoff, I. Timtcheva, F. Fratev, St. Minchev. 1983. J. Photochem. 23, 61.
4. Timtcheva, I., P. Nikolov, J.Zechner, N. Getoff, to be published.
5. Sofroniev, N., St. Minchev,Dokl. Bolg. Akad. Nauk. (in press)

Participants

Prof. Dr. B. Aleksiev
Higher Institute of Chemical Technology
1156 SOFIA
BULGARIA

Prof. M.J.S.A. Amaral Trigo
Faculdade de Ciencias do Porto
Departamento de Quimica
4000 PORTO
PORTUGAL

Dr A. J. Andersen
Group Manager
Development Department
331 Havrevaenget
DK-2980 KOKKEDAL
DENMARK

Dr R.P. Andrews
LKB Biochrom Ltd.
Cambridge Science Park
Milton Road
CAMBRIDGE CB4 4FJ
U. K.

Dr T.M. Andronova
Shemyakin Institute of Bioorganic Chemistry
USSR Academy of Sciences
Ul. Miklukho-Maklava 16/10
117871 MOSCOW V-437
U. S. S. R.

Dr. S. Arielly
KabiVitrum AB
Taljegårdsgatan 3
S-431 33 MOLNDAL
SWEDEN

Dr Ch. Ashton
Applied Biosystems Ltd.
Kelvin Close
Birchwood Science Park North
Warrington - Cheshire
UNITED KINGDOM

Dr A. Aubry
Laboratoire de Minéralogie et Cristallographie
Université de Nancy I
Boîte Postale 239
54506 VANDOEUVRE LES NANCY CEDEX
FRANCE

Dr L. Balaspiri
Department of Medical Chemistry
Medical University of Szeged
Dóm Tér 8
6720 SZEGED
HUNGARY

Dr I. T. Barnish
Pfizer Central Research
Sandwich
KENT CT13 9NJ
U. K.

Prof. E. Bayer
Institute for Organic Chemistry
University Tübingen
Auf der Morgenstelle 18
7400 TUBINGEN
F. R. G.

Dr Th. Beisswenger
Degussa AG
Rodenbacher Chaussee 4
Postfach 1345
6450 Hanau 1
F. R. G.

Prof. C. Di Bello
Univers. di Padova
Istituto Chimica Industriale
9, Via Marzolo
35131 PADOVA
ITALY

Dr W. Bender
FE-EP-CWL-Pharma
Bayer AG
Postfach 10 17 09
5600 WUPPERTAL 1
F. R. G.

Prof. E. Benedetti
Dipartimento di Chimica
Univ. di Napoli
Via Mezzocannone 4
80134 NAPOLI
ITALY

Prof. N. L. Benoiton
University of Ottawa
Department of Biochemistry
OTTAWA - ONTARIO
CANADA K1H 8M5

Dr G. Benz
FE-EP-CWL-Pharma
Bayer AG
5600 WUPPERTAL 1
F. R. G.

Doc. Dr M. Bienert
Institute of Drug Research
Institut f. Wirkstoforschung der AdW d. DDR
Alfred-Kowalke-Str. 4
DDR-1136 BERLIN
G. D. R.

Prof. G. Van Binst
Vrije Universiteit Brussel
Pleinlaan 2
B-1050 BRUSSELS
BELGIUM

Prof. Ch. Birr
Organogen GmbH
Czerny-Ring 22
D-6900 HEIDELBERG 1
F. R. G.

Dr Ch. Bladon
Medical Research Council
Brain Metabolism Unit
University Department of Pharmacology
1 George Square
EDINBURG EH8 9S2
Scotland
U. K.

Dr K. Bláha
Institute of Organic Chemistry and Biochemistry
Czechoslovak Academy of Sciences
Flemingovo nam. 2
166 10 PRAGUE 6
CZECHOSLOVAKIA

Dr D. Blanot
C. N. R. S.
UA 04-1131
Bâtiment 432
Université de Paris-sud
91405 ORSAY
FRANCE

Prof. T. Blundell
Department of Crystallography
Division of Molecular Biology
Birbeck College
Malet Street
LONDON WC1 FHX
U. K.

Dr K. Bode
Sandoz Ltd.
Pharmaceutical Division
Chemical Production
Lichtstr. 35
CH-4002 BASLE
SWITZERLAND

Prof. G. Borin
Biopolymer Research Center
Department of Organich Chemistry
Via Marzolo 1
35131 PADOVA
ITALY

Dr A. Brack
Centre de Biophysique Moléculaire
C. N. R. S.
1A, Avenue de la Recherche Scientifique
45071 ORLEANS Cedex 2
FRANCE

Dr A. F. Bradbury
National Institute for Medical Research
The Ridgeway
Mill Hill
LONDON NW7 1AA
U. K.

Prof. D. Brandenburg
Deutsches Wollforschungsinstitut
Veltmanplatz 8
D-51 AACHEN
F. R. G.

Prof. M. Brenner
Institute of Organic Chemistry
University of Basle
St. Johannsring 19
CH-4056 BASLE
SWITZERLAND

Dr H. Brückner
Institute of Food Technology
University of Hohenheim
Garbenstrasse 25
D-7000 STUTTGART 70
F. R. G.

Prof. K. Brunfeldt
Instituttet for Kemiindustri
The Technical University of Denmark
Building 227
DK 2800 LYNGBY
DENMARK

Dr A. Buku
Center of Polypeptide and Membrane Research
The Mount Sinai Med. Center
One Gustave L. Levy Place
NEW YORK, N.Y. 10029
U. S. A.

Dr R. Callens
Solvay and Co.
Central Research Laboratory
310 Rue de Ransbeek
B-1120 BRUSSELS
BELGIUM

Dr L. Caporale
Merck Sharp and Dohme Research Laboratories
W 26-208
WEST POINT, P.A. 19486
U.S.A.

Dr F. Cardinaux
Sandoz Ltd.
Pharmaceutical Div.
Preclinical Res.
P. O. Box
CH-4002 BASLE
SWITZERLAND

Dr R. De Castiglione
Farmitalia - Carlo Erba
Chemical Research and Development
Via dei Gracchi 35
20146 MILAN
ITALY

Dr G. Chassaing
Laboratoire de Chimie Organique Biologique
C. N. R. S.
Université Pierre et Marie Curie
4, Place Jussieu
PARIS 75005
FRANCE

Dr M. Chorev
The Hebrew University
School of Pharmacy
P. O. Box 12065
JERUSALEM 91120
ISRAEL

Prof. P. Cordopatis
University of Patras
Lab. of Organic Chemistry
PATRAS
GREECE

Prof. A. Cosmatos
39, D. Gounaris Str.
151 24 ATHENS
GREECE

Prof. D. H. Coy
Peptide Research Laboratories
Dept. of Medicine
Tulane University School of Medicine
1430 Tulane Avenue
NEW ORLEANS, LA 70112
U. S. A.

Dr J. S. Davies
Chem. Dept.
University College of Swansea
Singleton Park
SWANSEA SA2 8PP
U. K.

Prof. Ch. Deber
University of Toronto
Research Institute
Hospital for Sick Children
555 University Avenue
TORONTO, ONTARIO
CANADA M5G 1X8

Dr Derrien
Institut Choay
10, Rue Morel
F-92120 Montrouge
FRANCE

Dr J. Diaz
Centre de Recherches
Clin-Midy-Sanofi
Rue du Prof. Joseph Blayac
34082 MONTPELLIER Cedex
FRANCE

Dr C. N. C. Drey
King Alfred's College
WINCHESTER S022 4NR
U. K.

Dr J. P. Durieux
Novabiochem A.G.
Weidenmattweg. 4
CH-4448 LAUFELFINGEN
SWITZERLAND

Dr A. S. Dutta
ICI Pharmaceuticals Division
Mereside - Alderley Park
Macclesfield
CHESHIRE SK10 4TG
U. K.

Prof. R. Epton
Wolverhampton Polytechnic
School of Applied Sciences
WOLVERHAMPTON WV1 1SB
U. K.

Dr J. A. Eriksen
Nycomed AS
Nycoveien 1-2
P. O. Box 4220 Torshov
N-0401 OSLO 4
NORWAY

Prof. E. Escher
Lab. of Pharmacology and Biochemistry
Medical School
University of Sherbrooke
SHERBROOKE,
QUEBEC J1H 5N4
CANADA

Dr G. Evin
I.N.S.E.R.M.
Unité 36
17, Rue du Fer-à-Moulin
75005 PARIS
FRANCE

Doc. Dr F. Fahrenholz
Max-Planck-Institut für Biophysik
Kennedyallee 70
D-6000 FRANKFURT /M. 70
F. R. G.

Dr A. M. Felix
Peptide Research Dept.
Hoffmann-La Roche Inc.
NUTLEY, N. J. 07110
U. S. A.

Dr S. Fermandjian
Département de Biologie
Centre d'Etudes Nucléaires
de Saclay
91191 GIF-SUR-YVETTE Cedex
FRANCE

Mr N. Fertherigos
BioHellas S. A.
Hellenic Biotechnology Co.
10 Parnithos Street
154 52 ATHENS
GREECE

Dr M. Flegel
Institute of Pharmacology
Czechoslovak Academy of Sciences
Albertov 4
12800 PRAGUE 2
CZECHOSLOVAKIA

Dr G. Fölsch
Institute of Medical
Biochemistry
University of Göteborg
Box 33031
S-40033 GÖTEBORG
SWEDEN

Dr L. Forni
University of Mons
Faculty of Sciences
Laboratory of Biological
Chemistry
21, Avenue Maistriav
B-7000 MONS
BELGIUM

Dr M. C. Fournié - Zaluski
Départment de Chimie
Organique
Faculté de Pharmacie
4 Avenue de l'Observatoire
75006 PARIS
FRANCE

Dr I. Frič
Inst. of Organic Chemistry
and Biochemistry
C. S. A. V.
Flemingovo nám. 2
16610 PRAGUE 6
CZECHOSLOVAKIA

Prof. M. Fridkin
Department of Organic
Chemistry
Weizmann Institute of
Science
REHOVOT 76100
ISRAEL

Dr P. Fromageot
Institut de Recherche
Fondamentale
Centre d'Etudes
Nucléaires de Saclay
91191 GIF-SUR-YVETTE
Cedex
FRANCE

Prof. J. S. Fruton
Yale University
123 York Street
NEW HAVEN, CT 06511
U. S. A.

Dr N. Galakatos
Massachusetts Inst.
of Technology
Department of Chemistry
CAMBRIDGE,
MASSACHUSETTS 021390
U. S. A.

Dr M. Galantino
Farmitalia Carlo Erba
Via Dei Gracchi 35
20146 MILANO
ITALY

Dr I. J. Galpin
University of Liverpool
Robert Robinson
Laboratories
P. O. Box 147
LIVERPOOL L69 3BX
U. K.

Dr J. Gante
E. Merck Darmstadt
Pha Fo Chem HK
Frankfurter Str. 250
D-6100 DARMSTADT.
F. R. G.

Dr J. M. Garcia Antón
Centro de Investigasión y
Desarrolo de Barcelona
(C.S.I.C.)
Jorge Girona Salgado 18-26
08034 BARCELONA
SPAIN

Dr D. Gazis
Mount Sinai School of
Medicine
Center for Polypeptide and
Membrane Research
NEW YORK, NY 10029
U. S. A.

Prof. R. Geiger
Hoechst AG
Pharma-Synthese
Postfach 800 320
D-6230 FRANKFURT/M. 80
F. R. G.

Dr T. Geiser
Peptide Synthesis Project
Applied Biosystems, Inc.
850 Lincoln Centre Drive
POSTER CITY, CA 94404
U. S. A.

Prof. I. Gerothanassis
Department of Chemistry
University of Ioannina
451 10 IOANNINA
GREECE

Dr S. Gestrelius
Ferring AB
P. O. Box 30561
S-200 61 MALMÖ
SWEDEN

Prof. W. A. Gibbons
Pharmaceutical Chemistry
The School of Pharmacy
University of London
29/39 Brunswick Square
LONDON WC1N 1AX
U. K.

Prof. L. Gierasch
Department of Chemistry
University of Delaware
NEWARK, DELAWARE 19716
U.S.A.

Dr D. Gillessen
F. Hoffmann-La Roche
and Co. Ltd.
ZFE Bau 71/304
CH-4002 BASEL
SWITZERLAND

Dr Ch. Gilon
The Hebrew University of
Jerusalem
JERUSALEM 91904
ISRAEL

Prof. E. Giralt
Dept. Quimica Organica
Facultat de Quimica
Marti Franqués 1
08028 BARCELONA
SPAIN

Dr J. D. Glass
Mount Sinai School
of Medicine
5th Avenue at 100th Street
NEW YORK, NY 10029
U. S. A.

Dr W. Göhring
Max-Planck-Institut
f. Biochemie
Abtl. Peptidchemie
Am Klopferspitz
D-8033 MARTINSRIED
F. R. G.

Prof. M. Goodman
Department of Chemistry
University of California
SAN DIEGO,
LA JOLLA, CA 92093
U. S. A.

Dr L. Grehn
Institute of Biochemistry
University of Uppsala
Biomedical Center
Box 576
S-751 23 UPPSALA
SWEDEN

Dr P. Grogg
Bachem AG
Hauptstrasse 144
CH-4416 BUBENDORF
SWITZERLAND

Prof. Z. Grzonka
Institute of Chemistry
University of Gdansk
Ul. Sobieskiego 18
80-952 GDANSK
POLAND

Dr D. Hadfield
Applied Biosystems Ltd.
Kelvin Close
Birchwood Science Park North
Warrington - Cheshire
UNITED KINGDOM

Dr J. Halstrøm
Novo Research Institute
Novo Allé
DK-2880 BAGSVAERD
DENMARK

Dr G. W. Hardy
Department of Medicinal
Chemistry
The Welcome Research
Laboratories
Langley Court
Beckenham
KENT BR3 3BS
U. K.

Dr A. Hartmann
Ciba-Geigy AG
Postfach K-122.1.28
CH-4002 BASEL
SWITZERLAND

Prof. E. Haslinger
Institut f. Organische Chemie
Universität Wien
Währinger Strasse 38
A-1090 WIEN
AUSTRIA

Dr R. F. Hirschmann
Merck Sharp and Dohme
Research Laboratories
P. O. Box 2000
RAHWAY, NJ 07065
U. S. A.

Dr J. Hlavácek
Institute of Organic
Chemistry and
Biochemistry
Flemingovo n. 2
166 10 PRAGUE 6
CZECHOSLOVAKIA

Dr S. M. Hoare
Boots Peptides
D 61 – The Boots Co. PLC
NOTTINGHAM NG2 3AA
U. K.

Prof. A. Holm
University of Copenhagen
Universitetsparken 5
2100 COPENHAGEN Ø
DENMARK

Dr P. Hoogerhout
Gorlaeus Laboratories
Leiden State University
P. O. Box 9502
2300 RA LEIDEN
THE NETHERLANDS

Prof. V. J. Hruby
University of Arizona
Department of Chemistry
TUCSON, ARIZONA 85721
U. S. A.

Dr J. Humphries
Beecham Pharmaceuticals
The Pinnacles
Harlow
ESSEX CM19 5AD
U. K.

Dr H. Immer
Novabiochem AG
Postfach
Weidenmattweg 4
CH-4448 LÄUFELFINGEN
SWITZERLAND

Dr J. Izdebski
University of Warsaw
Faculty of Chemistry
Pasteura 1
02-093 WARSAW
POLAND

Prof. H. - D. Jakubke
Karl-Marx-Univerität
Sektion Biowissenschaften
Talstrasse 33
DDR-7010 LEIPZIG
G. D. R.

Dr J. T. Johansen
Carlsberg Biotechnology Ltd.
16 Tagensvej
DK-2200 COPENHAGEN N
DENMARK

Dr J. H. Johansen
Nycomed A/S
Nycovn 2
N-0485 OSLO 4
NORWAY

Dr N. L. Johansen
Novo Industri A/S
Novo Allé
DK-2880 BAGSVAERD
DENMARK

Dr W. H. Johnson
Roche Products Ltd.
P. O. Box 8
Welwyn Garden City
HERTS AL7 3AY
U. K.

Dr P. A. Jouin
C.N.R.S. - C.C.I.P.E.
Rue de la Cardonille
34 094-MONTPELLIER Cedex
FRANCE

Dr A. Juhász
Institute for Drug Research
Szabadśagharcosok u. 47
1045 BUDAPEST
HUNGARY

Prof. G. Jung
Inst. of Organic Chemistry
University of Tübingen
Auf der Morgenstelle 18
D-7400 TÜBINGEN
F. R. G.

Dr Gr. Kalopissis
80, Boulevard Bineau
92200 NEUILLY-SUR-SEINE
FRANCE

Dr B. Kamber
Ciba-Geigy AG
CH-4002 BASEL
SWITZERLAND

Prof. P. G. Katsoyannis
Department of Biochemistry
Mount Sinai School
of Medicine
of the City University
of New York
1 Gustave L. Levy Place
NEW YORK, NY 10029
U. S. A.

Dr O. Keller
Fluka AG
CH-9470 BUCHS
SWITZERLAND

Dr G. Kéri
1st Institute of Biochemistry
Semmelweis University
Medical School
P. O. Box 260
1444 BUDAPEST 8
HUNGARY

Prof. H. Kessler
Institute of Organic
Chemistry
J. W. Goethe - University
D-6000 FRANKFURT 50
F. R. G.

Prof. L. Kisfaludy
Chemical Works of Gedeon
Richter Ltd.
P. O. Box 27
1475 BUDAPEST 10
HUNGARY

Dr Ch. Klein
Boehringer Mannheim GmbH
Biochemie Werk Tutzing
D-8132 TUTZING
F. R. G.

Dr Y. Kobayashi
Institute for Protein Research
Osaka University
3 - 2 Yamadaoka
Suita
OSAKA 565
JAPAN

Dr E. Kojro
Max- Planck-Institut
für Biophysik
Kennedyallee 70
D-6000 FRANKFURT AM
MAIN 70
F. R. G.

Dr A. Kolodziejczyk
Technical University
of Gdansk
Dept. Organ. Chem.
Ul. Majakowskiego 11
80-952 GDANSK
POLAND

Prof. W. A. König
Institut für Organische
Chemie
der Universität Hamburg
Martin-Luther-King-Platz 6
D-2000 HAMBURG 13
F. R. G.

Dr D. Konopinska
Institute of Chemistry
Wroclaw University
Joliot-Curie 14
50035 WROCLAW
POLAND

Prof. K. D. Kopple
Smith Kline and French
Laboratories
Dept. of AP &S Chemistry F-90
1500 Spring Garden Street
PHILADELPHIA, PA 19101
U. S. A.

Dr A. Kozhich
Shemyakin Institute of
Bioorganic Chemistry
USSR Academy of Sciences
Ul. Miklukho-Maklaya 16/10
117871 MOSCOW V-437
U. S. S. R.

Prof. H. Kunz
Institut für Organische
Chemie
Universität Mainz
J. - J. - Becher - Weg 18 - 20
D-6500 MAINZ
F. R. G.

Prof. G. Kupryszewski
Institute of Chemistry
University of Gdansk
Ul. Sobieskiego 18
80-952 GDANSK
POLAND

Mr J. Kyriakides
J. M. Kyriakides S.A.
479, Messogion Street
115 26 ATHENS
GREECE

Prof. L. Lapatsanis
Lab. of Organic Chemistry
University of Athens
13A, Navarinou Street
106 80 ATHENS
GREECE

Dr L. - E. Larsson
Pharmacia AB
S-75182 UPPSALA
SWEDEN

Dr M. Lebl
Inst. of Organic Chemistry
and Biochemistry
C. S. A. V.
Flemingovo n. 2
16610 PRAGUE 6
CZECHOSLOVAKIA

Prof. M. T. Leplawy
Institute of Organic
Chemistry
Technical University
Ul. Zwirki 36
90-924 LODZ
POLAND

Dr G. Lindeberg
Institute of
Biochemistry
University of Uppsala
Box 576
S-751 23 UPPSALA
SWEDEN

Dr Al. Lobbia
Diamalt AG
Georg-Reismüller Str. 3236
D-8000 MUNCHEN 50
F. R. G.

A. Loffet
Propeptide
B. P. 12
F-91710 VERT-LE-PETIT
FRANCE

Dr M. Löw
Chemical Works
of Gedeon Richter Ltd
P. O. Box 27
H-1475 BUDAPEST
HUNGARY

Dr R. Lundin
Pharmacia AB
S-75182 UPPSALA
SWEDEN

Dr J. A. MacLaren .
C. S. I. R. O.
Division of Protein
Chemistry
343 Royal Parade
Parkville
VICTORIA 3052
AUSTRALIA

Prof. H. L. S. Maia
Universidade do Minho
Largo do Paço
4719 BRAGA Codex
PORTUGAL

Dr Ch. - P. Mak
Sandoz Research Institute
Brunnerstrasse 59
A-1235 VIENNA
AUSTRIA

Dr R. Makineni
Bachem Inc.
3132 Kashiwa Street
TORRANCE, CA 90505
U. S. A.

Prof. M. Manning
Medical College of Ohio
Department of Biochemistry
C. S. Suite 10008
TOLEDO, OHIO 43699
U. S. A.

Dr J. Markussen
Novo Research Institute
Novo Allé
DK-2880 BAGSVAERD
DENMARK

Dr M. Marraud
L. C. P. M.
ENSIC - INPL
1, Rue Grandville
54042 NANCY Cedex
FRANCE

Prof. G. R. Marshall
Washington University
School of Medicine
Dept. of Physiology and

Biophysics
ST. LOUIS, MO 63110
U. S. A.

Dr J. Martinez
Centre de Recherche
C.N.R.S.-I.N.S.E.R.M.
de Pharmacologie -
Endocrinologie
Rue de la Cardonille
B. P. 5055
34033 MONTPELLIER Cedex
FRANCE

Prof. J. M. Matsoukas
University of Patras
Dept. of Chemistry
PATRAS
GREECE

Prof. H. Matsuo
Dept. Biochem.
Miyazaki Medical College
Kiyotake
MIYAZAKI 889-16
JAPAN

Prof. K. Medzihradszky
Central Research Institute
of Chemistry
Hungarian Academy of
Sciences
Pusztaszeri Ut 57
H-1025 BUDAPEST
HUNGARY

Dr J. Meienhofer
Berlex Laboratories, Inc.
110 East Hanover Ave.
CEDAR KNOLLS, NJ 07927
U. S. A.

Dr M. Meldal
Institute of Organic
Chemistry
The Technical University of
Denmark
DK-2800 LYNGBY
DENMARK

Prof. B. Merrifield
Rockefeller University
1230 York Avenue
NEW YORK, NY 10021
U. S. A.

Dr A. Michel
University of Mons
Faculty of Sciences
Laboratory of Biological
Chemistry
21, Avenue Maistriav
B-7000 MONS
BELGIUM

Dr I. I. Mikhaleva
Shemyakin Inst. of
Bioorganic Chemistry
USSR Academy of Sciences
Ul. Miklukho-Maklaya 16/10
117871 MOSCOW V-437
U. S. S. R.

Doc. Dr S. Minchev
Higher Pedagogical Institute
9700 SHOUMEN
BULGARIA

Doc. Dr L. Mladenova-Orlinova
Higher Institute of Chemical
Technology
1156 SOFIA
BULGARIA

Dr Sv. Mojsov
Howard Hughes Med.
Institute
Harvard Medical School
Massachusetts Gen. Hospital
Blossom Street, Wellman 3
BOSTON, MA 02114
U. S. A.

Dr S. Moore
Peninsula Laboratory
Europe Ltd.
P. O. Box 62
17K Westside Jackson Str.
St Helens
MERSEYSIDE WA9 3AJ
U. K.

Dr B. A. Morgan
Sterling-Winthrop Research
Institute
81 Columbia Turnpike
RENSSELAER, NY 12144
U. S. A.

Dr J. - L. Morgat
Institut de Recherche
Fondamentale
Service de biochimie du
Départment de Biologie
Centre d'Etudes Nucléaires
de Saclay
91191 GIF-SUR-YVETTE Cedex
FRANCE

Prof. M. Mutter
Institute of Organic
Chemistry
University of Basel
St. Johanns-Ring 19
CH-4056 BASEL
SWITZERLAND

Dr V. K. Naithani
Deutsches
Wollforshungsinstitut
an der Th. Aachen
Veltmanplatz 8
51 AACHEN
F. R. G.

Dr J. J. Nestor
Syntex Research
Institute of Bio-organic
Chemistry
3401 Hillview Avenue
PALO ALTO, CA 94304
U. S. A.

Dr J. W. F. M. Van Nispen
Organon Int. B.V.
Organon Scientific
Development Group
P. O. Box 20
5340 BH OSS
THE NETHERLANDS

Prof. R. F. Offord
Biochimie Médicale
Faculté de Médecine
C. M. U.
9, Avenue de Champel
1211 GENEVA
SWITZERLAND

Prof. A. C. M. Paiva
Escola Paulista de Medicina
Caixa Postal 20.388
04034 SAO PAULO, SP
BRAZIL

Prof. A. Patchornik
The Weizmann Institute
of Science
REHOVOT 76100
ISRAEL

Dr Fr. Paulsen
Ferring Pharmaceuticals
Ferring Malmo
P. O. Box 30561
SWEDEN

Prof. A. A. Pavia
University of Avignon
33, Rue Pasteur
84000 AVIGNON
FRANCE

Prof. E. Pedroso
Dept. Quimica Organica
Facultat de Quimica
Marti Franqués 1
08028 BARCELONA
SPAIN

Prof. E. Peggion
Department of Organic
Chemistry
University of Padua
Via Marzolo 1
35131 PADUA
ITALY

Prof. B. Penke
Institute of Medical
Chemistry
Szeged Medical School
H-6720 SZEGED
HUNGARY

Dr D. Petkov
Bulgarian Academy of

Sciences
Institute of Organic
Chemistry
1040 SOFIA
BULGARIA

Dr R. Pipkorn
Zentrum fuer Molekulare
Biologie
Im Neuenheimer Feld 282
D-6900 Heidelberg
F. R. G.

Prof. C. Poulos
University of Patras
Department of Chemistry
PATRAS
GREECE

Dr G. Precigoux
Lab. Cristallographie
Université Bordeaux I
33405 TALENCE
FRANCE

Dr G. Priestley
Cambridge Research
Biochemicals Ltd
Button Gud
Harston
CAMBRIDGE CB2 5NX
U. K.

Dr J. Przybylski
Institute of Chemistry
University of Gdansk
Ul. Sobieskiego 18
80-952 GDANSK
POLAND

Prof. U. Ragnarsson
Institute of Biochemistry
University of Uppsala
Biomedical Center
Box 576
S-751 23 UPPSALA
SWEDEN

Dr S. Rakhit
Bio-Méga, Inc.
2100 Cunard
LAVAL, QUEBEC H7S 2G5
CANADA

Prof. R. Ramage
University of Edinburgh
West Mains Road
EDINBURGH EH9 3JJ
Scotland
U. K.

Dr K. - L. Reichelt
Pediatric Research Institute
National Hospital of Norway
0027 OSLO 1
NORWAY

Dr Fr. Reig-Isart
Centro de Investigación y
Desarrolo de Barcelona
(C.S.I.C.)
Jorge Girona Salgado 18-26
08034 BARCELONA
SPAIN

Prof. S. Reissmann
Friedrich-Schiller-
Universität
Department of Biology
Goetheallee 26
6900 JENA
G. D. R.

Dr J. Van Rietschoten
CNRS - INSERM
Laboratoire de Biochimie
Faculté de Médecine
Secteur Nord
13326 MARSEILLE Cedex 15
FRANCE

Dr H. Rink
Ciba-Geigy AG
CH-4002 BASEL
SWITZERLAND

Dr P. Rivaille
INSERM U. 113
C.H.U. Saint-Antoine
27, Rue Chaligny
75571 PARIS Cedex 12
FRANCE

Dr J. Rivier
Clayton Foundation
Laboratories for
Peptide Biology
The Salk Institute
10010 N. Torrey Pines Road
LA JOLLA, CA 92037
U. S. A.

Prof. R. Rocchi
University of Padova
Dept. of Organic Chemistry
Via Marzolo 1
35131 PADOVA
ITALY

Dr S. Romani
Max-Planck-Institut
f. Biochemie
Abtl. Peptidchemie
Am Klopferspitz
D-8033 MARTINSRIED
F. R. G.

Prof. B. P. Roques
Département de Chimie
Organique
Université R. Descartes
4, Avenue de l 'Observatoire
75006 PARIS
FRANCE

Dr K. Rose
University of Geneva
Biochimie Médicale
C. M. U.
9, Avenue Champel
1211 GENEVA 4
SWITZERLAND

Prof. M. Rothe
Lehrstuhl Organische
Chemie II
University of Ulm
Oberer Eselsberg
7900 ULM
F. R. G.

Prof. S. Sakakibara
Peptide Institute
Protein Research
Foundation
Minoh-shi
OSAKA 562
JAPAN

Prof. C. Sakarellos
University of Ioannina
Lab. of Organic Chemistry
451 10 IOANNINA
GREECE

Prof. M. Sakarellos-Daitsiotis
Department of Chemistry
University of Ioannina
451 10 IOANNINA
GREECE

Prof. S. Salvadori
Dipartimento Scienze
Farmaceutiche
Via Scandiana 21
44100 FERRARA
ITALY

Dr H. Saneii
Biomolecular Research
International Corp.
P. O. Box 1403
LOUISVILLE, KY 40401
U. S. A.

Dr D. Sarantakis
Wyeth Laboratories Inc.
Research Department
Medicinal Chemistry
P. O. Box
PHILADELPHIA, PA 19380
U. S. A.

Dr N. A. Sasaki
Institut de Chimie des
Substances Naturelles
C. N. R. S.
91190 GIF-SUR-YVETTE
FRANCE

Dr A. Scarso
UCB - Bioproducts S.A.
Chemin du Foriest
B-1420 BRAINE L'ALLEUD
BELGIUM

Prof. A. Scatturin
Dipartimento di Scienze

Farmaceutiche
Via Scandiana 21
44100 FERRARA
ITALY

Dr W. M. M. Schaaper
Central Veterinary Institute
Edelhertweg 15
8219 PH LELYSTAD
THE NETHERLANDS

Prof. P. W. Schiller
Clinical Research Institute
of Montreal
110 Pine Avenue West
MONTREAL - QUEBEC
CANADA H2W 1R7

Prof. C. H. Schneider
Institute for Clinical
Immunology
Inselspital
CH-3010 BERN
SWITZERLAND

Dr. I. Schön
Chemical Works of Gedeon
Richter Ltd.
P. O. Box 27
H-1475 BUDAPEST
HUNGARY

Prof. R. Schwyzer
Swiss Federal Institute of
Technology (ETH)
Dept. of Molecular Biology
CH-8093 ZURICH
SWITZERLAND

Dr J. Seprődi
1st Institute of Biochemistry
Semmelweis University
Medical School
P. O. Box 260
H-1444 BUDAPEST 8
HUNGARY

Prof. R. C. Sheppard
Medical Research Council
Laboratory for Molecular
Biology
Hills Road
CAMBRIDGE CB2 2QH
U. K.

Prof. T. Shiba
Dept. of Chemistry
Faculty of Sciences
Osaka University
Machikaneyama-cho 1-1
Toyonaka
OSAKA 560
JAPAN

Dr A. J. Slotboom
Laboratory of Biochemistry
State University of Utrecht
Transitorium III
Pedualaan 8
NL-3485 CH UTRECHT
THE NETHERLANDS

Dr C. W. Smith
Biopolymer Chemistry
The Upjohn Company
KALAMAZOO, MI 49001
U. S. A.

Prof. A. F. Spatola
Department of Chemistry
University o Louisville
LOUISVILLE, KY 40292
U. S. A.

Prof. G. Stavropoulos
Department of Chemistry
University of Patras
PATRAS
GREECE

Prof. G. Stelakatos
Lab. of Organic Chemistry
University of Athens
13A, Navarinou Street
106 80 ATHENS
GREECE

Prof. J. M. Stewart
Dept. of Biochemistry-B 126
University of Colorado
Medical School
DENVER, CO 80262
U. S. A.

Prof. M. Szelke
Department of Chemical
Pathology
Royal Postgraduate Medical
School
Hammersmith Hospital
LONDON W12 OHS
U. K.

Prof. A. Tartar
Faculté de Pharmacie
Lab. de Chimie Organique
3, Rue du Professeur
Laguesse
59045 LILLE Cedex
FRANCE

Prof. P. A. Temussi
Universitá di Napoli
Via Mezzocannone 4
80134 NAPOLI
ITALY

Dr I. Teplán
Institute of Chemistry and
Biochemistry
Semmelweis University
Medical School
Puskin u. 9
1088 BUDAPEST
HUNGARY

Prof. G. I. Tesser
Laboratory for Organic
Chemistry
Toernooiveld
NL-6525 ED NIJMEGEN
THE NETHERLANDS

Prof. D. Theodoropoulos
University of Patras
Lab. of Organic Chemistry
PATRAS
GREECE

Dr J. Trojnar
Ferring AB
Box 30561
S-20062 MALMÖ
SWEDEN

Dr Ch. Tzougraki
University of Athens
Department of Chemistry
13A, Navarinou Street
106 80 ATHENS
GREECE

Prof. I. Ugi
Lehrstuhl I
Organische Chemie
Technische Universität
München
Lichtenbergstr. 4
D-8046 GARCHING
F. R. G.

Prof. F. Vandesande
Zoological Institute
K. U. L.
Naamsestraat 59
3000 LEUVEN
BELGIUM

Dr R. J. Vavrek
Biochemistry Department
University of Colorado
Medical School
4200 East 9th Avenue
DENVER, CO 80220
U. S. A.

Dr M. Verlander
Bachem Inc.
P. O. Box 3426
TORRANCE, CA 90510
U. S. A.

Prof. G. P. Vlasov
Institute of High Molecular
Compounds
USSR Academy of Sciences
Bolshoi Prospect 31
LENINGRAD 199004
U. S. S. R.

Prof. W. Voelter
Abteilung f. Physikal.
Biochemie
Physiologisch-Chemisches
Institut
Tübingen University

Hoppe-Seyler-Str. 1
D-7400 TÜBINGEN
F. R. G.

Prof. J. Wade
Howard Florey Institute
University of Melbourne
Parkville
VICTORIA 3052
AUSTRALIA

Dr R. Wade
Ciba-Geigy Pharmaceuticals
Wimblehurst Road
Horsham
WEST SUSSEX RH12 4AB
U. K.

Dr P. Ward
Glaxo Group Research Ltd.
Greenford Road
Greenford
MIDDLESEX UB6 OHE
U. K.

Dr G. Westin-Sjödahl
Dept. of Organic Chemistry
KabiVitrum AB
S-112 87 STOCKHOLM
SWEDEN

Dr Br. J. Williams
Merck Sharp and Dohme
Neuroscience Research
Centre
Eastwick Road
HARLOW-ESSEX CM20 2QR
U. K.

Dr M. Von Wilm
Beckman Instruments
International
17, Rue de Pierres-Du-Niton
B. P. 76
CH-1211 GENEVA 6
SWITZERLAND

Prof. A. Yaron
The Weizmann Institute of
Science
Department of Biophysics
REHOVOT 76100
ISRAEL

Prof. A. Yiotakis
University of Athens
Lab. of Organic Chemistry
13A, Navarinou Street
106 80 ATHENS
GREECE

Prof. M. Zaoral
Institute of Organic
Chemistry and
Biochemistry
Czechoslovak Academy
of Sciences
Flemingovo n. 2
166 10 PRAHA 6
CZECHOSLOVAKIA

Dr G. Zanotti
Centro per lo Studio della
Chimica del Farmago del
CVR
Istituto di Chimica
Farmaceutica
Universitá "La Sapienza"
P. le Aldo Moro
00185 ROMA
ITALY

Authors' Index

Subject Index

FSC
www.fsc.org
MIX
Papier | Fördert
gute Waldnutzung
FSC® C083411

Kolibri 360 GmbH
Ferdinand-Jühlke-Straße 7
99095 Erfurt, Deutschland
produktsicherheit@kolibri360.de